Microfluidics and Nanophotonics: Science and Engineering

Volume 2

This Springer book series focuses on the science and engineering of nanophotonics and miniaturized systems. It covers topics such as metamaterials, metasurfaces, biophotonics, microfluidics, Microelectromechanical systems (MEMS), and nano/microdevices. The series explores the translation of scientific breakthroughs in nanophotonics, metamaterials, and microfluidics into practical engineering applications, including metalens, micro-spectrometers, MEMS devices, biomedical devices and instrumentations, etc., to name a few. By featuring the latest research from leading international researchers and engineers, the series serves as a valuable resource for researchers, industry professionals, and undergraduate and graduate students interested in fundamental nanophotonics, cutting-edge micro/nanodevices, and related applications. More specifically, it includes the following themes:

- Metamaterials and Metasurfaces
- Nanoplasmonic Biosensing
- Biomedical Devices and Instrumentation
- Microsystems and MEMS
- Nanophotonics in Quantum Engineering

Hong-Xiang Sun · Yong Ge · Shou-Qi Yuan

Acoustic Metamaterials

Asymmetric Transmission, Focusing, and Asymmetric Focusing

Springer

Hong-Xiang Sun
School of Physics and Electronic Engineering
Research Center of Fluid Machinery Engineering and Technology
Jiangsu University
Zhenjiang, China

Yong Ge
School of Physics and Electronic Engineering
Research Center of Fluid Machinery Engineering and Technology
Jiangsu University
Zhenjiang, China

Shou-Qi Yuan
Research Center of Fluid Machinery Engineering and Technology
Jiangsu University
Zhenjiang, China

ISSN 2730-9029 ISSN 2730-9037 (electronic)
Microfluidics and Nanophotonics: Science and Engineering
ISBN 978-981-96-8523-3 ISBN 978-981-96-8524-0 (eBook)
https://doi.org/10.1007/978-981-96-8524-0

This work was supported by the National Key Research and Development Program of China under Grant No. 2020YFC1512403, and the National Natural Science Foundation of China under Grants No. 12274183, and No. 12174159.

This Springer imprint is published by the registered company Springer Nature Singapore Pte Ltd.
The registered company address is: 152 Beach Road, #21-01/04 Gateway East, Singapore 189721, Singapore

Preface I

As the former president of Jiangsu University, it is with immense pride and enthusiasm that I introduce this significant contribution to the field of acoustic metamaterials. The book, *Acoustic Metamaterials—Asymmetric Transmission, Focusing, and Asymmetric Focusing*, represents a remarkable achievement in the ongoing exploration of materials with unique acoustic properties. Over the past two decades, the field has grown exponentially, offering new possibilities for wave control, energy focusing, and signal processing. This book stands as a testament to the innovative spirit and technical expertise that drive advancements in modern physics and engineering.

Our group at Jiangsu University has been at the forefront of research in acoustic metamaterials, and our work has not only advanced the fundamental understanding of acoustic wave phenomena but also opened new avenues for practical applications. Our investigations into asymmetric transmission, acoustic focusing, and asymmetric focusing have the potential to revolutionize fields such as medical ultrasound, noise control, architectural acoustics, and acoustic communication. By consolidating our research in this volume, we provide a comprehensive and detailed overview of the fundamental principles, design methods, and experimental results that underpin these exciting developments.

During my tenure at Jiangsu University, I had the privilege of leading a dedicated team of scholars who have made significant contributions to the field of acoustic metamaterials. Our ability to bridge theoretical insights with experimental validation has been a hallmark of our work, and our commitment to pushing the boundaries of knowledge is truly commendable. This book reflects our deep understanding of the subject and desire to share our insights with the broader scientific community.

For students, researchers, and practitioners in acoustics and metamaterials, this book serves as an invaluable resource. It not only elucidates complex phenomena but also highlights the practical potential of acoustic metamaterials in addressing real-world challenges. Our emphasis on asymmetric wave manipulation and focusing is particularly relevant, as these concepts hold great promise for advancing technologies in areas such as ultrasonic therapy, sonar detection, acoustic sensing, etc.

I am confident that this book will become a key reference in the study of acoustic metamaterials, inspiring further research and innovation in the field. It is with great admiration that I recommend this work to readers, and I look forward to its impact on the scientific community and beyond.

Zhenjiang, China
February 2025

Prof. Dr. Shou-Qi Yuan

Preface II

The realm of acoustics has long been a cornerstone of human civilization, enabling communication, exploration, and the appreciation of the arts. From the harmonious melodies of music to the clarity of spoken language, and from the precision of ultrasonic imaging to the reach of sonar detection, acoustic waves permeate and enrich our modern existence. Yet, traditional materials have offered limited control over these waves, constraining the potential for innovation in acoustic applications. The advent of acoustic metamaterials has emerged as a transformative solution, providing unprecedented manipulation of acoustic waves and heralding a new era in acoustic technology.

Acoustic metamaterials are artificially structured composites with subwavelength dimensions, whose acoustic properties are determined not by the intrinsic characteristics of their constituent materials, but by the design of their structural units. Through meticulous engineering of these units and their configurations, acoustic metamaterials can precisely control the propagation of sound, transcending the limitations of conventional materials and enabling a variety of novel phenomena and functionalities.

In recent years, the field of acoustic metamaterials has witnessed remarkable advancements, yielding a wealth of groundbreaking research. Among these, three key research areas have attracted significant attention and demonstrated immense potential for practical applications: **acoustic asymmetric transmission, acoustic focusing, and acoustic asymmetric focusing**.

Acoustic asymmetric transmission challenges the conventional reciprocity of sound propagation in acoustic systems, enabling acoustic waves to travel asymmetrically in opposite directions. This characteristic, akin to an electronic diode, facilitates acoustic one-way transmission and holds promise for applications in acoustic diodes and isolators. For instance, leveraging asymmetric transmission can lead to the development of efficient, compact acoustic isolators that effectively mitigate reflections and crosstalk in acoustic systems, thereby enhancing their performance and stability.

Acoustic focusing aims to concentrate sound energy within a specific region, surpassing the diffraction limits of traditional acoustic lenses and achieving subwavelength-scale sound manipulation. Conventional acoustic lenses are constrained by diffraction limits, which hinder high-resolution acoustic focusing. Acoustic metamaterials, however, can precisely modulate acoustic wavefronts through specially designed unit cells, overcoming diffraction limits and enabling subwavelength focusing of sound. This breakthrough has revolutionized fields such as high-resolution acoustic imaging and precise acoustic therapy. For example, acoustic metamaterial lenses can facilitate ultra-high-resolution ultrasonic imaging of human tissues, providing a more accurate foundation for early disease diagnosis.

Acoustic asymmetric focusing combines the advantages of asymmetric transmission and focusing, enhancing acoustic wave concentration in a particular direction while suppressing propagation in the opposite direction. This property holds potential for applications in directional acoustic communication and high-intensity focused ultrasound therapy. For instance, asymmetric focusing can be utilized to construct directional acoustic antennas for long-distance, high-fidelity acoustic communication. It can also address challenges in focused ultrasound therapy, such as the damaging effects of high-intensity reflected sound waves on transducers and surrounding healthy tissues, thereby reducing treatment costs and improving accuracy.

Regarding the research background, with the support of the National Key Research and Development Program of China under Grant No. 2020YFC1512403, and the National Natural Science Foundation of China under Grant Nos. 12274183 and 12174159, our group has summarized and organized its work over the past decade on acoustic asymmetric transmission, acoustic focusing, acoustic asymmetric focusing, and related applications based on acoustic metamaterials. Mature research findings have been carefully selected, and the latest advancements in the field have been incorporated to compile this book. It is intended to serve as a reference for professionals in physics, acoustics, medical ultrasound, non-destructive testing, and power mechanical engineering.

The book is organized into ten chapters. Chapter 1, **Introduction**, provides an overview of the fundamental concepts, historical development, current research hotspots, and future trajectories of acoustic metamaterials and their key branches. Chapter 2, **Basic Theory**, introduces a series of fundamental principles and methods for sound manipulation in 2D and 3D spaces based on acoustic metamaterials, including the interference theory, effective medium theory, finite element method, generalized Snell's law, mode conversion theory, and band theory of sonic crystals. Chapters 3 and 4, **Acoustic Asymmetric Transmission**, present several AAT systems designed with underwater plate-like metamaterials and airborne metamaterials, and discuss their potential applications in architectural acoustics, noise control, and medical ultrasound in detail. Chapters 5–7, **Acoustic Focusing**, showcase a series of AF lenses designed with ring-shaped metamaterials, phase-controlled metamaterials, and thermoacoustic phase-controlled metamaterials, exploring their applications in underwater signal detection, medical ultrasound therapy, acoustic communication, and acoustic energy harvesting. Chapter 8, **Acoustic Asymmetric Focusing**, demonstrates a variety of AAF lenses designed with asymmetric modulation metamaterials and discusses their potential applications in focused ultrasound therapy and medical ultrasound imaging. Chapter 9, **Three-Dimensional Acoustic Focusing and Asymmetric Focusing**, presents several types of 2D acoustic lenses designed with phase-controlled metamaterials. These 2D lenses can simultaneously achieve a combination of focusing and other effects in 3D space, promoting their practical applications in non-contact manipulation for trapping particles, medical ultrasound imaging and treatment. Chapter 10, **Conclusions, Challenges, and Future Prospects**, summarizes key findings in acoustic metamaterials, covering AAT, AF, and AAF systems. It highlights advancements in sound manipulation, identifies challenges, and outlines future research directions, including broadband devices, 3D wavefront control, and machine learning.

During the process of conducting relevant research and writing this book, we benefited from the guidance and assistance of several respected individuals, including Prof. Shu-yi Zhang, Prof. Xiao-jun Liu, Prof. Yun Lai, and Prof. Xiaozhou Liu from Nanjing University; Prof. Baile Zhang from Nanyang Technological University; Prof. Zhengyou Liu from Wuhan University; Prof. Menglu Qian and Associate Prof. Zheng Xu from Tongji University; Assistant Prof. Haoran Xue from the Chinese University of Hong Kong; Prof. Yihao Yang from Zhejiang University; Associate Prof. Yi-jun Guan, Dr. Ding Jia, and Prof. Qiao-rui Si from Jiangsu University. We are also grateful for the contributions of Ph.D. student Yin Wang; postgraduate students Jian-ping Xia, Jiao Qian, and Hong-yu Zou; and undergraduate students Chen Liu, Yu-lei Huang, Wen-ting Gao, Jia-he Chen, An-ru Hou, Tian-chi Zhang, Jia-li Yin, and others who participated in various aspects of this work. We would like to express our deepest gratitude to all of them.

The author has strived to maintain a serious and rigorous approach in writing this book. However, we acknowledge that there may still be shortcomings. We warmly welcome readers' feedback, criticisms, and corrections. Together, may we contribute our modest efforts to advancing the field of acoustic metamaterials, particularly in the areas of asymmetric transmission, focusing, and asymmetric focusing.

Zhenjiang, China

Hong-Xiang Sun
Yong Ge
Shou-Qi Yuan

Competing Interests The authors have no competing interests to declare that are relevant to the content of this manuscript.

Contents

Abbreviations

1D	One-dimensional
2D	Two-dimensional
3D	Three-dimensional
AAF	Acoustic asymmetric focusing
AAT	Acoustic asymmetric transmission
AAVF	Acoustic asymmetric vortex focusing
ABR	Acoustic blind region
AF	Acoustic focusing
AM	Acoustic metasurface
AOT	Acoustic one-way transmission
AVF	Acoustic vortex focusing
BI	Bottom incidence
DDB	Downward deflected branch
EM	Effective medium
FE	Finite element
FLHM	Full length at half maximum
HIFU	High-intensity focused ultrasound
ICCD	Intensification charge-coupled device
Im	Imaginary part
LI	Left incidence
MB	Middle branch
ME	Measured
MMR	Monopole Mie resonance
PP	Plane plate
Re	Real part
RI	Right incidence
SC	Sonic crystal
SI	Simulated
TI	Top incidence

TPA	Thermoacoustic phased array
UDB	Upward deflected branch
WS	Wigner-Seitz

Chapter 1
Introduction

1.1 Acoustic Metamaterials

Acoustic wave is a type of common energy and information carrier in nature, and the realization of sound manipulation has always been widely studied due to its application prospects in medical ultrasound, information processing, and architectural acoustics, etc. However, the realization of accurate control of sound still poses a great challenge. In recent years, the emergence and development of acoustic metamaterials have provided feasibility for achieving sound manipulation with high performance [1]. Generally, the acoustic metamaterials consist of periodically arranged subwavelength artificial structures, and can realize special characteristics that natural materials do not possess.

By analogizing the equivalent mass density (ρ) and the reciprocal bulk modulus ($1/\kappa$) of acoustic materials with the dielectric constant (ε) and the magnetic permeability (μ) of electromagnetic materials [2], respectively, a theoretical basis is established for designing various acoustic metamaterials with negative equivalent parameters. In 2000, Liu et al. proposed a type of acoustic metamaterial based on a localized resonant sonic crystal (SC) with a cubic lattice [3]. As shown in Fig. 1.1a, a basic unit cell of the SC consists of a shot wrapped in a layer of silicone rubber material, which is then embedded in a cubic epoxy resin matrix. The SC is achieved by periodically arranging the basic unit cells (Fig. 1.1b) and exhibits a localized resonance mode, resulting in a bandgap in the low-frequency region. The frequency of this bandgap is much lower than that of Bragg scattering SCs (Fig. 1.1c) and appears as a trough in the pressure amplitude spectrum (Fig. 1.1d). To better understand the bandgap of the SC, the basic unit cell can be modeled as a spring-damped resonant system [4], where the outer layer of silicone rubber material is simplified as a spring connecting the matrix and the shot (Fig. 1.1e). When an external excitation (F) is applied to the matrix within the bandgap, the motion of both the matrix and the shot becomes out of phase. According to Newton's second law, the ratio of F to the system's acceleration

H. Sun et al., *Acoustic Metamaterials*, Microfluidics and Nanophotonics: Science and Engineering 2, https://doi.org/10.1007/978-981-96-8524-0_1

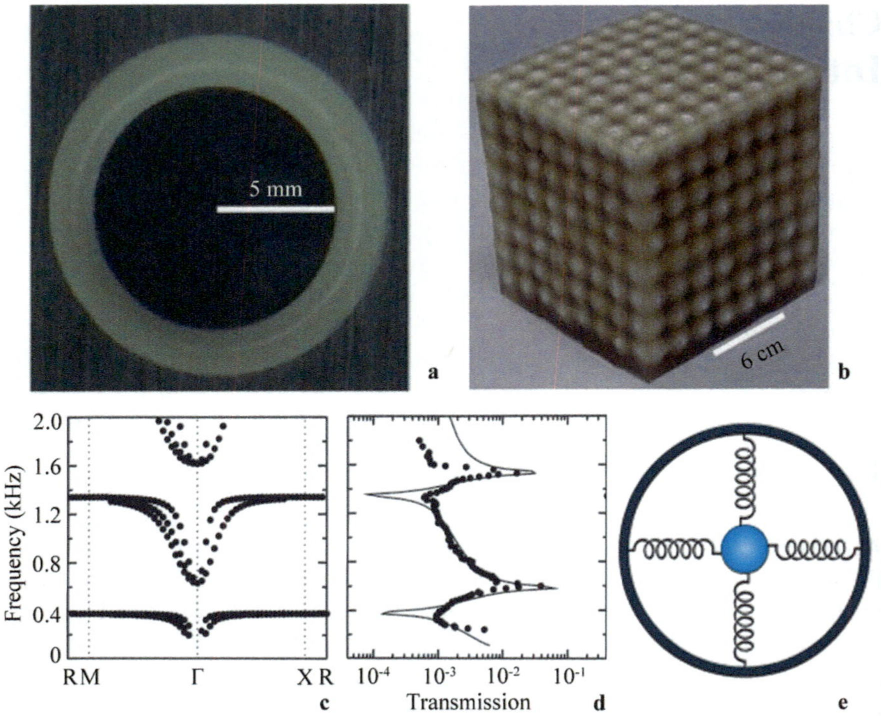

Fig. 1.1 Photographs of **a** a basic unit cell and **b** a type of localized resonant SC. **c** Calculated band structure of the SC. **d** Calculated and measured transmission spectra of the SC. **e** Schematic of a spring-damped resonant system for the basic unit cell

(a) changes, indicating a change in the equivalent mass density (ρ_{eff}). Notably, when the directions of F and a are opposite, the real part of ρ_{eff} becomes negative.

In addition to the development of acoustic metamaterials with negative mass density [3–7], those with negative elastic modulus have also been designed subsequently [8, 9]. As shown in Fig. 1.2a, b, Fang et al. designed a one-dimensional (1D) array of Helmholtz resonators filled with fluid [8]. When an acoustic wave is incident on the structure, the acoustic pressure inside the channel changes, causing the fluid in the resonant cavities to oscillate back and forth, leading to expansion and compression. It is noteworthy that when the frequency of the acoustic wave sweeps through the resonance frequency of the structure (33 kHz in Fig. 1.2c), the process of fluid expansion and compression inside the structure is in opposite phase with the incident wave, indicating a negative bulk elastic modulus response.

Furthermore, by coupling negative equivalent mass density and elastic modulus, double-negative metamaterials at specific frequencies can be achieved. In 2004, Li et al. designed a type of Mie-resonance structure made of silicone rubber material in water [10], where the sound velocity of the resonator is much lower than that of water. The results show that the resonator can achieve double-negative equivalent

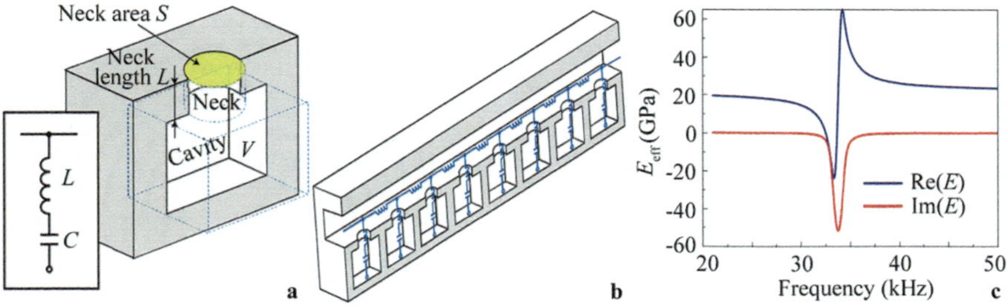

Fig. 1.2 **a** Schematic of a Helmholtz resonator and **b** a 1D array of Helmholtz resonators. **c** Real and imaginary parts of the bulk elastic modulus spectrum

parameters, which are closely related to its low-frequency resonance modes. Specifically, the negative bulk elastic modulus is determined by the monopole mode of the resonator, while the negative mass density is related to the dipole mode. At a certain frequency, both monopole and dipole modes exist simultaneously, and the resonator can be equivalent to a double-negative equivalent parameter material, meaning both the equivalent elastic modulus and mass density are negative, thereby realizing an acoustic negative refractive index material.

In experimental studies, Fok et al. observed negative refraction of sound in an acoustic system composed of a Helmholtz resonator with a monopole resonance mode and a spring resonator with a dipole resonance mode [11]. In 2015, Cheng et al. designed a type of Mie resonator with low equivalent sound velocity and high refractive index based on a coiling-up space structure [12]. The sample photograph and cross-section are shown in Fig. 1.3a, b, respectively. As a type of double-negative metamaterial, there are two frequency bands for both negative equivalent parameters: one corresponding to the negative bulk modulus caused by the monopole Mie resonant mode (the light blue region in Fig. 1.3c), and the other corresponding to the negative mass density supported by the dipole Mie resonance (the light blue region in Fig. 1.3d).

Another milestone in acoustic metamaterials is the development of coiling-up space structures [13–16]. In 2012, Liang et al. designed an acoustic metamaterial with a high refractive index and near-zero mass density, referred to as a coiling-up space structure [13]. As shown in Fig. 1.4a, the acoustic wave propagates along a narrow zigzag path within the structure, resulting in a total propagation distance much longer than the physical size of the structure. This enables the manipulation of the refractive index and phase delay of sound. By tuning the length of the zigzag path, the phase and group velocities of the coiling-up space structure can be modulated, altering the equivalent refractive index and dispersion relationship. Consequently, various sound effects, such as negative refraction and tunneling, can be achieved. As shown in Fig. 1.4b, by stacking coiling-up space structures in a triangular shape, an acoustic prism with negative mass density and negative bulk modulus can be realized. When an acoustic wave is incident from the bottom side of the prism, the sound energy can transmit horizontally through the oblique edge of the prism,

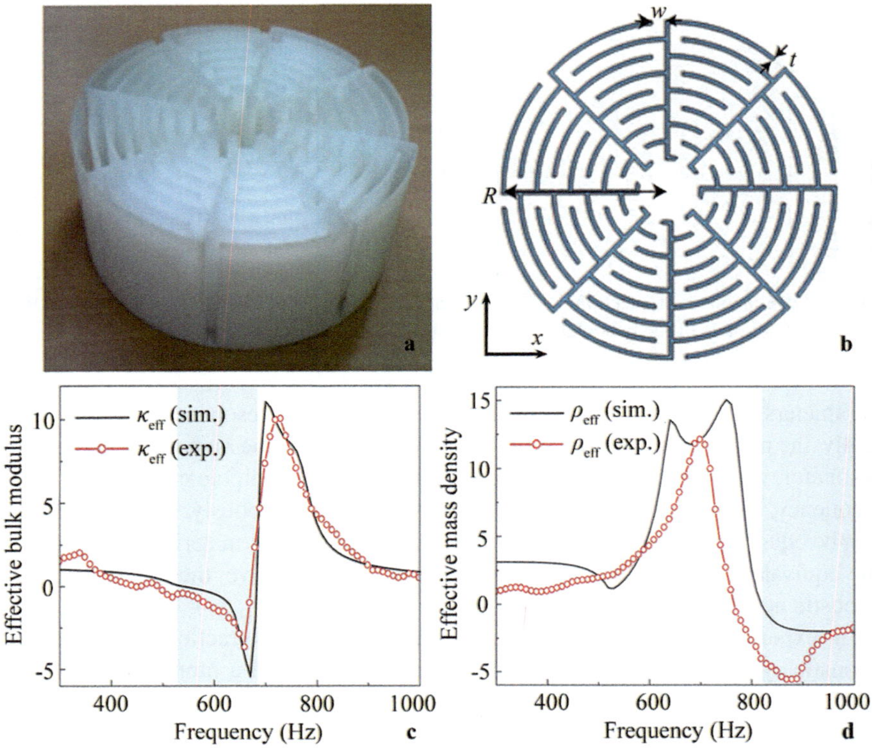

Fig. 1.3 **a** Photograph of the sample and **b** schematic of the cross-section of a Mie resonator. Measured and simulated **c** bulk modulus and **d** mass density spectra of the Mie resonator

demonstrating negative refraction. Additionally, by tuning the path length of the coiling-up space structure, the equivalent mass density of the structure approaches zero in the low-frequency region, enabling tunneling effects. Furthermore, as shown in Fig. 1.4c, when an acoustic wave is incident from the left side of a waveguide, the sound energy is severely scattered by a wall. However, by surrounding the wall with a density-near-zero metamaterial, the incident acoustic wave can enter the narrow channel of the coiling-up space structure and transmit through the wall with high efficiency and a constant waveform.

Later, the coiling-up space structure was widely used in the design of metamaterials and metasurfaces [17–25]. By tuning the propagation distance within the unit cell, a phase delay of sound across the full 2π range can be achieved. Based on the generalized Snell's law [26], acoustic metasurfaces (AMs) composed of a series of phased unit cells with different parameters can be designed to achieve precise sound manipulation, enabling various effects such as focusing [17, 18, 20, 21, 24, 27–29], self-bending beams [30–33], and invisibility [34–36].

Despite the rapid development of acoustic metamaterials in recent years, most works have focused on linear, passive, and static structures. To further broaden

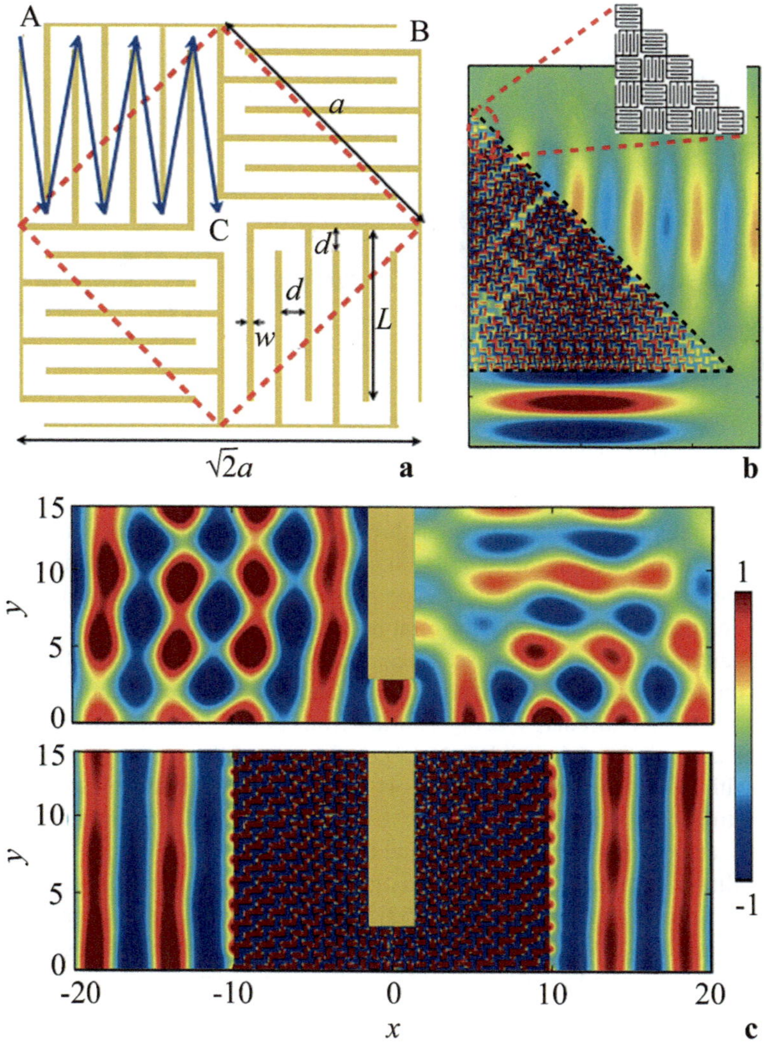

Fig. 1.4 **a** Schematic of a coiling-up space structure. **b** Simulated pressure distribution of acoustic negative refraction caused by an acoustic prism composed of coiling-up space structures. **c** Simulated pressure distributions of acoustic tunneling through a waveguide without and with the density-near-zero metamaterial

their applicability, active acoustic metamaterials composed of piezoelectric materials, speakers, fans, and other components have been designed [37–44]. Compared to passive acoustic metamaterials, active ones offer advantages such as reconfigurability, real-time control, and sound energy gain.

In 2014, Fleury et al. achieved non-reciprocal sound transmission by introducing a rotating airflow in a three-port acoustic circulator [38]. Figure 1.5a schematically

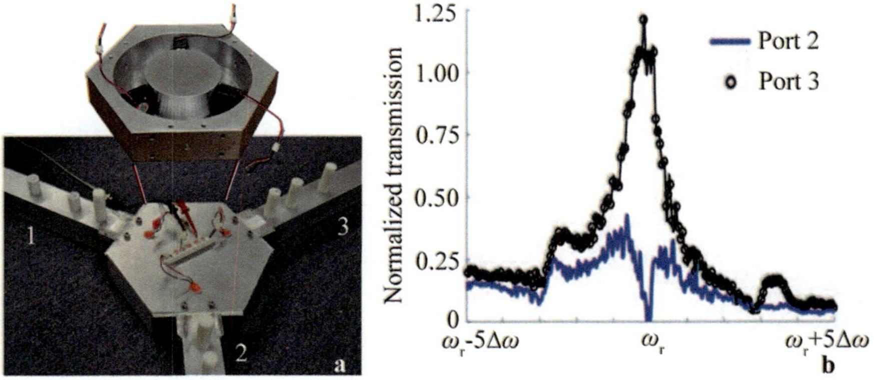

Fig. 1.5 **a** Photograph of a three-port acoustic circulator. The top inset shows the circulator without the cover. **b** Measured transmission spectra at ports 2 and 3 with the incident wave at port 1

shows the circulator structure, in which three actively controlled fans with rotational symmetry are installed in the annular channel. When the fans are turned off, two degenerate resonant modes with opposite propagation directions exist in the circulator. When the fans are turned on, a rotating airflow is introduced, causing a split in the resonant frequency. In this case, when an acoustic wave is incident from port 1, the two modes interfere destructively at port 2 and constructively at port 3, resulting in most of the sound energy being emitted from port 3 (Fig. 1.5 b). Conversely, when the acoustic wave is incident from port 3, most of the energy is emitted from port 2, with almost no transmission at port 1. Thus, the designed acoustic circulator with rotating airflow breaks the inherent reciprocity of sound propagation in natural materials, achieving non-reciprocal sound transmission.

In addition to rotating airflow, acoustic amplification circuits have been used to design metamaterial structures [45–52]. Zhang et al. utilized an acoustic unidirectional amplifier to achieve non-reciprocal coupling between two resonant cavities [45]. As shown in Fig. 1.6a, two identical acoustic resonators (labeled "1" and "2") are connected by two waveguides, providing reciprocal coupling between the resonators. To introduce non-reciprocal coupling, an acoustic amplification circuit, including an acoustic unidirectional amplifier, a loudspeaker, and a microphone, is externally connected to the two resonant cavities. When an acoustic wave is incident from resonator 1, the forward transmitted pressure spectrum $|s_{12}|$ can be detected in resonator 2. Similarly, the pressure spectrum $|s_{21}|$ can be obtained in resonator 1 when the acoustic wave is incident from resonator 2. Figure 1.6b shows the experimentally measured and theoretically fitted results of $|s_{12}|$ and $|s_{21}|$ for the acoustic system without and with the amplifier, respectively. Without the amplifier, both spectra $|s_{12}|$ and $|s_{21}|$ are nearly identical, with two resonance peaks due to reciprocal coupling. However, when the unidirectional amplifier is introduced, $|s_{12}| > |s_{21}|$, and the maximum $|s_{12}|$ exceeds 1, with resonance frequencies differing from those without the amplifier, demonstrating non-reciprocal coupling.

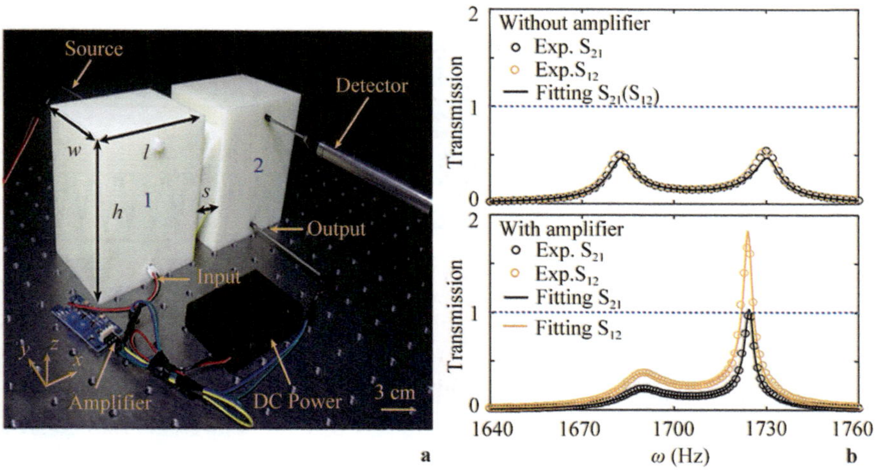

Fig. 1.6 **a** Photograph of two acoustic resonators connected by a pair of cross-linked narrow waveguides and a unidirectional amplifier. **b** Measured and fitted transmission spectra of acoustic systems without and with the unidirectional amplifier

1.2 Acoustic Metasurfaces

AMs are characterized by their planar structure, subwavelength thickness, and high-performance sound manipulation capabilities [53–59], falling within the scope of acoustic metamaterials. Generally, AMs are composed of a series of ultrathin unit cells. By tuning structural parameters of these unit cells, a phase delay of sound across the full 2π range can be achieved. This allows for arbitrary shaping of acoustic wavefronts under the theoretical framework of the generalized Snell's law. Based on this characteristic, AMs can be applied to the design of various phased control devices with exotic phenomena, including arbitrary sound reflection and transmission, focusing, invisibility, and absorption. These applications have broad prospects in fields such as medical ultrasound, architectural acoustics, nondestructive testing, and noise reduction, and they also provide new pathways for the development of acoustic materials and technologies. In recent years, with the rapid development of AMs, a variety of metasurface structures with diverse functions have been proposed, constructed using different types of unit cells, including coiling-up space structures [18–25, 60–63], Helmholtz resonators [31, 64–67], and groove structures [68–70], among others.

AMs based on coiling-up space structures with subwavelength dimensions force acoustic waves to propagate along curled channels (zigzags, labyrinths, and spirals), which are significantly longer than the actual physical dimension of the structure. The extended propagation distance enables the manipulation of reflected waves and allows the reflected phase delay to be tailored across the full 2π range. As shown in Fig. 1.7a, a reflection-type unit cell consists of a zigzag structure with a sound-hard boundary on the left side [18, 20]. By adjusting the width of the unit cell,

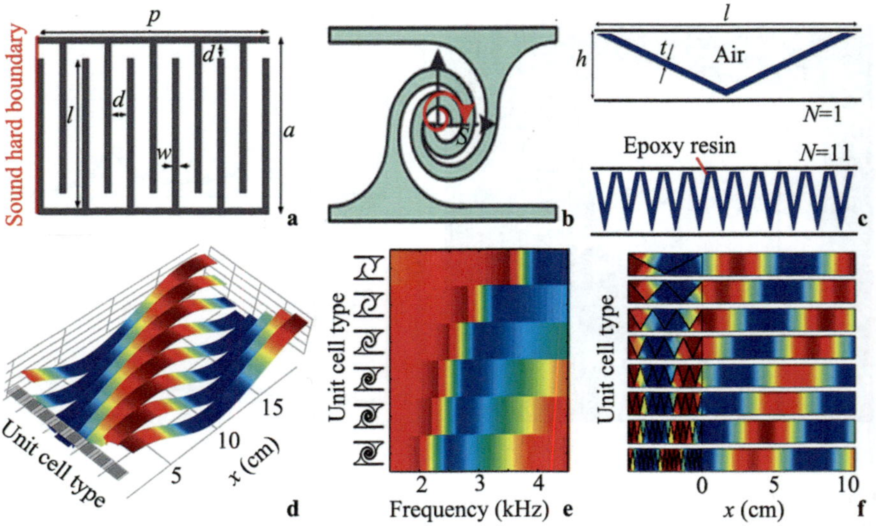

Fig. 1.7 Schematic of three types of metasurface unit cells composed of **a** the zigzag structure, **b** labyrinthine structure, and **c** V-shaped structure. **d–f** Simulated pressure distributions caused by the unit cells with different structural parameters in (**a–c**), demonstrating phase delays of sound covering the full 2π range

the propagation distance of sound can be precisely controlled, achieving a reflected phase delay across the full 2π range for unit cells of the same thickness (Fig. 1.7d). In addition to reflection-type unit cells, transmission-type metasurface unit cells can also be designed using other coiled structures without sound-hard boundaries, such as labyrinthine [19] and V-shaped structures [71] (Fig. 1.7b, c). By tuning the coiled length of the labyrinthine structure and the number of V-shaped structures, the transmitted phase delay of sound can be adjusted across the 2π range, as shown in Fig. 1.7e, f, respectively. However, these coiling-up space structures must be carefully designed to minimize viscous loss and ensure optimal impedance matching.

Coiled AMs can also be used to design low-frequency, ultrathin, near-perfect sound absorbers [72–76]. As shown in Fig. 1.8a, b, both metasurface unit cells consist of a centered perforated plate and a spiral coplanar air chamber [72, 73]. Sound energy is absorbed into the spiral chamber due to the resonant states of the structure and near-perfect impedance matching with the external air. The absorbed sound energy propagates through the coiled channel, increasing the total propagation distance of the sound, which results in low velocity and a high refractive index. This phenomenon promotes the dissipation of sound energy through conversion into heat, as shown in Fig. 1.8c, d. Importantly, the total thicknesses of both metasurfaces are as small as $\lambda/50$ and $\lambda/223$, respectively.

To simultaneously achieve a high refractive index and high transmission, a transmission-type metasurface unit cell can be constructed using Helmholtz

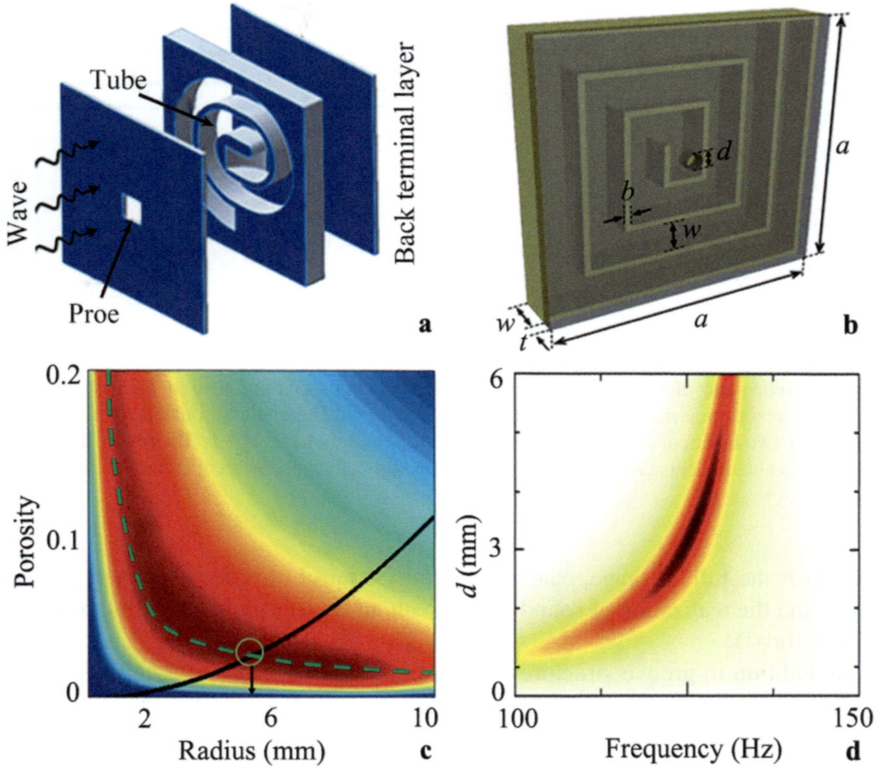

Fig. 1.8 **a**, **b** Schematic of two types of metasurface unit cells composed of a centred perforated plate and a spiral coplanar air chamber. **c** Simulated absorption coefficients caused by the unit cells with different values of panel porosity and tube radius in (**a**). **d** Simulated absorption spectra caused by the unit cells with different values of *d* in (**b**)

resonators for flexible phase control and a straight channel to introduce hybrid resonances that compensate for impedance mismatch. The corresponding metasurface unit cells (Fig. 1.9a–c) can achieve tunable phase velocity and near-perfect transmission efficiency [31, 64, 65]. Additionally, the subwavelength width of the unit cell enables fine spatial resolution of phase distributions. Similar to coiling-up space structures, by simultaneously tuning the heights of the Helmholtz resonators and the straight channel, the metasurface unit cells can achieve phase delays across the full 2π range and maintain high transmission, as shown in Fig. 1.9d–f, respectively. These metasurface unit cells can be used to design acoustic lenses with functionalities such as focusing, asymmetric focusing, and vortex generation.

Given the resonant characteristics of coiling-up space structures and Helmholtz resonators, these unit cells inherently have a limited working bandwidth. This limitation can be overcome by using 1D and two-dimensional (2D) reflection-type metasurfaces composed of grooves with subwavelength widths [77–79], as shown in Fig. 1.10a–c. The reflected phase delays caused by grooves of different depths

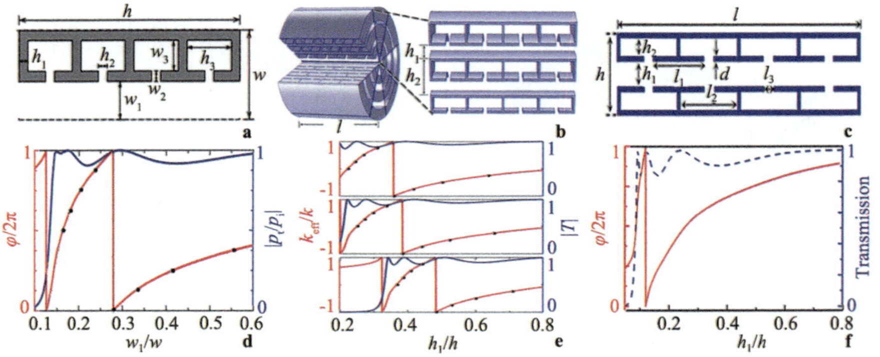

Fig. 1.9 Schematic of three types of metasurface unit cells composed of **a** a single layer of Helmholtz resonators with a straight channel, **b** three layers of Helmholtz resonators with two channels, and **c** two layers of Helmholtz resonators with a central channel. **d–f** The phase delays and transmission coefficients caused by the metasurface unit cells with different structural parameters in (**a–c**)

can cover the full 2π range, enabling arbitrary control of reflected acoustic wavefronts and the realization of sound effects without significant bandwidth limitations (Fig. 1.10d–f).

In addition to groove structures, the design of metasurface unit cells by filling them with gases of different volume ratios or air at different temperatures can also effectively overcome bandwidth limitations [80–82]. This is because the gas inside the unit cells can achieve near-perfect impedance matching with the external air. As shown in Fig. 1.11a, by filling the unit cells with two types of gases (argon and xenon) at different volume ratios [80], the phase delay of sound can cover the full 2π

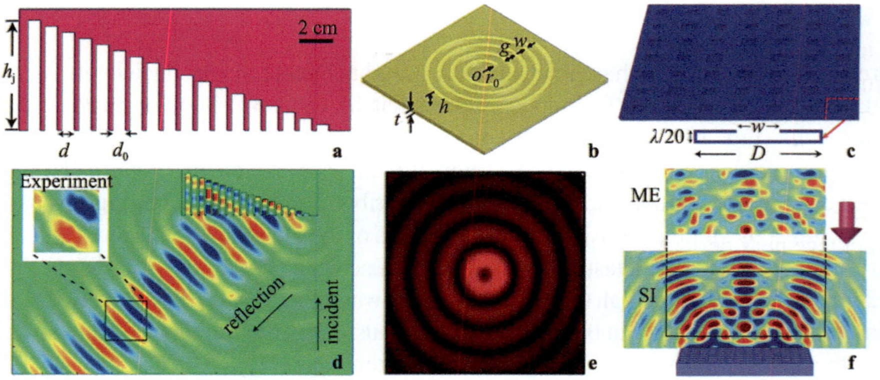

Fig. 1.10 Schematic of **a** a 1D reflection-type metasurface composed of grooves with periodic depth, **b** an Archimedean spiral-shaped grating engraved on a back surface, and **c** a 2D metasurface-based Schroeder diffuser. Measured and simulated field distributions of **d** abnormal sound refraction, **e** spiral sound beam, and **f** scattered acoustic wave caused by the metasurfaces in (**a–c**)

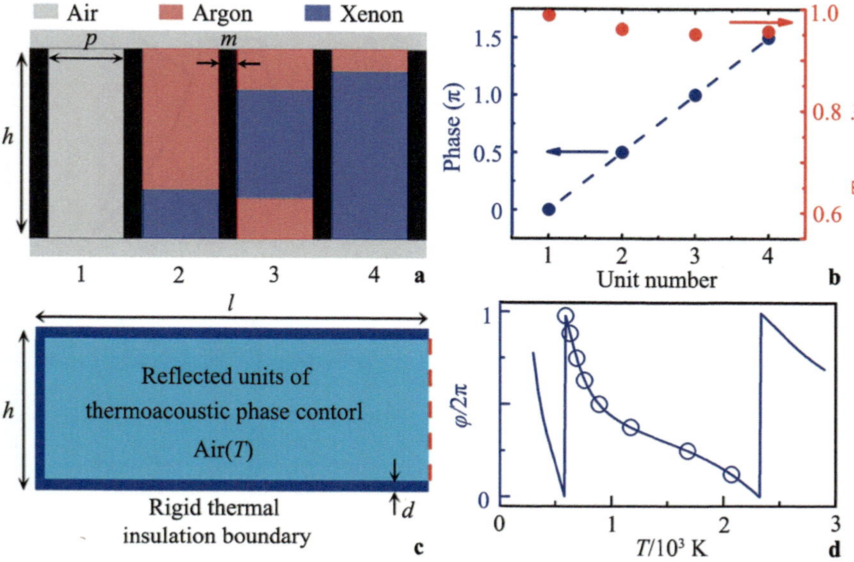

Fig. 1.11 **a** Schematic of metasurface unit cells filled with two types of gases (argon and xenon) at different volume ratios, and **b** simulated phase delays and transmittances caused by different unit cells. **c** Schematic of a reflection-type unit cell filled with air at different temperatures, and **d** the corresponding phase delays caused by the unit cells at different temperatures

range, and high transmission can be achieved, as shown in Fig. 1.11b. Additionally, as shown in Fig. 1.11c, a reflection-type unit cell filled with air [82] can achieve reflected phase delays across the 2π range by tuning the air temperature (Fig. 1.11d). Both types of metasurface unit cells exhibit broadband characteristics.

Finally, to realize a no-reflection surface, an impedance-matched metasurface with hybrid resonances can be designed. When an acoustically reflecting surface is impedance-matched to incident waves, no sound reflection is generated, and sound transmission is also prevented. As shown in Fig. 1.12a, the impedance-matched metasurface unit cell consists of a decorated membrane, sealed gas, and a reflecting hard surface [83]. The decorated membrane exhibits two resonant modes (Fig. 1.12b), which hybridize due to the presence of the sealed gas and the reflecting sound surface. The impedance-matched metasurface, with its deep subwavelength thickness, can achieve near-perfect low-frequency sound absorption (Fig. 1.12c).

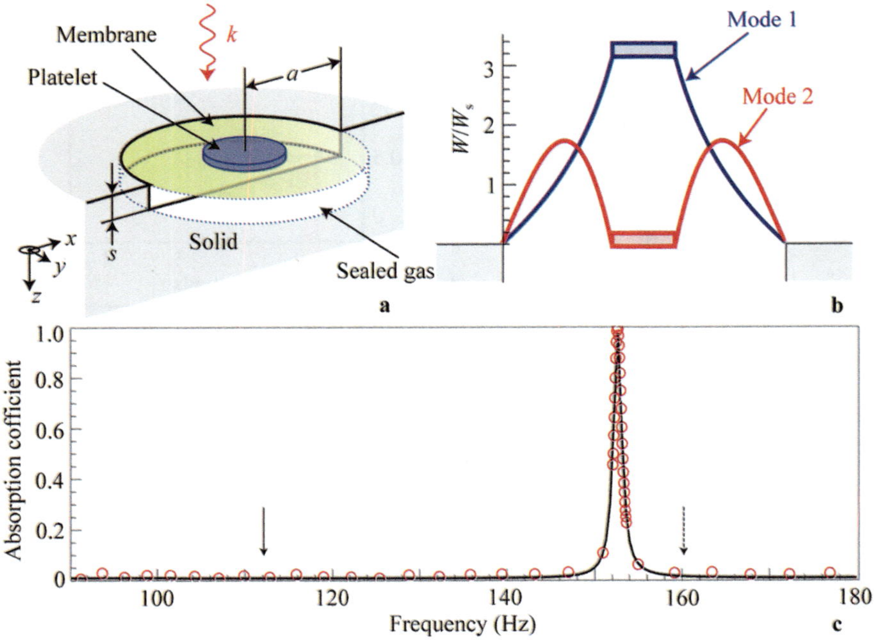

Fig. 1.12 **a** Schematic of an impedance-matched metasurface, and **b** the two lowest-frequency eigenmodes of a decorated membrane. **c** Measured absorption coefficient as a function of frequency

1.3 Acoustic One-Way Transmission

In nature, sound propagation follows the reciprocity law, meaning that acoustic waves can symmetrically propagate through a medium in two opposite directions. In recent years, there has been growing interest in achieving non-reciprocity in sound propagation due to the increasing demand for various acoustic non-reciprocal devices, such as acoustic diodes, acoustic circulators, and acoustic rectifiers. The rapid development of acoustic metamaterials has made it feasible to achieve acoustic one-way transmission (AOT).

First, the symmetry of time reversal can be broken using nonlinear media [84–87]. For example, by combining a nonlinear medium (such as a microbubble suspension) and a 1D SC (Fig. 1.13a), an acoustic diode with one-way transmission of sound can be achieved by breaking the time reversal symmetry [84, 85]. In this system, when an acoustic wave with frequency ω is incident from the nonlinear medium, the frequency of the second harmonic generated by the nonlinear effect becomes 2ω, which falls within the pass band of the SC. As a result, the sound can pass through the acoustic diode. In this case, the acoustic diode is conductive, and its conductivity increases gradually with the sound intensity. However, when the acoustic wave with the same frequency propagates in the opposite direction, the sound energy is completely reflected by the SC due to the bandgap, preventing it from passing through

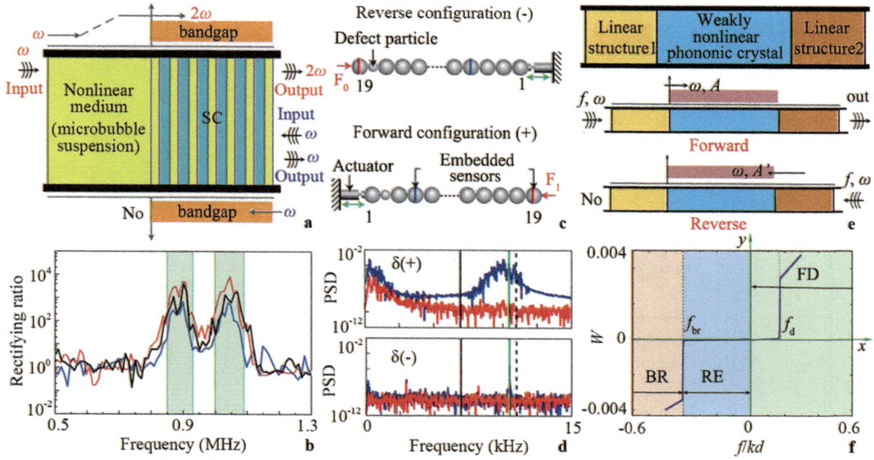

Fig. 1.13 **a** Schematic of an acoustic diode composed of a nonlinear medium and a SC, and **b** measured rectifying ratio spectra. **c** Schematic of a granular crystal composed of a statically compressed array of particles with a light mass defect, and **d** its power spectral density spectra. **e** Schematic of an acoustic diode consisting of two types of linear structures and a nonlinear SC, and **f** its output power versus driving force

the acoustic diode. The rectifying ratio of this system can reach approximately 10^4 (Fig. 1.13b).

Similarly, a granular crystal composed of a statically compressed array of particles with a light mass defect near a boundary (Fig. 1.13c), and an acoustic diode consisting of two types of linear structures and a nonlinear SC (Fig. 1.13e) can also be used to achieve AOT [86, 87]. These systems exhibit obvious AOT characteristics (Fig. 1.13d) and share several similarities with electrical diodes (Fig. 1.13f). However, the sound transmission efficiency of these systems is relatively low due to the conversion efficiency of the nonlinear media.

Second, as shown in Fig. 1.14a, by designing a non-reciprocal active unit cell composed of a single piezoelectric membrane augmented by a nonlinear electronic circuit and sandwiched between two highly subwavelength Helmholtz cavities with different resonance frequencies [88], AOT can be achieved (Fig. 1.14b). In this system, the piezoelectric membrane excites a strong second harmonic, and the two Helmholtz cavities with different frequencies create the asymmetry necessary for non-reciprocal behavior. Inspired by this mechanism, a type of AOT system consisting of a nonlinear medium, acoustic gain, and lossy materials (Fig. 1.14c) has been realized [89]. Based on the nonlinear medium's characteristic of responding nonlinearly to input sound signals with different amplitudes, the designed system can achieve AOT with an invariant frequency identical to that of the incident wave (Fig. 1.14d).

In electronic systems, the symmetry of time reversal can be broken using magnetic fields. However, this mechanism cannot be applied to acoustics because sound is magnetically inert. To overcome this, circulating airflow and active circuits have

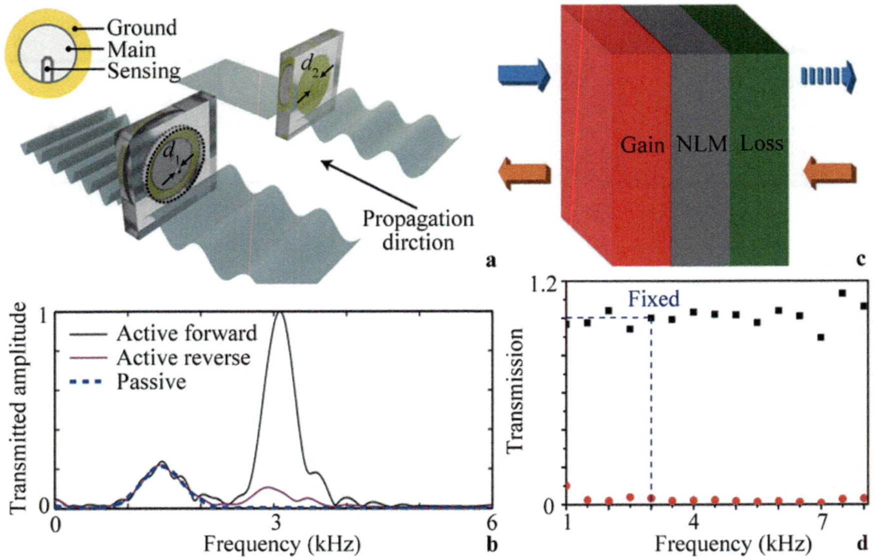

Fig. 1.14 **a** Schematic of an acoustic non-reciprocal system composed of an active unit cell and two Helmholtz cavities, **b** and its transmitted amplitude spectra. **c** Schematic of an acoustic non-reciprocal system consisting of a nonlinear medium and acoustic gain and lossy materials, **d** and its transmission spectra

been introduced into acoustic systems to create an analogous magnetic field for sound, which can break time-reversal symmetry. As shown in Fig. 1.15a, a biased circulator is composed of a subwavelength resonant ring cavity and three fans [38]. When the fans are turned on, a circulating airflow is generated in the ring cavity, and the resonant frequency ω_0 splits into two frequencies $\omega_{\pm}$, analogous to the Zeeman effect caused by a magnetic field. This analogous magnetic field can direct sound propagation to one of the output ports (Fig. 1.15e), achieving the AOT effect. Additionally, similar to the resonant cavity, time-reversal symmetry can be broken by introducing circulating airflow around each unit cell of the SCs [90–92]. As shown in Fig. 1.15b, when circulating airflow is introduced into a hexagonal-lattice SC [90], edge states with AOT localized on its boundary can be obtained (Fig. 1.15f). This phenomenon can also be realized in honeycomb-lattice SCs [91, 92] with circulating airflow (Fig. 1.15c, d), and the propagation of edge states is shown in Fig. 1.15g, h.

However, in experiments, non-synchronous rotation and flow instabilities of the circulating airflow can significantly affect the performance of AOT, posing a major challenge. To address this, a rotating chiral structure composed of ring resonators and chiral-structured fins with rotational symmetry has been developed [43], as shown in Fig. 1.16a. A sample photograph of this structure is shown in Fig. 1.16b. This system generates a stable and uniform airflow even at low rotation speeds of the fins. When an acoustic wave is incident from port P_a, the transmission coefficients at port P_b are larger than those at port P_c at different frequencies (Fig. 1.16c), demonstrating the AOT effect.

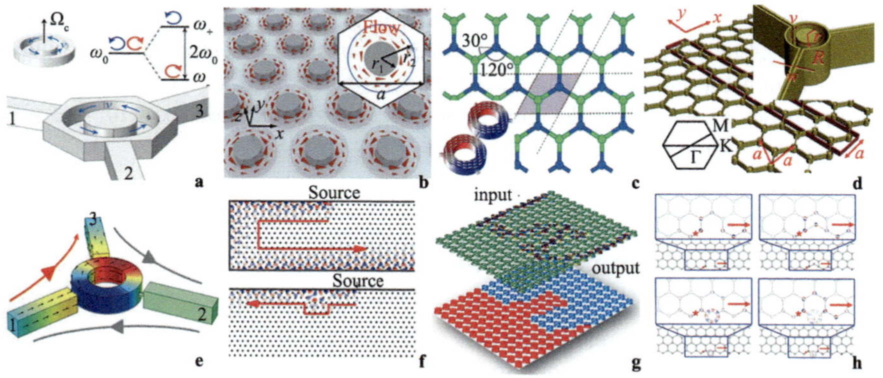

Fig. 1.15 **a** Schematic of a biased circulator composed of a subwavelength resonant ring cavity and three fans. Schematic of **b** a hexagonal-lattice SC and **c**, **d** two types of honeycomb-lattice SCs with circulating airflow. **e** Simulated field distribution of AOT caused by the biased circulator in (**a**), and **f–h** simulated field distributions of edge states with the AOT propagating along the boundaries of the SCs in (**b–d**), respectively

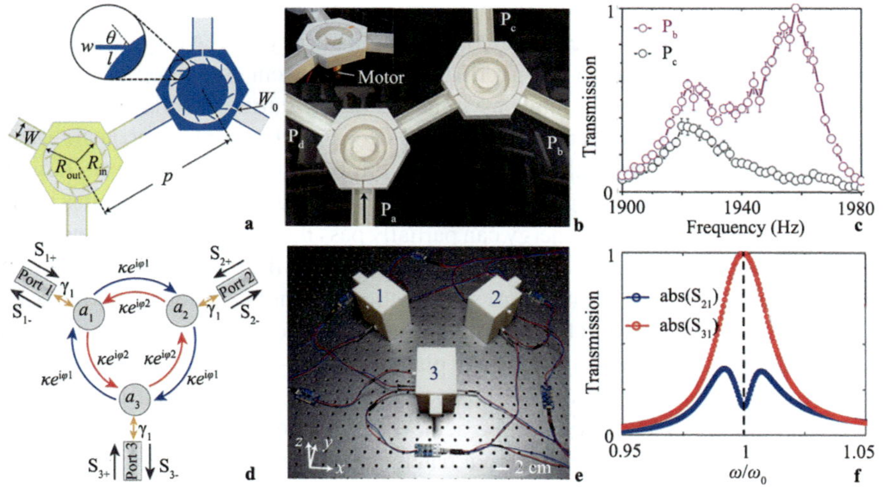

Fig. 1.16 **a** Schematic of a rotating chiral structure composed of a ring resonator and a chiral-structured fin, and **b** its sample photograph. **c** Measured transmission spectra at output ports P_b and P_c of the rotating chiral structure in (**b**). **d** Schematic of the tight-binding model of an acoustic circulator, and **e** its sample photograph. **f** Measured transmission spectra at two output ports of the acoustic circulator

In addition to circulating airflow, asymmetric Peierls phases caused by active circuits also provide a means to achieve AOT [52]. Inspired by non-Hermitian acoustics, an acoustic circulator composed of three resonant cavities and six unidirectional amplifiers has been designed to achieve AOT. The tight-binding model and a sample photograph of this circulator are shown in Fig. 1.16d, e, respectively. The

Peierls phases are defined as the phases of couplings between each site (φ_1 and φ_2 in Fig. 1.16d). Figure 1.16f shows the measured transmission spectra at different output ports, demonstrating the AOT characteristics of the acoustic circulator. This work also provides a method for designing compact, integratable devices with one-way sound transmission.

1.4 Acoustic Asymmetric Transmission

Over the past decade, the realization of acoustic asymmetric transmission (AAT) has attracted significant attention due to its wide range of potential applications. The AAT effect, caused by asymmetric linear structures, can be observed along specific propagation paths. Unlike AOT, AAT does not violate the classical reciprocity principle. AAT systems can significantly improve the conductivity of sound energy and expand the working bandwidth, effectively enhancing the practical applicability of AOT systems [93]. As a result, a series of theoretical and experimental studies on AAT effects based on various asymmetric linear systems have been conducted.

First, by using the intrinsic characteristics of SCs, such as the directionality of bandgaps and asymmetric mode conversion, a variety of AAT systems can be designed [94–99]. As shown in Fig. 1.17a, an AAT system composed of a SC and a diffraction structure [94] exhibits asymmetric sound transmission in two working bands (Fig. 1.17d). This phenomenon occurs because when a plane acoustic wave with frequency ω is vertically incident from the SC, the sound energy is largely reflected due to the SC's bandgap. However, when the acoustic wave propagates in the opposite direction, the sound energy can partially pass through the AAT system. This is because the sound energy is converted into high-order diffractions by the diffraction structure, and some of these high-order diffractions are not constrained by the SC's bandgap, allowing the sound to transmit through the system. Additionally, as shown in Fig. 1.17b, c, by designing a 1D SC plate with anti-symmetric and symmetric meta-atoms [95] and a directional acoustic waveguide with a built-in SC [96], high-performance AAT effects can be achieved. These effects are based on asymmetric mode conversion and directional bandgaps, as demonstrated in Fig. 1.17e, f, respectively.

Second, the AAT effect can also be achieved by exploiting the unique characteristics of acoustic metamaterials [100–109]. As shown in Fig. 1.18a, an asymmetric steel grating structure [100] achieves the AAT effect through different slit periods on the two surfaces of the grating. The working band of this system can be easily determined by the grating periods (Fig. 1.18d). Furthermore, as shown in Fig. 1.18b, an AAT prism composed of coiling-up space unit cells with near-zero refractive index [14] allows energy transmission only for sound waves incident normally on the prism's surface due to the inherent tunneling effect. When a plane acoustic wave is incident from the left inclined surface of the prism, the sound energy cannot pass through. However, when the wave is incident from the right side, it can transmit

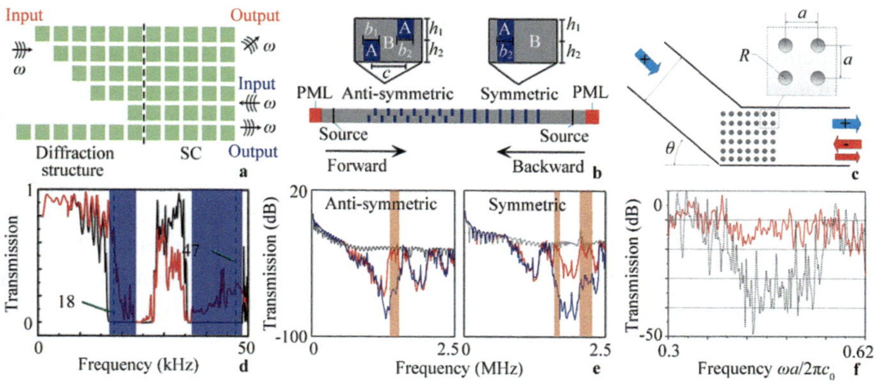

Fig. 1.17 Schematic of **a** an AAT system composed of a SC and a diffraction structure, **b** a 1D SC with symmetric and anti-symmetric meta-atoms, and **c** a directional acoustic waveguide with a built-in SC. **d**–**f** Measured and simulated transmission spectra for both incident directions of the AAT systems in (**a**–**c**)

through the prism due to its vertical incidence on the surface (Fig. 1.18e). Additionally, the propagation path of sound can be controlled using a metamaterial with a refractive index gradient [105], enabling high-performance AAT (Fig. 1.18c, f).

To further improve the bandwidth of AAT systems, it is necessary to achieve asymmetric distributions of refractive index gradients within a single medium. This approach effectively overcomes the large impedance difference between the lens material and the surrounding medium. As shown in Fig. 1.19a, by arranging six

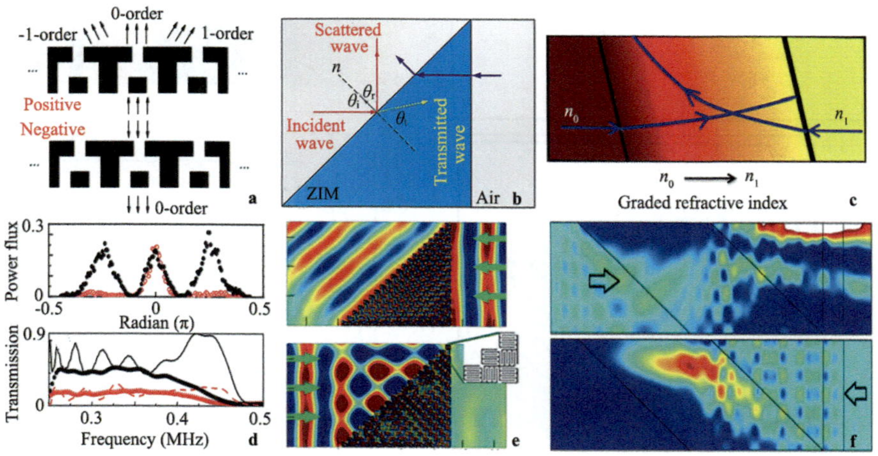

Fig. 1.18 Schematic of **a** an asymmetric steel grating structure, **b** an AAT prism composed of coiling-up space unit cells with near-zero refractive index, and **c** a metamaterial with refractive index gradient. **d**–**f** Simulated and measured intensity distributions caused by the AAT structures in (**a**–**c**)

heat sources at different temperatures in an asymmetric array [110], asymmetric distributions of temperature and refractive index gradients in air can be achieved simultaneously (Fig. 1.19b). The AAT effect exists above 2.5 kHz, with a contrast ratio greater than 0.9 above 5.0 kHz (Fig. 1.19c, d), demonstrating the broadband characteristics of this thermoacoustic AAT system.

Finally, by using the phase control capabilities of AMs, AAT systems with subwavelength thickness and planar structures can be realized [111–123]. As shown in Fig. 1.20a, an AAT tunnel can be designed using two pairs of AMs attached to both sides of a waveguide [111]. The propagation path of sound can be controlled based on the generalized Snell's law, enabling the AAT effect (Fig. 1.20d). Moreover, a dual-layer AAT lens composed of a near-zero index structure and a metasurface [16] achieves the AAT effect by manipulating sound waves through the AMs and leveraging the directional selectivity of the near-zero refractive index material (Fig. 1.20b, e). Furthermore, by carefully tailoring losses, a robust and tunable AAT effect can be achieved using gradient-index AMs [115], as shown in Fig. 1.20c, f. Theoretically, the AAT effect arises from the loss-induced suppression of high-order diffraction, which expands the potential applications of lossy AMs.

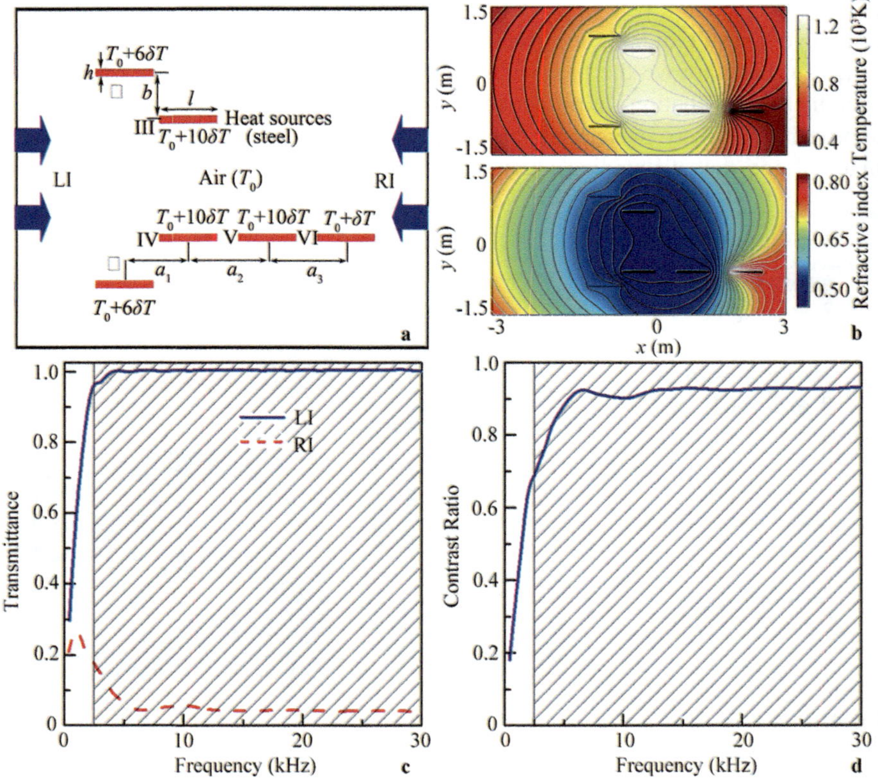

Fig. 1.19 Schematic of **a** a thermoacoustic AAT system composed of six heat sources at different temperatures. **b** Simulated distributions of temperature and refractive index gradients caused by the system. Simulated **c** transmission spectra and **d** contrast ratio for the thermoacoustic AAT system

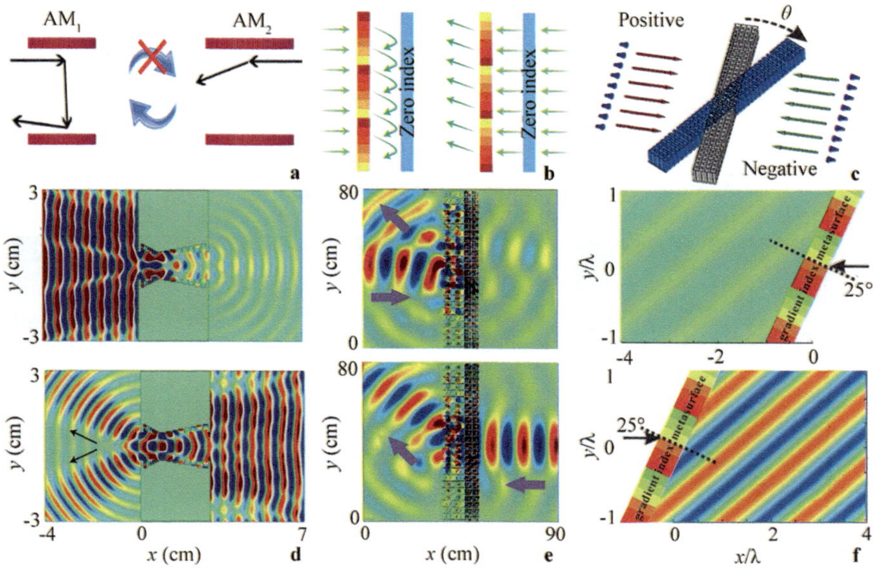

Fig. 1.20 Schematic of **a** an AAT tunnel with two groups of AMs attached on both sides, **b** a dual-layer AAT lens composed of near-zero index structure and metasurface, and **c** a gradient-index AM with a rotation angle of θ. **d–f** Simulated field distributions through the AAT systems with the AMs in (**a–c**)

1.5 Acoustic Focusing

Acoustic focusing (AF) has long been a hot topic due to its practical applications in several important fields [124–133], such as biomedical imaging, nondestructive testing, and ultrasonic medical treatment. Traditionally, AF lenses made of natural materials have been characterized by their large size and curved shape, which limit their practical applications. However, the emergence of acoustic metamaterials with unique sound properties has effectively overcome these limitations, enabling the design of various high-performance AF lenses.

The first type of AF lens is achieved using SCs [134–143]. In this approach, a negative refractive index for sound can be realized by carefully selecting the parameters of the SCs in the direction transverse to the sound propagation, as illustrated in Fig. 1.21. By gradually tuning the refractive index gradient of the SC [135], sound propagation can be effectively manipulated, resulting in high-performance AF effects (Fig. 1.21a, d). Additionally, as shown in Fig. 1.21b, c, high-performance AF lenses can also be achieved by designing SCs with lattice spacing and radius gradients [138, 139], as demonstrated in Fig. 1.21e, f. However, in the low-frequency region, the size of AF lenses composed of SCs tends to be large, as the lattice constant of the SC is comparable to or larger than the wavelength.

Another type of AF lens is constructed using acoustic metamaterials [144–149]. In this case, the unit cells typically have a higher refractive index and a smaller

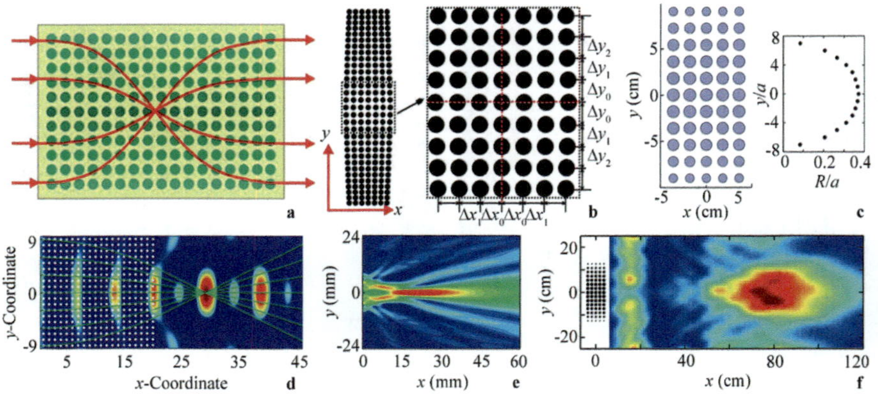

Fig. 1.21 Schematic of three types of AF lenses composed of SCs with **a** refractive index, **b** lattice spacing, and **c** radius gradients. **d–f** Simulated acoustic intensity distributions caused by the AF lenses in (**a–c**)

size, effectively overcoming the limitations of SC-based AF lenses. As shown in Fig. 1.22a–c, the unit cells are designed as cross structures [144], coiled-up spaces [17], and Hilbert fractal structures [145], respectively. By designing AF lenses composed of a series of unit cells with a specific size gradient, a refractive index gradient for sound can be achieved, resulting in AF lenses with high efficiency and subwavelength thickness, as shown in Fig. 1.22d–f.

As a special case of acoustic metamaterials, AMs also show great potential in advanced AF lenses due to their subwavelength thickness and planar structure [54]. Based on the generalized Snell's law [26], several types of metasurface-based AF

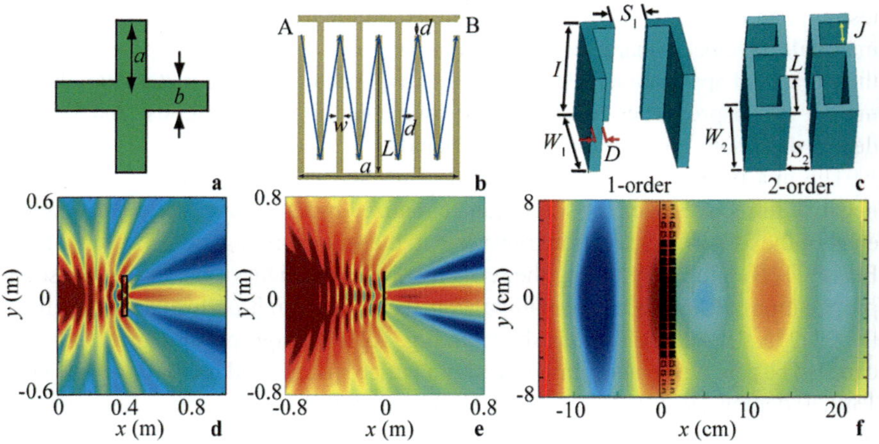

Fig. 1.22 Schematic of three types of unit cells composed of **a** coiled-up space, **b** cross structure, and **c** Hilbert fractal structure. **d–f** Simulated intensity distributions caused by the AF lenses composed of the unit cells in (**a–c**)

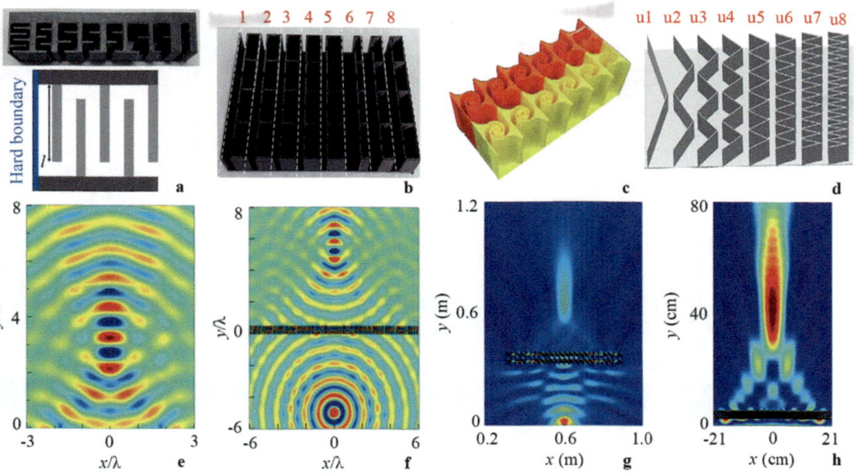

Fig. 1.23 Schematic of four types of metasurface-based unit cells composed of **a** coiled-up space, **b** Helmholtz resonant cavities, **c** labyrinthine structures, and **d** V-shaped meta-atoms. **e**–**h** Simulated acoustic intensity distributions caused by the AF lenses composed of the phased unit cells in (**a**–**d**)

lenses have been demonstrated [18, 20, 21, 24, 27–29, 71, 150–152]. As shown in Fig. 1.23a, by tuning the parameters of the unit cell consisting of coiled-up space [20], eight types of phased unit cells with the same thickness and an equally spaced phase delay from 0 to 2π are realized. A reflected AF lens can then be designed using these eight types of phased unit cells (Fig. 1.23e). Moreover, phased unit cells composed of Helmholtz resonant cavities [150], labyrinthine structures [21], and V-shaped meta-atoms [71] have also been proposed for designing AF lenses (Fig. 1.23b–d). High-performance AF effects can be achieved using these metasurface-based lenses, as shown in Fig. 1.23f–h.

The aforementioned lenses have achieved highly efficient sound focusing. However, their working bands are narrow, often limited to a single frequency. This limitation arises from the complexity of the refractive index gradient of the lens and the large impedance difference between the lens material and the surrounding medium. To achieve broadband AF lenses, it is generally necessary to realize the refractive index gradient in a single medium. To address this, researchers have used two heat sources and a curved thermal insulation interface [153] to control sound propagation through temperature gradients (Fig. 1.24a, b). This approach results in a working bandwidth exceeding 15 kHz (Fig. 1.24c). Additionally, by placing 20 heat sources of different temperatures in a symmetrical array [154], an AF lens can be designed without the need for a thermal insulation interface (Fig. 1.24d, e). The working bandwidth of the AF effect caused by the temperature gradient exceeds 3.5 kHz (Fig. 1.24f). These designs demonstrate the feasibility of realizing broadband acoustic lenses in a single medium.

In addition to 2D AF effects, the realization of AF in three-dimensional (3D) space has attracted significant attention due to its practical applications and value

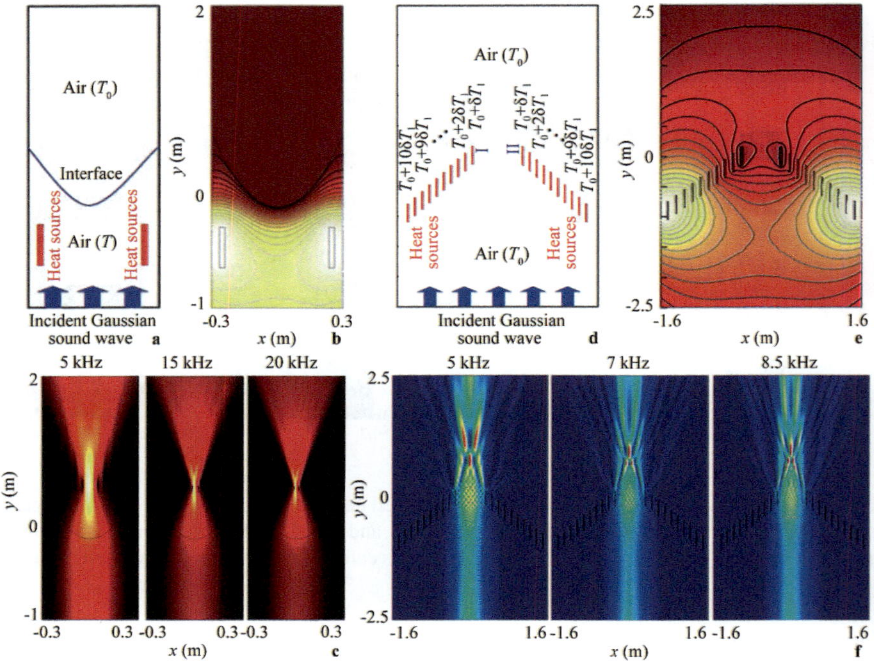

Fig. 1.24 **a** Schematic of the AF lens composed of two heat sources and a curved thermal insulation interface, and simulated distributions of **b** temperature gradient and **c** sound intensity caused by the lens. **d** Schematic of the AF lens composed of twenty heat sources, and simulated distributions of **e** temperature gradient and **f** sound intensity caused by the lens

[155–158]. As shown in Fig. 1.25a, by designing a 2D ultrathin planar acoustic lens composed of hybrid labyrinthine unit cells [155], the AF effect in 3D space can be observed. The sound energy at the focus can be enhanced by up to 15 dB, and the lens thickness is only about 1/6 of the wavelength (Fig. 1.25c). Furthermore, by simply using two types of test tubes with and without lids (Fig. 1.25b), a 2D reflected AF lens with binary phases can be designed [157]. The reflected intensity distributions at 2500, 3500, and 4000 Hz demonstrate broadband AF with high efficiency (Fig. 1.25d). The measured results agree well with the simulated results, showcasing a maximally simplified configuration and strong robustness of the AF effect in 3D space.

1.6 Overview of Main Research in This Book

In the aforementioned works, we observe that the rapid development of AOT, AAT, and AF based on acoustic metamaterials has been achieved over the past two decades. During this time, a series of physical mechanisms and design methods have been

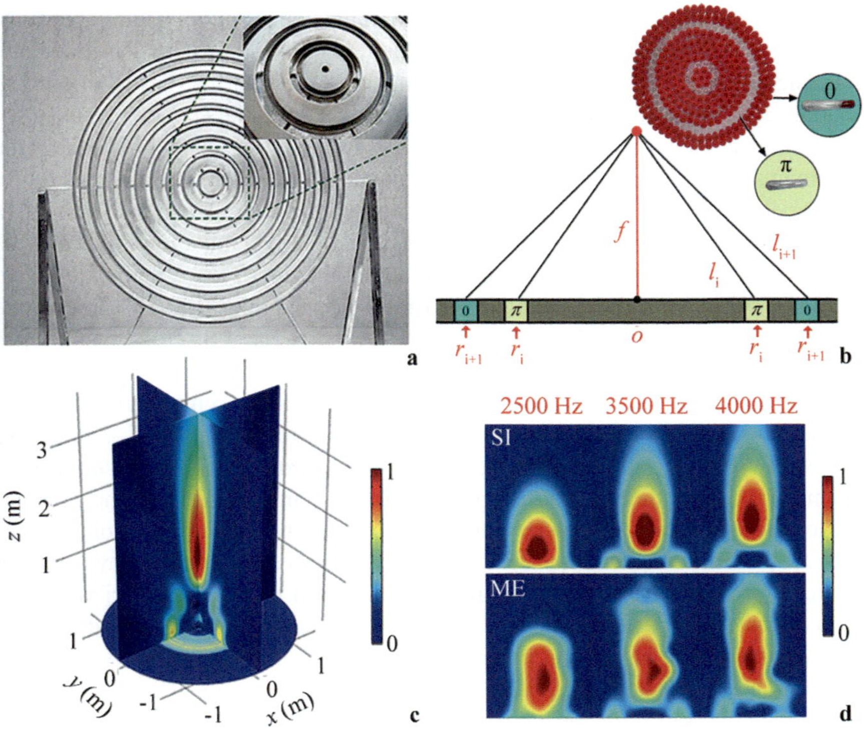

Fig. 1.25 Schematic of two types of 2D AF lenses composed of **a** hybrid labyrinthine unit cells and **b** two types of test tubes with/without lids. **c**, **d** Simulated and measured acoustic intensity distributions caused by both types of AF lenses in (**a**, **b**)

successively developed. However, in some specific application scenarios, the existing mechanisms, performance optimization, and modulation of AAT and AF based on acoustic metamaterials still require in-depth study. Additionally, the design of multifunctional acoustic effects with special applications, based on the coupling of multiple mechanisms, remains a challenge. For example, the effect of acoustic asymmetric focusing (AAF), which is based on the coupling of both AAT and AF effects, holds great promise for applications in medical focused ultrasound therapy. Furthermore, most of these metamaterial systems are 1D structures that achieve corresponding sound effects in 2D space, leading to significant limitations in practical applications. The design of 2D metamaterial systems to achieve sound effects in 3D space still faces considerable challenges.

In this book, based on the research foundation in acoustic metamaterials established by our group, we conduct theoretical and experimental studies on the physical mechanisms and performance modulation of acoustic metamaterials with asymmetric transmission and focusing effects. Our aim is to develop design methods, preparation techniques, and performance modulation strategies for high-performance AAT, AF, and AAF structures. We also explore their practical applications in medical ultrasound, underwater detection, acoustic communication, architectural acoustics, and noise reduction. The achievements of this research not only hold fundamental academic significance in the field of acoustic metamaterials but also have the potential for application in various important fields, offering substantial economic value.

The main content of this book is organized as follows: this chapter introduces the fundamental concepts, historical development, current research hotspots, and future trajectories of acoustic metamaterials and their key branches. Chapter 2 presents a series of fundamental principles and methods for sound manipulation in 2D and 3D spaces based on acoustic metamaterials, including the interference theory, effective medium (EM) theory, finite element (FE) method, generalized Snell's law, mode conversion theory, and band theory of SCs. Chapters 3 and 4 discuss several types of AAT systems designed using underwater plate-like metamaterials and airborne metamaterials. The potential applications of these systems in architectural acoustics, noise control, and medical ultrasound are explored in detail. Chapters 5–7 showcase a series of AF lenses designed using ring-shaped metamaterials, phase-controlled metamaterials, and thermoacoustic phase-controlled metamaterials. The applications of these AF lenses in underwater signal detection, medical ultrasound treatment, acoustic communication, and acoustic energy harvesting are also investigated. Chapter 8 presents five types of AAF lenses designed using asymmetric modulation metamaterials. These lenses have practical applications in focused ultrasound therapy and medical ultrasound imaging. Chapter 9 demonstrates four types of 2D acoustic lenses designed using phase-controlled metamaterials. These lenses can achieve the effects of AF, acoustic vortex focusing (AVF), AAF, and acoustic asymmetric vortex focusing (AAVF) in 3D space. These lenses have potential applications in non-contact manipulation for trapping particles, medical ultrasound imaging and treatment. Chapter 10 summarizes acoustic metamaterial advancements in AAT, AF, and AAF, addressing challenges and future directions like broadband devices, 3D wavefront control, and machine learning, with transformative applications in medical ultrasound, communication, and environmental acoustics.

References

1. Cummer SA, Christensen J, Alù A (2016) Controlling sound with acoustic metamaterials. Nat Rev Mater 1(3):16001. https://doi.org/10.1038/natrevmats.2016.1
2. Kushwaha MS, Halevi P, Dobrzynski L, Djafari-Rouhani B (1993) Acoustic band structure of periodic elastic composites. Phys Rev Lett 71(13):2022. https://doi.org/10.1103/PhysRevLett.71.2022
3. Liu ZY, Zhang XX, Mao YW, Zhu YY, Yang ZY, Chan CT, Sheng P (2000) Locally resonant sonic materials. Science 289(5485):1734–1736. https://doi.org/10.1126/science.289.5485.1734
4. Liu ZY, Chan CT, Sheng P (2005) Analytic model of phononic crystals with local resonances. Phys Rev B 71(1):014103. https://doi.org/10.1103/PhysRevB.71.014103
5. Milton GW, Willis JR (2007) On modifications of Newton's second law and linear continuum elastodynamics. Proc R Soc A 463(2079):855–880. https://doi.org/10.1098/rspa.2006.1795

6. Yao SS, Zhou XM, Hu GK (2008) Experimental study on negative effective mass in a 1D mass-spring system. New J Phys 10:043020. https://doi.org/10.1088/1367-2630/10/4/043020
7. Yang Z, Mei J, Yang M, Chan NH, Sheng P (2008) Membrane-type acoustic metamaterial with negative dynamic mass. Phys Rev Lett 101(20):204301. https://doi.org/10.1103/PhysRevLett.101.204301
8. Fang N, Xi DJ, Xu JY, Ambati M, Srituravanich W, Sun C, Zhang X (2006) Ultrasonic metamaterials with negative modulus. Nat Mater 5(6):452–456. https://doi.org/10.1038/nmat1644
9. Lee SH, Park CM, Seo YM, Wang ZG, Kim CK (2009) Acoustic metamaterial with negative modulus. J Phys-Condens Mat 21(17):175704. https://doi.org/10.1088/0953-8984/21/17/175704
10. Li J, Chan CT (2004) Double-negative acoustic metamaterial. Phys Rev E 70(5):055602. https://doi.org/10.1103/PhysRevE.70.055602
11. Fok L, Zhang X (2011) Negative acoustic index metamaterial. Phys Rev B 83(21):214304. https://doi.org/10.1103/PhysRevB.83.214304
12. Cheng Y, Zhou C, Yuan BG, Wu DJ, Wei Q, Liu XJ (2015) Ultra-sparse metasurface for high reflection of low-frequency sound based on artificial Mie resonances. Nat Mater 14(10):1013–1019. https://doi.org/10.1038/NMAT4393
13. Liang ZX, Li J (2012) Extreme acoustic metamaterials by coiling up space. Phys Rev Lett 108(11):114301. https://doi.org/10.1103/PhysRevLett.108.114301
14. Li Y, Liang B, Gu ZM, Zou XY, Cheng JC (2013) Unidirectional acoustic transmission through a prism with near-zero refractive index. Appl Phys Lett 103(5):053505. https://doi.org/10.1063/1.4817249
15. Li Y, Liang B, Zou XY, Cheng JC (2013) Extraordinary acoustic transmission through ultrathin acoustic metamaterials by coiling up space. Appl Phys Lett 103(6):063509. https://doi.org/10.1063/1.4817925
16. Shen C, Xie YB, Li JF, Cummer SA, Jing Y (2016) Asymmetric acoustic transmission through near-zero-index and gradient-index metasurfaces. Appl Phys Lett 108(22):223502. https://doi.org/10.1063/1.4953264
17. Li Y, Liang B, Tao X, Zhu XF, Zou XY, Cheng JC (2012) Acoustic focusing by coiling up space. Appl Phys Lett 101(23):233508. https://doi.org/10.1063/1.4769984
18. Li Y, Liang B, Gu ZM, Zou XY, Cheng JC (2013) Reflected wavefront manipulation based on ultrathin planar acoustic metasurfaces. Sci Rep 3:2546. https://doi.org/10.1038/srep02546
19. Xie YB, Wang WQ, Chen HY, Konneker A, Popa BI, Cummer SA (2014) Wavefront modulation and subwavelength diffractive acoustics with an acoustic metasurface. Nat Commun 5:5553. https://doi.org/10.1038/ncomms6553
20. Li Y, Jiang X, Li RQ, Liang B, Zou XY, Yin LL, Cheng JC (2014) Experimental realization of full control of reflected waves with subwavelength acoustic metasurfaces. Phys Rev Appl 2(6):064002. https://doi.org/10.1103/PhysRevApplied.2.064002
21. Wang WQ, Xie YB, Konneker A, Popa BI, Cummer SA (2014) Design and demonstration of broadband thin planar diffractive acoustic lenses. Appl Phys Lett 105(10):101904. https://doi.org/10.1063/1.4895619
22. Xie YB, Konneker A, Popa BI, Cummer SA (2013) Tapered labyrinthine acoustic metamaterials for broadband impedance matching. Appl Phys Lett 103(20):201906. https://doi.org/10.1063/1.4831770
23. Tang K, Qiu CY, Ke MZ, Lu JY, Ye YT, Liu ZY (2014) Anomalous refraction of airborne sound through ultrathin metasurfaces. Sci Rep 4:6517. https://doi.org/10.1038/srep06517
24. Yuan BG, Cheng Y, Liu XJ (2015) Conversion of sound radiation pattern via gradient acoustic metasurface with space-coiling structure. Appl Phys Express 8(2):027301. https://doi.org/10.7567/APEX.8.027301
25. Fu YY, Shen C, Cao YY, Gao L, Chen HY, Chan CT, Cummer SA, Xu YD (2019) Reversal of transmission and reflection based on acoustic metagratings with integer parity design. Nat Commun 10:2326. https://doi.org/10.1038/s41467-019-10377-9

26. Yu NF, Genevet P, Kats MA, Aieta F, Tetienne JP, Capasso F, Gaburro Z (2011) Light propagation with phase discontinuities: generalized laws of reflection and refraction. Science 334(6054):333–337. https://doi.org/10.1126/science.1210713
27. Xie HF, Hou ZL (2021) Nonlocal metasurface for acoustic focusing. Phys Rev Appl 15(3):034054. https://doi.org/10.1103/PhysRevApplied.15.034054
28. Zeng LS, Li ZM, Lin ZB, Wu H, Peng YG, Zhu XF (2022) Far-field super-resolution focusing with weak side lobes and defect detection via an ultrasonic meta-lens of sharp-edge apertures. Appl Phys Lett 120(20):202202. https://doi.org/10.1063/5.0094606
29. Ren J, Hou ZL (2024) Highly efficient nonlocal metasurface for transmission-type acoustic focusing. Phys Rev Appl 22(1):014040. https://doi.org/10.1103/PhysRevApplied.22.014040
30. Zhang P, Li TC, Zhu J, Zhu XF, Yang S, Wang Y, Yin XB, Zhang X (2014) Generation of acoustic self-bending and bottle beams by phase engineering. Nat Commun 5:4316. https://doi.org/10.1038/ncomms5316
31. Li Y, Jiang X, Liang B, Cheng JC, Zhang LK (2015) Metascreen-based acoustic passive phased array. Phys Rev Appl 4(2):024003. https://doi.org/10.1103/PhysRevApplied.4.024003
32. Ren XM, Zhou QX, Xu Z, Liu XJ (2021) Particle trapping in arbitrary trajectories using first-order Bessel-like acoustic beams. Phys Rev Appl 15(5):054041. https://doi.org/10.1103/PhysRevApplied.15.054041
33. Li XS, Zhou HT, Wang YF, Wang YS (2011) Modulation of acoustic self-accelerating beams with tunable curved metasurfaces. Appl Phys Lett 118(2):023503. https://doi.org/10.1063/5.0035286
34. Dubois M, Shi CZ, Wang Y, Zhang X (2017) A thin and conformal metasurface for illusion acoustics of rapidly changing profiles. Appl Phys Lett 110(15):151902. https://doi.org/10.1063/1.4979978
35. Wang X, Mao DX, Li Y (2017) Broadband acoustic skin cloak based on spiral metasurfaces. Sci Rep 7:11604. https://doi.org/10.1038/s41598-017-11846-1
36. Wang YH, Cheng Y, Liu XJ (2019) Ultrathin acoustic cloaking by a conformal hybrid metasurface. Sci Rep 9:12700. https://doi.org/10.1038/s41598-019-49148-3
37. Popa BI, Zigoneanu L, Cummer SA (2013) Tunable active acoustic metamaterials. Phys Rev B 88(2):024303. https://doi.org/10.1103/PhysRevB.88.024303
38. Fleury R, Sounas DL, Sieck CF, Haberman MR, Alù A (2014) Sound isolation and giant linear nonreciprocity in a compact acoustic circulator. Science 343(6170):516–519. https://doi.org/10.1126/science.1246957
39. Bergamini A, Delpero T, Simoni LD, Lillo LD, Ruzzene M, Ermanni P (2014) Phononic crystal with adaptive connectivity. Adv Mater 26(9):1343–1347. https://doi.org/10.1002/adma.201305280
40. Popa BI, Shinde D, Konneker A, Cummer SA (2015) Active acoustic metamaterials reconfigurable in real time. Phys Rev B 91(22):220303. https://doi.org/10.1103/PhysRevB.91.220303
41. Xie BY, Tang K, Cheng H, Liu ZY, Chen SQ, Tian JG (2017) Coding acoustic metasurfaces. Adv Mater 29(6):1603507. https://doi.org/10.1002/adma.201603507
42. Xia JP, Jia D, Sun HX, Yuan SQ, Ge Y, Si QR, Liu XJ (2018) Programmable coding acoustic topological insulator. Adv Mater 30(46):1805002. https://doi.org/10.1002/adma.201805002
43. Ding YJ, Peng YG, Zhu YF, Fan XD, Yang J, Liang B, Zhu XF, Wan XG, Cheng JC (2019) Experimental demonstration of acoustic Chern insulators. Phys Rev Lett 122(1):014302. https://doi.org/10.1103/PhysRevLett.122.014302
44. Tian ZH, Shen C, Li JF, Reit E, Gu YY, Fu H, Cummer SA, Huang TJ (2019) Programmable acoustic metasurfaces. Adv Funct Mater 29(13):1808489. https://doi.org/10.1002/adfm.201808489
45. Zhang L, Yang YH, Ge Y, Guan YJ, Chen QL, Yan QH, Chen FJ, Xi R, Li YZ, Jia D, Yuan SQ, Sun HX, Chen HS, Zhang BL (2021) Acoustic non-Hermitian skin effect from twisted winding topology. Nat Commun 12(1):6297. https://doi.org/10.1038/s41467-021-26619-8
46. Zhang QC, Li YT, Sun HF, Liu X, Zhao LK, Feng XL, Fan XY, Qiu CY (2023) Observation of acoustic non-Hermitian Bloch braids and associated topological phase transitions. Phys Rev Lett 130(1):017201. https://doi.org/10.1103/PhysRevLett.130.017201

47. Zhang QC, Zhao LK, Liu X, Feng XL, Xiong LW, Wu WQ, Qiu CY (2023) Experimental characterization of three-band braid relations in non-Hermitian acoustic lattices. Phys Rev Res 5(2):L022050. https://doi.org/10.1103/PhysRevResearch.5.L022050
48. Chen ZX, Li ZW, Weng JK, Liang B, Lu YQ, Cheng JC, Alù A (2023) Sound non-reciprocity based on synthetic magnetism. Sci Bull 68(19):2164–2169. https://doi.org/10.1016/j.scib.2023.08.013
49. Chen ZX, Peng YG, Chen ZG, Liu Y, Chen P, Zhu XF, Lu YQ (2024) Robust temporal adiabatic passage with perfect frequency conversion between detuned acoustic cavities. Nat Commun 15(1):1478. https://doi.org/10.1038/s41467-024-45932-6
50. Chen ZX, Ma LL, Ge SJ, Chen ZG, Lu MH, Chen YF, Lu YQ (2024) Transient logic operations in acoustics through dynamic modulation. Phys Rev Appl 21(1):L011001. https://doi.org/10.1103/PhysRevApplied.21.L011001
51. Chen ZX, Chen A, Peng YG, Li ZW, Liang B, Yang J, Zhu XF, Lu YQ, Cheng JC (2024) Observation of acoustic Floquet π modes in a time-varying lattice. Phys Rev B 109(2):L020302. https://doi.org/10.1103/PhysRevB.109.L020302
52. Zhang L, Ge Y, Guan YJ, Chen FJ, Han N, Chen QL, Pan Y, Jia D, Yuan SQ, Sun HX, Christensen J, Chen HS, Yang YH (2024) Nonreciprocal acoustic devices with asymmetric Peierls phases. Phys Rev Lett 133(13):136601. https://doi.org/10.1103/PhysRevLett.133.136601
53. Pacchioni G (2017) Acoustic metasurfaces: ultrathin sound diffusers. Nat Rev Mater 2:17047. https://doi.org/10.1038/natrevmats.2017.47
54. Assouar B, Liang B, Wu Y, Li Y, Cheng JC, Jing Y (2018) Acoustic metasurfaces. Nat Rev Mater 3(12):460–472. https://doi.org/10.1038/s41578-018-0061-4
55. Li A, Singh S, Sievenpiper D (2018) Metasurfaces and their applications. Nanophotonics 7(6):989–1011. https://doi.org/10.1515/nanoph-2017-0120
56. Liang B, Cheng JC, Qiu CW (2018) Wavefront manipulation by acoustic metasurfaces: from physics and applications. Nanophotonics 7(6):1191–1205. https://doi.org/10.1515/nanoph-2017-0122
57. Chen AL, Wang YS, Wang YF, Zhou HT, Yuan SM (2022) Design of acoustic/elastic phase gradient metasurfaces: principles, functional elements, tunability, and coding. Appl Mech Rev 74(2):020801. https://doi.org/10.1115/1.4054629
58. Liu DA, Hao LM, Zhu WR, Yang X, Yan XL, Guan C, Xie Y, Pang SF, Chen Z (2023) Recent progress in resonant acoustic metasurfaces. Materials 16(21):7044. https://doi.org/10.3390/ma16217044
59. Zabihi A, Ellouzi C, Shen C (2023) Tunable, reconfigurable, and programmable acoustic metasurfaces: a review. Front Mater 10:1132585. https://doi.org/10.3389/fmats.2023.1132585
60. Qi SB, Li Y, Assouar B (2017) Acoustic focusing and energy confinement based on multilateral metasurfaces. Phys Rev Appl 7(5):054006. https://doi.org/10.1103/PhysRevApplied.7.054006
61. Fan SW, Wang WQ, Dong HW, Liu J, Qi HB, Wang YS (2023) Broadband high-efficiency acoustic vortices via a topology-optimized space-coiling-cavity metasurface. Appl Phys Lett 123(10):127891. https://doi.org/10.1063/5.0154688
62. Cui X, Shi J, Liu X, Lai Y (2021) A panel acoustic energy harvester based on the integration of acoustic metasurface and Helmholtz resonator. Appl Phys Lett 119(25):253903. https://doi.org/10.1063/5.0074701
63. Zou H, Xu Z, Hu Y, Du Q, Peng P (2022) Reflected continuously tunable acoustic metasurface with rotatable space coiling-up structure. Phys Lett A 426:127891. https://doi.org/10.1016/j.physleta.2021.127891
64. Jiang X, Li Y, Liang B, Cheng JC, Zhang LK (2016) Convert acoustic resonances to orbital angular momentum. Phys Rev Lett 117(3):034301. https://doi.org/10.1103/PhysRevLett.117.034301
65. Xia JP, Zhang XT, Sun HX, Yuan SQ, Qian J, Ge Y (2018) Broadband tunable acoustic asymmetric focusing lens from dual-layer metasurfaces. Phys Rev Appl 10(1):014016. https://doi.org/10.1103/PhysRevApplied.10.014016

66. Qian J, Xia JP, Sun HX, Wang Y, Ge Y, Yuan SQ, Yang YH, Liu XJ, Zhang BL (2020) Aperiodic metagratings for high-performance multifunctional acoustic lenses. Adv Mater Technol 5(12):2000542. https://doi.org/10.1002/admt.202000542
67. Feng XD, Wang Y, Shi LJ, Zou HY, Lu YJ, Jia D, Ge Y, Guan YJ, Yuan SQ, Sun HX, Liu XJ (2023) Mode converter of vortex beams by phase-gradient acoustic metagratings. J Appl Phys 133(3):034502. https://doi.org/10.1063/5.0130015
68. Zhu YF, Assouar B (2019) Multifunctional acoustic metasurface based on an array of Helmholtz resonators. Phys Rev B 99(17):174109. https://doi.org/10.1103/PhysRevB.99.174109
69. Chen J, Sun ZQ, Fan Z (2019) Groove-structured meta-surface for patterned sub-diffraction sound focusing. Appl Phys Lett 114(25):254102. https://doi.org/10.1063/1.5096258
70. Xu WK, Zhang M, Lin ZB, Liu CL, Qi WC, Wang W (2019) Anomalous refraction manipulation of Lamb waves using single-groove metasurfaces. Phys Scr 94(10):105807. https://doi.org/10.1088/1402-4896/ab2b01
71. Lan J, Li YF, Liu XZ (2017) Broadband manipulation of refracted wavefronts by gradient acoustic metasurface with V-shape structure. Appl Phys Lett 111(26):263501. https://doi.org/10.1063/1.5005950
72. Cai XB, Guo QQ, Hu GK, Yang J (2014) Ultrathin low-frequency sound absorbing panels based on coplanar spiral tubes or coplanar Helmholtz resonators. Appl Phys Lett 105(12):121901. https://doi.org/10.1063/1.4895617
73. Li Y, Assouar B (2016) Acoustic metasurface-based perfect absorber with deep subwavelength thickness. Appl Phys Lett 108(6):063502. https://doi.org/10.1063/1.4941338
74. Du J, Luo Y, Zhao X, Sun X, Song Y, Hu X (2021) Bilayer ventilated labyrinthine metasurfaces with high sound absorption and tunable bandwidth. Sci Rep 11(1):5829. https://doi.org/10.1038/s41598-021-84986-0
75. Chen JS, Chung YT, Wang CY, Lo WY, Liu CH, Yu CH, Chang IL, Lin TR (2023) Ultrathin arch-like labyrinthine acoustic metasurface for low-frequency sound absorption. Appl Acoust 202:109142. https://doi.org/10.1016/j.apacoust.2022.109142
76. Qi HB, Fan SW, Jiang M, Tang XL, Wang YS (2024) Low-frequency ultra-broadband ventilated muffler based on a resonance-labyrinthine metamaterial. Extreme Mech Lett 67:102120. https://doi.org/10.1016/j.eml.2023.102120
77. Zhu YF, Zou XY, Li RQ, Jiang X, Tu J, Liang B, Cheng JC (2015) Dispersionless manipulation of reflected acoustic wavefront by subwavelength corrugated surface. Sci Rep 5:10966. https://doi.org/10.1038/srep10966
78. Wang T, Ke MZ, Li WP, Yang Q, Qiu CY, Liu ZY (2016) Particle manipulation with acoustic vortex beam induced by a brass plate with spiral shape structure. Appl Phys Lett 109(12):123506. https://doi.org/10.1063/1.4963185
79. Zhu YF, Fan XD, Liang B, Cheng JC, Jing Y (2017) Ultrathin acoustic metasurface-based Schroeder diffuser. Phys Rev X 7(2):021034. https://doi.org/10.1103/PhysRevX.7.021034
80. Mei J, Wu Y (2014) Controllable transmission and total reflection through an impedance-matched acoustic metasurface. New J Phys 16:123007. https://doi.org/10.1088/1367-2630/16/12/123007
81. Song AL, Chen TN, Wang XP, Wan LL (2016) Waveform-preserved unidirectional acoustic transmission based on impedance-matched acoustic metasurface and phononic crystal. J Appl Phys 120(8):085106. https://doi.org/10.1063/1.4961659
82. Hou AR, Gao WT, Qian J, Sun HX, Ge Y, Yuan SQ, Si QR, Liu XJ (2018) Thermoacoustic-reflected focusing lens based on acoustic Bessel-like beam with phase manipulation. Chin Phys B 27(12):124301. https://doi.org/10.1088/1674-1056/27/12/124301
83. Ma GC, Yang M, Xiao SW, Yang ZY, Sheng P (2014) Acoustic metasurface with hybrid resonances. Nat Mater 13(9):873–878. https://doi.org/10.1038/NMAT3994
84. Liang B, Yuan B, Cheng JC (2009) Acoustic diode: rectification of acoustic energy flux in one-dimensional systems. Phys Rev Lett 103(10):104301. https://doi.org/10.1103/PhysRevLett.103.104301

85. Liang B, Guo XS, Tu J, Zhang D, Cheng JC (2010) An acoustic rectifier. Nat Mater 9(12):989–992. https://doi.org/10.1038/NMAT2881
86. Boechler N, Theocharis G, Daraio C (2011) Bifurcation-based acoustic switching and rectification. Nat Mater 10(9):665–668. https://doi.org/10.1038/NMAT3072
87. Liu C, Du ZL, Sun Z, Gao HJ, Guo X (2015) Frequency-preserved acoustic diode model with high forward-power-transmission rate. Phys Rev Appl 3(6):064014. https://doi.org/10.1103/PhysRevApplied.3.064014
88. Popa BI, Cummer SA (2014) Non-reciprocal and highly nonlinear active acoustic metamaterials. Nat Commun 5:3398. https://doi.org/10.1038/ncomms4398
89. Gu ZM, Hu J, Liang B, Zou XY, Cheng JC (2016) Broadband non-reciprocal transmission of sound with invariant frequency. Sci Rep 6:19824. https://doi.org/10.1038/srep19824
90. Yang ZJ, Gao F, Shi XH, Lin X, Gao Z, Chong YD, Zhang BL (2015) Topological acoustics. Phys Rev Lett 114(11):114301. https://doi.org/10.1103/PhysRevLett.114.114301
91. Khanikaev AB, Fleury R, Mousavi SH, Alù A (2015) Topologically robust sound propagation in an angular-momentum-biased graphene-like resonator lattice. Nat Commun 6:8260. https://doi.org/10.1038/ncomms9260
92. Ni X, He C, Sun XC, Liu XP, Lu MH, Feng L, Chen YF (2015) Topologically protected one-way edge mode in networks of acoustic resonators with circulating air flow. New J Phys 17:053016. https://doi.org/10.1088/1367-2630/17/5/053016
93. Liang B, Yuan Y, Cheng JC (2015) Recent advances in acoustic one-way manipulation. Acta Phys Sin 64(9):094305. https://doi.org/10.7498/aps.64.094305
94. Li XF, Ni X, Feng L, Lu MH, He C, Chen YF (2011) Tunable unidirectional sound propagation through a sonic-crystal-based acoustic diode. Phys Rev Lett 106(8):084301. https://doi.org/10.1103/PhysRevLett.106.084301
95. Zhu XF, Zou XY, Liang B, Cheng JC (2010) One-way mode transmission in one-dimensional phononic crystal plates. J Appl Phys 108(12):124909. https://doi.org/10.1063/1.3520491
96. Yuan B, Liang B, Tao JC, Zou XY, Cheng JC (2012) Broadband directional acoustic waveguide with high efficiency. Appl Phys Lett 101(4):043503. https://doi.org/10.1063/1.4739081
97. Cicek A, Kaya OA, Ulug B (2012) Refraction-type sonic crystal junction diode. Appl Phys Lett 100(11):111905. https://doi.org/10.1063/1.3694020
98. Oh JH, Kim HW, Ma PS, Seung HM, Kim YY (2012) Inverted bi-prism phononic crystals for one-sided elastic wave transmission applications. Appl Phys Lett 100(21):213503. https://doi.org/10.1063/1.4721485
99. Ouyang SL, He HL, He ZJ, Deng K, Zhao HP (2016) Acoustic one-way mode conversion and transmission by sonic crystal waveguides. J Appl Phys 120(10):104504. https://doi.org/10.1063/1.4962201
100. He ZJ, Peng SS, Ye YT, Dai ZW, Qiu CY, Ke MZ, Liu ZY (2012) Asymmetric acoustic gratings. Appl Phys Lett 98(8):083505. https://doi.org/10.1063/1.3562306
101. Li Y, Tu J, Liang B, Guo XS, Zhang D, Cheng JC (2012) Unidirectional acoustic transmission based on source pattern reconstruction. J Appl Phys 112(6):064504. https://doi.org/10.1063/1.4752407
102. Xu SJ, Qiu CY, Liu ZY (2012) Acoustic transmission through asymmetric grating structures made of cylinders. J Appl Phys 111(9):094505. https://doi.org/10.1063/1.4709730
103. Jia H, Ke MZ, Li CH, Qiu CY, Liu ZY (2013) Unidirectional transmission of acoustic waves based on asymmetric excitation of Lamb waves. Appl Phys Lett 102(15):153508. https://doi.org/10.1063/1.4802254
104. Li CH, Ke MZ, Ye TY, Xu SJ, Qiu CY, Liu ZY (2014) Broadband asymmetric acoustic transmission by a plate with quasi-periodic surface ridges. Appl Phys Lett 105(2):023511. https://doi.org/10.1063/1.4890721
105. Li RQ, Liang B, Li Y, Kan WW, Zou XY, Cheng JC (2012) Broadband asymmetric acoustic transmission in a gradient-index structure. Appl Phys Lett 101(26):263502. https://doi.org/10.1063/1.4773481
106. Gu ZM, Liang B, Zou XY, Yang J, Li Y, Yang J, Cheng JC (2015) One-way acoustic mirror based on anisotropic zero-index media. Appl Phys Lett 107(21):213503. https://doi.org/10.1063/1.4936373

107. Song AL, Li JF, Shen C, Peng XY, Zhu XH, Chen TN, Cummer SA (2019) Broadband high-index prism for asymmetric acoustic transmission. Appl Phys Lett 114(12):121902. https://doi.org/10.1063/1.5092125
108. Cao WK, Wu LT, Zhang C, Ke JC, Cheng Q, Cui TJ, Jing Y (2019) Asymmetric transmission of acoustic waves in a waveguide via gradient index metamaterials. Sci Bull 64(12):808–813. https://doi.org/10.1016/j.scib.2019.01.002
109. Wu HW, Zhou QJ, Sun BY, Cao YY, Xu YD (2023) Asymmetric angular selected transmission in phase gradient metagratings and zero index metamaterials. J Appl Phys 133(22):223103. https://doi.org/10.1063/5.0153044
110. Guan YJ, Sun HX, Xia JP, Yuan SQ (2017) Broadband asymmetric acoustic transmission in a single medium by an array of heat sources. J Phys D: Appl Phys 50(16):165102. https://doi.org/10.1088/1361-6463/aa6101
111. Zhu YF, Zou XY, Liang B, Cheng JC (2015) Acoustic one-way open tunnel by using metasurface. Appl Phys Lett 107(11):113501. https://doi.org/10.1063/1.4930300
112. Wang XP, Wan LL, Chen TN, Liang QX, Song AL (2016) Broadband acoustic diode by using two structured impedance-matched acoustic metasurfaces. Appl Phys Lett 109(4):044102. https://doi.org/10.1063/1.4960019
113. Zhu YF, Gu ZM, Liang B, Yang J, Yang J, Yin LL, Cheng JC (2016) Asymmetric sound transmission in a passive non-blocking structure with multiple ports. Appl Phys Lett 109(10):103504. https://doi.org/10.1063/1.4962435
114. Jiang X, Liang B, Zou XY, Yang J, Yin LL, Yang J, Cheng JC (2016) Acoustic one-way metasurfaces: asymmetric phase modulation of sound by subwavelength layer. Sci Rep 6:28023. https://doi.org/10.1038/srep28023
115. Li Y, Shen C, Xie YB, Li JF, Wang WQ, Cummer SA, Jing Y (2017) Tunable asymmetric transmission via lossy acoustic metasurfaces. Phys Rev Lett 119(3):035501. https://doi.org/10.1103/PhysRevLett.119.035501
116. Zhang HL, Zhu YF, Liang B, Yang J, Yang J, Cheng JC (2017) Sound insulation in a hollow pipe with subwavelength thickness. Sci Rep 7:44106. https://doi.org/10.1038/srep44106
117. Ju FF, Tian Y, Cheng Y, Liu XJ (2018) Asymmetric acoustic transmission with a lossy gradient-index metasurface. Appl Phys Lett 113(12):121901. https://doi.org/10.1063/1.5032263
118. Liu BY, Jiang YY (2018) Controllable asymmetric transmission via gap-tunable acoustic metasurface. Appl Phys Lett 112(17):173503. https://doi.org/10.1063/1.5023852
119. Gu ZM, Fang XS, Liu T, Gao H, Liang SJ, Li Y, Liang B, Cheng JC, Zhu J (2021) Tunable asymmetric acoustic transmission via binary metasurface and zero-index metamaterials. Appl Phys Lett 118(11):113501. https://doi.org/10.1063/5.0046756
120. Tang S, Ren B, Feng YX, Song J, Jiang YY (2021) Asymmetric acoustic beam shaping based on monolayer binary metasurfaces. Appl Phys Express 14(8):085504. https://doi.org/10.35848/1882-0786/ac15bf
121. Ju FF, Zou X, Liu ZG, Qian SY (2022) Efficient asymmetric acoustic transmission based on the combination of a binary metasurface and a gradient index metasurface. Appl Phys Express 15(11):114001. https://doi.org/10.35848/1882-0786/ac97db
122. Hao ZL, Chen HJ, Yin YH, Qiu CW, Zhu S, Chen HY (2023) Asymmetric conversion of arbitrary vortex fields via acoustic metasurface. Appl Phys Lett 123(20):201702. https://doi.org/10.1063/5.0171813
123. Quan JQ, Sun BY, Fu YY, Gao L, Xu YD (2024) Acoustic bilayer gradient metasurfaces for perfect and asymmetric beam splitting. Chin Phys Lett 41(1):014301. https://doi.org/10.1088/0256-307X/41/1/014301
124. Li JS, Fok L, Yin XB, Bartal G, Zhang X (2009) Experimental demonstration of an acoustic magnifying hyperlens. Nat Mater 8(12):931–934. https://doi.org/10.1038/NMAT2561
125. Yan X, Zhu R, Huang GL, Yuan FG (2013) Focusing guided waves using surface bonded elastic metamaterials. Appl Phys Lett 103(12):121901. https://doi.org/10.1063/1.4821258
126. Zhao JJ, Ye HP, Huang K, Chen ZN, Li BW, Qiu CW (2014) Manipulation of acoustic focusing with an active and configurable planar metasurface transducer. Sci Rep 4:6257. https://doi.org/10.1038/srep06257

127. Gu Y, Cheng Y, Liu XJ (2015) Acoustic planar hyperlens based on anisotropic density-near-zero metamaterials. Appl Phys Lett 107(13):133503. https://doi.org/10.1063/1.4932097
128. Karthick S, Sen AK (2016) Role of shear induced diffusion in acoustophoretic focusing of dense suspensions. Appl Phys Lett 109(1):014101. https://doi.org/10.1063/1.4955274
129. Augustsson P, Karlsen JT, Su HW, Bruus H, Voldman J (2016) Iso-acoustic focusing of cells for size-insensitive acousto-mechanical phenotyping. Nat Commun 7:11556. https://doi.org/10.1038/ncomms11556
130. Miniaci M, Gliozzi AS, Morvan B, Krushynska A, Bosia F, Scalerandi M, Pugno NM (2017) Proof of concept for an ultrasensitive technique to detect and localize sources of elastic nonlinearity using phononic crystals. Phys Rev Lett 118(21):214301. https://doi.org/10.1103/PhysRevLett.118.214301
131. Li WB, Meng F, Huang XD (2020) Coding metalens with helical-structured units for acoustic focusing and splitting. Appl Phys Lett 117(2):021901. https://doi.org/10.1063/5.0012784
132. Li XR, Jia YR, Luo YC, Yao J, Wu DJ (2021) Mixed focused-acoustic-vortices generated by an artificial structure plate engraved with discrete rectangular holes. Appl Phys Lett 118(4):043503. https://doi.org/10.1063/5.0038892
133. Ma FY, Huang Z, Liu CR, Wu JH (2022) Acoustic focusing and imaging via phononic crystal and acoustic metamaterials. J Appl Phys 131(1):011103. https://doi.org/10.1063/5.0074503
134. Torrent D, Sánchez-Dehesa J (2007) Acoustic metamaterials for new two-dimensional sonic devices. New J Phys 9:323. https://doi.org/10.1088/1367-2630/9/9/323
135. Lin SCS, Huang TJ, Sun JH, Wu TT (2009) Gradient-index phononic crystals. Phys Rev B 79(9):094302. https://doi.org/10.1103/PhysRevB.79.094302
136. Deng K, Ding YQ, He ZJ, Zhao HP, Shi J, Liu ZY (2009) Graded negative index lens with designable focal length by phononic crystal. J Phys D: Appl Phys 42(18):185505. https://doi.org/10.1088/0022-3727/42/18/185505
137. Climente A, Torrent D, Sanchez-Dehesa J (2010) Sound focusing by gradient index sonic lenses. Appl Phys Lett 97(10):104103. https://doi.org/10.1063/1.3488349
138. Peng SS, He ZJ, Jia H, Zhang AQ, Qiu CY, Ke MZ, Liu ZY (2010) Acoustic far-field focusing effect for two-dimensional graded negative refractive-index sonic crystals. Appl Phys Lett 96(26):263502. https://doi.org/10.1063/1.3457447
139. Martin TP, Nicholas M, Orris GJ, Cai LW, Torrent D, Sánchez-Dehesa J (2010) Sonic gradient index lens for aqueous applications. Appl Phys Lett 97(11):113503. https://doi.org/10.1063/1.3489373
140. Martin TP, Layman CN, Moore KM, Orris GJ (2012) Elastic shells with high-contrast material properties as acoustic metamaterial components. Phys Rev B 85(16):161103. https://doi.org/10.1103/PhysRevB.85.161103
141. Xue XW, Li P, Jin F (2019) Refraction behavior investigation and focusing control of phononic crystals under external magnetic fields. Ultrasonics 96:261–266. https://doi.org/10.1016/j.ultras.2018.12.007
142. Li ZX, Yang SH, Wang DF, Shan H, Chen DD, Fei CL, Xiao M, Yang YT (2021) Focus of ultrasonic underwater sound with 3D printed phononic crystal. Appl Phys Lett 119(7):073501. https://doi.org/10.1063/5.0058415
143. Dasila S, Krishnamurthy CV, Subramanian V (2024) Acoustic Bessel-like beam generation using phononic crystals. J Appl Phys 135(2):024903. https://doi.org/10.1063/5.0182429
144. Zigoneanu L, Popa BI, Cummer SA (2011) Design and measurements of a broadband two-dimensional acoustic lens. Phys Rev B 84(2):024305. https://doi.org/10.1103/PhysRevB.84.024305
145. Song GY, Huang B, Dong HY, Cheng Q, Cui TJ (2016) Broadband focusing acoustic lens based on fractal metamaterials. Sci Rep 6:35929. https://doi.org/10.1038/srep35929
146. Jiang X, Liang B, Yang J, Yang J, Cheng JC (2018) Acoustic planar antireflective focusing lens with sub-diffraction-limit resolution based on metamaterials. J Appl Phys 123(9):091717. https://doi.org/10.1063/1.5001305
147. Liao GX, Wang ZW, Luan CC, Liu JP, Yao XH, Fu JZ (2021) Broadband controllable acoustic focusing and asymmetric focusing by acoustic metamaterials. Smart Mater Struct 30(4):045021. https://doi.org/10.1088/1361-665X/abeb2e

148. Zhang LL, Liu XJ (2022) Acoustic negative refraction and planar focusing based on purely imaginary metamaterials. Appl Sci 12(12):5962. https://doi.org/10.3390/app12125962
149. Gao SL, Zeng QL, Gong MY, Lan J, Liu XZ (2022) An acoustic flat lens for broadband focusing via cross-shape structure. Micromachines-Basel 14(1):12. https://doi.org/10.3390/mi14010012
150. Li Y, Qi SB, Assouar B (2016) Theory of metascreen-based acoustic passive phased array. New J Phys 18:043024. https://doi.org/10.1088/1367-2630/18/4/043024
151. Song AL, Bai YZ, Sun CY, Xiang YX, Xuan FZ (2023) A reconfigurable acoustic coding metasurface for tunable and broadband sound focusing. J Appl Phys 134(24):243104. https://doi.org/10.1063/5.0178338
152. Huang SZ, Wen GL, Yin S, Lv ZY, Liu J, Pan ZB, Jian LJ (2024) Tunable acoustic metasurface for broadband asymmetric focusing based on Helmholtz resonator. J Sound Vib 591:118628. https://doi.org/10.1016/j.jsv.2024.118628
153. Liu C, Sun HX, Yuan SQ, Xia JP (2016) Broadband acoustic focusing effect based on temperature gradient distribution. Acta Phys Sin 65(4):044303. https://doi.org/10.7498/aps.65.044303
154. Ge Y, Sun HX, Liu C, Qian J, Yuan SQ, Xia JP, Guan YJ, Zhang SY (2016) Acoustic focusing by an array of heat sources in air. Appl Phys Express 9(6):066701. https://doi.org/10.7567/APEX.9.066701
155. Li Y, Yu GK, Liang B, Zou XY, Li GY, Cheng S, Cheng JC (2014) Three-dimensional ultrathin planar lenses by acoustic metamaterials. Sci Rep 4:6830. https://doi.org/10.1038/srep06830
156. Molerón M, Serra-Garcia M, Daraio C (2014) Acoustic Fresnel lenses with extraordinary transmission. Appl Phys Lett 105(11):114109. https://doi.org/10.1063/1.4896276
157. Fan XD, Zhu YF, Liang B, Yang J, Cheng JC (2016) Broadband convergence of acoustic energy with binary reflected phases on planar surface. Appl Phys Lett 109(24):243501. https://doi.org/10.1063/1.4971795
158. Cao J, Lu CY, Ma QJ, Hu XP, You CZ, Xu Z, Wu DW (2022) Coding acoustic metasurfaces for tunable focusing in three-dimensional space. Appl Phys Express 15(5):057001. https://doi.org/10.35848/1882-0786/ac678e

Chapter 2
Basic Theory of Sound Manipulation Based on Acoustic Metamaterials

2.1 Introduction

Over the past two decades, the rapid development of acoustic metamaterials has led to their widespread application in various advanced sound devices and structures. Generally, acoustic metamaterials enable precise control of sound propagation and exhibit characteristics such as sub-wavelength thickness, broad bandwidth, simple structure, and ease of modulation. These properties make them highly promising for applications in medical ultrasound, noise control, architectural acoustics, nondestructive testing, and more. To analyze the propagation characteristics of sound and the mechanisms behind various sound phenomena caused by different types of acoustic metamaterials, it is essential to understand the basic theoretical frameworks and modeling methods of sound.

In this chapter, we introduce the fundamental theories of sound manipulation in 2D and 3D spaces based on acoustic metamaterials. These include the interference theory, EM theory, FE method, generalized Snell's law, mode conversion theory, and band theory of SCs. These theories not only provide methodological support for investigating the effects of AAT, AF, and AAF caused by acoustic metamaterials but also offer a theoretical basis for in-depth analysis of the underlying physical mechanisms.

2.2 Interference Theory

The linear interference of sound is closely related to the propagation and interaction of acoustic waves. When two or more acoustic waves reach a certain point simultaneously, their pressures superimpose based on the principle of linear superposition, enabling the enhancement or cancellation of sound through interference.

H. Sun et al., *Acoustic Metamaterials*, Microfluidics and Nanophotonics: Science and Engineering 2, https://doi.org/10.1007/978-981-96-8524-0_2

This phenomenon has significant potential in fields such as noise control, source localization, and acoustic imaging.

Consider a sound source with amplitude A, initial phase φ, and frequency f. The incident acoustic wave can be expressed as [1]:

$$p(t) = A\cos(\omega t - \varphi), \tag{2.1}$$

where $\omega = 2\pi f$ is the angular frequency, and t is time. By keeping the frequency constant and varying the initial phase and amplitude, we obtain two sound sources with different amplitudes (A_1, A_2) and initial phases (φ_1, φ_2). The two incident acoustic waves can then be written as [1]:

$$p_1(t) = A_1\cos(\omega t - \varphi_1), \tag{2.2}$$

$$p_2(t) = A_2\cos(\omega t - \varphi_2). \tag{2.3}$$

Based on the principle of linear superposition, the total pressure at a given point is the sum of the pressures from the two waves [1]:

$$p_{\mathrm{t}}(t) = p_1(t) + p_2(t) = A_{\mathrm{t}}\cos(\omega t - \varphi_{\mathrm{t}}), \tag{2.4}$$

where

$$A_{\mathrm{t}} = \sqrt{A_1^2 + A_2^2 + 2A_1A_2\cos(\varphi_2 - \varphi_1)}, \tag{2.5}$$

$$\varphi_{\mathrm{t}} = \arctan\frac{A_1\sin(\varphi_1) + A_2\sin(\varphi_2)}{A_1\cos(\varphi_1) + A_2\cos(\varphi_2)}, \tag{2.6}$$

In Eqs. (2.5) and (2.6), A_{t} is determined by the phase difference between φ_2 and φ_1, and φ_{t} represents the phase of the superimposed acoustic wave. When the phase difference between the two waves is an even multiple of π (i.e. $\varphi_2 - \varphi_1 = 2n\pi$, $n = 0, \pm1, \pm2, \pm3\ldots$), the amplitude of the total pressure is maximized ($A_{\mathrm{t}} = A_1 + A_2$), indicating constructive interference. Conversely, when the phase difference is an odd multiple of π (i.e. $\varphi_2 - \varphi_1 = (2n+1)\pi$, $n = 0, \pm1, \pm2, \pm3\ldots$), the amplitude of the total pressure is minimized ($A_{\mathrm{t}} = |A_1 - A_2|$), indicating destructive interference.

2.3 Effective Medium Theory

The EM theory is primarily used to derive the effective parameters of acoustic metamaterials. The core idea is to treat the metamaterial as a uniform medium [2]. As shown in Fig. 2.1a, an acoustic metamaterial is replaced by a planar EM that provides

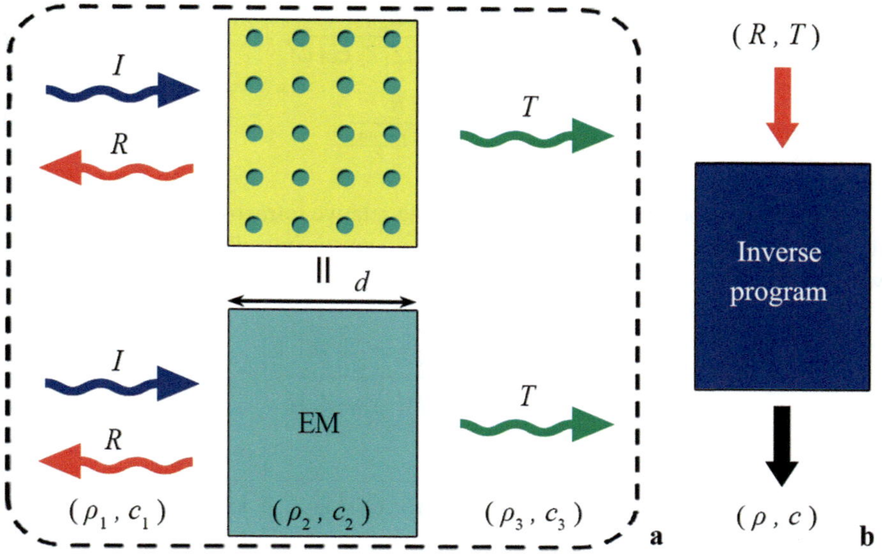

Fig. 2.1 **a** Schematic of an incident acoustic wave passing through an acoustic metamaterial and an EM. **b** The inverse program for calculating effective parameters based on transmission and reflection coefficients

the same amplitude and phase for reflection and transmission coefficients. Additionally, the effective refractive index and impedance can be obtained from the reflection and transmission coefficients (R and T) for the normal incidence of an acoustic wave on the acoustic metamaterial (Fig. 2.1b). The effective density ρ and sound velocity c are then calculated based on the effective refractive index and impedance, as discussed below.

Next, consider an EM model with thickness d placed between two media with densities ρ_1 and ρ_3 and sound velocities c_1 and c_3. When an acoustic wave is incident on the EM with density ρ_2 and sound velocity c_2, the reflection and transmission coefficients can be written as [2]:

$$R = \frac{(Z_1 + Z_2)(Z_2 - Z_3)e^{-2i\varphi} + (Z_1 - Z_2)(Z_2 + Z_3)}{(Z_1 + Z_2)(Z_2 - Z_3)e^{-2i\varphi} + (Z_1 - Z_2)(Z_2 - Z_3)}, \tag{2.7}$$

$$T = \frac{4Z_1Z_2}{(Z_1 - Z_2)(Z_2 - Z_3)e^{i\varphi} + (Z_1 + Z_2)(Z_1 + Z_3)}, \tag{2.8}$$

where $Z_i = \rho_i c_i / \cos\theta_i$ is the acoustic impedance, θ_i is the angle between the wave vector and the normal line of the EM, and $\varphi = \pi f d \cos\theta_2 / c_2$ is the phase delay of sound through the EM. For normal incidence and identical media on both sides, the reflection and transmission coefficients can be simplified as [2]:

$$R = \frac{Z_2^2 - Z_1^2}{Z_2^2 + Z_1^2 - 2iZ_1Z_2 \cot\varphi}, \tag{2.9}$$

$$T = \frac{1 + R}{\cos\varphi - \frac{Z_2 i \sin\varphi}{Z_1}}. \tag{2.10}$$

In Eqs. (2.9) and (2.10), by introducing the relative refractive index $n_\text{r} = c_1 / c_2$, the wave number $k = 2\pi f / c_1$, and the relative acoustic impedance $Z_r = \rho_2 c_2 / \rho_1 c_1$, we obtain [2]:

$$R = \frac{\tan(n_\text{r}kd)\left(\frac{1}{Z_r} - Z_r\right)i}{2 - \tan(n_\text{r}kd)\left(\frac{1}{Z_r} + Z_r\right)i}, \tag{2.11}$$

$$T = \frac{2}{\cos(n_\text{r}kd)\left[2 - \tan(n_\text{r}kd)\left(\frac{1}{Z_r} + Z_r\right)\right]}. \tag{2.12}$$

By inverting Eqs. (2.11) and (2.12), we can calculate the relative acoustic impedance Z_r and the refractive index n_r as follows [2]:

$$Z_\text{r} = \pm\sqrt{\frac{(1 + R)^2 - T^2}{(1 - R)^2 - T^2}}, \tag{2.13}$$

$$n_\text{r} = \frac{\pm\cos^{-1}\left(\frac{1}{2}[1 - (R^2 - T^2)]\right)}{kd} + \frac{2\pi m}{kd}, \tag{2.14}$$

where m is the branch number of the $\cos^{-1}$ function. In Eqs. (2.13) and (2.14), both Z_r and n_r are complex functions of complex variables. Mathematically, any combination of signs in these equations and any value of m can yield the same reflection and transmission coefficients.

To resolve these ambiguities, additional constraints must be imposed on the characteristics of the acoustic metamaterial. Generally, the real part of Z_r must be positive, which determines the sign in Eq. (2.13). Additionally, a positive imaginary sound speed component is required, leading to a negative imaginary part of n_r. Note that these signs are interrelated in Eqs. (2.13) and (2.14). When $\text{Re}(Z_\text{r})$ or $\text{Im}(n)$ is close to zero, errors in the measurement or calculation of reflection and transmission coefficients may lead to incorrect sign combinations, causing discontinuities in Z_r and n with frequency. To address this, we rewrite Eqs. (2.13) and (2.14) as [2]:

$$Z_\text{r} = \frac{r}{1 - 2R + R^2 - T^2}, \quad r = \mp\sqrt{(R^2 - T^2 - 1)^2 - 4T^2}, \tag{2.15}$$

$$n_\text{r} = \frac{-i\log x + 2\pi m}{kd}, \quad x = \frac{(1 - R^2 + T^2 + r)}{2T}. \tag{2.16}$$

We first solve for r in Eq. (2.15) and select the root that yields a positive solution for $\mathrm{Re}(Z_\mathrm{r})$. Then, the expression for x uses the value of r to eliminate ambiguity in the expression for n_r in Eq. (2.16). However, determining $\mathrm{Re}(n_r)$ from Eq. (2.11) is complicated by the need to select the proper value of m. This issue can be avoided by determining the effective parameters for a minimum thickness of the acoustic metamaterials, where m is zero. By using this approach, the sign and branch number for increasing frequency can be determined by the requirement of continuous n_r. Based on Z_r and n_r, the effective density and sound velocity of the EM can be calculated as [2]:

$$\rho_2 = \rho_1 n_\mathrm{r} Z_\mathrm{r}, \tag{2.17}$$

$$c_2 = \frac{c_1}{n_\mathrm{r}}. \tag{2.18}$$

2.4 Finite Element Method

The FE method is a widely used numerical technique in scientific research and engineering applications. It originated in the 1940s from the ideas proposed by Hrennikoff and Courant for solving problems related to elastic structures in mechanical mechanics. However, its practical application was initially limited due to the enormous computational load. In recent years, with the rapid advancement of computer technology, the FE method has become a powerful tool for solving a wide range of physical problems.

The fundamental idea behind the FE method is to convert the problem of solving differential equations into the problem of finding the extremum of a functional. This is achieved by dividing the solution domain into a series of interconnected elements, which are linked by nodes. This discrete model approximates the actual physical system. Based on the variational principle, an algebraic system of equations is established to solve for the field function at the nodes. Finally, the solution at any point within any element can be calculated using interpolation.

Solving a problem using the FE method typically involves the following steps:

(1) Domain discretization: the solution domain is divided into a finite number of interconnected but non-overlapping elements, connected by nodes. The nodes are usually located at the vertices of the elements. For a 2D domain, the elements are typically triangular, while for a 3D domain, tetrahedral elements are commonly used.
(2) Selection of interpolation functions: an appropriate interpolation function is chosen as the basis function for the elements, based on the number of nodes and the desired solution accuracy. The field function for each element is then derived.

(3) Formulation of element equations: the field function is substituted into the integral equation, and by integrating over the element region, the FE equation for the element is obtained.
(4) Assembly of global equations: the FE equations for all elements are summed to form the global FE equation for the entire structure. By incorporating the continuity conditions at the nodes and the boundary conditions of the solution domain, numerical methods such as Gaussian elimination or iterative methods can be used to solve the system of equations.
(5) Solution and interpolation: the physical variables at any point in the solution domain can be approximated using interpolation functions based on the solutions at the nodes.

Throughout this book, the propagation characteristics of sound are numerically simulated using the FE software COMSOL Multiphysics. Typically, the pressure acoustics module is used to design the FE model. In this module, sound propagation is simulated in a fluid under static background conditions in the frequency domain, which satisfies the Helmholtz equation. The FE form of the Helmholtz equation can be expressed as [3]:

$$\nabla \cdot \left(-\frac{1}{\rho_0} \nabla p \right) - \frac{\omega^2 p}{\rho_0 c_0^2} = 0, \tag{2.19}$$

where p represents the total acoustic pressure.

The incident sound source can be set as either a plane wave or a cylindrical wave. For plane acoustic wave, the incident pressure distribution p_{i} is located at the left boundary and is expressed as [3]:

$$p_{\mathrm{i}} = p_1 e^{-ikx}, \tag{2.20}$$

where p_1 represents the pressure amplitude, and $k = \omega / c_0$ is the wavenumber in air. The incident pressure distribution p_{i} satisfies the FE equation [3]:

$$Q_{\mathrm{i}} = i\frac{kp_{\mathrm{i}}}{\rho_0} + \frac{i}{2k\rho_0} \Delta_{||} p_{\mathrm{i}} + \vec{\mathbf{n}} \cdot \frac{1}{\rho_0} \nabla p_{\mathrm{i}}, \tag{2.21}$$

where Q_{i} represents the incident sound source, $\Delta_{||}$ indicates the Laplace operator in the cutting plane. For cylindrical acoustic wave, a point source is set at the position of sound source $\vec{\mathbf{x}}_0$, satisfying the FE equation [3]:

$$\nabla \cdot \left(-\frac{1}{\rho_0} \nabla p \right) - \frac{kp}{\rho_0} = \frac{4\pi}{\rho_0} S \delta(\vec{\mathbf{x}} - \vec{\mathbf{x}}_0), \tag{2.22}$$

where S represents the Monopole amplitude.

The boundaries of the model can be set as open boundaries, sound-hard boundaries, or Floquet periodic boundaries. For an open boundary, the boundary can be set as plane or cylindrical wave radiation, corresponding to the plane or cylindrical sound source. The FE equations for the two boundaries are expressed as [3]:

$$-\vec{\mathbf{n}}\cdot\left(-\frac{1}{\rho_0}\nabla p\right)+i\frac{kp}{\rho_0}+\frac{i}{2k\rho_0}\Delta_{||}p=Q_{\mathrm{i}},\tag{2.23}$$

$$-\vec{\mathbf{n}}\cdot\left(-\frac{1}{\rho_0}\nabla p\right)+\left(ik+\frac{1}{2r}-\frac{1}{8r(1+ikr)}\right)\frac{p_{\mathrm{r}}}{\rho_0}+\frac{r\Delta_{||}p_{\mathrm{r}}}{2(1+ikr)\rho_0}=Q_{\mathrm{i}},\tag{2.24}$$

where $\vec{\mathrm{n}}$ is the unit normal vector, r is the shortest distance from the point on the boundary to the sound source, and p_{r} is the reflected acoustic pressure. For a sound-hard boundary, the normal component of the acceleration is zero, satisfying the FE equation [3]:

$$-\vec{\mathbf{n}}\cdot\left(-\frac{1}{\rho_0}\nabla p\right)=0.\tag{2.25}$$

Floquet periodicity is used to reduce the model size by exploiting symmetries and periodicities in the geometry and physics interfaces. The FE equations for Floquet periodicity are expressed as [3]:

$$p_s=p_x e^{-i\vec{\mathbf{k}}_{\mathrm{F}}\cdot(\vec{\mathbf{r}}_s-\vec{\mathbf{r}}_x)},\tag{2.26}$$

$$-\vec{\mathbf{n}}_s\cdot\left(-\frac{1}{\rho_0}\nabla p\right)_s=\vec{\mathbf{n}}_x\cdot\left(-\frac{1}{\rho_0}\nabla p\right)_x e^{-i\vec{\mathbf{k}}_{\mathrm{F}}\cdot(\vec{\mathbf{r}}_s-\vec{\mathbf{r}}_x)},\tag{2.27}$$

where the subscripts "s" and "x" represent a set of symmetrical boundaries, and $\vec{\mathbf{k}}_{\mathrm{F}}$ is the Floquet periodic wave vector.

Finally, for solid structures in the model, the solid mechanics module is typically used. The interfaces between the solid and fluid domains are set as acoustic-structure coupling boundaries. The FE equations for the external and internal boundaries are expressed as [3]:

$$-\vec{\mathbf{n}}\cdot\left(-\frac{1}{\rho_0}\nabla p\right)=-\vec{\mathbf{n}}\cdot\vec{\mathbf{u}}_{\mathrm{tt}},\tag{2.28}$$

$$\vec{\mathbf{F}}_{\mathrm{A}}=\vec{\mathbf{n}}p,\tag{2.29}$$

where $\vec{\mathbf{u}}_{\mathrm{tt}}$ represents the structural acceleration, and $\vec{\mathbf{F}}_{\mathrm{A}}$ is the force applied per unit area.

$$-\vec{\mathbf{n}}\cdot\left(-\frac{1}{\rho_0}\nabla p\right)_{\mathrm{up}}=-\vec{\mathbf{n}}\cdot\vec{\mathbf{u}}_{\mathrm{tt}},\tag{2.30}$$

$$-\vec{\mathbf{n}} \cdot \left(-\frac{1}{\rho_0}\nabla p\right)_{\text{down}} = -\vec{\mathbf{n}} \cdot \vec{\mathbf{u}}_{\text{tt}}, \tag{2.31}$$

$$\vec{\mathbf{F}}_{\text{A}} = \vec{\mathbf{n}}\left(p_{\text{down}} - p_{\text{up}}\right), \tag{2.32}$$

where the subscripts "up" and "down" represent the two sides of the internal boundary.

Finally, the solution domain is divided into multiple basic elements by meshing the geometric model. The field distributions across the entire solution domain are obtained through parameter calculations using the COMSOL Multiphysics software.

2.5 Generalized Snell's Law

2.5.1 1D Generalized Snell's Law

As shown in Fig. 2.2a, to derive the 1D generalized Snell's law for sound transmission, we design a 1D transmitted AM with a phase gradient of $\varphi(x)$. Consider an incident wave with an angle θ_{i} (measured from the +z direction, where $0 \leq \theta_{\text{i}} \leq \pi$). The acoustic wave passes through the metasurface and is transmitted at an angle θ_{t} (measured from the +z direction, where $0 \leq \theta_{\text{t}} \leq \pi$). As shown in Fig. 2.2b, assuming that two specified propagation paths, A_1–A_2–A_3 and B_1–B_2–B_3, are infinitesimally close to the actual acoustic paths, we can derive the phase delay differences between points A_1 and A_2, A_3 and A_4, B_1 and B_2, and B_3 and B_4 as follows:

$$\begin{cases} \varphi_{\text{A}_2} - \varphi_{\text{A}_1} = \varphi(x) \\ \varphi_{\text{A}_3} - \varphi_{\text{A}_2} = n_{\text{t}}k_0|\text{A}_2\text{A}_3| \end{cases}, \tag{2.33}$$

$$\begin{cases} \varphi_{\text{B}_2} - \varphi_{\text{B}_1} = n_{\text{i}}k_0|\text{B}_1\text{B}_2| \\ \varphi_{\text{B}_3} - \varphi_{\text{B}_2} = \varphi(x + \text{d}x) \end{cases}, \tag{2.34}$$

where $|\text{B}_1\text{B}_2| = d^{\text{i}}$ and $|\text{A}_2\text{A}_3| = d^{\text{t}}$ are the differences in propagation distances for the incident and transmitted acoustic waves, respectively. Here, $n_{\text{i/t}}$ is the refractive index of the incident/transmitted medium, and k_0 is the wave number in air.

Based on Fermat's principle, the difference in phase delays between the two paths is zero, which can be expressed as:

$$\varphi_{\text{A}_3} - \varphi_{\text{A}_1} = \varphi_{\text{B}_3} - \varphi_{\text{B}_1}. \tag{2.35}$$

Based on Eqs. (2.33)–(2.35), we obtain:

$$\varphi(x + \text{d}x) + n_{\text{i}}k_0 d^{\text{i}} = \varphi(x) + n_{\text{t}}k_0 d^{\text{t}}. \tag{2.36}$$

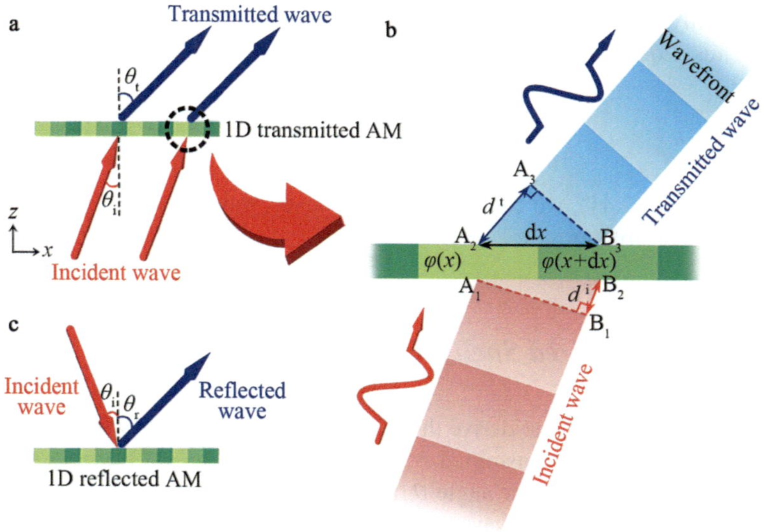

Fig. 2.2 **a** Schematic of a 1D transmitted AM. **b** Schematic of incident acoustic waves passing through the 1D transmitted AM. **c** Schematic of a 1D reflected AM

The phase delay difference between positions $x + \mathrm{d}x$ and x on the metasurface is given by:

$$\mathrm{d}\varphi = \varphi(x + \mathrm{d}x) - \varphi(x). \quad (2.37)$$

The difference in propagation distance $d^{\mathrm{i/t}}$ can be expressed as:

$$d^{\mathrm{i/t}} = \sin\theta_{\mathrm{i/t}}\mathrm{d}x. \quad (2.38)$$

By substituting Eqs. (2.37) and (2.38) into (2.36), we obtain the 1D generalized Snell's law for sound transmission [4]:

$$n_{\mathrm{t}}\sin\theta_{\mathrm{t}} - n_{\mathrm{i}}\sin\theta_{\mathrm{i}} = \frac{1}{k_0}\frac{\mathrm{d}\varphi}{\mathrm{d}x}. \quad (2.39)$$

As shown in Fig. 2.2c, for a 1D reflected AM with a phase gradient of $\varphi(x)$ and an incident wave with angle θ_{i}, the sound is reflected at an angle θ_{r} (measured from the $-z$ direction, where $0 \le \theta_{\mathrm{r}} \le \pi$). Similar to the case of sound transmission, the 1D generalized Snell's law for sound reflection can be written as [4]:

$$\sin\theta_{\mathrm{r}} - \sin\theta_{\mathrm{i}} = \frac{1}{k_0 n_{\mathrm{i}}}\frac{\mathrm{d}\varphi}{\mathrm{d}x}. \quad (2.40)$$

Specifically, when both the incident and transmitted media are air, we have $n_\mathrm{i} = n_\mathrm{t} = 1$. Thus, Eqs. (2.39) and (2.40) can be simplified as [4]:

$$\sin\theta_{\mathrm{t/r}} - \sin\theta_\mathrm{i} = \frac{1}{k_0}\frac{\mathrm{d}\varphi}{\mathrm{d}x}. \tag{2.41}$$

Based on Eq. (2.41), we can manipulate acoustic wavefronts in 2D space using 1D metasurface structures.

2.5.2 2D Generalized Snell's Law

As illustrated in Fig. 2.3a, to derive the 2D generalized Snell's law for sound transmission, we design a 2D transmitted AM with a phase gradient of $\varphi(x, y)$. Consider an incident wave with a polar angle θ_i and an azimuth angle β_i impinging on the metasurface. The transmitted wave has angles θ_t and β_t. Here, θ_i and θ_t are measured from the $+z$ direction ($0 \leq \theta_{\mathrm{i/t}} \leq \pi$), and β_i and β_t are measured from the $+x$ direction in the x–y plane ($0 \leq \beta_{\mathrm{i/t}} \leq \pi$).

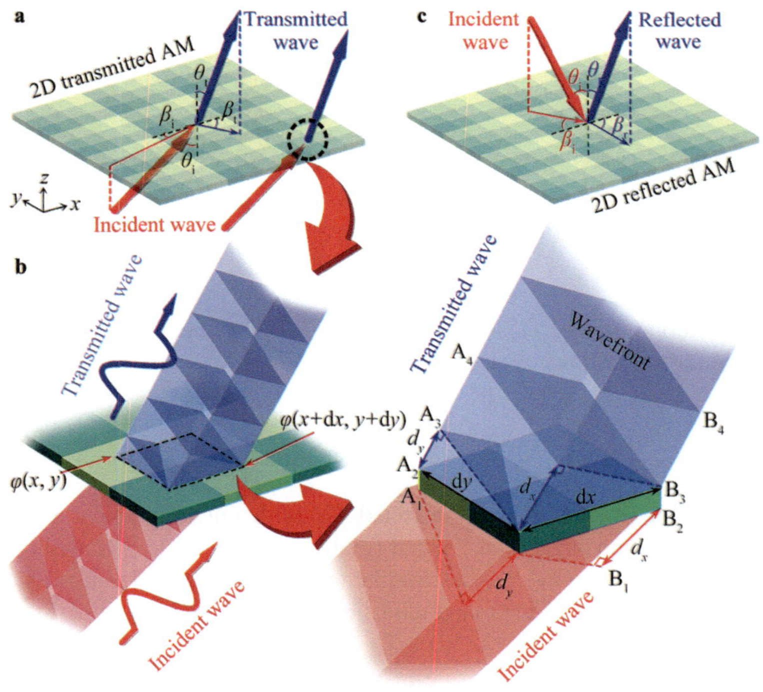

Fig. 2.3 **a** Schematic of a 2D transmitted AM. **b** Schematic of incident acoustic waves passing through the 2D transmitted AM. **c** Schematic of a 2D reflected AM

As shown in Fig. 2.3b, there are two specified propagation paths, A_1–A_2–A_3–A_4 and B_1–B_2–B_3–B_4, which are infinitesimally close to the actual acoustic paths. The phase delay differences between two adjacent points on these paths can be written as:

$$\begin{cases} \varphi_{A_2} - \varphi_{A_1} = \varphi(x, y) \\ \varphi_{A_3} - \varphi_{A_2} = n_t k_0 |A_2 A_3| \\ \varphi_{A_4} - \varphi_{A_3} = n_t k_0 |A_3 A_4| \end{cases}, \tag{2.42}$$

$$\begin{cases} \varphi_{B_2} - \varphi_{B_1} = n_i k_0 |B_1 B_2| \\ \varphi_{B_3} - \varphi_{B_2} = n_i k_0 |B_2 B_3| \\ \varphi_{B_4} - \varphi_{B_3} = \varphi(x + dx, y + dy) \end{cases}, \tag{2.43}$$

where $|A_3A_4| = d_x^t$ and $|A_2A_3| = d_y^t$ are the differences in propagation distances for the transmitted acoustic waves in the x and y directions, respectively, and $|B_2B_3| = d_x^i$ and $|B_1B_2| = d_y^i$ are those for the incident acoustic waves in both directions. Based on Fermat's principle, we have:

$$\varphi_{A_4} - \varphi_{A_1} = \varphi_{B_4} - \varphi_{B_1}. \tag{2.44}$$

By substituting Eqs. (2.42) and (2.43) into (2.44), we obtain:

$$\varphi(x + dx, y + dy) + n_i k_0 \left(d_x^i + d_y^i\right) = \varphi(x, y) + n_t k_0 \left(d_x^t + d_y^t\right). \tag{2.45}$$

The phase delay difference between the positions $(x + dx, y + dy)$ and (x, y) on the metasurface is given by:

$$d\varphi = \varphi(x + dx, y + dy) - \varphi(x, y). \tag{2.46}$$

The differences in propagation distances $d_x^{i/t}$ and $d_y^{i/t}$ can be expressed as:

$$\left(d_x^{i/t}, d_y^{i/t}\right) = \sin\theta_{i/t}(\cos\beta_{i/t} dx, \sin\beta_{i/t} dy). \tag{2.47}$$

By substituting Eqs. (2.46) and (2.47) into (2.45), we obtain the 2D generalized Snell's law for sound transmission [5]:

$$n_t \sin\theta_t(\cos\beta_t dx + \sin\beta_t dy) - n_i \sin\theta_i(\cos\beta_i dx + \sin\beta_i dy) = \frac{d\varphi}{k_0}. \tag{2.48}$$

As shown in Fig. 2.3c, for a 2D reflected AM with a phase gradient of $\varphi(x, y)$ and an incident wave with polar angle θ_i and azimuth angle β_i, the sound is reflected at angles θ_r $(0 \le \theta_r \le \pi)$ and β_r $(0 \le \beta_r \le \pi)$, measured from the $-z$ direction and the $+x$ direction in the x–y plane, respectively. Similar to the case of sound transmission, the 2D generalized Snell's law for sound reflection can be written as [5]:

$$\sin\theta_{\mathrm{r}}(\cos\beta_{\mathrm{r}}\mathrm{d}x + \sin\beta_{\mathrm{r}}\mathrm{d}y) - \sin\theta_{\mathrm{i}}(\cos\beta_{\mathrm{i}}\mathrm{d}x + \sin\beta_{\mathrm{i}}\mathrm{d}y) = \frac{\mathrm{d}\varphi}{k_0 n_{\mathrm{i}}}. \tag{2.49}$$

For the special case where the AM is immersed in air, we have $n_{\mathrm{i}} = n_{\mathrm{t}} = 1$. Thus, the 2D generalized Snell's law for sound transmission/reflection can be simplified as [5]:

$$\sin\theta_{\mathrm{t/r}}(\cos\beta_{\mathrm{t/r}}\mathrm{d}x + \sin\beta_{\mathrm{t/r}}\mathrm{d}y) - \sin\theta_{\mathrm{i}}(\cos\beta_{\mathrm{i}}\mathrm{d}x + \sin\beta_{\mathrm{i}}\mathrm{d}y) = \frac{\mathrm{d}\varphi}{k_0}. \tag{2.50}$$

Based on Eq. (2.50), we can arbitrarily shape acoustic wavefronts in 3D space using 2D metasurface structures.

2.6 Mode Conversion Theory

2.6.1 Mode Conversion Between Zero-Order and First-Order Waves

In this section, we theoretically derive the mode conversion between zero-order and first-order waves. As illustrated in Fig. 2.4, a mode-conversion meta-atom is placed in a straight waveguide with a width of w_y^1 in the y direction. When a zero-order wave impinges on the meta-atom, the sound energy can transmit through it and be converted into a first-order wave. The pressure distributions of the zero-order and first-order waves in the waveguide can be expressed as [1]:

$$p_0(x, y) = A_0 e^{-\mathrm{i}k_0 x}, \tag{2.51}$$

$$p_1(x, y) = -A_1 e^{-\mathrm{i}k_x^1 x} \sin\frac{\pi y}{w_y^1}, \tag{2.52}$$

where A_0 and A_1 are the amplitudes of the zero-order and first-order waves, respectively, $k_0 = 2\pi f/c$ is the wave number in air, c is the sound velocity in air, f is the frequency, and $k_x^1 = \sqrt{k_0^2 - \left(\pi/w_y^1\right)^2}$ is the wave number of the first-order wave in the x direction.

It is important to note that the zero-order wave at any frequency can pass through the waveguide, while the first-order wave has a cutoff frequency $f_c^1 = c/\left(2w_{1y}\right)$. When the frequency of the first-order wave satisfies $f > f_c^1$, the first-order wave can propagate through the waveguide. Otherwise, the value of k_x^1 becomes imaginary, and the first-order wave cannot pass through the meta-atom. Therefore, the first-order wave can only transmit through the waveguide when $w_y^1 \geq \lambda/2$, where λ is the wavelength.

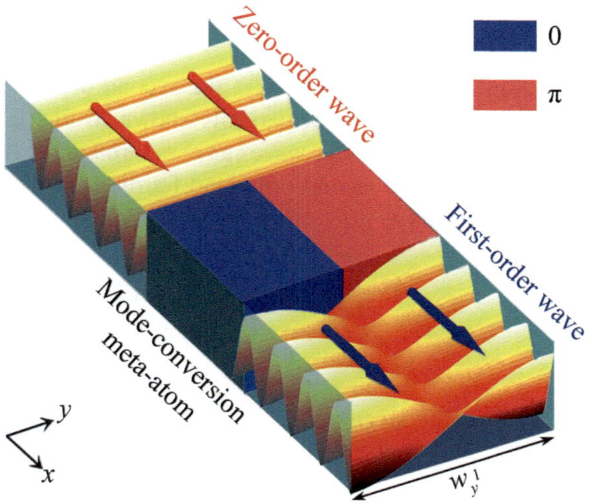

Fig. 2.4 Schematic of the mode conversion between the zero-order and first-order waves caused by a mode-conversion meta-atom in a straight waveguide. The colors of the meta-atom represent the theoretical phase distributions

From Eqs. (2.51) and (2.52), it can be observed that the phase distribution of the zero-order wave is constant, while that of the first-order wave is anti-symmetric with respect to $y = 0$. Therefore, to convert the zero-order wave into the first-order wave, the theoretical phase distribution of the mode-conversion meta-atom must satisfy [6, 7]:

$$\varphi(y) = \begin{cases} \pi, & y \in \left[-\frac{w_y^1}{2}, 0\right) \\ 0, & y \in \left[0, \frac{w_y^1}{2}\right] \end{cases}, \tag{2.53}$$

which is schematically represented by two types of colors in Fig. 2.4.

2.6.2 Mode Conversion Between Zero-Order and High-Order Waves

The zero-order wave can also be converted into other types of high-order waves using mode-conversion meta-atoms. The pressure distribution of the n-th-order wave in the straight waveguide is given by [1]:

$$p_n(x, y) = A_n e^{-\mathrm{i}k_x^n x} \cos n\pi\left(\frac{y}{w_y^n} + \frac{1}{2}\right), \tag{2.54}$$

where A_n is the amplitude of the n-th-order wave, $k_x^n = \sqrt{k_0^2 - \left(n\pi/w_y^n\right)^2}$ is the wave number of the n-th-order wave in the x direction, and w_y^n is the waveguide width which must satisfy $w_y^n \geq n\lambda/2$ to allow the propagation of the n-th-order wave [1].

As an example, we consider the second-order ($n = 2$) and fourth-order ($n = 4$) waves, which are schematically shown in Fig. 2.5a, b, respectively. Based on Eq. (2.54), the zero-order wave can pass through the two types of meta-atoms and be converted into the second-order and fourth-order waves. The corresponding phase distributions in the y direction are given by [7]:

$$\varphi(y) = \begin{cases} \pi, & \frac{y}{w_y^2} \in \left[-\frac{1}{2}, -\frac{1}{4}\right) \cup \left(\frac{1}{4}, \frac{1}{2}\right] \\ 0, & \frac{y}{w_y^2} \in \left[-\frac{1}{4}, \frac{1}{4}\right] \end{cases}, \tag{2.55}$$

$$\varphi(y) = \begin{cases} \pi, & \frac{y}{w_y^4} \in \left[-\frac{3}{8}, -\frac{1}{8}\right) \cup \left(\frac{1}{8}, \frac{3}{8}\right] \\ 0, & \frac{y}{w_y^4} \in \left[-\frac{1}{2}, -\frac{3}{8}\right) \cup \left[-\frac{1}{8}, \frac{1}{8}\right] \cup \left(\frac{3}{8}, \frac{1}{2}\right] \end{cases}. \tag{2.56}$$

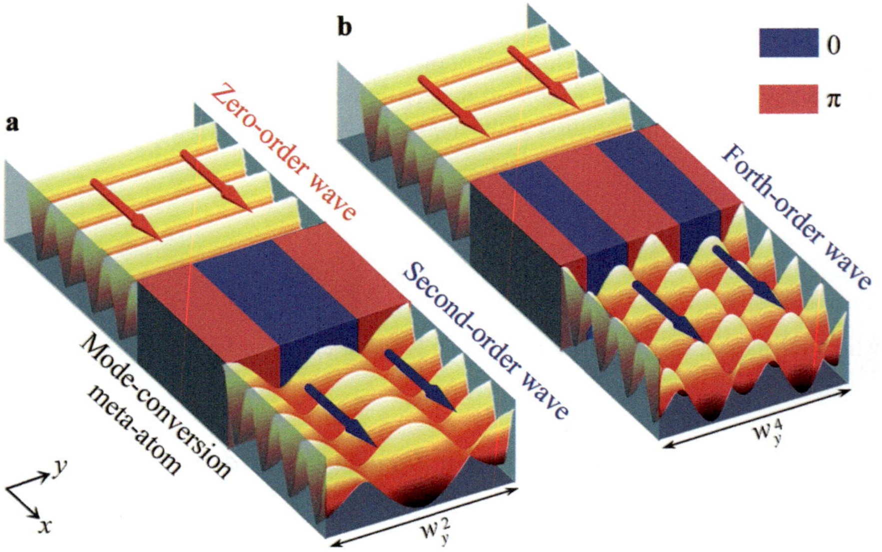

Fig. 2.5 Schematic of the mode conversion between **a** the zero-order and second-order waves, and **b** the zero-order and fourth-order waves caused by mode-conversion meta-atoms in straight waveguides. The colors of the meta-atoms represent their theoretical phase distributions

2.7 Band Theory of Sonic Crystal

2.7.1 Concept of SC

The concept of SC is derived from that of photonic crystals. In a periodic elastic system, the propagation direction of sound changes when its wavelength is close to the periodic length of the structure. This phenomenon is known as Bragg scattering. Due to Bragg scattering, the periodic elastic system forms a specific band structure called the phononic band. Acoustic waves with frequencies within the phononic band can propagate through the system. However, for frequencies outside the phononic band, sound transmission is forbidden due to the exponential decay of sound energy caused by interference and cancellation in the periodic elastic system, forming phononic bandgaps. A periodic elastic system with phononic bandgaps is defined as a SC, which is generally composed of two or more types of elastic media. The continuous and discontinuous physical phases are referred to as matrices and dispersions, respectively.

2.7.2 Lattice Periodicity

(1) Direct space

Similar to natural crystals, SCs also have a periodic structure and can be described using the concept of a crystal lattice. The periodically arranged atoms can be abstracted as lattice points, and the entire crystal can be described by a space lattice, referred to as a crystal lattice. The smallest repetitive structure of the crystal lattice is defined as the unit cell. To determine the position of each lattice point, we arbitrarily select a reference point, and the position vectors of the nearest three non-coplanar lattice points, $\vec{a}_1$, $\vec{a}_2$ and $\vec{a}_3$, are the direct basis vectors. These vectors can be used to construct the direct space. Therefore, the position vector of any lattice point can be expressed as a linear combination of the three direct basis vectors:

$$\vec{R}_l = l_1\vec{a}_1 + l_2\vec{a}_2 + l_3\vec{a}_3, \tag{2.57}$$

where l_1, l_2 and l_3 are arbitrary integers, and $\vec{R}_l$ is defined as the direct lattice vector. At this reference point, a parallelepiped can be constructed by the three direct basis vectors, and thus a unit cell can be obtained. Therefore, the crystal lattice can be obtained by repeating integer multiples of the unit cell along the direction of the basis vector.

Based on the symmetry of the crystal structure, the crystal lattice can be divided into fourteen types of lattices, known as Bravais lattices. As shown in Fig. 2.6, the 2D Bravais lattices mainly include the square and hexagonal lattices, while the 3D

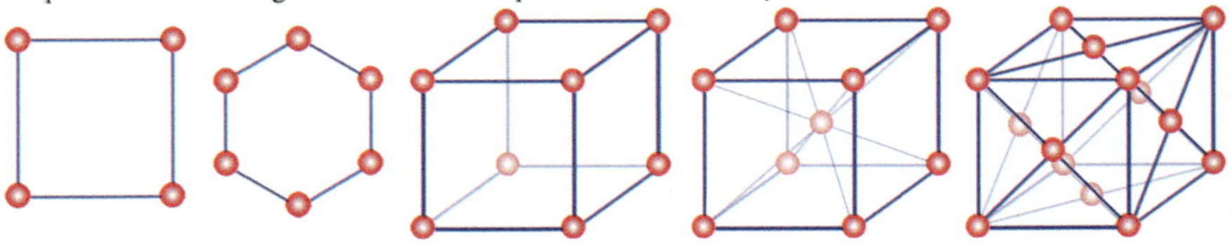

Fig. 2.6 Schematic of several types of Bravais lattices

ones include the simple cubic, body-centered cubic, and face-centered cubic lattices, among others.

Additionally, there is another method to determine the unit cell. As shown in Fig. 2.7, a lattice point is taken as the center, and a closed region surrounded by the perpendicular lines connecting the lattice point with its nearest neighboring lattice points is called the Wigner–Seitz unit cell, abbreviated as the WS unit cell. Note that the WS unit cell has the same symmetry as the corresponding Bravais lattice.

(2) Reciprocal space

To discuss the band structure of SCs, we introduce the concept of reciprocal space. The basis vectors in the reciprocal space can be derived from those in the direct space. The relationships between the direct and reciprocal basis vectors are as follows [8]:

$$\vec{b}_1 = 2\pi \frac{\vec{a}_2 \times \vec{a}_3}{\vec{a}_1 \cdot \left(\vec{a}_2 \times \vec{a}_3\right)}, \tag{2.58}$$

Fig. 2.7 Schematic of a WS unit cell

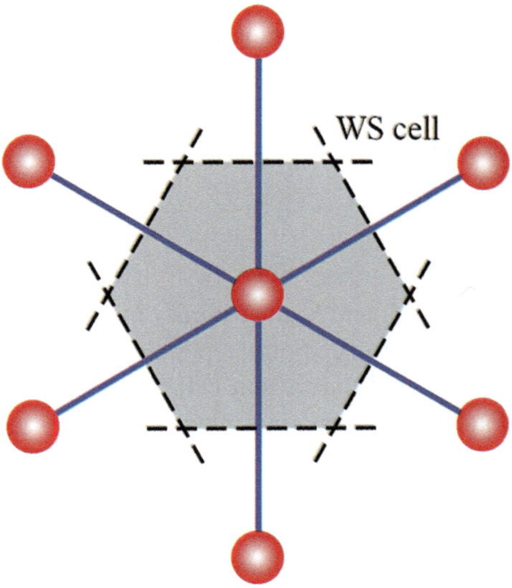

$$\vec{b}_2 = 2\pi \frac{\vec{a}_3 \times \vec{a}_1}{\vec{a}_1 \cdot \left(\vec{a}_2 \times \vec{a}_3\right)}, \tag{2.59}$$

$$\vec{b}_3 = 2\pi \frac{\vec{a}_1 \times \vec{a}_2}{\vec{a}_1 \cdot \left(\vec{a}_2 \times \vec{a}_3\right)}, \tag{2.60}$$

where $\vec{b}_1$, $\vec{b}_2$, and $\vec{b}_3$ are the reciprocal basis vectors. The direct and reciprocal basis vectors satisfy the orthogonal relation $\vec{a}_i \cdot \vec{b}_j = 2\pi\delta_{ij}$ $(i, j = 1, 2, 3)$, where δ_{ij} is the Kronecker symbol. Similarly, the position vector of an arbitrary lattice point in reciprocal space is expressed as:

$$\vec{G}_h = h_1\vec{b}_1 + h_2\vec{b}_2 + h_3\vec{b}_3, \tag{2.61}$$

where h_1, h_2, and h_3 are arbitrary integers, and $\vec{G}_h$ is defined as the reciprocal lattice vector.

In fact, each crystal structure can be described by both direct and reciprocal crystal lattices. It can be mathematically proven that there is a unique correspondence between direct and reciprocal lattices, and both of them are reciprocal to each other, accurately exhibiting the symmetry of the crystal structure.

2.7.3 Brillouin Zone

The symmetry of the band structure in SC is closely related to that of the crystal lattice. Therefore, calculating the band structure of the unit cell in reciprocal space can accurately characterize the symmetry of the band structure of the entire crystal lattice in reciprocal space. This method reduces computational load and improves efficiency. Generally, we select the WS unit cell in reciprocal space as the minimum repetitive structure, which is called the first Brillouin zone. Additionally, to further reduce the computational region in reciprocal space, we select a series of high-symmetry points within the first Brillouin zone to enclose a region with high symmetry, known as the irreducible Brillouin zone.

Next, we discuss the reciprocal lattice vectors and the first Brillouin zone of SCs with different lattice symmetries. Based on the dimension of the periodic arrangement of the unit cells in the SCs, we generally divide them into 1D, 2D, and 3D SCs.

(1) Brillouin zone of 1D SC

A 1D SC is composed of a periodic arrangement of unit cells in one direction. For a 1D crystal lattice with a lattice constant of a, the direct and reciprocal basis vectors can be expressed as $\vec{a} = a\vec{e}_x$ and $\vec{b} = \frac{2\pi}{a}\vec{e}_x$, respectively, where $\vec{e}_x$ is the unit vector

Fig. 2.8 Schematic of the **a** direct and **b** reciprocal crystal lattices of a 1D SC

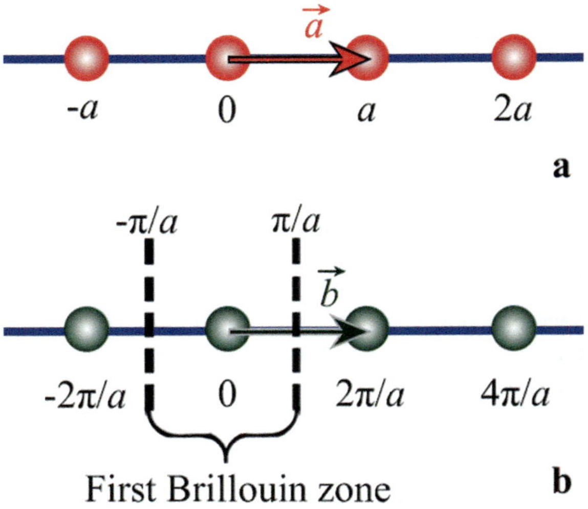

in the periodic direction of the SC. As shown in Fig. 2.8, by taking a lattice point in reciprocal space as a reference, the vectors from this point to the two nearest lattice points are $\vec{b}$ and $-\vec{b}$. The region between the perpendicular bisectors of both vectors is the first Brillouin zone, defined as $[-\pi/a, \pi/a]$.

(2) Brillouin zone of 2D SC

2D SCs can be mainly divided into two types: orthogonal and oblique lattices. Figure 2.9a schematically shows the orthogonal lattice, in which both direct basis vectors are orthogonal to each other. Assuming that the length and width of the WS unit cell are a and b, and the unit vectors in the x and y directions are $\vec{e}_x$ and $\vec{e}_y$, respectively, the direct basis vectors can be written as [8]:

$$\vec{a}_1 = a\vec{e}_x, \tag{2.62}$$

$$\vec{a}_2 = b\vec{e}_y. \tag{2.63}$$

Next, we calculate the reciprocal basis vectors of the crystal lattice. To do this, we introduce a direct basis vector $\vec{a}_3$ that is orthogonal with $\vec{a}_1$ and $\vec{a}_2$, which can be expressed as [8]:

$$\vec{a}_3 = \vec{e}_z, \tag{2.64}$$

where $\vec{e}_z$ is the unit vector in the z direction. Therefore, we can calculate the three reciprocal basis vectors as follows [8]:

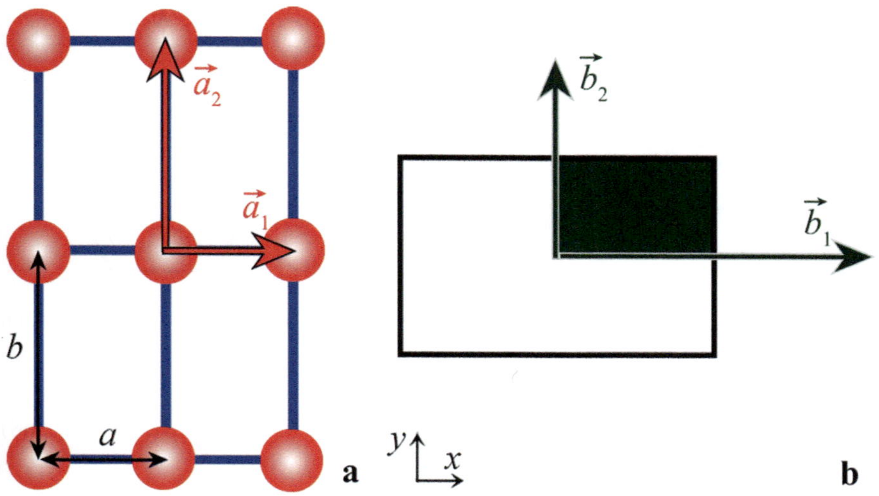

Fig. 2.9 **a** Schematic of the 2D orthogonal direct lattice and **b** its Brillouin zone

$$\vec{b}_1 = \frac{2\pi}{a}\vec{e}_x, \tag{2.65}$$

$$\vec{b}_2 = \frac{2\pi}{b}\vec{e}_y, \tag{2.66}$$

$$\vec{b}_3 = 2\pi\vec{e}_z. \tag{2.67}$$

The reciprocal basis vector in the z direction in Eq. (2.64) is redundant in the 2D lattice and can be ignored. As shown in Fig. 2.9b, the black open rectangle represents the first Brillouin zone of the crystal lattice, and the dark green shaded region is the irreducible Brillouin zone.

For the oblique lattice, its two basis vectors are not orthogonal with each other, and form a certain angle, usually 60°. There exist two basic structures in the oblique lattice system, such as triangular and honeycomb lattices, which are shown in Fig. 2.10a, b respectively. The unit cell of the triangular lattice only contains one atom, while that of the honeycomb lattice has two non-equivalent atoms. It is worthy of note that the direct basis vectors of the two oblique lattices are the same. Therefore, the reciprocal lattices as well as the first Brillouin zones are the same, as shown in Fig. 2.10c. Then, we take the honeycomb lattice as an example, and calculate its reciprocal basis vectors, and obtain the first Brillouin zone.

As shown in Fig. 2.10b, the unit cell of the 2D honeycomb lattice is a regular hexagon, and the distance between the nearest neighboring lattice points is a. Therefore, the direct basis vectors can be expressed as [8]:

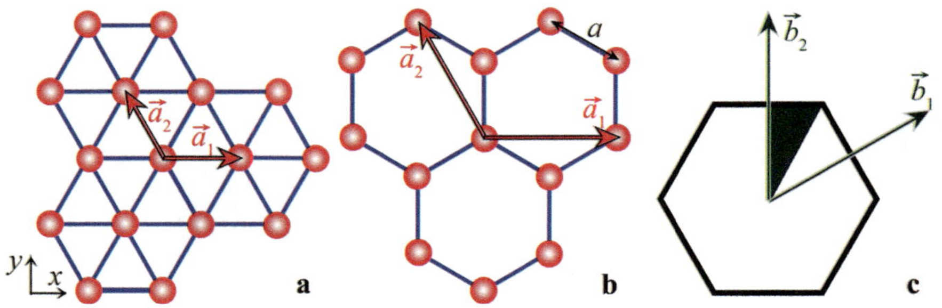

Fig. 2.10 Schematic of **a** a triangular lattice and **b** a honeycomb lattice, and **c** their first Brillouin zone

$$\vec{a}_1 = a\vec{e}_x, \tag{2.68}$$

$$\vec{a}_2 = \frac{a}{2}\left(\sqrt{3}\vec{e}_y - \vec{e}_x\right). \tag{2.69}$$

Similar to the 2D orthogonal lattice, by introducing the direct basis vector in the z direction in Eq. (2.64), we obtain the three reciprocal basis vectors based on Eqs. (2.58)–(2.60), which can be written as [8]:

$$\vec{b}_1 = \frac{2\pi}{a}\left(\vec{e}_x + \frac{\vec{e}_y}{\sqrt{3}}\right), \tag{2.70}$$

$$\vec{b}_2 = \frac{4\pi}{\sqrt{3}a}\vec{e}_y, \tag{2.71}$$

$$\vec{b}_3 = 2\pi\vec{e}_z. \tag{2.72}$$

The basis vector in the z direction in Eq. (2.72) is redundant and can also be ignored. As shown in Fig. 2.10c, the minimum region surrounded by the perpendicular bisectors of the two reciprocal basis vectors in Eqs. (2.70) and (2.71) is the first Brillouin zone.

2.7.4 Calculation of Band Structure of SC

When solving the problem of electronic band structure in crystals, it is generally transformed into solving the wave function and energy eigenvalues of electrons in the crystal under the potential field of a periodic spatial distribution of atoms. The effect of SCs on elastic waves is similar to the effect of a periodic potential field on electrons. Therefore, the band structure of the SCs can be calculated using a similar method. In the context of airborne sound, the characteristics of sound propagation in the SCs can be characterized based on pressure acoustics. For wave problems in the frequency domain, the Helmholtz equation is usually used to describe the propagation of sound, which is given as [1]:

$$\nabla \cdot \left[\frac{\nabla p(\vec{r})}{\rho(\vec{r})} \right] + \frac{1}{\rho(\vec{r})} \left(\frac{\omega}{c} \right)^2 p(\vec{r}) = 0, \tag{2.73}$$

where $\vec{r}$ is the position vector, $p(\vec{r})$ is the acoustic pressure, $\rho(\vec{r})$ is the density of the medium, and ω and c are the angular frequency and sound velocity, respectively. Based on the Bloch theorem, the pressure eigenfunction obtained by solving Eq. (2.73) can be expressed in a special form, known as the Bloch function [8]:

$$p(\vec{r}) = e^{i\vec{k}\cdot\vec{r}} u_{n,\vec{k}}(\vec{r}), \tag{2.74}$$

where $\vec{k}$ is the Bloch wave vector, n is the band index, and $u_{n,\vec{k}}(\vec{r})$ is the amplitude modulation function with translational symmetry, satisfying $u_{n,\vec{k}}(\vec{r}) = u_{n,\vec{k}}(\vec{r} + \vec{R}_l)$. By substituting Eq. (2.74) into (2.73), we obtain:

$$\begin{aligned} & i\vec{k} \cdot \left(\nabla \frac{u_{n,\vec{k}}(\vec{r})}{\rho(\vec{r})} + \frac{\nabla u_{n,\vec{k}}(\vec{r})}{\rho(\vec{r})} \right) + \nabla \cdot \frac{\nabla u_{n,\vec{k}}(\vec{r})}{\rho(\vec{r})} \\ & + \frac{u_{n,\vec{k}}(\vec{r})}{\rho(\vec{r})} \left[\left(\frac{\omega}{c} \right)^2 - \left| \vec{k} \right|^2 \right] = 0. \end{aligned} \tag{2.75}$$

Therefore, for any given wave vector $\vec{k}$, we can calculate a series of eigenvalues and eigenfunctions, which correspond to the energy bands and eigenmodes of the SC, respectively.

2.8 Conclusion

This chapter introduces the fundamental theories and modeling methods for sound manipulation using acoustic metamaterials in both 2D and 3D spaces. Key theories include the interference theory, EM theory, FE method, generalized Snell's law, mode conversion theory, and band theory of SCs. These theories provide a comprehensive framework for understanding and designing advanced acoustic devices.

References

1. Du GH, Zhu ZM, Gong XF (2012) Fundamentals of Acoustics. Nanjing University Press
2. Fokin V, Ambati M, Sun C, Zhang X (2007) Method for retrieving effective properties of locally resonant acoustic metamaterials. Phys Rev B 76(14):144302. https://doi.org/10.1103/PhysRevB.76.144302
3. COMSOL multiphysics user's guide, version 4.2. http://www.comsol.com
4. Yu NF, Genevet P, Kats MA, Aieta F, Tetienne JP, Capasso F, Gaburro Z (2011) Light propagation with phase discontinuities: generalized laws of reflection and refraction. Science 334(6054):333–337. https://doi.org/10.1126/science.1210713
5. Qian J, Wang Y, Yuan SQ, Sun HX, Liu XJ (2019) Reflected acoustic wavefront manipulation by an ultrathin metasurface based on three-dimensional generalized Snell's law. Appl Phys Express 12(9):094001. https://doi.org/10.7567/1882-0786/ab3492
6. Qian J, Wang Y, Xia JP, Ge Y, Yuan SQ, Sun HX, Liu XJ (2020) Broadband integrative acoustic asymmetric focusing lens based on mode-conversion meta-atoms. Appl Phys Lett 116(22):223505. https://doi.org/10.1063/5.0004579
7. Chen JH, Qian J, Guan YJ, Ge Y, Yuan SQ, Sun HX, Lai Y, Liu XJ (2021) Broadband bidirectional and multi-channel unidirectional acoustic insulation by mode-conversion phased units. Front Mater 8:766491. https://doi.org/10.3389/fmats.2021.766491
8. Jain VK (2022) Solid state physics. Springer Nature Switzerland AG

Chapter 3
Acoustic Asymmetric Transmission by Underwater Plate-Like Metamaterials

3.1 Introduction

Underwater metamaterials are a type of acoustic metamaterial designed to manipulate sound propagation in water. This is typically achieved by precisely designing metallic materials with special structures. Based on underwater metamaterials, the modulation of sound propagation paths, asymmetric transmission, focusing, and isolation can be realized in water. These capabilities are of great significance in fields such as underwater acoustic communication, sonar detection, and medical ultrasound imaging and treatment. To improve the performance of underwater communication devices and enhance the detection capability of sonar systems, it is essential to achieve asymmetric transmission of sound along specific propagation paths to avoid interference caused by sound energy in other directions. Therefore, the study of underwater AAT devices has significant application value and scientific importance.

In this chapter, we numerically design and experimentally demonstrate four types of AAT systems using underwater metamaterials. First, we design an AAT system composed of a thin brass plate with a single-sided periodic grating immersed in water [1]. The AAT effect is observed in two working bands below 1500 kHz, which can be systematically controlled by tuning the size of the unit cell. Second, we propose an enhanced AAT system composed of a periodic grating made of brass cylinders accompanied by multi-layer brass plates [2]. The enhanced AAT effect arises from the leaky A_0 mode in the brass plates, which is excited by the coupling of diffracted waves with resonant surface states localized at the interface of the grating and brass plate. The measured maximum transmittance and rectifying ratio reach 0.97 and 10^4, respectively. Third, we investigate the performance modulation of an AAT system composed of a brass plate and a periodic grating immersed in water [3]. The pass band and the output angle of sound in the AAT system are closely related to the angle distribution curves of the ± 1-order diffractions and the A_0 mode. Consequently, by adjusting the grating period and the plate thickness, the control of the passband and the output angle of sound can be achieved. Fourth, we design a multi-band AAT

H. Sun et al., *Acoustic Metamaterials*, Microfluidics and Nanophotonics: Science and Engineering 2, https://doi.org/10.1007/978-981-96-8524-0_3

system consisting of two periodic gratings with different periods placed on both sides of two brass plates [4]. The AAT effect is observed in four working bands below 1500 kHz, attributed to the interaction between various diffractions caused by the gratings with different periods and Lamb modes in the brass plates. Finally, we propose an ultra-broadband AAT system composed of a brass plate and a right triangle reflector in water [5]. The ultra-broadband AAT effect is attributed to the asymmetric excitation of the leaky A_0 mode in the brass plate caused by the incident angle of external waves. The bandwidth of the AAT can reach 2000 kHz with an incident angle of 48°. The designed underwater AAT systems have potential applications in medical ultrasound imaging, diagnosis, and acoustic functional components.

3.2 Acoustic Asymmetric Transmission by a Metal Plate with Single-Sided Periodic Gratings

3.2.1 Design and Performance of AAT System

As schematically shown in Fig. 3.1, the designed AAT system consists of a thin brass plate immersed in water, with a periodic array of rectangular gratings milled on the left surface of the plate [1]. The abbreviations LI and RI present the normal incidence of sound from the left and right sides of the system, respectively. The parameters of the unit cell are $a = 4$ mm, $h = 1$ mm, $a_1 = 1$ mm and $h_1 = 0.5$ mm. The characteristics of the AAT system are simulated using the COMSOL Multiphysics software, with material parameters selected as follows: density $\rho = 8400$ kg/m^3, longitudinal wave velocity $c_l = 4400$ m/s and transversal wave velocity $c_t = 2200$ m/s for the brass plates; $\rho = 1000$ kg/m^3 and $c = 1433$ m/s for water at 277 K.

Figure 3.2a–d show the simulated intensity distributions caused by the AAT system at 362 and 844 kHz for LI and RI, with the number of unit cells in the system set to 40. For comparison, a brass plane plate (PP) with the same length and thickness of 1.5 mm is also simulated (Fig. 3.2e, f). The intensity distribution in each figure is normalized by the maximum intensity of the incident wave. As shown in Fig. 3.2, the simulated intensity distributions exhibit obvious AAT characteristics. In the case of LI, strong sound energy is transmitted to the right region, primarily consisting of three branches: the middle branch (MB), the downward deflected branch (DDB), and the upward deflected branch (UDB). In the far field, the sound intensities of DDB and UDB are much stronger than that of MB. This phenomenon is attributed to sound diffraction caused by the grating. However, for RI and PP, the sound energy of MB is very weak, and DDB and UDB almost disappear.

To quantitatively display the AAT behavior, we simulate the transmission spectra for LI, RI, and PP, as shown in Fig. 3.3a. The transmission spectra exhibit remarkable AAT behavior, particularly in the two black shaded regions. We can see that LI has

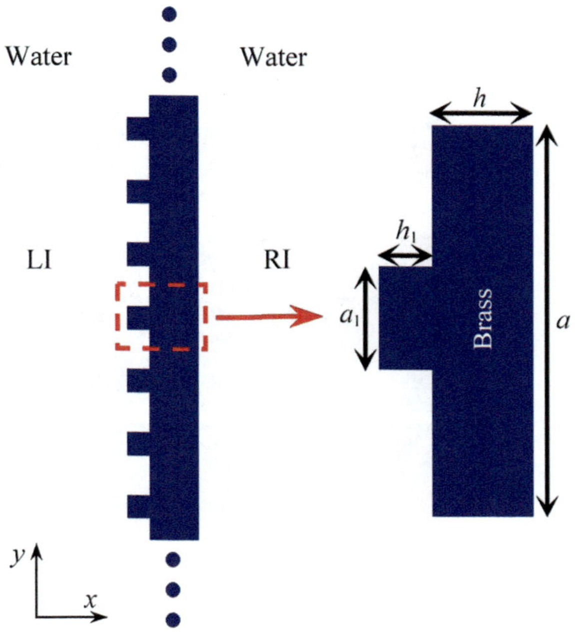

Fig. 3.1 Schematic of the AAT system

high transmission efficiency, while RI has no transmission in the ranges of 358–416 and 786–934 kHz. Additionally, the transmission for PP is rather low due to an extremely large acoustic impedance mismatch between brass and water.

To quantify the performance of the AAT system, we introduce the contrast ratio, defined as [6]:

$$R_c = \frac{|T_L - T_R|}{T_L + T_R}, \tag{3.1}$$

where T_L and T_R are the transmission coefficients for LI and RI, respectively. The absolute value of R_c represents the relative transmission weight between LI and RI. Figure 3.3b presents R_c calculated as a function of frequency. It is shown that the R_c is larger than 0.6 in both shaded regions, indicating that the system has good AAT performance.

3.2.2 *Experimental Measurement*

To experimentally verify the AAT system, we measure the transmission coefficients using the well-known ultrasonic transmission technique [7, 8]. In the experiment, the sample sizes are 16 cm and 8 cm in the y and z directions respectively, with the unit cell parameters matching those in Fig. 3.2. A plane brass plate with the same

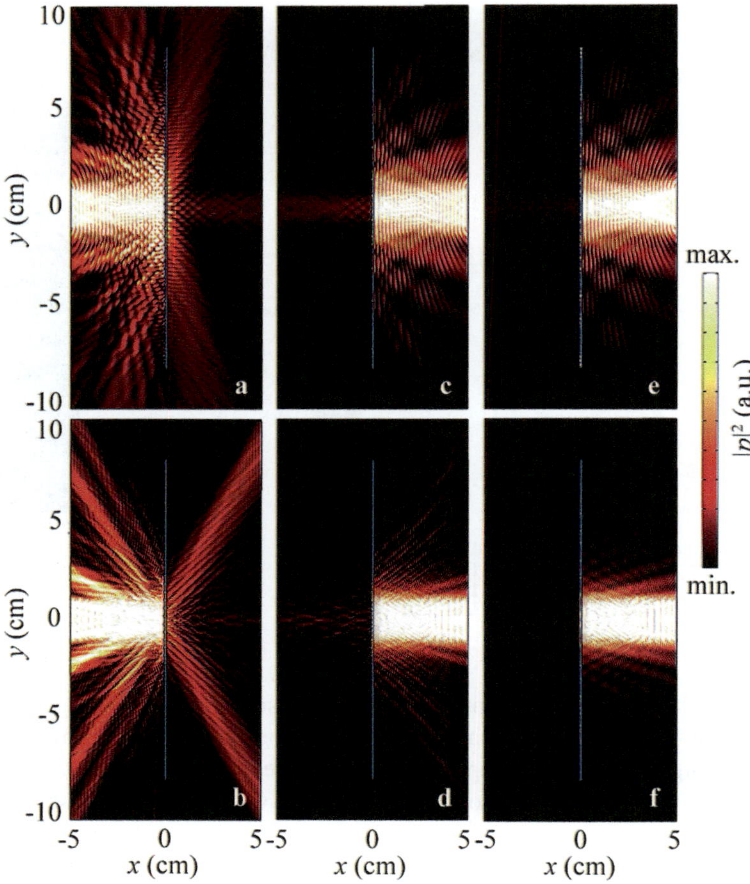

Fig. 3.2 Simulated intensity distributions caused by the AAT system for **a**, **b** LI, **c**, **d** RI, and **e**, **f** PP at 362 and 844 kHz

parameters as in the simulation is also measured. Figure 3.4 shows the schematic of the experimental set-up, which is immersed in a water tank. The sample is clamped and placed between a pair of specially designed ultrasonic generation and detection transducers (with a central frequency of 900 kHz, a bandwidth of 500–1300 kHz, and a diameter of 20 mm). The generation transducer is aligned normally towards the sample at a distance of 10 cm. On the other side, the detection transducer is positioned around the sample at the same distance. The average water temperature is 277 K.

Figure 3.5 shows the experimental angular transmission distributions for LI, RI, and PP. It is observed that only MB exists for RI and PP, whereas for LI, additional DDB and UDB appear apart from MB. The maximum intensities occur at angles of −40 and 40°, which agrees well with the simulations in Fig. 3.2b. This demonstrates that DDB and UDB dominate the transmitted sound energy.

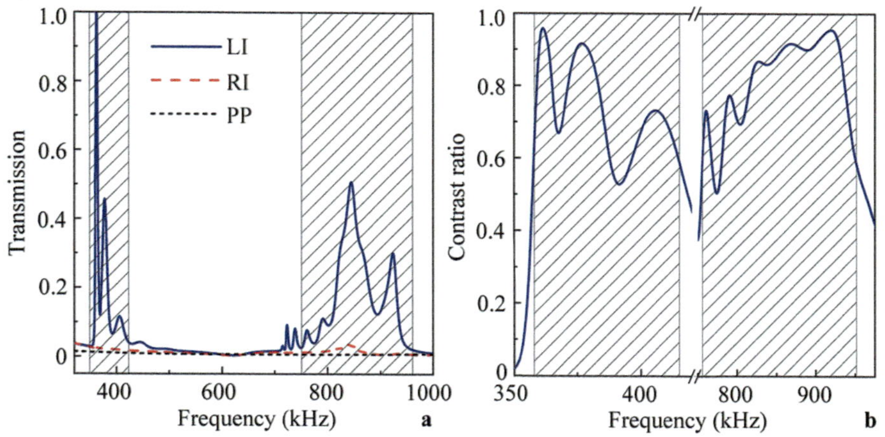

Fig. 3.3 Simulated **a** transmission spectra for LI, RI, and PP, and **b** contrast ratio of the AAT system

Fig. 3.4 Schematic of the experimental set-up for the AAT system in Fig. 3.1

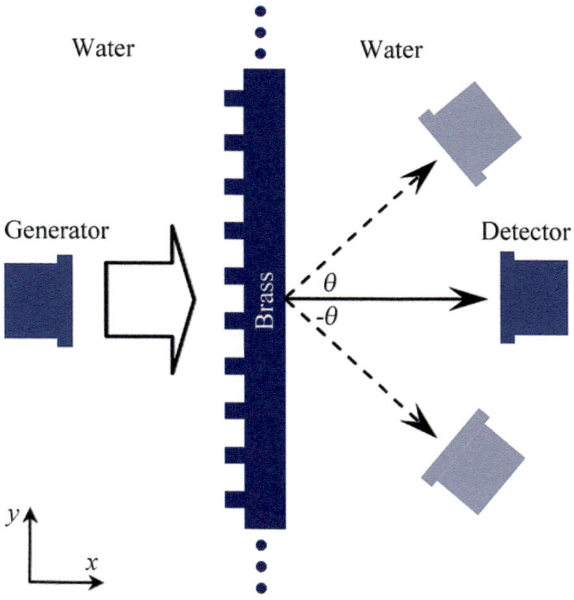

Next, we measure the transmission spectra by superposing all transmitted sound energy. Figure 3.6a shows the measured transmission spectra of the AAT system for LI, RI, and PP in the range of 600–1100 kHz, which agrees well with the simulated results. In particular, the transmission peak in the range of 800–950 kHz matches well between the measurement and simulation. Figure 3.6b presents the measured and simulated contrast ratio of the AAT system, showing good agreement in the AAT range.

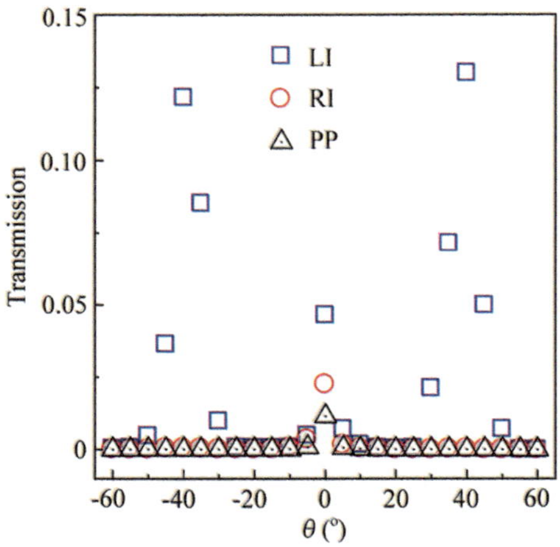

Fig. 3.5 Measured angular transmission distributions for LI, RI, and PP at 844 kHz

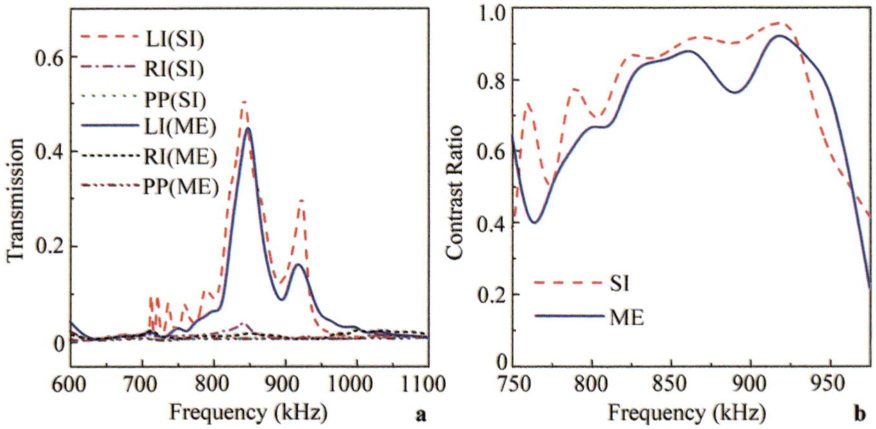

Fig. 3.6 **a** Measured and simulated transmission spectra for LI, RI, and PP, and **b** contrast ratio of the AAT system

3.2.3 Performance Modulation

We also find that the working bands of the AAT system can be controlled by tuning the size of the unit cell. In the simulations, we gradually change the size of the unit cell ($a = 4$ mm, $h = 1$ mm, $a_1 = 1$ mm, and $h_1 = 0.5$ mm) by scaling it by a factor n from 0.75 to 1.25, while keeping the configuration and material of the AAT system unchanged. Figure 3.7 plots the transmission spectra of the AAT system for different values of n. As shown in Fig. 3.7a–d, the working bands (the black shaded

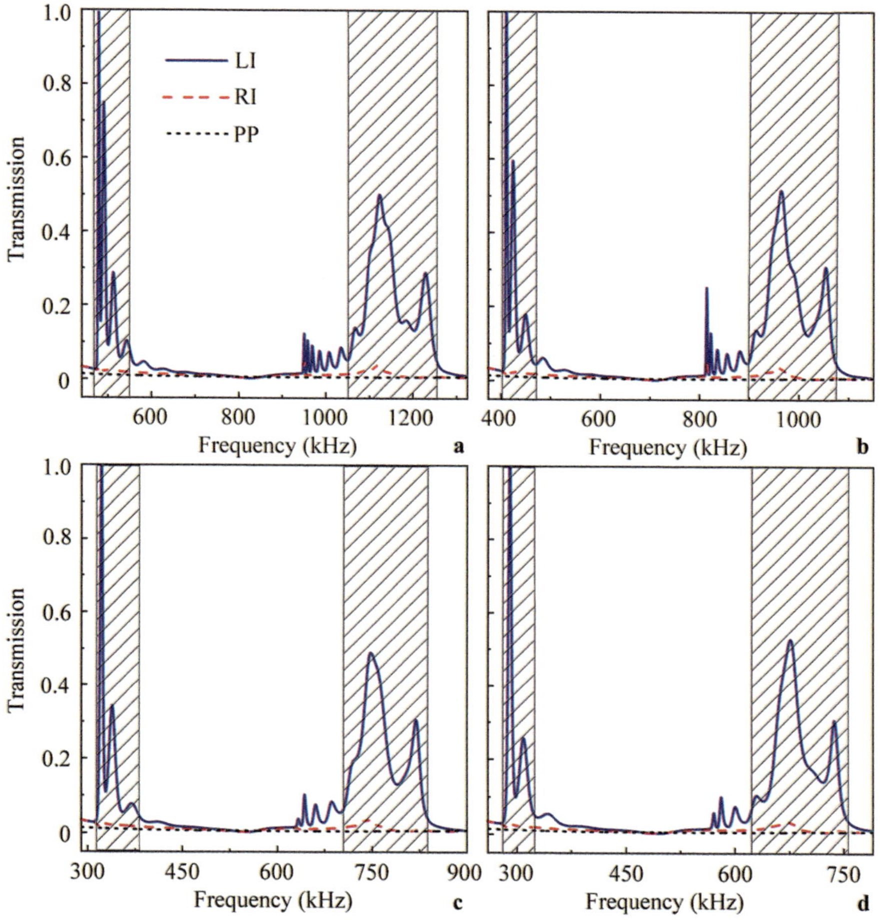

Fig. 3.7 Simulated transmission spectra of the AAT system with different values of n for LI, RI, and PP. **a** $n = 0.75$, working bands: 480–556 and 1062–1246 kHz; **b** $n = 0.875$, working bands: 408–460 and 908–1068 kHz; **c** $n = 1.125$, working bands: 320–374 and 740–830 kHz; **d** $n = 1.25$, working bands: 286–320 and 628–748 kHz

regions) change gradually as n is adjusted, while the shapes and magnitudes of the transmission spectra remain nearly constant. Thus, the working band of the AAT system can be systematically controlled by adjusting the size of the unit cell.

3.3 Enhanced Acoustic Asymmetric Transmission by Multi-Layer Metal Plates with Single-Sided Periodic Gratings

3.3.1 Design and Performance of Enhanced AAT System

As schematically shown in Fig. 3.8, the enhanced AAT system consists of a multi-layer brass plate (with a thickness $h = 1$ mm) immersed in a water background. A periodic array of brass cylinders (with a diameter $d = 0.8$ mm and a lattice constant $a = 2$ mm) is placed along the y direction on the left side of the multi-layer plates [2]. The brass plates are 20 cm long (covering $n = 100$ periods) and 10 cm wide in the y and z directions, respectively. The distance between the grating and the plate (b) and the distance between the plates (c) are both 0.5 mm. The abbreviations LI and RI refer to the incident waves emanating from the left and right sides of the AAT system, respectively. The software COMSOL Multiphysics is used to simulate the transmittance spectra, rectifying ratios, and intensity distributions. The material parameters are as follows: density $\rho = 8400$ kg/m^3, longitudinal wave velocity $c_l = 4400$ m/s, and transversal wave velocity $c_t = 2200$ m/s for brass; $\rho_w = 997$ kg/m^3 and $c_w = 1482$ m/s for water at 296 K. The width of the incident plane wave is set to 20 mm.

Figure 3.9 shows the simulated intensity distributions caused by the enhanced AAT systems with $m = 1$, 2, and 3. As shown in Fig. 3.9, the intensity distributions exhibit clear characteristics of AAT. For RI (Fig. 3.9b, d, and f), the sound energy is almost completely reflected with no transmission. However, in the case of LI (Fig. 3.9a, c, and e), the strong sound energy is transmitted to the right region and

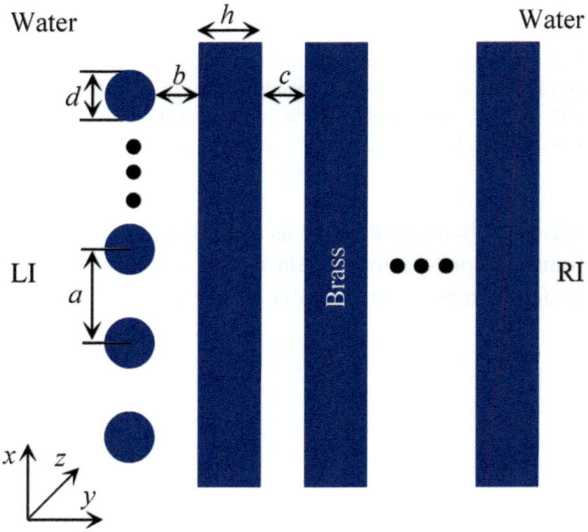

Fig. 3.8 Schematic of the enhanced AAT system

splits into two branches: UDB and DDB. Meanwhile, in the left region, in addition to the incident sound energy, the reflected sound energy is also divided into two parts that propagate in the opposite directions of UDB and DDB.

To verify this, we experimentally measure the transmittance spectra of the enhanced AAT system. The structure parameters of the sample are the same as those used in the simulations. The sample is clamped and placed between a pair of specially designed ultrasonic generation and detection transducers (with a central frequency of 900 kHz, a bandwidth of 500–1300 kHz, and a diameter of 20 mm). The entire assembly is immersed in a large water tank, and the water temperature is maintained at 296 K during the experiment.

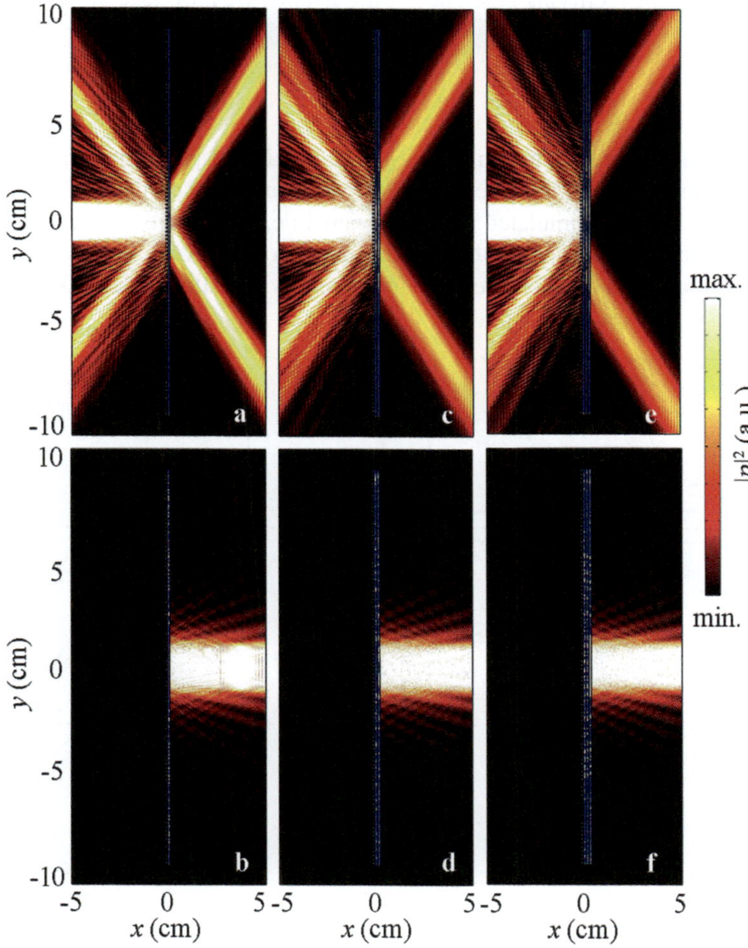

Fig. 3.9 Simulated intensity distributions caused by the enhanced AAT system at 900 kHz: **a** LI and **b** RI for $m = 1$; **c** LI and **d** RI for $m = 2$; **e** LI and **f** RI for $m = 3$

Figure 3.10a–c display the measured transmittance spectra by integrating the total transmitted sound energy for $m = 1, 2$, and 3, which agree well with the simulations. The working bands of the AAT systems with $m = 1, 2$, and 3 are 750–952 kHz, 776–952 kHz, and 802–952 kHz, respectively, as indicated by the black shaded regions. It is observed that the transmittance spectra exhibit remarkable AAT behavior. By comparing Fig. 3.10a–c, we observe that the transmittance for both LI and RI gradually decreases with the increase of m in the range of 750–870 kHz, especially between $m = 1$ and 2, due to the reflected sound energy and energy loss between the plates. However, in the range of 880–950 kHz, the transmittance for LI remains almost unchanged for $m = 2$ and 3, while that for RI continues to decrease (Fig. 3.10b, c), which can be used to further optimize the performance of the AAT system.

To quantify the rectifying efficiency of the sample, we introduce the ratio defined as [9]:

$$\delta = T_L / T_R, \tag{3.2}$$

Where T_L and T_R are the transmittance for LI and RI, respectively. The measured rectifying ratios of the AAT systems with $m = 1, 2$, and 3 are plotted in Fig. 3.10d–f, which agree with the simulated results. Meanwhile, as m increases, the rectifying ratios significantly increase in the working bands of the AAT systems (the black shaded regions), and the measured maximum rectifying ratio reaches 10^3 for $m = 3$

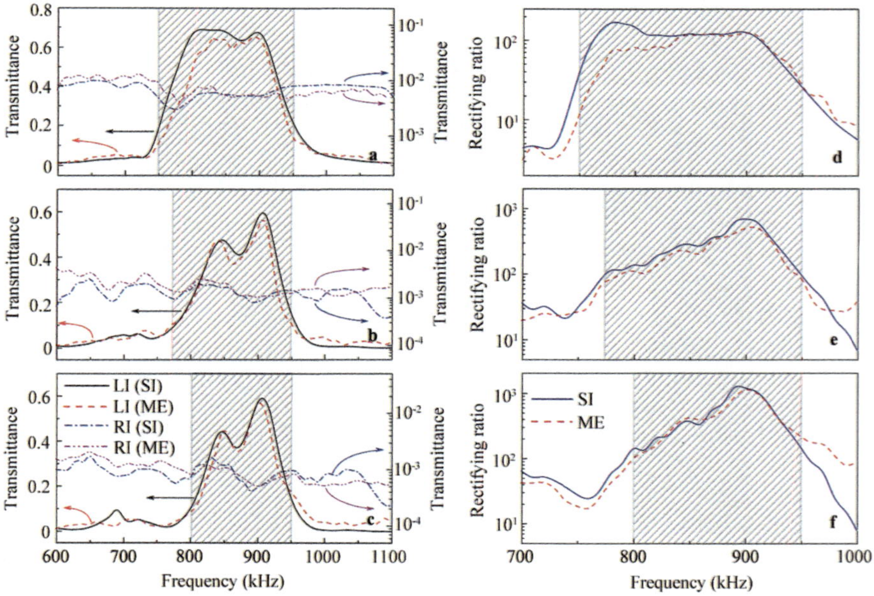

Fig. 3.10 Measured and simulated transmittance spectra of the enhanced AAT systems with **a** $m = 1$, **b** 2, and **c** 3, and rectifying ratios of the enhanced AAT systems with **d** $m = 1$, **e** 2, and **f** 3

at 900 kHz. It is therefore deduced that a higher rectifying ratio (more than 10^3) can be achieved in the AAT system with more brass plates.

3.3.2 Physical Mechanism

To gain insight into the mechanism of AAT, we investigate the diffracted waves caused by the grating structure without the brass plates. The diffraction angles can be calculated by $\theta = \arcsin\left(|k|c_{\mathrm{w}}/af\right)$, where k is the order of the diffracted wave, and f is the incident frequency. Figure 3.11a displays the angle distribution of the diffracted waves. It shows that, in addition to the 0-order, the ±1-orders are dominant below 1200 kHz, and the higher-order diffracted waves do not exist due to the smaller lattice constant of the grating. Moreover, the lower cutoff frequency of the ±1-orders is 741 kHz, which is almost the same as that of the AAT system with $m = 1$. Figure 3.11b shows the measured and simulated transmittance spectra of the diffracted waves through the grating, along with those caused by a uniform plate. As shown in Fig. 3.10a and 3.11b, the transmittance for LI is stronger than that for the ±1-orders but is almost the same as that for the 0- and ±1-orders. It is therefore deduced that the enhanced sound transmission is attributed to the coupling of the diffracted waves (0- and ±1-orders) with the resonant surface states localized at the interface between the grating and the brass plate [10]. However, this phenomenon cannot be observed for a single uniform plate without the grating. As shown in Fig. 3.11b, the transmittance is rather low for the uniform plate.

Furthermore, when the uniform brass plates are introduced into the AAT system, the transmitted waves through the system form intrinsic Lamb modes in the plates,

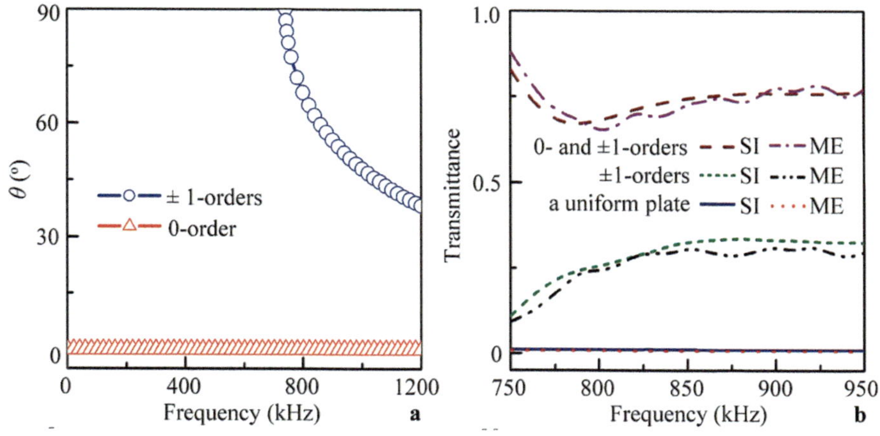

Fig. 3.11 **a** Angle distribution of diffracted waves through a single grating without a plate. **b** Simulated and measured transmittance spectra of the diffracted waves through the grating, along with those of sound through a uniform brass plate

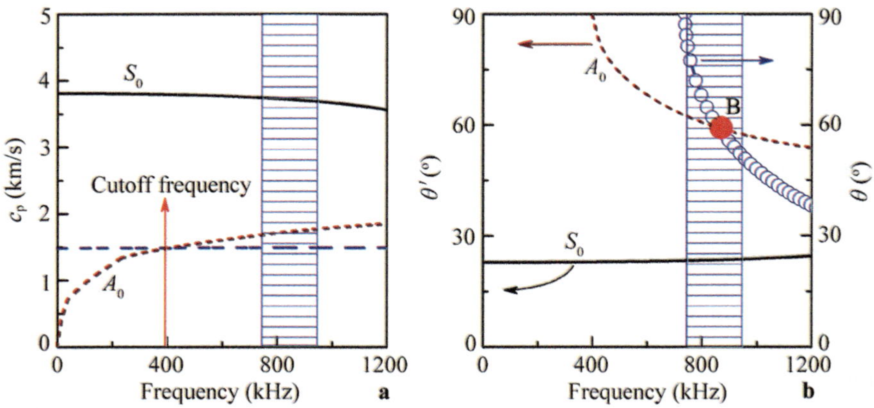

Fig. 3.12 **a** Dispersion curves of Lamb modes for a brass plate with a thickness of 1 mm immersed in water, along with the water line (the blue dashed line). The vertical red line denotes the cutoff frequency below which the A_0 mode is non-leaky. **b** Angle distribution of Lamb modes for the brass plate immersed in water, along with that of the ±1-orders (the blue open circles)

which can be excited by the diffracted waves. It is well known that when a plate is immersed in water, the energy of the Lamb modes will either leak or not, depending on their phase velocities (relative to that of water). Figure 3.12a shows the dispersion curve for a brass plate with a thickness of 1.0 mm immersed in water. We can see that the zero-order Lamb modes S_0 and A_0 are dominant in the AAT range (the blue shaded region), while the higher-order modes do not exist due to their lower cutoff frequencies. For the S_0 mode, the dispersion curve lies above the velocity line of water (the blue dashed line), making it leaky. However, for the A_0 mode, below a cutoff frequency, the dispersion curve lies below the water line: it is non-leaky and cannot be excited by the incident wave in any direction due to the momentum mismatch.

Figure 3.12b presents the output angle distribution of the Lamb modes at the interface between the brass plate and the water, calculated by $\theta' = \arcsin(c_w/c_p)$, where c_p is the phase velocity of the Lamb modes. As shown in Fig. 3.12b, the angle curve of the A_0 mode intersects that of the ±1-orders (the blue open circles) at point B in the AAT range (the blue shaded region), and the frequency of point B is close to 900 kHz, which corresponds to the frequency of maximum transmittance for $m = 3$. Moreover, as shown in Fig. 3.9a, c, and e, the propagation angles of both UDB and DDB are 58.4°, which matches that of the A_0 mode at 900 kHz in Fig. 3.12b. Therefore, it is straightforward to deduce that the enhanced AAT arises from the leaky A_0 mode in the brass plate, which is excited by the coupling of the diffracted waves with the resonant surface states localized at the interface between the grating and the plate.

3.3.3 Performance Optimization

To further enhance the transmittance and rectifying ratio, we optimize the AAT system by placing two inclined brass plates (with a thickness of 5 mm and a length of 50 mm) with x-axis symmetry on the left side of the system. Figure 3.13 shows the simulated intensity distributions caused by the AAT systems with $m = 1, 2$, and 3 at 900 kHz. The inclined angle between the plate and the incident wave is 50.2°, and the coordinate at the center of the top inclined plate is set as (−22, 40 mm). As shown in Fig. 3.13a, c, and e, the two reflected beams are reflected again by both inclined plates and transmit through the AAT systems. It is noteworthy that the transmitted sound energy of UDB and DDB is significantly enhanced. Meanwhile, for RI (Fig. 3.13b, d, and f), a portion of the transmitted sound energy is also reflected by both inclined plates and returns to the right side, which further enhances the rectifying ratio of the AAT systems.

Figure 3.14 presents the measured and simulated transmittance spectra and rectifying ratios for the optimized AAT systems with $m = 1, 2$, and 3. We observe that the measured and simulated transmittance spectra and rectifying ratios agree well with each other. Compared with Fig. 3.10a–c, the sound transmission is enhanced for LI but weakened for RI in Fig. 3.14a–c. The measured maximum transmittance reaches 0.97 for $m = 1$. Meanwhile, compared with Fig. 3.10d–f, the rectifying ratios are much higher in Fig. 3.14d–f, and the measured maximum rectifying ratio is as high as 10^4 for $m = 3$.

3.4 Performance Modulation of Acoustic Asymmetric Transmission by a Metal Plate with Single-Sided Periodic Gratings

3.4.1 Design of AAT System

The designed AAT system consists of a single-layer brass plate and a periodic brass rectangular grating immersed in water [3]. Photographs of the system from both sides are presented in Fig. 3.15a, b. As shown in Fig. 3.15a, the width (w) and length (l) of the grating (the red dashed rectangle) are 72 mm and 120 mm, respectively. On the opposite side (Fig. 3.15b), the width (w_1) and length (l_1) of the brass plate are 200 mm and 120 mm, respectively. Figure 3.15c illustrates the cross-section of the AAT system, where the plate thickness (h), the grating period (a), the width (s) and thickness (d) of the rectangular unit, and the distance between the grating and plate (b) are 1.0 mm, 2.4 mm, 0.6 mm, 1.0 mm, and 0.5 mm, respectively. The abbreviations LI and RI represent the incident plane waves emanating from the grating and brass plate sides, respectively. The COMSOL Multiphysics software is used to simulate the performance of the AAT system. The material parameters are as follows: density

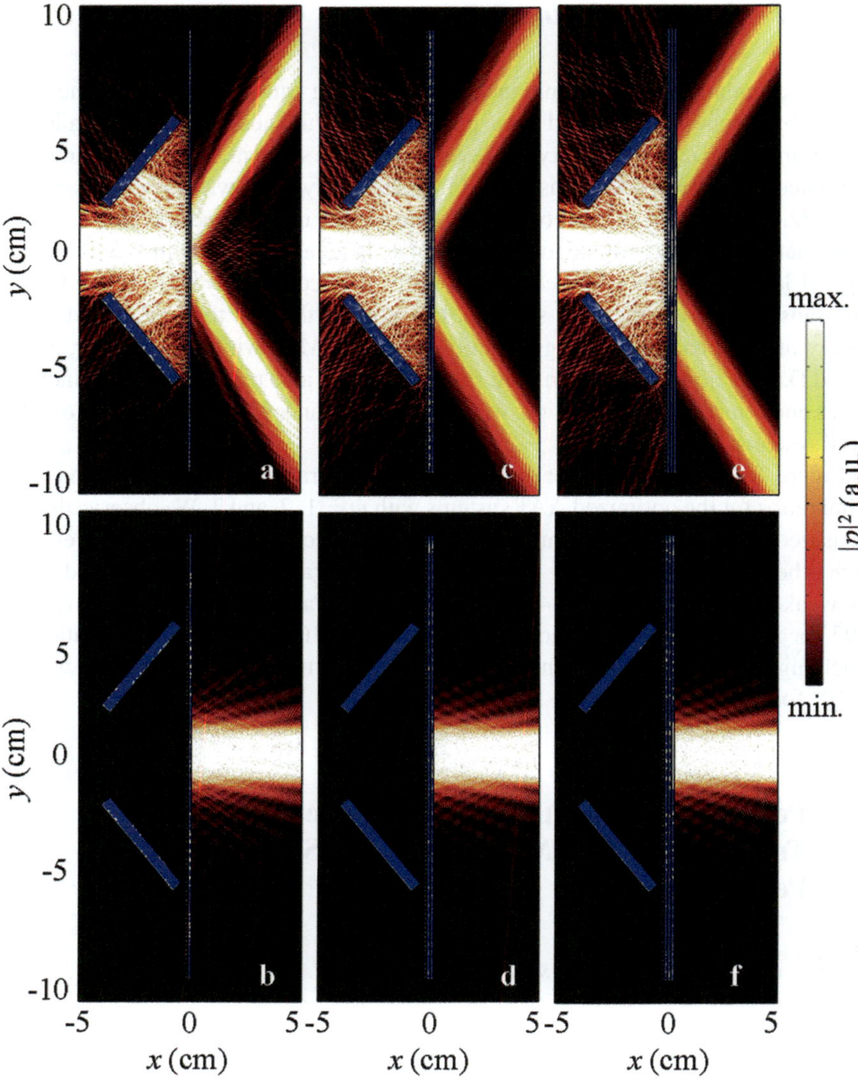

Fig. 3.13 Simulated intensity distributions caused by the optimized AAT systems at 900 kHz: **a** LI and **b** RI for $m = 1$; **c** LI and **d** RI for $m = 2$; **e** LI and **f** RI for $m = 3$

$\rho = 8400$ kg/m^3, longitudinal velocity $c_l = 4400$ m/s, and transversal velocity $c_t = 2200$ m/s for brass; and $\rho = 997$ kg/m^3 and $c = 1482$ m/s for water. Additionally, the width of the incident plane wave is set to 20 mm, matching the diameter of the excitation transducer used in the experiment, and the initial pressure amplitude ($|p_0|$) is selected as 1.0 Pa.

In the experiment, transmission spectra and intensity distributions are measured using the well-known ultrasonic transmission technique [7, 8]. As schematically

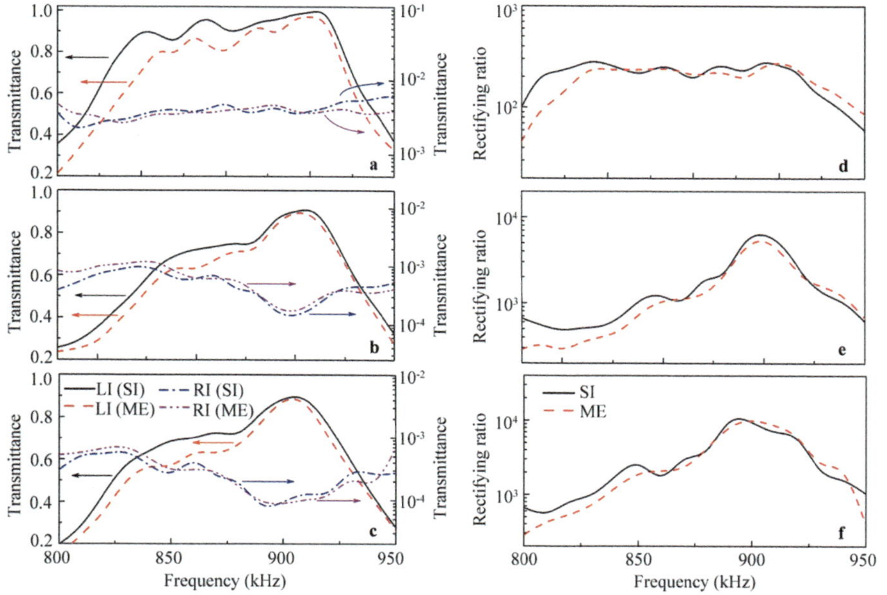

Fig. 3.14 Simulated and measured transmittance spectra of the optimized AAT systems with **a** $m = 1$, **b** 2, and **c** 3, and rectifying ratios of the optimized AAT systems with **d** $m = 1$, **e** 2, and **f** 3

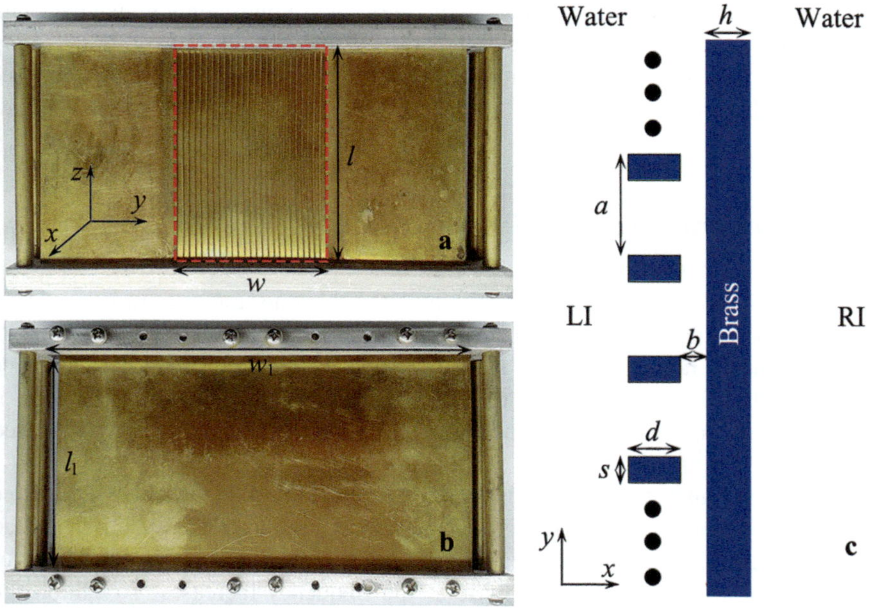

Fig. 3.15 Photographs of the AAT system from the **a** grating side and **b** plate side, and **c** schematic of the cross-section of the AAT system

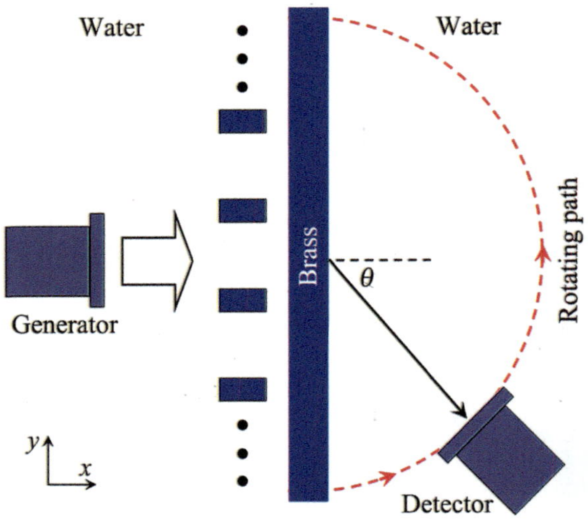

Fig. 3.16 Schematic of the experimental set-up for the AAT system in Fig. 3.15

shown in Fig. 3.16, the AAT system is placed between a pair of ultrasonic excitation and detection transducers (with central frequencies of 500 kHz, bandwidths of 250–1100 kHz, and diameters of 20 mm) and is immersed in a water tank. The excitation transducer is aligned normally toward the system at a distance of 12 cm. On the opposite side, the detecting transducer scans the system along a semicircular path with a radius of 5 cm, where the parameter θ represents the rotation angle of the detecting transducer. The measured transmittance spectra can be obtained by integrating the transmitted sound energy.

3.4.2 Performance of AAT System

Figure 3.17 shows the measured and simulated transmission spectra of the AAT system. A significant difference in the transmission spectra is observed for LI and RI in the frequency range of 608–774 kHz (the black shaded region). The transmittance is greater than 0.3 for LI and less than 0.05 for RI, indicating high performance of the AAT system. The measured spectrum is in good agreement with the simulated one.

Next, we simulate the intensity distributions through the AAT system for LI at three selected frequencies within the pass band, as shown in Fig. 3.18. Each transmitted wave (TW) is divided into two beams in the right region. The output angles (θ) of both beams are identical and decrease as the frequency increases. To exhibit the output angles of these TWs, we measure the intensity distributions along the same semicircular path on the right side for LI at 685, 705, and 725 kHz, as shown in Fig. 3.19. The simulated results are also plotted for comparison. The maximum values of the measured intensity occur at 65.9°, 62.9°, and 59.6° in Fig. 3.19a–c, respectively,

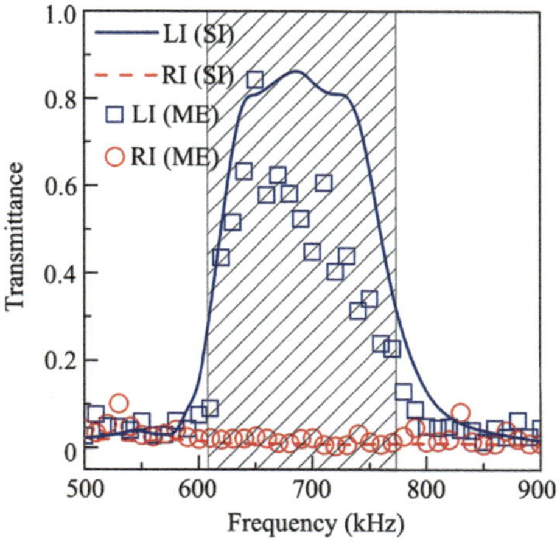

Fig. 3.17 Measured and simulated transmittance spectra of the AAT system

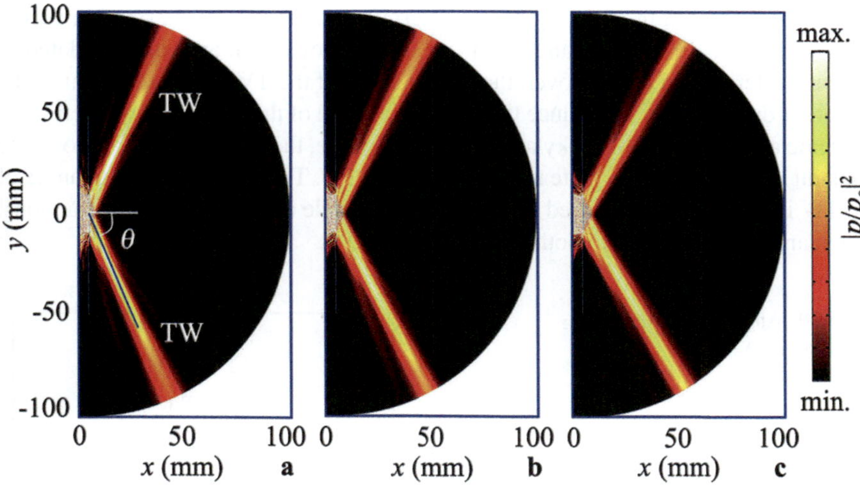

Fig. 3.18 Simulated intensity distributions ($|p/p_0|^2$) through the AAT system for LI at **a** 685 kHz, **b** 705 kHz, and **c** 725 kHz

which agree well with the simulated results. Furthermore, it is demonstrated that the output angles decrease as the frequency increases.

3.4.3 Physical Mechanism

To study the relationship between the output angle of the TWs and the parameters of the AAT system, we theoretically calculate the diffracted angles of the ± 1-order

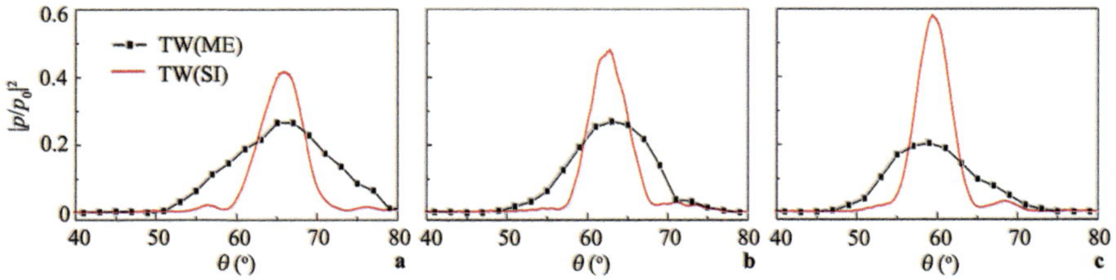

Fig. 3.19 Measured and simulated intensity distributions along the same semicircle path on the right side for LI at **a** 685 kHz, **b** 705 kHz, and **c** 725 kHz

diffractions caused by the grating with $a = 2.4$ mm and the leaky angles of the asymmetric zero-order Lamb mode (A_0) in the brass plate with $h = 1.0$ mm immersed in water, as shown in Fig. 3.20. Additionally, we plot the measured and simulated output angles of the TWs at 685, 705, and 725 kHz. The angle distribution curve of the A_0 mode intersects that of the ±1-order diffractions at 660 kHz, and the pass band of the AAT system is distributed on both sides of the intersection point. This indicates that the pass band of the AAT system is determined by the angle distribution curves of the ±1-order diffractions and the A_0 mode, both of which are closely related to the parameters a and h. Moreover, the output angle of the TWs is close to that of the ±1-order diffractions. We deduce that when the angle of the ±1-order diffractions is the same as or close to the leaky angle of the A_0 mode, the ±1-order diffractions can transmit through the brass plate in the form of the TW. Therefore, the output angle of the TW is primarily determined by the diffracted angle of the ±1-order diffractions, which are closely related to both parameters a and h.

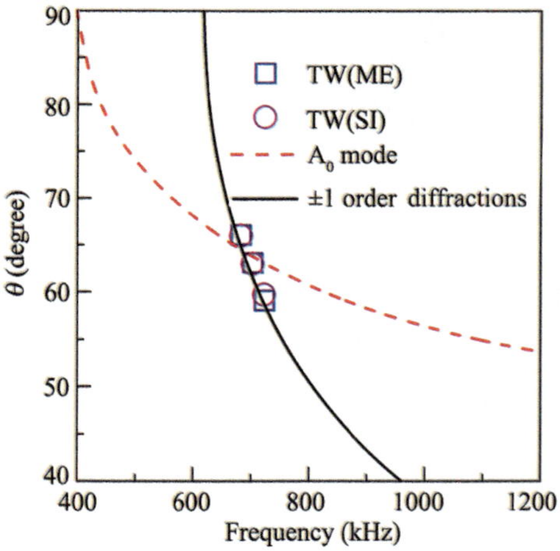

Fig. 3.20 Angle distribution curves of the A_0 mode, ±1-order diffractions and TWs

3.4.4 Performance Modulation

(1) Grating period a

First, we simulate the transmission spectra of the AAT systems by scaling the grating period (a) by a factor of n for LI, as shown in Fig. 3.21. In the simulations, $n(a)$ ranges from 0.5 to 2, while the other parameters remain constant. As shown in Fig. 3.21, increasing the value of a shifts the pass band of the AAT system to the low-frequency region, covering a frequency range of 450–1180 kHz. Figure 3.22 shows the transmission spectra of the AAT systems with $n(a) = 0.8$ and 1.2 for LI and RI. The pass bands are 785–990 kHz and 515–640 kHz for the AAT systems with $n(a) = 0.8$ and 1.2, respectively, and the bandwidth of the AAT system with $n(a) = 1.2$ is narrower than that with $n(a) = 0.8$. This suggests that the working band of the AAT system is closely related to the parameter a.

Furthermore, to exhibit the output angle and its variation of the TW, we simulate the intensity distributions through both AAT systems for LI. The incident frequencies are selected as 910, 930, and 950 kHz for the AAT system with $n(a) = 0.8$ (Fig. 3.23a–c), and 550, 570, and 590 kHz for the AAT system with $n(a) = 1.2$ (Fig. 3.23d–f). Compared with Fig. 3.23a–c, the output angles of the TWs are larger, and the corresponding variation is more obvious in Fig. 3.23d–f.

Figure 3.24 shows the measured and simulated intensity distributions along the same semicircle path on the right side for LI, which correspond to Fig. 3.23. We observe that the measured maximum intensities exist at 59.6°, 57.7°, 56.3°, 71.7°, 66.4° and 62.4° in Fig. 3.24a–f, respectively. The output angles of the TWs caused by the AAT system with $n(a) = 1.2$ (Fig. 3.24d–f) are larger than those with $n(a) = 0.8$

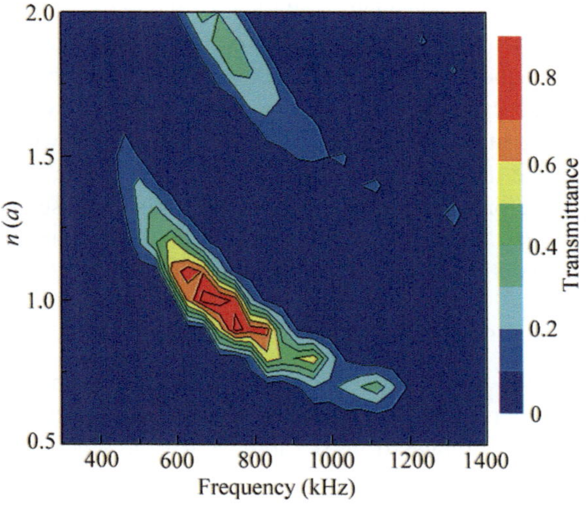

Fig. 3.21 Simulated transmittance spectra of the AAT systems with the parameter a scaled by a factor of n for LI

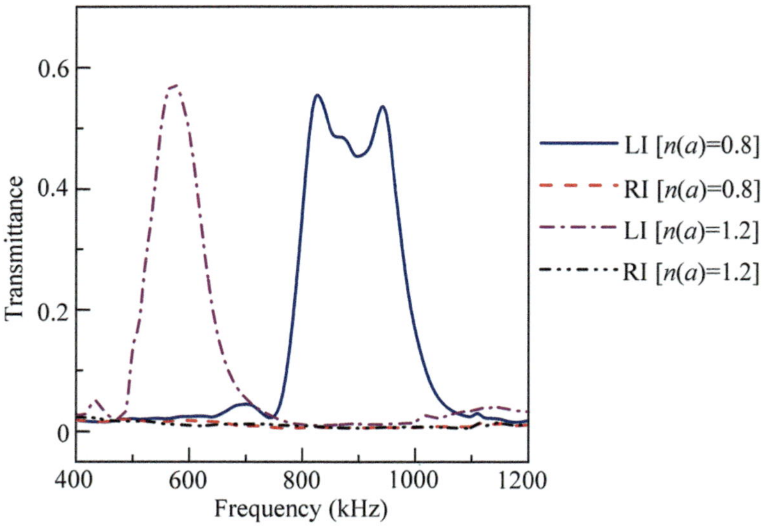

Fig. 3.22 Simulated transmittance spectra of the AAT systems with $n(a) = 0.8$ and 1.2 for LI and RI

(Fig. 3.24a–c), and the variation of the output angle is more obvious in Fig. 3.24d–f, which agree with the results in Fig. 3.23a–f.

In addition, we theoretically calculate the angle distribution curves of the A_0 mode and the ±1-order diffractions for the AAT systems with $n(a) = 0.8$ and 1.2, as shown in Fig. 3.25. The output angles of the TWs caused by the AAT systems with $n(a) =$ 0.8 and 1.2 are also plotted. With the increase of $n(a)$, the angle distribution curve of the A_0 mode remains unchanged, while that of the ±1-order diffractions shifts to the low-frequency region. Consequently, the intersection points of both curves move to the low-frequency region, causing the pass band of the AAT system and the excitation frequency of the TW to shift to the low-frequency region. Compared with the angle distribution curve of the ± 1-order diffractions with $n(a) = 0.8$, the diffracted angles and the curve gradient of the ± 1-order diffractions with $n(a) =$ 1.2 are larger. Therefore, the output angles of the TWs increase and change more significantly with $n(a) = 1.2$, and the pass band of the AAT system with $n(a) = 1.2$ is narrower than that with $n(a) = 0.8$, as also shown in Fig. 3.22.

(2) Grating period *a* and plate thickness *h*

Next, we simulate the transmission spectra of the AAT system by simultaneously scaling both parameters *a* and *h* by a factor of *n* for LI, as shown in Fig. 3.26. In the simulations, $n(a, h)$ ranges from 0.5 to 2, while the other parameters remain unchanged. The working band of the AAT system covers a frequency range of 400–1350 kHz for different values of $n(a, h)$. Compared with the results in Fig. 3.21, the working band of the AAT system becomes wider, and the transmittance in the high-frequency region is larger in Fig. 3.26. Figure 3.27 shows the transmission spectra of

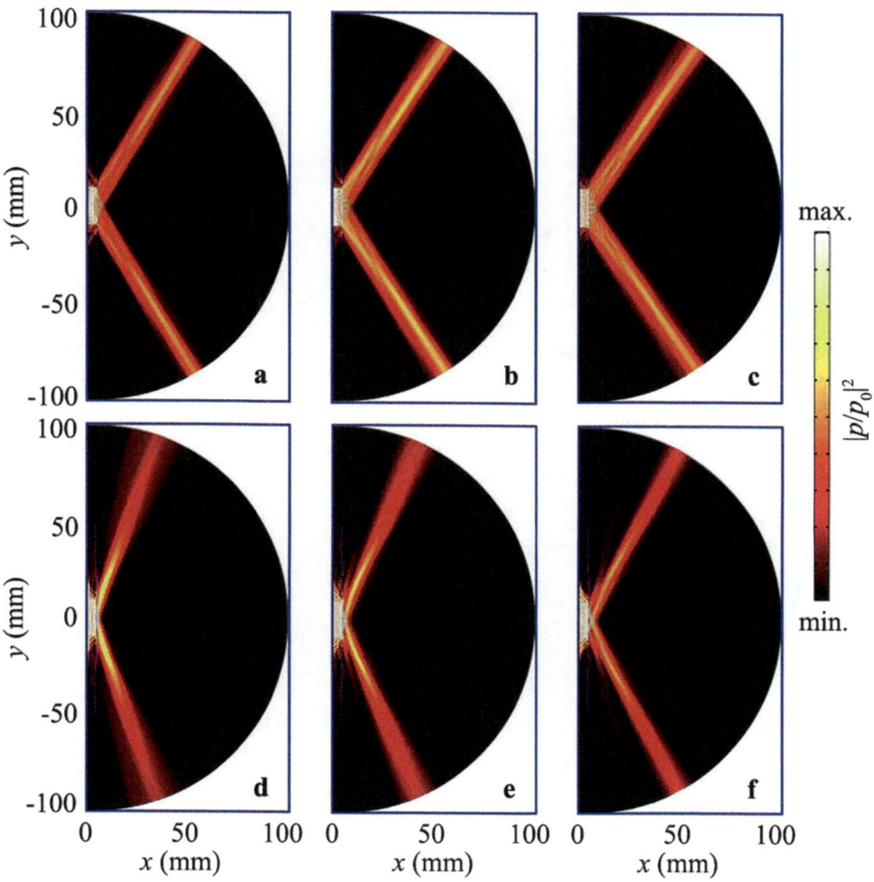

Fig. 3.23 Simulated intensity distributions through the AAT systems with $n(a) = 0.8$ at **a** 910 kHz, **b** 930 kHz, and **c** 950 kHz, and with $n(a) = 1.2$ at **d** 550 kHz, **e** 570 kHz, and **f** 590 kHz

the AAT systems with $n(a, h) = 0.8$ and 1.2 for LI and RI. The working band shifts to the low-frequency region as $n(a, h)$ increases, covering the ranges of 778–948 and 501–653 kHz for the AAT systems with $n(a, h) = 0.8$ and 1.2, respectively, similar to the results in Fig. 3.22. However, the transmittances for LI are larger than those in Fig. 3.22, especially for $n(a, h) = 0.8$.

Figure 3.28 shows the simulated intensity distributions through the AAT systems for LI, with incident frequencies selected as 855, 875, and 895 kHz for $n(a, h) = 0.8$ (Fig. 3.28a–c) and 570, 590, and 610 kHz for $n(a, h) = 1.2$ (Fig. 3.28d–f). The output angles of the TWs and their variation in Fig. 3.28a–c are similar to those in Fig. 3.28d–f.

Figure 3.29 shows the measured and simulated intensity distributions along the same semicircular path on the right side for LI, corresponding to Fig. 3.28, respectively. The maximum intensities occur at 65.7°, 63.7°, 61.3°, 66.7°, 62.5°, and 58.8°

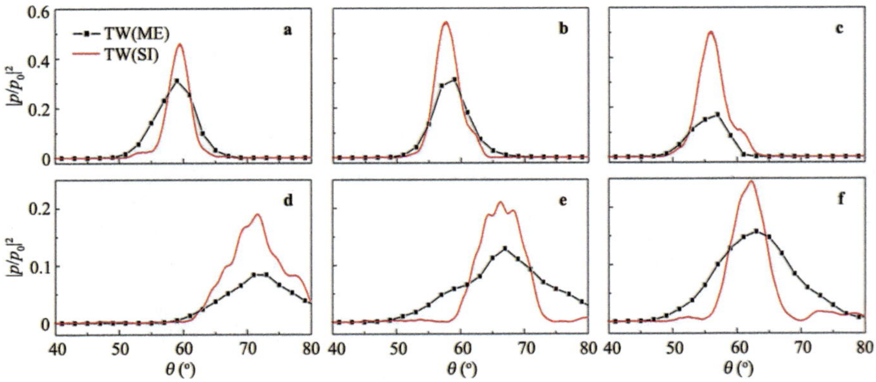

Fig. 3.24 Measured and simulated intensity distributions along the same semicircle path on the right side for LI caused by the AAT systems with $n(a) = 0.8$ at **a** 910 kHz, **b** 930 kHz, and **c** 950 kHz, and with $n(a) = 1.2$ at **d** 550 kHz, **e** 570 kHz, and **f** 590 kHz

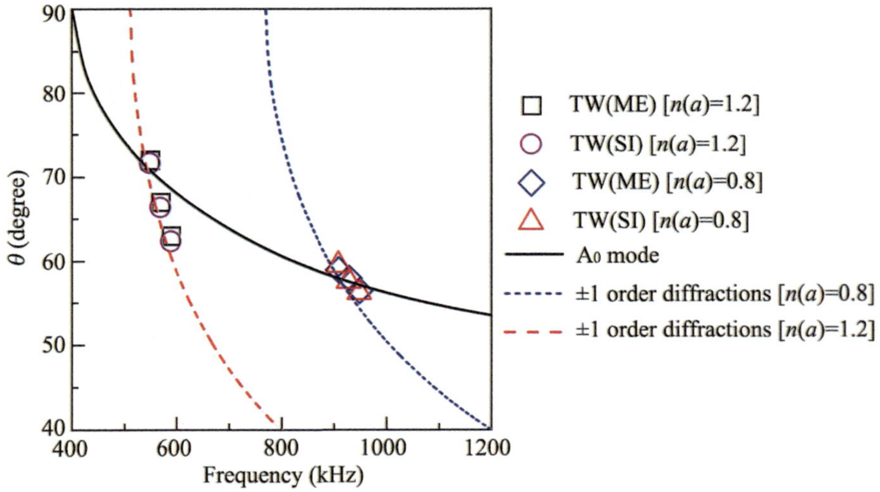

Fig. 3.25 Angle distribution curves of the A_0 mode, ±1-order diffractions, and TWs for the AAT systems with $n(a) = 0.8$ and 1.2

in Fig. 3.29a–f, respectively. The output angles of the TWs and their variation for the AAT system with $n(a, h) = 0.8$ (Fig. 3.29a–c) are similar to those with $n(a, h) = 1.2$ (Fig. 3.29d–f), consistent with the results in Fig. 3.28a–f. However, compared with Fig. 3.24a–c, the output angles of the TWs and their variation are larger in Fig. 3.29a–c, resulting from the simultaneous variation of both parameters *a* and *h*.

Figure 3.30 illustrates the angle distribution curves of the ±1-order diffractions and the A_0 mode for the AAT systems with $n(a, h) = 0.8$ and 1.2. The output angles of the TWs generated by both AAT systems are also presented. It can be observed

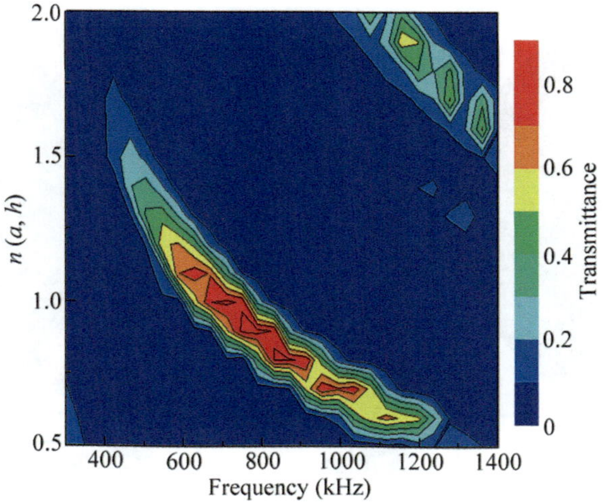

Fig. 3.26 Simulated transmittance spectra of the AAT systems with the parameters a and h simultaneously scaled by a factor of n for LI

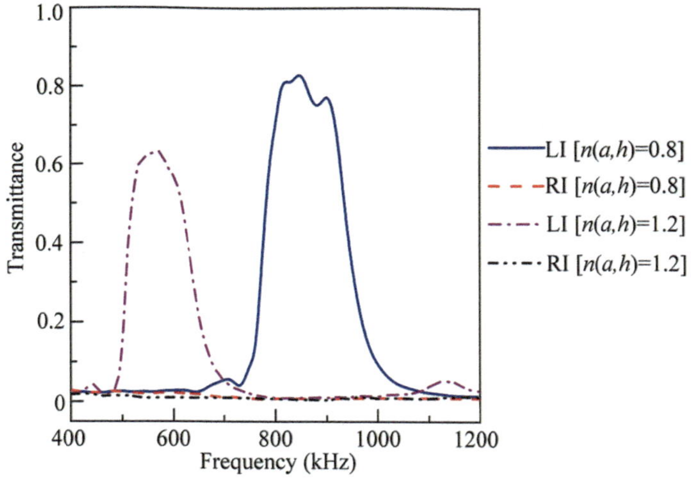

Fig. 3.27 Simulated transmittance spectra of the AAT systems with $n(a, h) = 0.8$ and 1.2 for LI and RI

that by simultaneously increasing both parameters a and h, the angle distribution curves of the ±1-order diffractions and the A_0 mode shift to the low-frequency region concurrently. Additionally, the diffracted angles and the curve gradient of the ±1-order diffractions for the AAT systems with $n(a, h) = 0.8$ and 1.2 remain identical. As a result, the output angles of the TWs and their variation are very similar, and the bandwidth of the AAT systems with $n(a, h) = 0.8$ and 1.2 is also comparable.

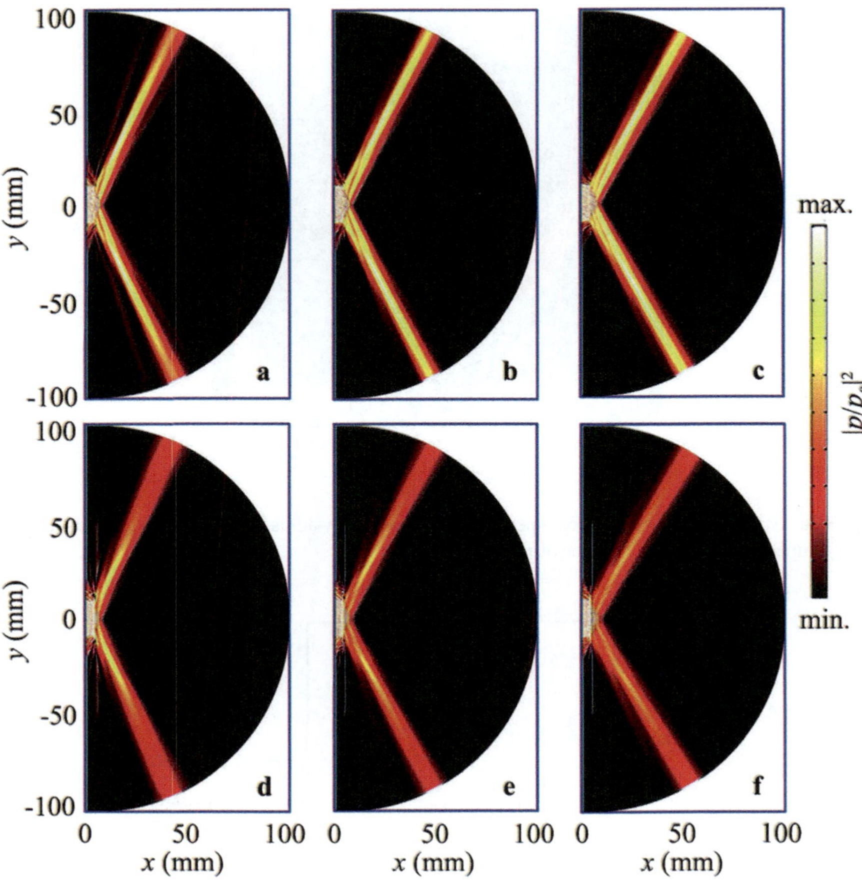

Fig. 3.28 Simulated intensity distributions through the AAT systems with $n(a, h) = 0.8$ at **a** 855 kHz, **b** 875 kHz, and **c** 895 kHz, and with $n(a, h) = 1.2$ at **d** 570 kHz, **e** 590 kHz, and **f** 610 kHz

3.5 Multi-Band Acoustic Asymmetric Transmission by Dual-Layer Metal Plates with Two-Sided Gratings

3.5.1 Design and Performance of Multi-band AAT System

The multi-band AAT system consists of two brass plates immersed in water, with two periodic rectangular gratings of different periods placed on both sides of the dual-layer plates [4]. The left and right sides of the system are schematically illustrated in Fig. 3.31a, b, respectively. The characteristics of the AAT system are simulated using the COMSOL Multiphysics software. The structure parameters are as follows: $a_1 = 3.6$ mm, $a_2 = 2.4$ mm, $b = 3.0$ mm, $c = 0.8$ mm, $h = 1.0$ mm, $d_1 = 0.4$ mm,

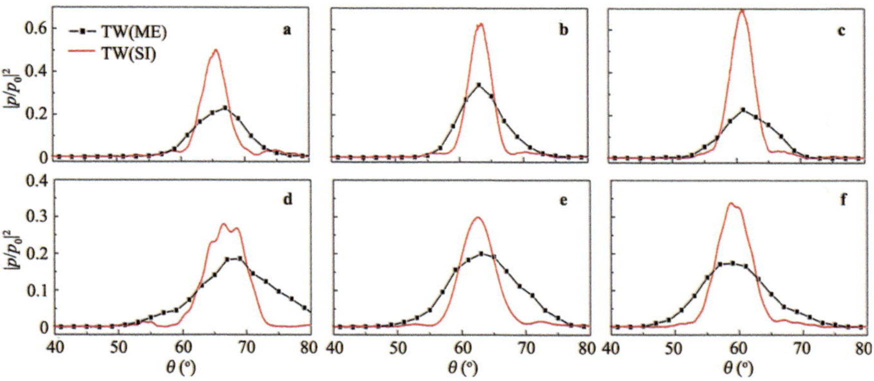

Fig. 3.29 Measured and simulated intensity distributions along the same semicircle path on the right side for LI caused by the AAT systems with $n(a, h) = 0.8$ at **a** 855 kHz, **b** 875 kHz, and **c** 895 kHz, and with $n(a, h) = 1.2$ at **d** 570 kHz, **e** 590 kHz, and **f** 610 kHz

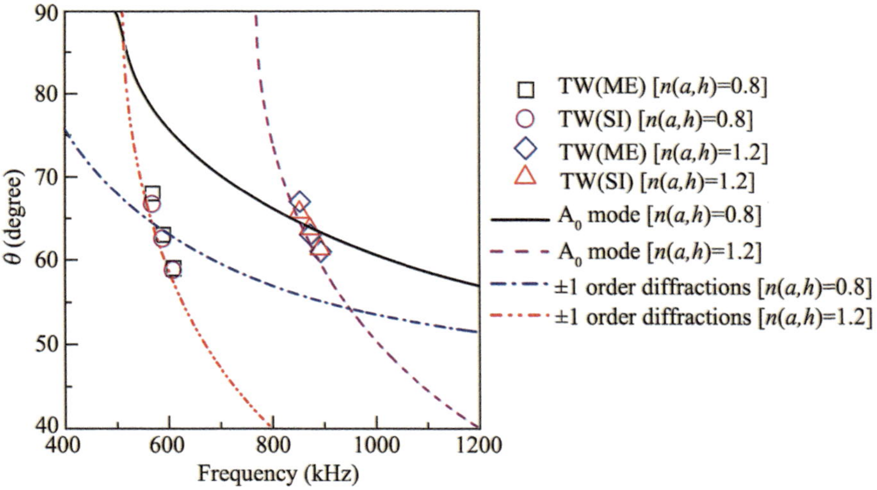

Fig. 3.30 Angle distribution curves of the A_0 mode, ±1-order diffractions, and TWs for the AAT systems with $n(a, h) = 0.8$ and 1.2

and $d_2 = 0.4$ mm. The material parameters are as follows: density $\rho = 8400$ kg/m^3, longitudinal wave velocity $c_l = 4400$ m/s, and transversal wave velocity $c_t = 2200$ m/s for brass; $\rho = 998$ kg/m^3 and $c = 1483$ m/s for water. Throughout this work, the normal incidence of sound from the left and right sides of the system is defined as left incidence (LI) and right incidence (RI), respectively.

Figure 3.32a presents the simulated transmittance spectra of the AAT system for LI and RI. It shows that the transmittance spectra exhibit a significant difference between LI and RI, and the AAT phenomenon exists in four bands below 1500 kHz: I: 431–506 kHz, II: 635–768 kHz, III: 955–1064 kHz, and IV: 1346–1430 kHz,

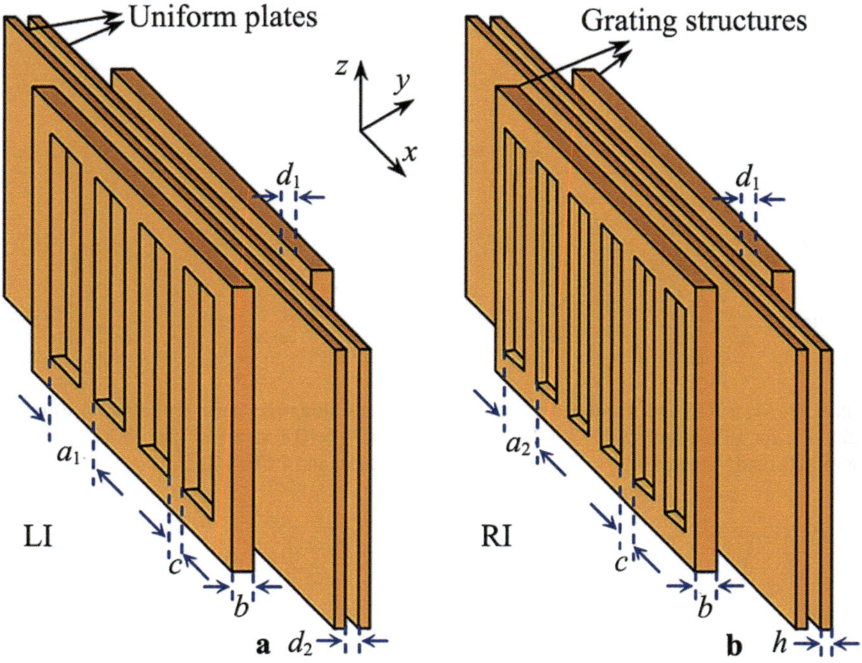

Fig. 3.31 Schematic of the **a** left and **b** right sides of the multi-band AAT system

as indicated by the black shaded regions. We can see that in bands I and III, the transmittance for LI is high, while that for RI is very low. Additionally, the maximum transmittance for LI is about 0.8 at 480 kHz in band I, which is larger than that in band III. Conversely, in bands II and IV, RI is associated with high transmittance, while for LI, the transmittance is close to zero. Moreover, the maximum transmittance for RI reaches 0.90 at 670 kHz in band II, which is larger than that in band IV.

To evaluate the performance of the AAT system, we calculate the contrast ratio based on Eq. (3.1). Figure 3.32b presents the contrast ratio of the system. It is noted that the contrast ratios of the multi-band AAT system are larger than 0.8 (the grey shaded regions) and can reach up to 0.95 in the ranges 640–730, 980–1060, and 1350–1390 kHz, indicating that the system has good multi-band AAT performance.

3.5.2 Physical Mechanism

To investigate the mechanism of multi-band AAT, we examine the Lamb modes in the brass plate and the diffracted waves excited by both types of gratings. We analytically calculate the leaky angles of the Lamb modes in a brass plate with a thickness of 1.0 mm and the angles of diffracted waves from the two gratings with periods of 3.6 (Fig. 3.33a) and 2.4 mm (Fig. 3.33b) immersed in water [11]. The black shaded

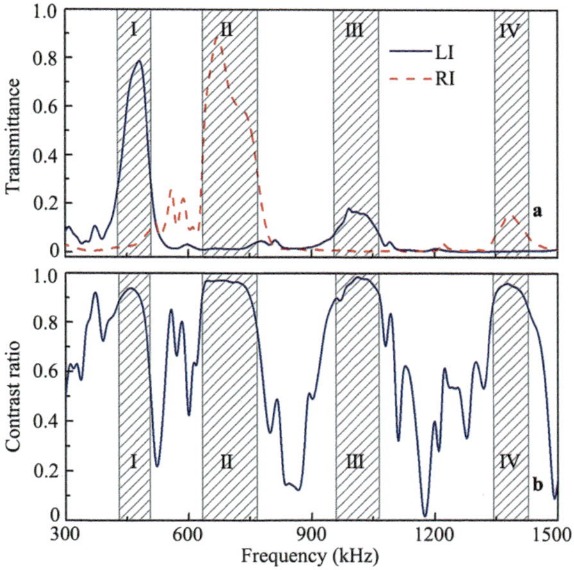

Fig. 3.32 **a** Simulated transmittance spectra of the multi-band AAT system for LI and RI, and **b** its contrast ratios

regions represent the AAT bands. As shown in Fig. 3.33a, the angle curve of the ±1-order diffracted waves intersects that of the A_0 mode at point A (425 kHz) and that of the S_0 mode at point C (1020 kHz). Additionally, the angle curve of the ±2-order diffracted waves intersects that of the A_0 mode at point B (990 kHz). It is worth noting that band I distributes on both sides of point A, and band III distributes on both sides of points B and C. As shown in Fig. 3.33b, the grating period is decreased from 3.6 to 2.4 mm, but the plate thickness remains constant. We can see that the angle curve of the ±1-order diffracted waves intersects those of the A_0 mode and S_0 mode at points D (682 kHz) and E (1392 kHz), respectively. Thus, bands II and IV distribute on both sides of points D and E, respectively.

Based on the aforementioned results, when the angles of the diffracted waves are the same as or close to the leaky angles of the Lamb modes, the diffraction can transmit through the system. Meanwhile, on the other side, the angles of the diffracted waves and the Lamb modes differ significantly, and the diffractions do not exist. Thus, little sound energy can transmit through the system, forming the AAT phenomenon. Additionally, the transmittance in bands I and II is larger than that in bands III and IV due to the difference in sound energy for different diffractions. Therefore, the multi-band AAT is attributed to the interaction of various diffracted waves generated from the two-sided gratings with different periods and the Lamb modes in the brass plate.

Next, we simulate the intensity distributions caused by a unit cell with a width of 7.2 mm in the AAT system and the total displacement distributions in the brass plates. The unit cell exactly covers two and three periods of the gratings on the left and right sides, respectively. As shown in Fig. 3.34a–c, the incident frequencies are selected in bands I and III, such as 468 (Fig. 3.34a), 988 (Fig. 3.34b), and 1038 kHz

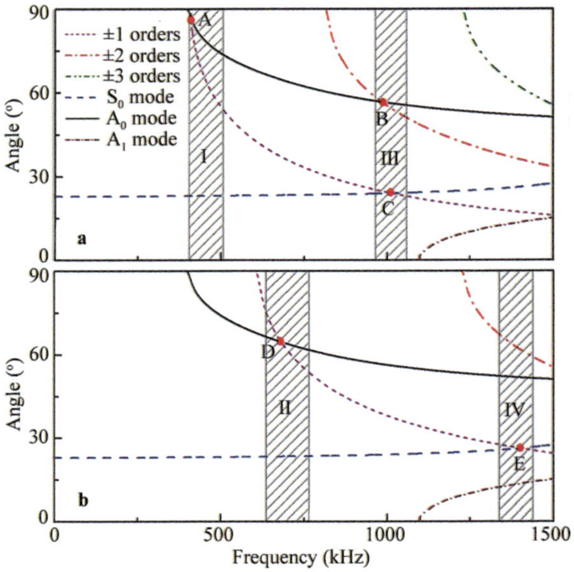

Fig. 3.33 Theoretically calculated angle curves of Lamb modes in a brass plate with a thickness of 1.0 mm, and those of diffracted waves excited by the gratings with periods of **a** 3.6 mm and **b** 2.4 mm immersed in water

(Fig. 3.34c). It is seen that most of the sound energy can transmit through the system for LI at these frequencies, and the transmitted energy at 468 kHz is significantly larger than that at 988 and 1038 kHz. Conversely, negligible transmission is observed for RI, as little sound energy can transmit through the system. Figure 3.34d–f present the total displacement distributions in the brass plate, corresponding to Fig. 3.34a–c, respectively. The vibration configurations in the brass plates demonstrate that the excited Lamb modes are a typical A_0 mode at 468 and 988 kHz, but a typical S_0 mode at 1038 kHz, indicating that band I stems from the A_0 mode, and band III is closely related to both the A_0 and S_0 modes, which agrees with the results in Fig. 3.33a.

Additionally, as presented in Fig. 3.35a, b, the incident frequencies are selected as 660 kHz (band II) and 1380 kHz (band IV). Compared with Fig. 3.34a–c, the AAT phenomenon exhibits an opposite effect for LI and RI in Fig. 3.35a, b. RI is associated with high transmitted acoustic energy, while for LI, the transmitted energy is quite weak. Furthermore, it is found from Fig. 3.35c, d that the excited Lamb modes are the A_0 mode at 660 kHz and the S_0 mode at 1380 kHz, respectively, indicating that band II arises from the A_0 mode and band IV stems from the S_0 mode. This agrees well with the result in Fig. 3.33b.

3.5.3 Experimental Measurement

To verify the results, we experimentally measure the transmittance spectra of the AAT system. The photographs of the sample on both sides are shown in Fig. 3.36a, b, respectively. The width (w) and length (l) of both gratings (the red dashed open

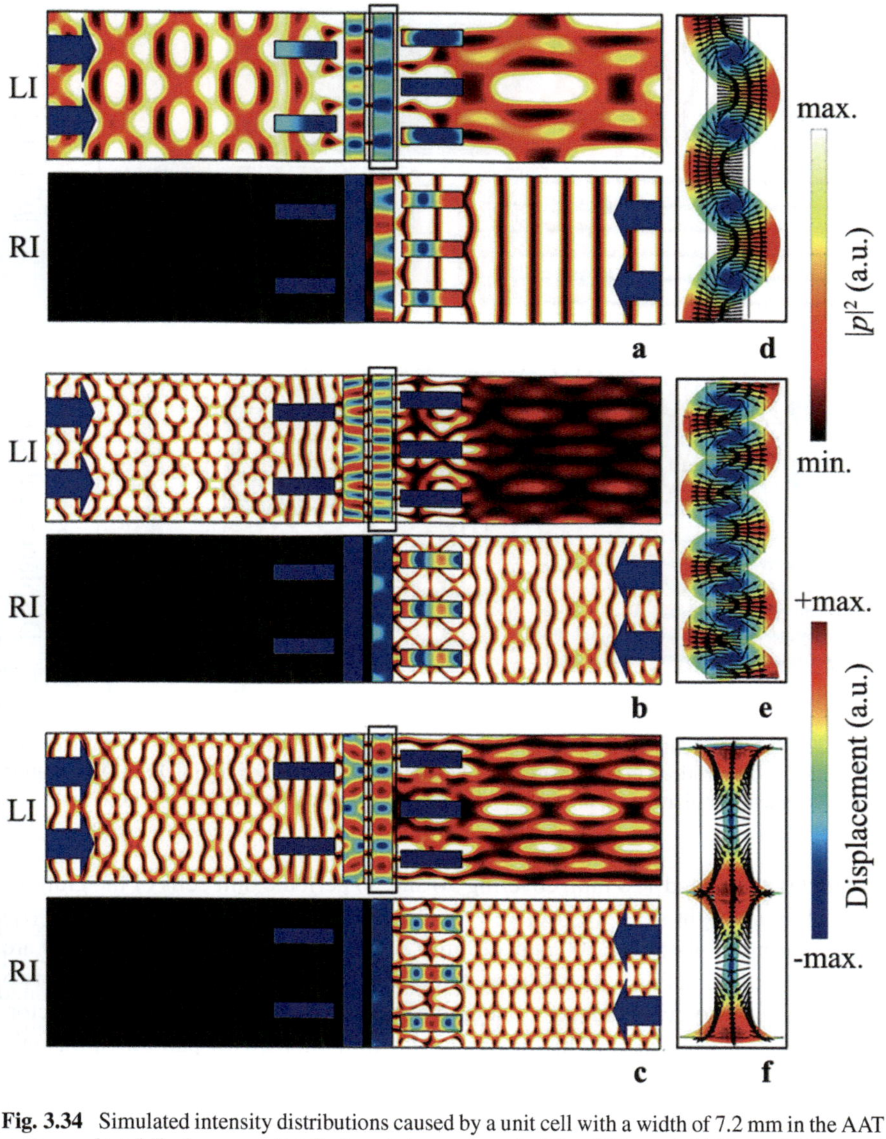

Fig. 3.34 Simulated intensity distributions caused by a unit cell with a width of 7.2 mm in the AAT system and total displacement distributions in brass plates for LI and RI at **a** 468 kHz, **b** 988 kHz, and **c** 1038 kHz. The blue arrows indicate the incident waves. Simulated deformation of displacement distributions in the brass plate for LI at **d** 468 kHz, **e** 988 kHz, and **f** 1038 kHz. The black arrows represent the vibration directions of the displacement

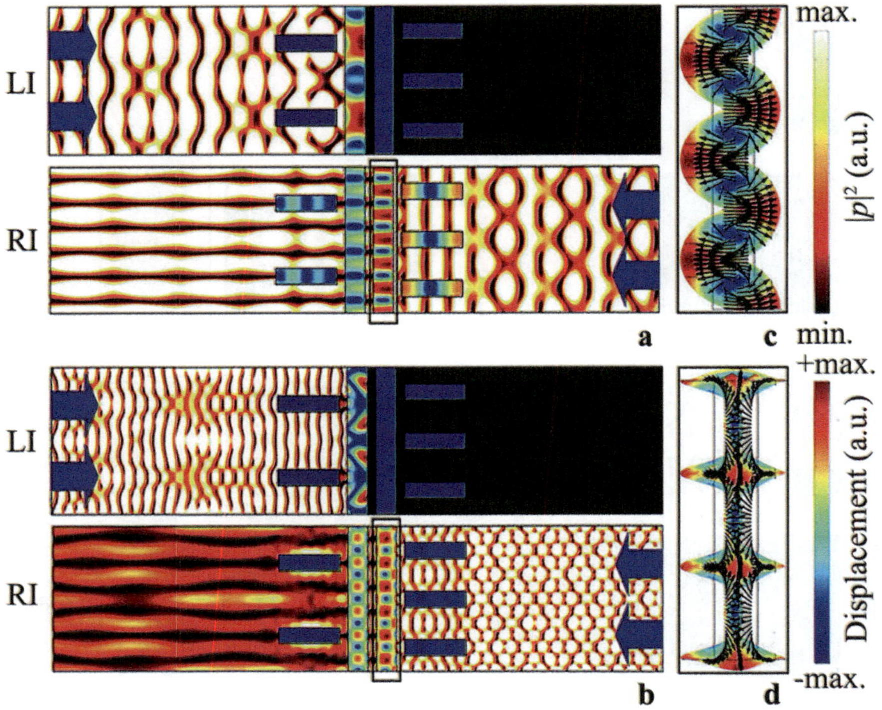

Fig. 3.35 Simulated intensity distributions caused by a unit cell in the AAT system and total displacement distributions in brass plates for LI and RI at **a** 660 kHz and **b** 1380 kHz. Deformation of displacement distributions in the brass plate for RI at **c** 660 kHz and **d** 1380 kHz

rectangle) are 72 and 120 mm, covering 20 and 30 periodic unit cells of the gratings on the left and right sides, respectively. Additionally, the width (w_1) and length (l_1) of the two brass plates are 200 and 120 mm, and the other structure parameters are the same as those in Fig. 3.31. In the measurement, the sample is placed between two broadband ultrasonic transducers, one as the generator and the other as the detector, immersed in a water tank. Due to bandwidth limitations, two pairs of ultrasonic transducers (with diameters of 20 mm) are adopted to fully cover the working bands of the AAT system from 300 to 1500 kHz. One pair operates at a central frequency of 500 kHz and a bandwidth of 250–1100 kHz, and the other pair at a central frequency of 1000 kHz and a bandwidth of 700–1600 kHz. The generation transducer is aligned normally towards the sample at a distance of 12 cm. On the other side, the detection transducer is set around the sample at the same distance. The measured transmittance spectra are obtained by integrating the total transmitted sound energy.

Figure 3.37a shows the measured transmittance spectra for LI and RI in the frequency range of 400–1500 kHz. The corresponding simulations are plotted for comparison. The structure parameters are the same as those in Fig. 3.36. It is observed that the measured transmittance spectra agree with the simulated results.

Fig. 3.36 Photographs of the **a** left and **b** right sides of the sample

However, there is a small difference in amplitude between the two types of results in bands II–IV, as the geometrical and material parameters of the sample are not precisely determined, and part of the loss and absorption is inevitably caused by the water environment, especially in the high-frequency region (bands III and IV). Figure 3.37b exhibits the measured and simulated contrast ratios of the system, and the measurements still agree with the simulations in the four AAT bands.

3.6 Ultra-Broadband Acoustic Asymmetric Transmission by a Metal Plate with a Right Triangle Reflector

3.6.1 Design and Performance of Ultra-Broadband AAT System

As schematically shown in Fig. 3.38, the ultra-broadband AAT system consists of a brass plate and a right triangle reflector immersed in water [5]. The abbreviations LI and RI indicate that the acoustic waves are normally incident from the left and right sides of the system, respectively. For LI, the incident wave is first reflected by the triangle reflector with a base angle θ, and then reaches the brass plate with an incident angle $\alpha = 180° - 2\theta$. However, for RI, the incident wave impinges directly on the brass plate with $\alpha = 0°$. Throughout this work, the COMSOL Multiphysics

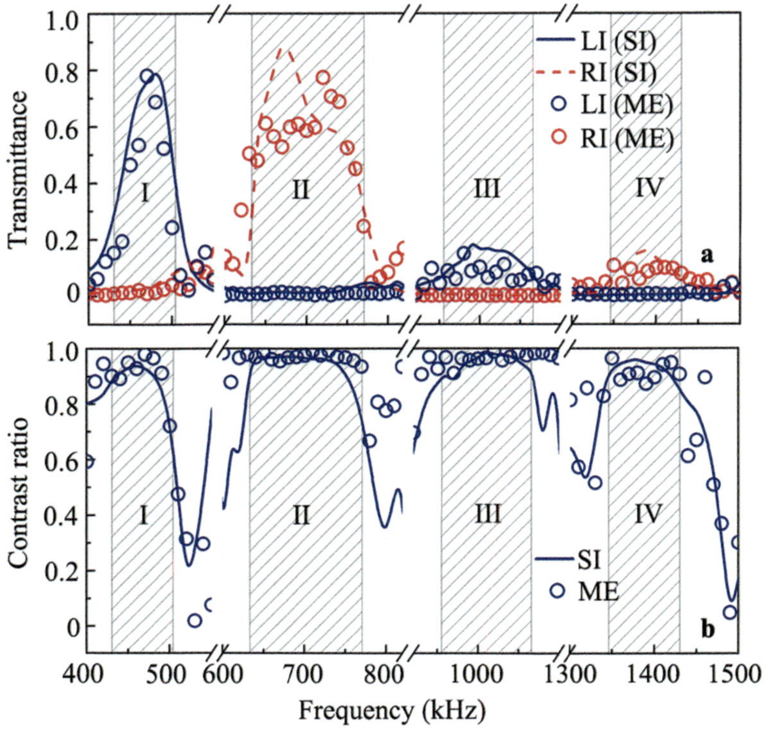

Fig. 3.37 **a** Measured and simulated transmittance spectra of the multi-band AAT system for LI and RI and **b** its contrast ratios

software is used to simulate the AAT characteristics. The structural parameters are d = 3.0 mm, l = 30.0 mm, and h = 1.0 mm. The material parameters are as follows: density $\rho = 8400$ kg/m^3, longitudinal wave velocity $c_l = 4400$ m/s, and transversal wave velocity $c_t = 2200$ m/s for brass; $\rho = 998$ kg/m^3 and $c = 1483$ m/s for water at 296 K. Additionally, the width of the plane acoustic wave is 20 mm.

Figure 3.39a–c show the simulated transmittance spectra of the AAT systems with base angles $\theta = 58°$, $60°$, and $62°$ for LI and RI, corresponding to $\alpha = 64°$, $60°$, and $56°$, respectively. The AAT effect is observed in the frequency ranges of 428–901 kHz, 547–1038 kHz, and 720–1267 kHz for the systems with $\theta = 58°$, $60°$, and $62°$, respectively, as indicated by the black shaded regions. The maximum bandwidth of the AAT system with $\theta = 62°$ reaches 550 kHz, which is significantly wider than previous broadband AAT systems [12–16]. Additionally, the peaks of the transmittance spectra for LI are located at 693 kHz, 819 kHz, and 1018 kHz for the systems with $\theta = 58°$, $60°$, and $62°$, respectively. This indicates that as θ increases, the working band of the AAT system becomes wider and shifts to higher frequencies. To quantify the AAT performance, we calculate the contrast ratio of the AAT system based on Eq. (3.1). Figure 3.39d–f present the simulated contrast ratios for the AAT systems with different values of θ, corresponding to Fig. 3.39a–c, respectively. In

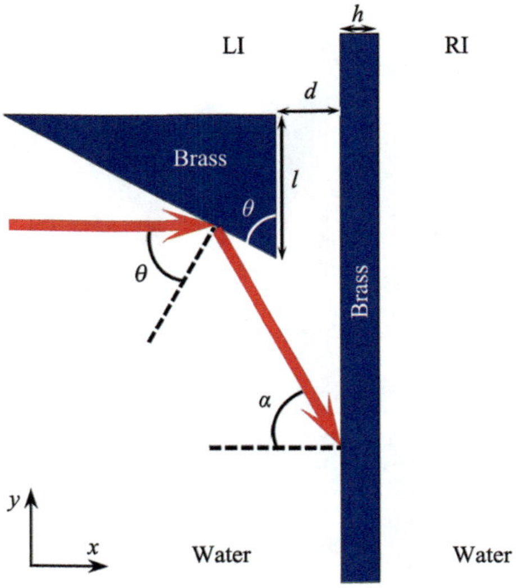

Fig. 3.38 Schematic of the ultra-broadband AAT system

the black shaded regions, most of the contrast ratios are close to 1.0, indicating that the systems exhibit high AAT performance.

Figure 3.40 shows the intensity distributions caused by the AAT systems with $\theta = 58°$, $60°$, and $62°$, where the incident frequencies are selected as 693, 819, and

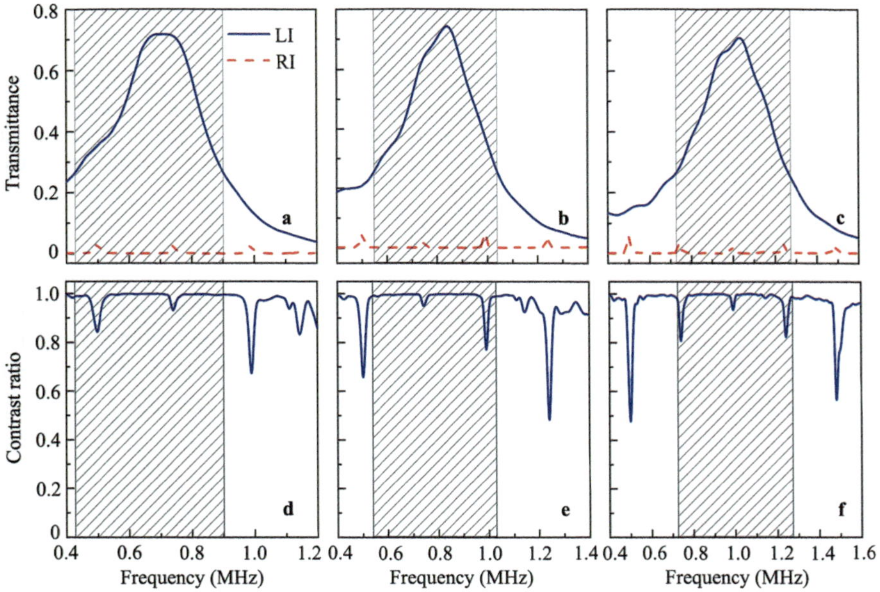

Fig. 3.39 Simulated transmittance spectra of the ultra-broadband AAT systems for **a** $\theta = 58°$, **b** 60°, and **c** 62°, and contrast ratios for **d** $\theta = 58°$, **e** 60°, and **f** 62°

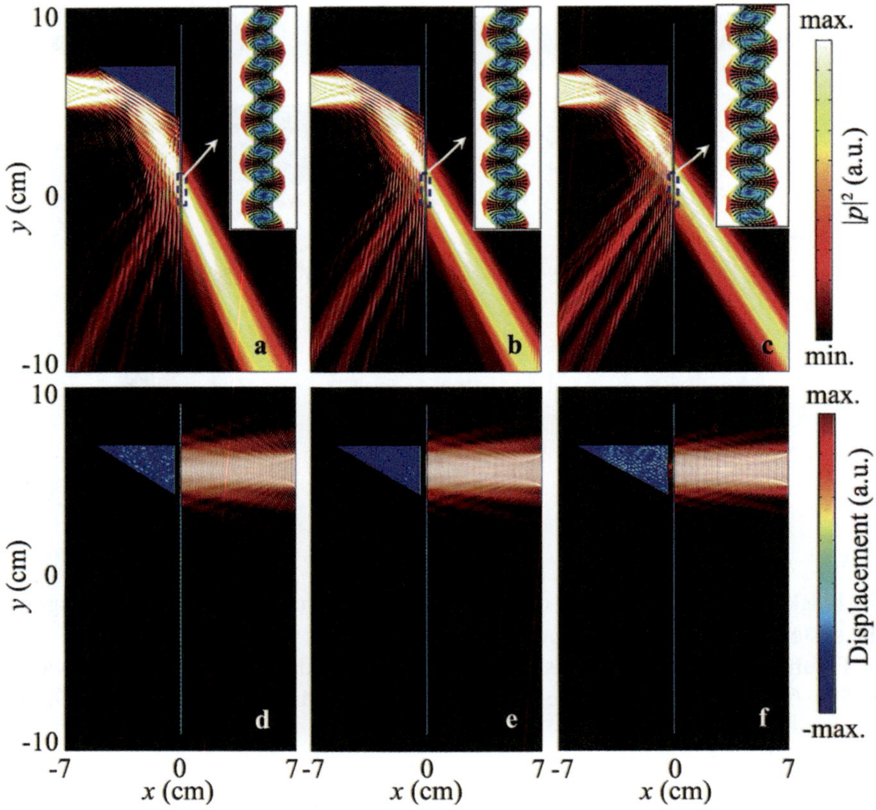

Fig. 3.40 Simulated intensity distributions caused by the ultra-broadband AAT systems with **a** $\theta = 58°$ at 693 kHz, **b** $\theta = 60°$ at 819 kHz, and **c** $\theta = 62°$ at 1018 kHz for LI, and **d–f** the corresponding intensity distributions for RI. Insets in (**a–c**) show the simulated deformation of displacement distributions in the brass plate for LI. The black arrows represent the vibration directions of the displacement

1018 kHz, respectively. For LI (Fig. 3.40a–c), the incident waves are first reflected by the triangle reflectors with $\theta = 58°$, 60°, and 62°, and then reach the brass plates with $\alpha = 64°$, 60°, and 56°, respectively. Finally, the waves transmit through the brass plates, forming single sound beams. However, for RI (Fig. 3.40d–f), the incident angle $\alpha = 0°$, and the incident sound energy is almost entirely reflected.

3.6.2 Physical Mechanism

As shown in insets of Fig. 3.40a–c, the deformation of displacement distributions in the brass plate are also simulated. Note that the asymmetric zero-order Lamb wave (A_0 mode) is excited for LI. Generally, when the phase velocity c_p of the Lamb wave

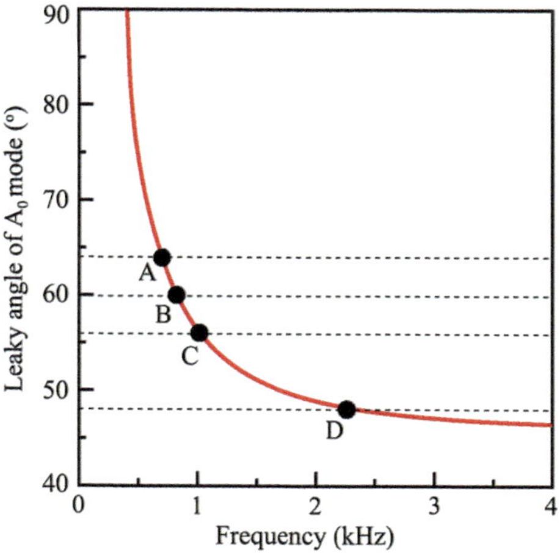

Fig. 3.41 Theoretically calculated leaky angle distribution of the A_0 mode in a brass plate with a thickness of 1.0 mm in water

in the brass plate is greater than the bulk velocity c_w of water, the Lamb wave can couple sound energy into water, leading to a leaky bulk wave at a critical angle. The leakage angle $\theta = \sin^{-1}(c_w/c_p)$ [3, 17].

Figure 3.41 shows the leaky angle distribution of the A_0 mode, where the frequencies at points A, B, and C are 693, 819, and 1018 kHz, corresponding to the leaky angles of the A_0 mode at 64°, 60°, and 56°, respectively. These frequencies and angles match the values of α in Fig. 3.40a–c. Additionally, the A_0 mode in the brass plate can be excited when the incident angle of the external wave is the same as or close to the leaky angle at a certain frequency. Therefore, sound energy can pass through the brass plate, as shown in Fig. 3.40a–c. However, in Fig. 3.40d–f, there is a significant difference between the incident angles and the leaky angles, resulting in little sound energy transmitting through the brass plate. Furthermore, as shown in Fig. 3.41, the gradient of the leaky angle curve of the A_0 mode at point C is smaller than those at points A and B, which explains why the working band of the AAT system becomes wider as α decreases. It is deduced that the AAT effect is attributed to the asymmetric excitation of the A_0 mode in the brass plate caused by the angle of the external wave, and the bandwidth of the AAT can be determined by the gradient of the leaky angle curve.

3.6.3 Experimental Measurement

In the AAT system, the right triangle reflector is used to adjust the incident angle of sound for LI, and thus it is replaced by a rotating transducer in the experiment. Figure 3.42 shows the schematic of the experimental set-up. The brass plate, with a thickness of 1.0 mm, is placed between two broadband ultrasonic transducers—one as a generator and the other as a detector—immersed in a water tank. Due to the bandwidth limitation, two pairs of ultrasonic transducers with diameters of 20 mm are used

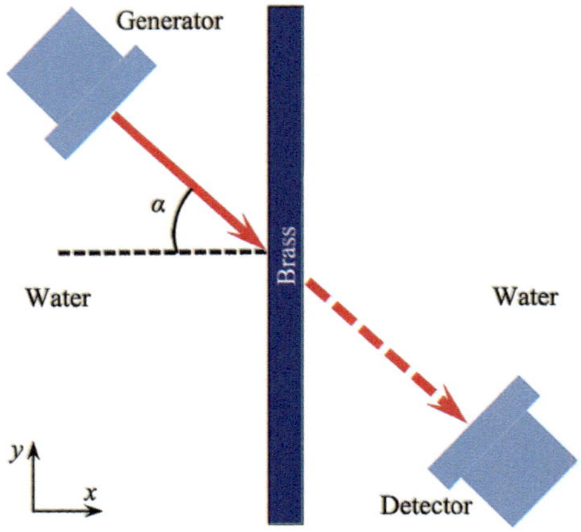

Fig. 3.42 Schematic of the experimental set-up for the AAT system in Fig. 3.38. The right triangle reflector is replaced by a rotating transducer

to fully cover the working range of the AAT from 300 to 1500 kHz. One pair operates with a bandwidth of 250–1100 kHz and a central frequency of 500 kHz, while the other pair operates with a bandwidth of 700–1600 kHz and a central frequency of 1000 kHz. The generation and detection transducers are obliquely positioned towards the center of the brass plate at the same distance of 10 cm, and the incident angles of the generation transducer are set to 64°, 60°, and 56°, corresponding to the values of α for LI in Fig. 3.40a–c, respectively. Additionally, the incident angle of 0° is also measured, corresponding to RI.

Figure 3.43 shows the measured transmittance spectra of the brass plate for different incident angles, with the corresponding simulations plotted for comparison. The measured transmittance spectra agree well with the simulated results. Moreover, as shown in the black shaded regions, the transmittance spectra agree with the working bands of the AAT in Fig. 3.39a–c, demonstrating the performance of the AAT system.

3.6.4 Performance Optimization

Based on the mechanism of AAT, we achieve an ultra-broadband AAT system with a single transmitted beam under the excitation of an external wave with $\alpha = 48°$. Figure 3.44a, b show the simulated transmittance spectra and contrast ratios of the ultra-broadband AAT system with $\theta = 66°$, respectively. Note that the working band of the AAT is 1320–3320 kHz (the black shaded region), and the bandwidth can reach up to 2000 kHz. Additionally, the maximum transmittance for LI is 0.65 at 2320 kHz, while the transmittance for RI is close to zero. Furthermore, as shown in Fig. 3.44b, most of the contrast ratios are close to 1.0 in the black shaded region, indicating that the system exhibits good ultra-broadband AAT performance.

As shown in Fig. 3.41, the frequency and the leaky angle at point D are 2320 kHz and 48°, respectively. The leaky angle of the A_0 mode and the frequency at point D

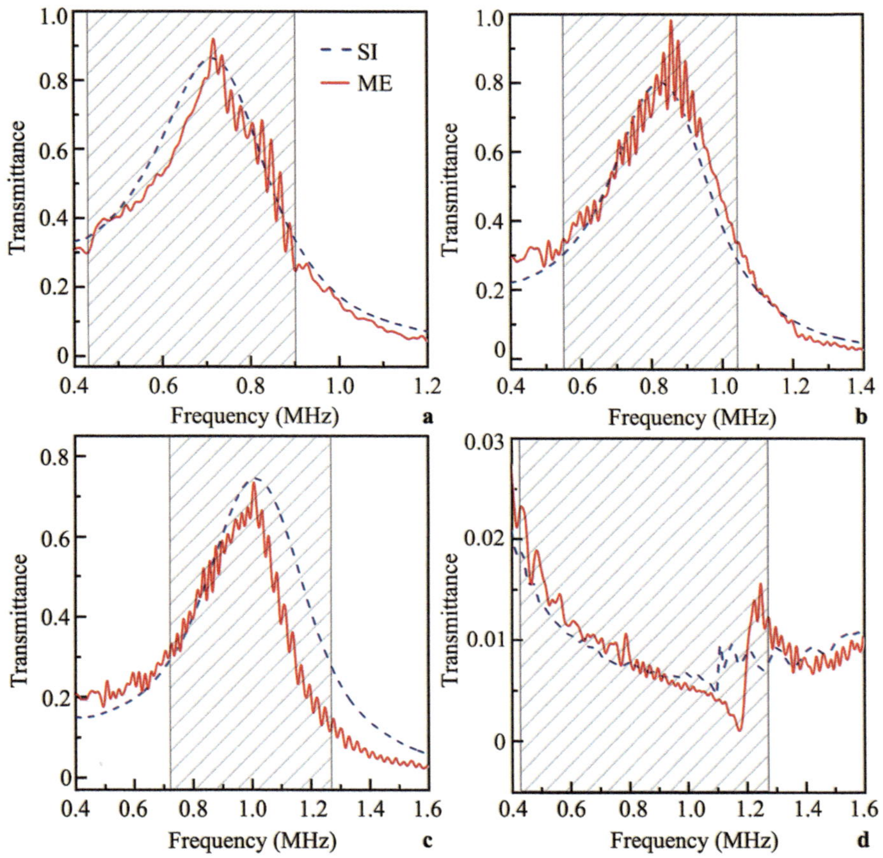

Fig. 3.43 Measured and simulated transmittance spectra of the brass plate for **a** $\alpha = 64°$, **b** 60°, **c** 56°, and **d** 0°

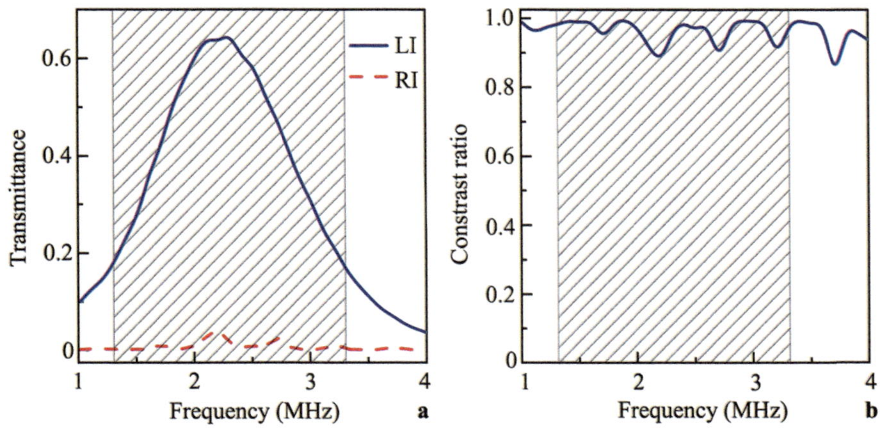

Fig. 3.44 Simulated **a** transmittance spectra and **b** contrast ratios of the AAT system with $\theta = 66°$

match the incident angle and the frequency of the maximum transmittance for LI in Fig. 3.44a. Therefore, the ultra-broadband AAT arises from the excitation of the A_0 mode. Moreover, the gradient of the leaky angle curve of the A_0 mode at point D is much smaller than that at points A, B, and C. Consequently, the working band of the AAT reaches 2000 kHz, which is significantly wider than previous broadband AAT systems.

3.6.5 Performance Modulation

Figure 3.45a, b show the simulated transmittance spectra of the AAT systems with plate thicknesses $h = 0.8$ mm and 1.2 mm, respectively, where the other parameters are the same as those in Fig. 3.39c. As shown in Fig. 3.45a, b, the working bands of the AAT are 895–1600 kHz and 585–1068 kHz, respectively. Note that as h increases, the working band of the AAT becomes narrower and shifts to lower frequencies.

Figure 3.46 displays the leaky angle curve of the A_0 mode in brass plates with thicknesses of 0.8 and 1.2 mm, where the leaky angles at points E and F are both 56°. As h increases, the leaky angle curve of the A_0 mode shifts to lower frequencies, and thus the frequencies at points E and F are 835 kHz and 1252 kHz, respectively. Additionally, the AAT effect occurs when the incident angle of the external wave is the same as or close to the leaky angle of the A_0 mode. Therefore, the working bands of the AAT systems are 895–1600 kHz and 585–1068 kHz for $h = 0.8$ mm and 1.2 mm, respectively. Furthermore, the gradient of the leaky angle curves of the A_0 mode at point F is lower than that at point E, which explains why the working band of the AAT becomes narrower as h increases. Therefore, in addition to the incident angle, the working band of AAT is also closely related to the plate thickness.

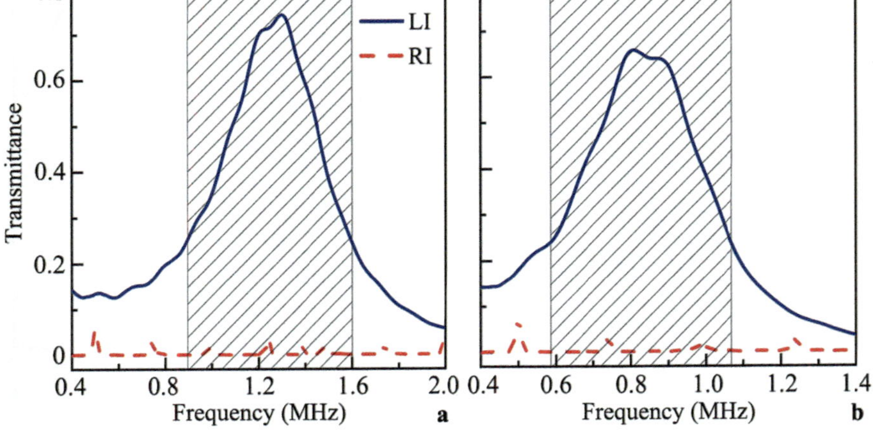

Fig. 3.45 Simulated transmittance spectra of the AAT systems with **a** $h = 0.8$ mm and **b** 1.2 mm

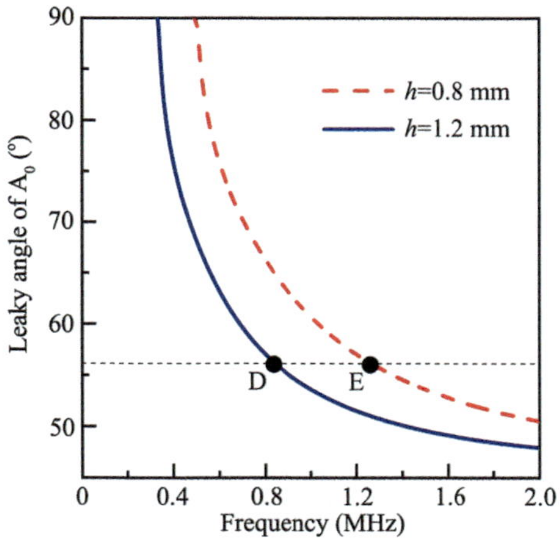

Fig. 3.46 Theoretically calculated leaky angle distribution of the A_0 mode in brass plates with $h =$ 0.8 mm and 1.2 mm in water

3.7 Conclusion

This chapter focuses on the design and experimental demonstration of underwater AAT systems using plate-like metamaterials. Four types of AAT systems are discussed: a single-sided periodic grating, multi-layer plates with periodic gratings, a brass plate with a periodic grating, and a multi-band AAT system with dual-layer plates. The AAT effect is achieved through the interaction of diffracted waves and Lamb modes in the brass plates, leading to high transmission in one direction and low transmission in the opposite direction. The systems are numerically simulated and experimentally validated, showing high rectifying ratios and tunable working bands. The physical mechanisms, such as the excitation of leaky A_0 modes and the role of grating periods, are thoroughly analyzed. The chapter concludes with the design of an ultra-broadband AAT system, demonstrating the potential applications of these systems in medical ultrasound imaging and underwater communication.

References

1. Sun HX, Zhang SY, Shui XJ (2012) A tunable acoustic diode made by a metal plate with periodical structure. Appl Phys Lett 100(10):103507. https://doi.org/10.1063/1.3693374
2. Sun HX, Zhang SY (2013) Enhancement of asymmetric acoustic transmission. Appl Phys Lett 102(11):113511. https://doi.org/10.1063/1.4798277
3. Sun HX, Zhang SY, Yuan SQ, Xia JP (2016) Experimental verification of manipulating propagation directions of transmitted waves in asymmetric acoustic transmission. Appl Phys A 122(4):328. https://doi.org/10.1007/s00339-016-9907-z
4. Sun HX, Yuan SQ, Zhang SY (2015) Asymmetric acoustic transmission in multiple frequency bands. Appl Phys Lett 107(21):213505. https://doi.org/10.1063/1.4936609
5. Jia D, Sun HX, Yuan SQ, Ge Y (2017) Ultra-broadband asymmetric acoustic transmission with single transmitted beam. Chin Phys B 26(2):024302. https://doi.org/10.1088/1674-1056/26/2/024302

6. Li XF, Ni X, Feng L, Lu MH, He C, Chen YF (2011) Tunable unidirectional sound propagation through a sonic-crystal-based acoustic diode. Phys Rev Lett 106(8):084301. https://doi.org/10.1103/PhysRevLett.106.084301
7. Hou B, Mei J, Ke MZ, Liu ZY, Shi J, Wen WJ (2008) Experimental determination for resonance-induced transmission of acoustic waves through subwavelength hole arrays. J Appl Phys 104(1):014909. https://doi.org/10.1063/1.2951457
8. Estrada H, de Abajo FJG, Candelas P, Uris A, Belmar F, Meseguer F (2009) Angle-dependent ultrasonic transmission through plates with subwavelength hole arrays. Phys Rev Lett 102(14):144301. https://doi.org/10.1103/PhysRevLett.102.144301
9. Liang B, Guo XS, Tu J, Zhang D, Cheng JC (2010) An acoustic rectifier. Nat Mater 9(12):989–992. https://doi.org/10.1038/NMAT2881
10. Ke MZ, He ZJ, Peng SS, Liu ZY, Shi J (2007) Surface resonant-states-enhanced acoustic wave tunneling in two-dimensional phononic crystals. Phys Rev Lett 99(4):044301. https://doi.org/10.1103/PhysRevLett.99.044301
11. Twersky V (1962) On scattering of waves by the infinite grating of circular cylinders. IEEE T Antenn Propag 10(6):737–765. https://doi.org/10.1109/tap.1962.1137940
12. He ZJ, Peng SS, Ye YT, Dai ZW, Qiu CY, Ke MZ, Liu ZY (2012) Asymmetric acoustic gratings. Appl Phys Lett 98(8):083505. https://doi.org/10.1063/1.3562306
13. Li Y, Tu J, Liang B, Guo XS, Zhang D, Cheng JC (2012) Unidirectional acoustic transmission based on source pattern reconstruction. J Appl Phys 112(6):064504. https://doi.org/10.1063/1.4752407
14. Xu SJ, Qiu CY, Liu ZY (2012) Acoustic transmission through asymmetric grating structures made of cylinders. J Appl Phys 111(9):094505. https://doi.org/10.1063/1.4709730
15. Jia H, Ke MZ, Li CH, Qiu CY, Liu ZY (2013) Unidirectional transmission of acoustic waves based on asymmetric excitation of Lamb waves. Appl Phys Lett 102(15):153508. https://doi.org/10.1063/1.4802254
16. Li CH, Ke MZ, Ye YT, Xu SJ, Qiu CY, Liu ZY (2014) Broadband asymmetric acoustic transmission by a plate with quasi-periodic surface ridges. Appl Phys Lett 105(2):023511. https://doi.org/10.1063/1.4890721
17. Sun HX, Zhang SY, Yuan SQ (2016) Control of pass-bands in asymmetric acoustic transmission. Chin Phys B 25(12):124313. https://doi.org/10.1088/1674-1056/25/12/124313

Chapter 4
Acoustic Asymmetric Transmission by Airborne Metamaterials

4.1 Introduction

Acoustic airborne metamaterials are artificial structures designed based on acoustic theory, exhibiting specific acoustic characteristics that differ from natural materials. These characteristics include AAT, acoustic invisibility, AF, and sound absorption and isolation. By precisely designing the shape and size of airborne metamaterials, sound manipulation and these exotic phenomena can be achieved. In practical applications, certain scenarios require open structures (such as windows and tunnels) with the AAT effect. For example, in hospitals, doctors need to hear patients' voices in the ward, while patients should not hear external sounds. Thus, the design of airborne AAT window/tunnel structures holds great potential in architectural acoustics and mechanical engineering.

In this chapter, we numerically design and experimentally demonstrate four types of AAT systems using airborne metamaterials. First, we design a broadband AAT tunnel/window with ultrathin metasurfaces [1]. The AAT effect arises from asymmetric sound manipulation caused by the metasurfaces, i.e., multiple reflections of sound and an acoustic blind region (ABR) for the left and right sides, respectively. The fractional bandwidth (the ratio of the bandwidth to the center frequency) of the AAT tunnel/window is 0.36, showing a broadband characteristic. Second, we propose an AAT tunnel attached with two pairs of triangle cavities of different sizes on both sides [2]. The AAT effect arises from asymmetric multiple scatterings caused by the triangle cavities. Moreover, the AAT tunnel exhibits high robustness against the excitation of a cylindrical sound source. Third, we design a multi-channel AAT lens composed of twelve mode-conversion meta-atoms [3]. The AAT effect arises from the cutoff frequency of the first-order wave caused by the step waveguide of the meta-atom, with a fractional bandwidth of 0.37. Interestingly, the AAT effect can be reversed under the incidence of the first-order wave. Finally, we propose an AAT prism composed of 31 phased unit cells, each consisting of n-layer meta-atoms with different discrete phase delays [4]. The fractional bandwidth of the AAT prism is 0.26,

H. Sun et al., *Acoustic Metamaterials*, Microfluidics and Nanophotonics: Science and Engineering 2, https://doi.org/10.1007/978-981-96-8524-0_4

attributed to asymmetric phase distributions caused by the non-parallel characteristic of exit interfaces on both sides of the prism. Furthermore, by tuning the parameters of the meta-atoms, a reversed AAT prism is also demonstrated. The designed AAT systems hold great potential in various applications, including architectural acoustics, noise control, and acoustic functional components.

4.2 Broadband Acoustic Asymmetric Transmission by Open Tunnel and Window Structures with Ultrathin Metasurfaces

4.2.1 Design and Performance of AMs

Figure 4.1a shows the propagation paths of incident and reflected waves through an AM with a phase gradient $-k < d\varphi/dx < 0$, where $k = 2\pi/\lambda$ is the wave vector, and λ is the wavelength. The incident (reflected) angle is defined as the angle between the direction of the incident (reflected) wave and the normal line of the metasurface. The signs + and − of the incident (reflected) angle correspond to anti-clockwise and clockwise directions, respectively. Based on Eq. (2.41), we can calculate the reflected angle θ_r caused by the metasurface under the excitation of an acoustic wave with an incident angle of θ_i.

As shown in Fig. 4.1a, under the effect of the AM, the reflected waves R_1 and R_2 with angles θ_{r1} and θ_{r2} are generated from the incident waves I_1 and I_2 with angles $\pi/2$ and 0, respectively. Note that as the incident angle decreases from $\pi/2$ to 0, the corresponding reflected angle changes from θ_{r1} to θ_{r2} ($\theta_{r2} < 0$). Therefore, the acoustic wave from the left side cannot reach the red shaded region on the right side, defined as ABR. Meanwhile, based on the reciprocity principle, when the incident wave from the right side is in the ABR (from $-\pi/2$ to $-\theta_{r1}$), the acoustic wave is

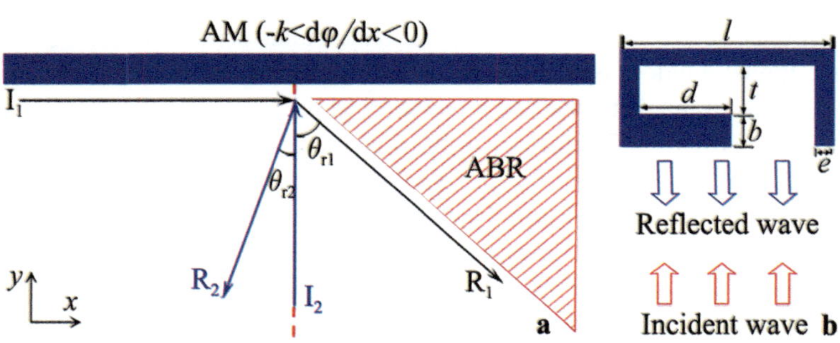

Fig. 4.1 **a** Schematic of different reflection paths caused by an AM with $-k < d\varphi/dx < 0$, and **b** a hooklike meta-atom

reflected back and cannot reach the left side. The existence of the ABR provides more possibilities for manipulating acoustic propagation paths and realizing AAT.

Figure 4.1b shows the structure of an ultrathin hook-like meta-atom made of epoxy resin. The length of the meta-atom, the height of the cavity, the thickness of the wall, and the thickness and length of the hook are l, b, e, t, and d, respectively. In the hook-like meta-atom, the parameter d is tailored to yield the desired phase delays, while the parameters l, b, e, and t are fixed and selected as 42.0 mm, 6.0 mm, 1.5 mm, and 3.0 mm, respectively. The reflected wave is significantly delayed after the incident wave normally impinges on the hook-like meta-atom. Here, we use the COMSOL Multiphysics software to simulate acoustic propagation. In the model, the material parameters are as follows: density $\rho = 1180$ kg/m^3, longitudinal wave velocity $c_l = 2720$ m/s, and transversal wave velocity $c_t = 1460$ m/s for epoxy resin; and $\rho = 1.21$ kg/m^3 and $c = 343$ m/s for air.

Figure 4.2 shows the reflected phase delays caused by the hook-like meta-atoms with different values of d. It is noted that as d increases, the reflected phase delay covers a range of 1.7π. Here, we select eight values of d (the red open circles): 9.0, 15.9, 17.9, 19.1, 20.0, 20.7, 21.4, and 22.2 mm, which can realize discrete phase delays from −0.75π to 0.30π in steps of 0.15π. To present the reflected phase delays of the hook-like meta-atoms, we simulate the reflected pressure distributions caused by the selected eight meta-atoms. The hook-like meta-atoms can achieve a gradual change of the reflected phase delay over a large range, as shown in Fig. 4.3.

It is important to note that for the control of sound reflections in Fig. 4.1, the most critical factor is a constant phase gradient of the metasurface rather than full coverage of 2π. Based on the hook-like meta-atoms, the thickness of the designed metasurfaces ($b + t = 9.0$ mm) is about λ/12, which is much thinner than previously

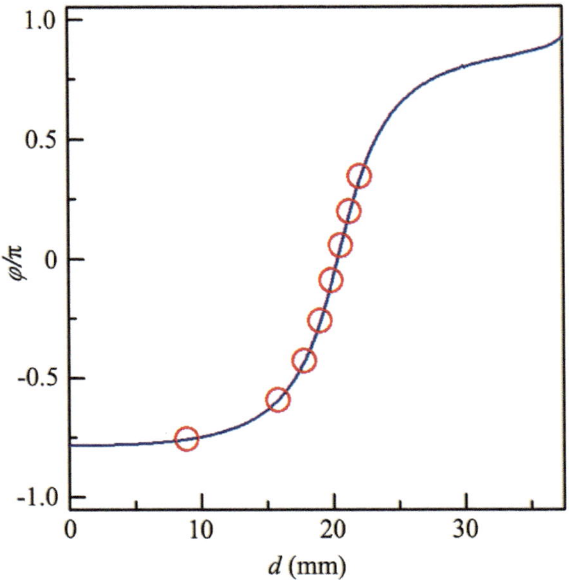

Fig. 4.2 Simulated reflected phase delays of hooklike meta-atoms with different values of d at 3.0 kHz

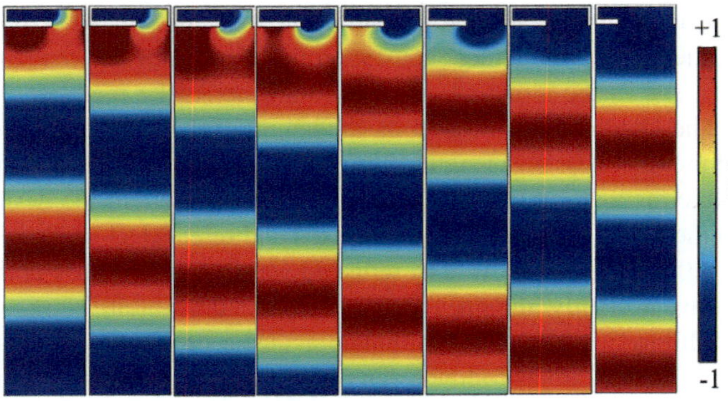

Fig. 4.3 Simulated reflected pressure distributions caused by the eight selected hooklike meta-atoms at 3.0 kHz

applied metasurfaces [5–7] and triangular cavities [2] in tunnel and window structures. Therefore, the metasurfaces composed of hook-like meta-atoms can be used to design AAT windows with ultrathin blades.

Based on Eq. (2.41), we can manipulate the reflected angle θ_r of sound by designing a metasurface with a desired phase gradient along the x direction. Here, we select the reflected angles $\theta_r = 60°$ and 45° with the incident angle $\theta_i = 90°$ at 3.0 kHz as examples, and use eight and seven hook-like meta-atoms to design the AMs with $\theta_r = 60°$ and 45°, respectively. The discrete phase delays of the metasurfaces corresponding to the theoretically continuous phase gradients $d\varphi(x)/dx = -7.36$ and -16.10 are presented in Fig. 4.4a, b, respectively.

Figure 4.5a schematically shows the reflection paths caused by the AM with $\theta_i = 90°$. In this case, the incident wave is reflected by the AM with the angle θ_{r1}. Fig. 4.5b, c show the simulated pressure distributions caused by the metasurfaces with $d\varphi(x)/dx = -7.36$ and -16.10 with $\theta_i = 90°$, respectively, where the black open arrows refer to the theoretical sound reflections with $\theta_{r1} = 60°$ and 45°. It shows that the reflected waves propagate along the black open arrows, which agree well with the theoretical predictions. Furthermore, for the normal incidence of sound ($\theta_i = 0°$), the reflection paths caused by the metasurface are presented in Fig. 4.5d. The theoretical reflected angles θ_{r2} can be calculated as $-7.7°$ and $-17.0°$ for the metasurfaces with $d\varphi(x)/dx = -7.36$ and -16.10, respectively. The reflected waves also agree with the theoretical values, as shown in Fig. 4.5e, f. These results indicate that the metasurfaces composed of hook-like meta-atoms can effectively manipulate acoustic reflection paths.

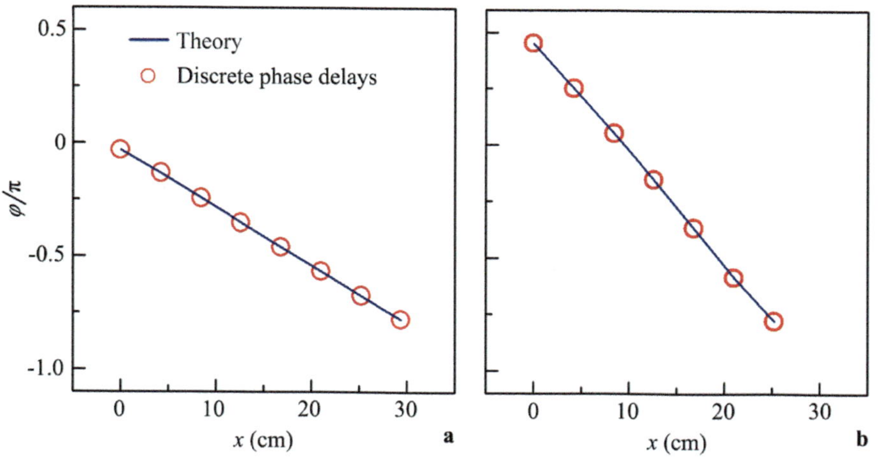

Fig. 4.4 Theoretically continuous phase distributions and discrete phase delays of the metasurfaces with **a** $d\varphi(x)/dx = -7.36$ and **b** -16.10. The values of the parameter d are 4.0, 13.9, 16.3, 17.6, 18.5, 19.2, 19.8, and 20.3 mm for the metasurface with $d\varphi(x)/dx = -7.36$, and 4.0, 14.6, 18.0, 19.6, 20.7, 21.7, and 23.0 mm for the metasurface with $d\varphi(x)/dx = -16.10$

4.2.2 Design and Performance of AAT Tunnel

Using the metasurface with the phase gradient $d\varphi(x)/dx = -7.36$ in Fig. 4.4a, we design a broadband AAT tunnel [1], as shown in Fig. 4.6. A pair of metasurfaces are attached symmetrically on both sides of a straight tunnel with a width $w = 13$ cm, which is close to the center wavelength of 11.4 cm. The normal incidence of acoustic waves from the left and right sides of the AAT tunnel (the red arrows) are defined as LI and RI, respectively.

Figure 4.7 shows the simulated transmittance spectra of the AAT tunnel for LI and RI. In the range 2.74–3.39 kHz (the black shaded region), the transmittance spectra exhibit an obvious AAT phenomenon, where the transmittance is greater than 0.6 for LI but lower than 0.2 for RI. Thus, the fractional bandwidth is calculated as 0.21. Figure 4.8a, b show the simulated pressure distributions caused by the AAT tunnel for LI and RI, respectively. There is a significant difference between the transmitted pressure distributions for LI and RI. The acoustic wave can transmit through the AAT tunnel for LI, while the sound energy is mostly reflected back, resulting in negligible transmission for RI.

4.2.3 Physical Mechanism of AAT Tunnel

Next, we discuss the mechanism of the AAT tunnel. Figure 4.9a shows the propagation path for LI. To simplify the analysis, we assume that the incident wave

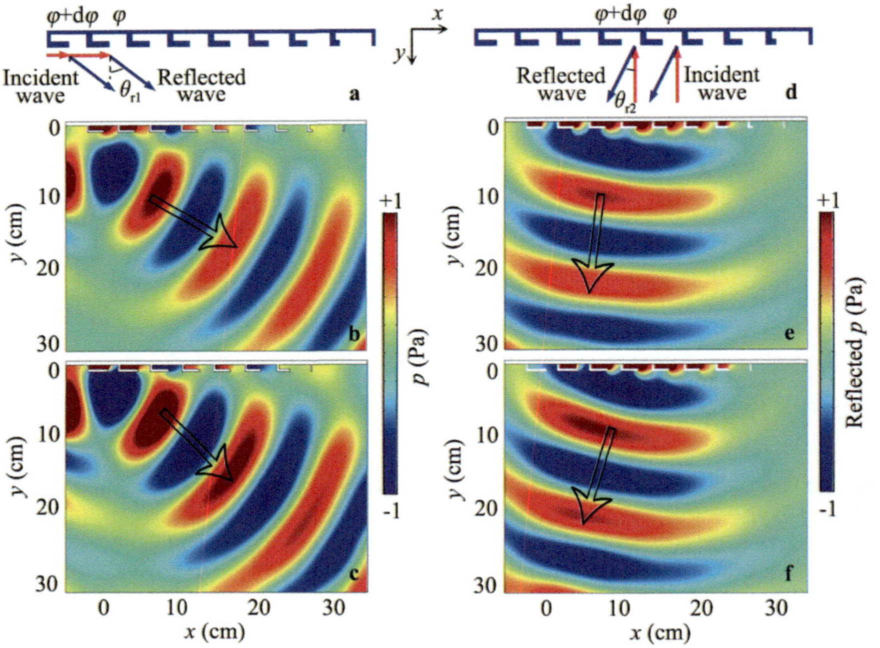

Fig. 4.5 **a** Schematic of reflection paths caused by the metasurface for the grazing incidence of sound, and simulated pressure distributions caused by the metasurfaces with **b** $d\varphi(x)/dx = -7.36$ and **c** -16.10 at 3.0 kHz. **d** Schematic of reflection paths caused by the metasurface for the normal incidence of sound, and simulated pressure distributions caused by the metasurfaces with **e** $d\varphi(x)/dx = -7.36$ and **f** -16.10 at 3.0 kHz. The black open arrows indicate the theoretical directions of sound reflection

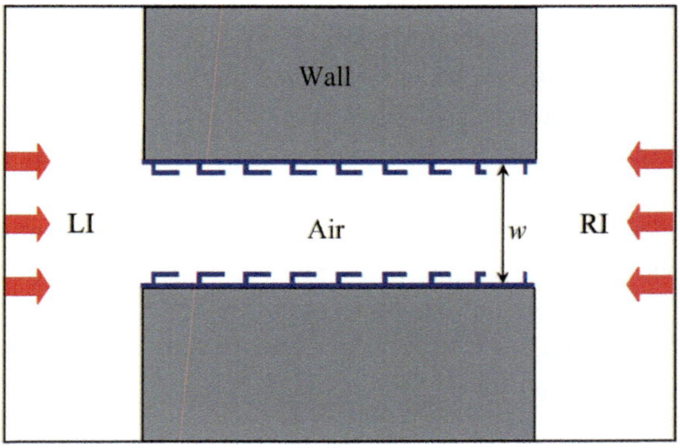

Fig. 4.6 Schematic of the AAT tunnel. The red arrows represent incident acoustic waves

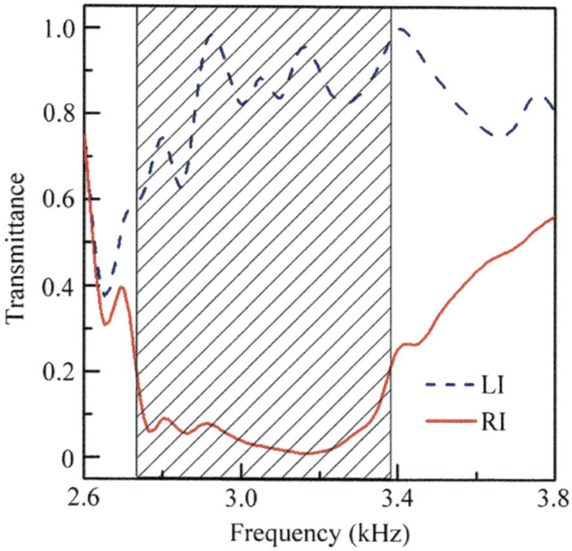

Fig. 4.7 Simulated transmittance spectra of the AAT tunnel for LI and RI in Fig. 4.6

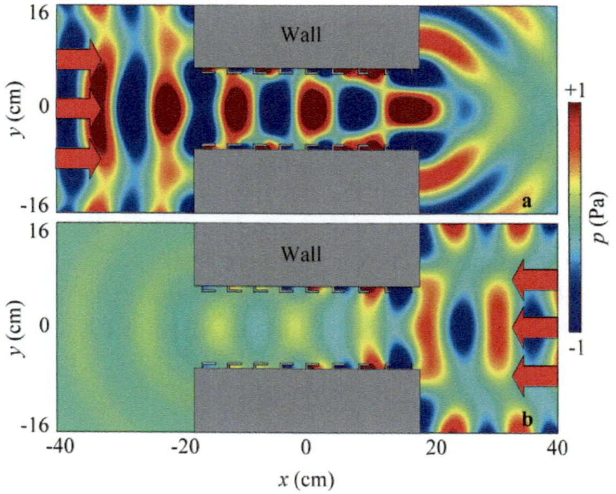

Fig. 4.8 Simulated pressure distributions caused by the AAT tunnel for **a** LI and **b** RI at 3.2 kHz

propagates along the upper metasurface. Based on the generalized Snell's law, the first reflected wave from the upper metasurface reaches the lower metasurface and undergoes further reflection, leading to multiple reflections between both metasurfaces. The multiple reflected angles can be calculated based on Eq. (2.41), as shown in Fig. 4.9c, where I_1 and R_i represent the incident wave and the i-th reflected wave, respectively. For the metasurface with $d\varphi(x)/dx = -7.36$, the angles of I_1, R_1, R_2, R_3, and R_4 are calculated as 90°, 60°, 47°, 36.7°, and 27.6°, respectively. All reflected angles are positive, indicating that the acoustic wave can pass through the AAT tunnel

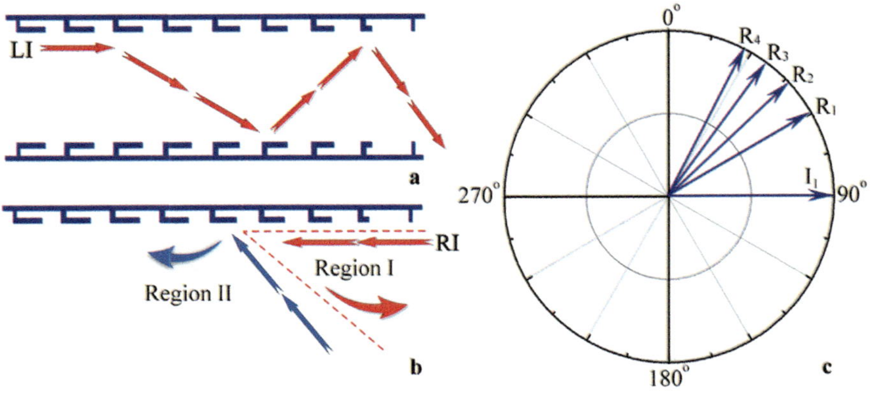

Fig. 4.9 Schematic of acoustic propagation paths for **a** LI and **b** RI, and **c** reflected directions of i-th reflections for LI

with a limited length. Conversely, the incident wave in the opposite direction cannot reach the left side for RI due to the ABR, as shown by the red arrows in Fig. 4.9b.

To demonstrate this, we simulate the reflectance spectra with different incident angles for RI, as shown in Fig. 4.10. For the metasurface with $d\varphi(x)/dx = -7.36$, the range of the ABR is from −60° to −90°. Thus, we select a series of incident angles: −90°, −80°, and −70° in region I (inside the ABR); and −60° and −45° in region II (outside the ABR). In the cases of −90°, −80°, and −70°, the reflectances are greater than 0.5 in the range 2.75–3.40 kHz. Especially at −90°, the reflectance is greater than 0.8 in the range 2.75–3.17 kHz and reaches about 1.0 at 2.80 kHz, indicating that most of the sound energy for RI is reflected back directly. However, the reflectance is close to zero for −60° and −45°, indicating that the effect of the ABR is negligible. Therefore, we deduce that the AAT phenomenon arises from multiple reflections for LI and the ABR for RI.

4.2.4 Experimental Measurement of AAT Tunnel

To further verify the AAT effect, we experimentally measure the transmittance spectra and the transmitted pressure distributions for LI and RI. The experimental set-up is shown in Fig. 4.11a. The tunnel attached with the metasurfaces is extended along the right to concentrate the transmitted sound energy. A loudspeaker array is located 1.6 m away from the sample to generate the incident plane wave. Two 0.25 inch-diameter Brüel & Kjær type-4961 microphones are used to detect acoustic pressure distributions. Microphone 1 is located in the scanning region, and microphone 2 is fixed near the loudspeaker array as a reference to obtain the phase of the sound signals. Using the software PULSE Labshop, we measure the pressure distributions by recording the pressure magnitude and phase at each position in the scanning

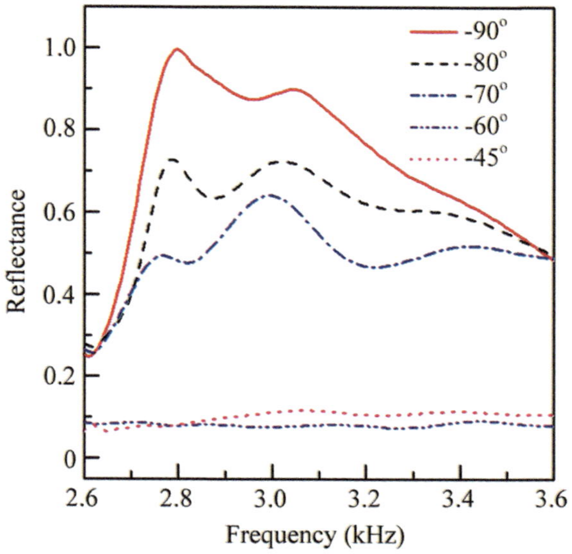

Fig. 4.10 Simulated reflectance spectra of the AAT tunnel with different incident angles for RI

region. In the measurement of the transmittance spectra, the transmittance is obtained by integrating the acoustic intensities at each position in the scanning region with and without the metasurfaces. Figure 4.11b shows the photograph of the sample, in which the selected values of *e*, *t*, *b*, *l*, and *w* are the same as those in Fig. 4.4a. Figure 4.12 presents the measured and simulated transmittance spectra of the sample for LI and RI. The measured spectra agree with the simulated ones in the range 2.74–3.39 kHz. However, there is a slight difference between the two results due to undesired reflected sound energy in the measurement.

Figure 4.13a, b show the comparison between the measured and simulated pressure distributions through the sample for LI and RI, respectively. In the measurement, we scan two square regions with a size of 10×10 cm^2, as marked by two red open squares (R1 and R2) in Fig. 4.13. As shown in Fig. 4.13a, the measured pressure distribution for LI in the scanning region is clear and large, which agrees well with the simulated result. For RI, both the measured and simulated pressures are close to zero, as shown in Fig. 4.13b. The measured pressure distributions further demonstrate the characteristic of AAT.

4.2.5 Design and Performance of AAT Window

Based on the AAT tunnel, we design an AAT window composed of eight blades with ultrathin metasurfaces attached on both sides and seven straight tunnels with a width *w* of 13 cm [1], as shown in Fig. 4.14. The phase gradient of the metasurfaces is the same as that in Fig. 4.4a.

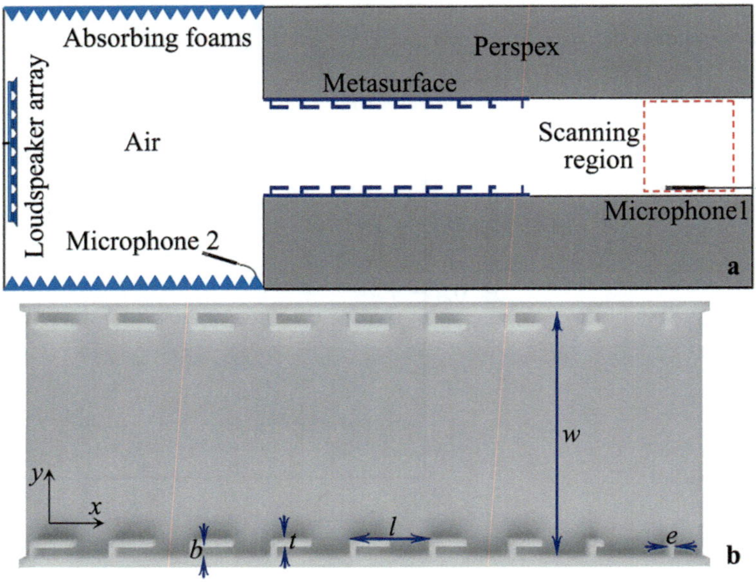

Fig. 4.11 **a** Schematic of the experimental set-up and **b** photograph of sample

Fig. 4.12 Measured and simulated transmittance spectra of the sample for LI and RI

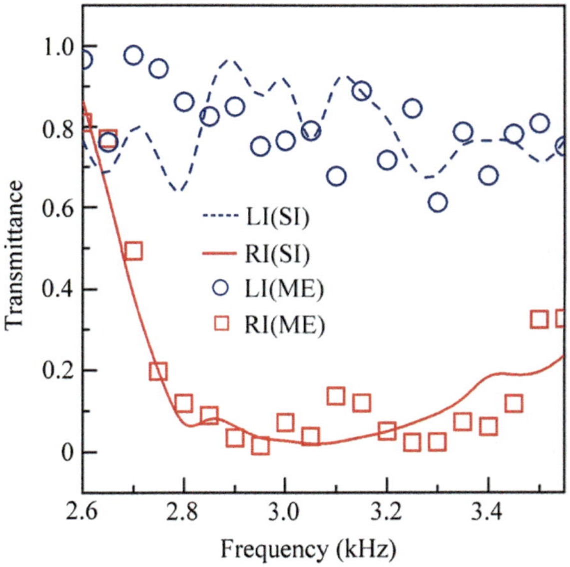

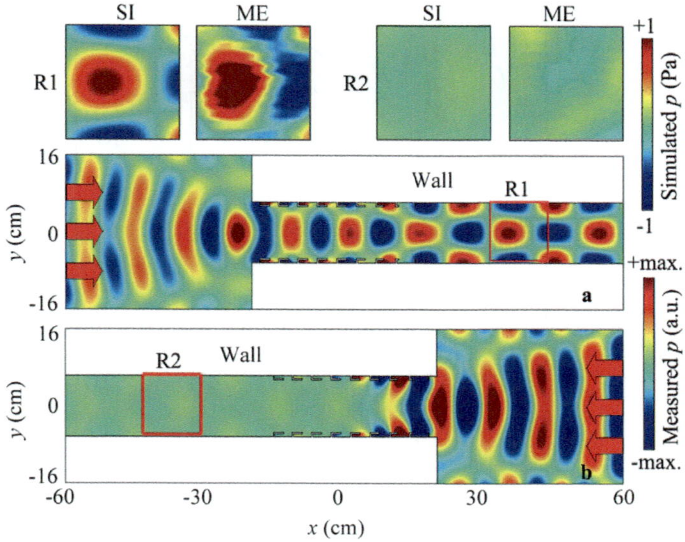

Fig. 4.13 Measured and simulated pressure distributions caused by the sample for **a** LI and **b** RI at 3.2 kHz

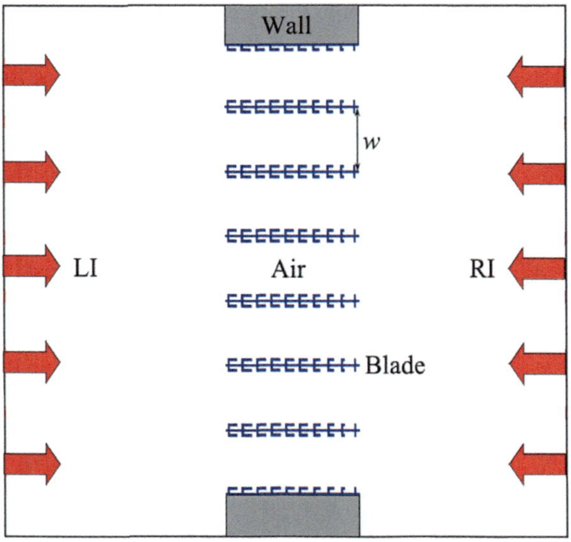

Fig. 4.14 Schematic of the AAT window

Figure 4.15a shows the transmittance spectra of the AAT window for LI and RI. Similar to the AAT tunnel, in the range 2.89–3.51 kHz (the black shaded region), the transmittance for LI is greater than 0.9, while that for RI is lower than 0.2. The working bandwidth of the AAT window reaches about 0.6 kHz.

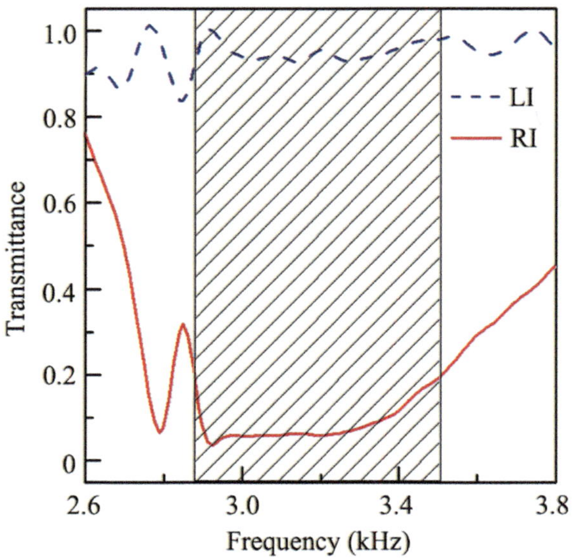

Fig. 4.15 Simulated transmittance spectra of the AAT window for LI and RI

Figure 4.16a, b show the pressure distributions caused by the AAT window for LI and RI, respectively. For LI, the incident wave is divided into several beams by the tunnels, and each sound beam can pass through the tunnel. Note that the transmitted beam is very wide due to the superposition of the transmitted acoustic waves from the seven tunnels of the AAT window. However, for RI, each sound beam cannot pass through the tunnel due to the ABR, and thus the transmitted sound energy for RI is very weak, as shown in Fig. 4.16b. The width of the AAT window is close to 1.0 m and can be further enlarged by introducing a larger number of blades. The AAT window is more beneficial for the exchange of media (including air and light) compared to a single AAT tunnel.

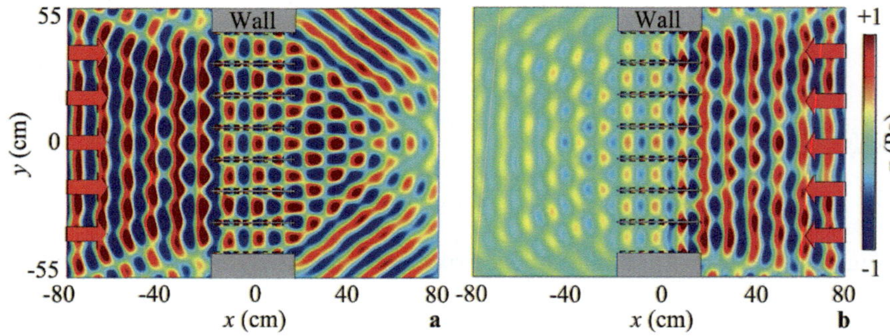

Fig. 4.16 Simulated pressure distributions caused by the AAT window for **a** LI and **b** RI at 3.2 kHz

4.2.6 Experimental Measurement of AAT Window

To experimentally demonstrate the performance of the AAT window, we measure the transmittance spectra of the AAT window for LI and RI. Figure 4.17 shows the photograph of the AAT window with five blades, in which the sizes of the meta-surfaces and the width w of the tunnels are reduced by twice compared to those in Fig. 4.14 due to the size limitation of the experimental platform.

Figure 4.18 shows the measured and simulated transmittance spectra of the AAT window. It is obvious that the measured spectra agree with the simulated ones. The slight difference between the measured and simulated results for LI may arise from the viscous loss of air in the measurement.

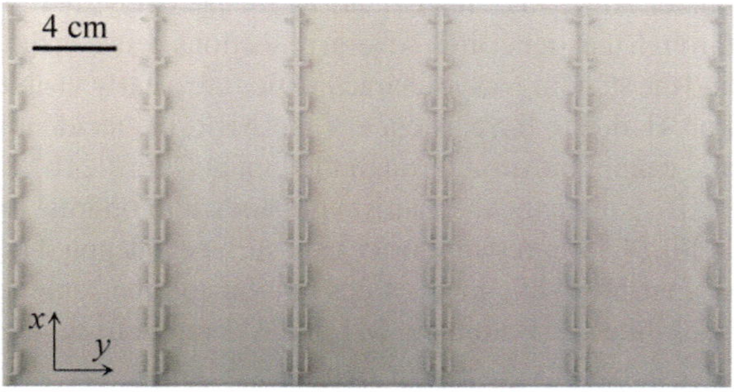

Fig. 4.17 Photograph of the AAT window

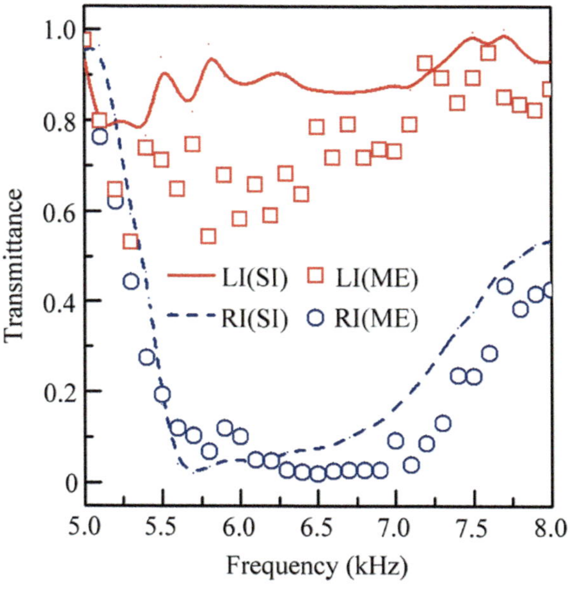

Fig. 4.18 Measured and simulated transmittance spectra of the AAT window

4.2.7 Performance Optimization of AAT Tunnel and Window

Furthermore, the working bandwidth of the designed AAT tunnel and window can be optimized by tuning the phase distribution of the metasurface. Unlike the metasurface in Fig. 4.4a, the values of d for the hook-like meta-atoms in the optimized metasurface range from 4.0 to 21.5 mm in steps of 2.5 mm, and the distribution of the phase delay is shown in Fig. 4.19, where $d\varphi(x)/dx$ is not constant.

Figure 4.20 shows the transmittance spectra of the optimized AAT tunnel. The transmittance spectra for LI and RI exhibit an obvious difference in the range 2.8–3.8 kHz (the black shaded region), and the working bandwidth of the optimized AAT tunnel reaches 1.0 kHz.

Figure 4.21a, b present the pressure distributions through the optimized AAT tunnel at 3.2 kHz for LI and RI, respectively. For LI, the acoustic wave can pass through the tunnel after several reflections between the upper and lower metasurfaces, and the standing wave characteristics are clearly visible inside the tunnel (Fig. 4.21a). For RI, due to the existence of the ABR, the incident wave is entirely reflected back and cannot transmit through the tunnel (Fig. 4.21b).

To explain the mechanism of bandwidth optimization, we divide the metasurface in Fig. 4.19 into two shorter metasurfaces (denoted as M1 and M2), each consisting of four hook-like meta-atoms and having the same length. The values of d for the hook-like meta-atoms in M1 and M2 range from 4.0 to 11.5 mm and from 14.0 to 21.5 mm, respectively, in steps of 2.5 mm.

Figure 4.22 shows the transmittance spectra of two types of AAT tunnels with a pair of M1 and M2 for RI. In the range 2.80–3.38 kHz (the blue shaded region), the transmittance is lower than 0.2 for M2. Additionally, in the high-frequency range

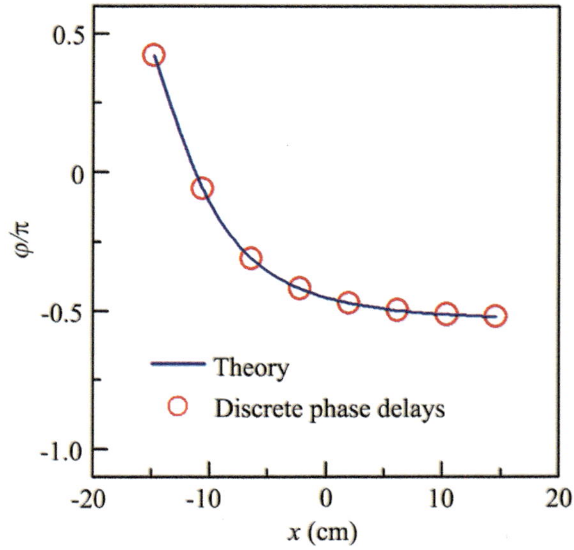

Fig. 4.19 Theoretical continuous phase distribution and discrete phase delays of the optimized metasurface, and the values of d for the eight red open circles are 4.0, 6.5, 9.0, 11.5, 14.0, 16.5, 19.0, and 21.5 mm

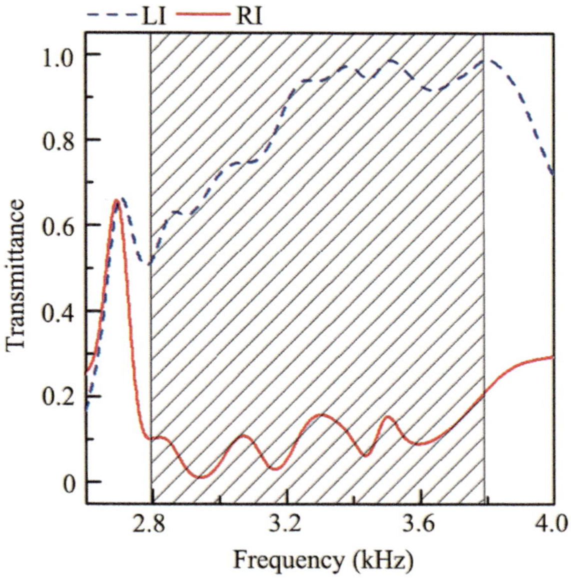

Fig. 4.20 Simulated transmittance spectra of the optimized AAT tunnel for LI and RI

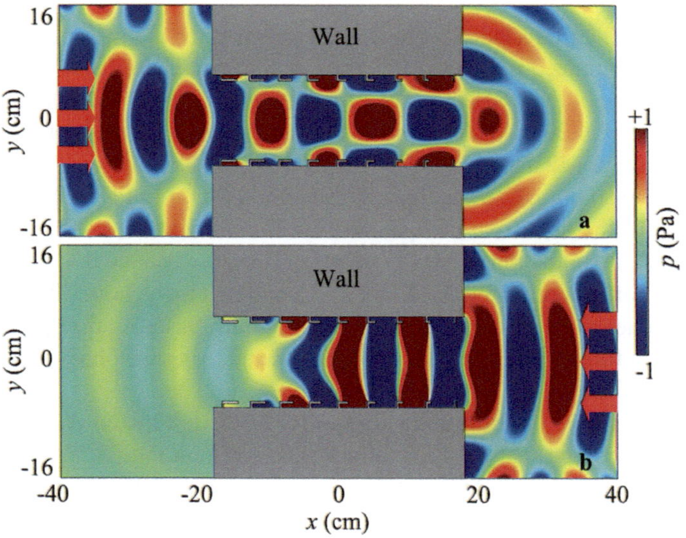

Fig. 4.21 Simulated pressure distributions caused by the optimized AAT tunnel for **a** LI and **b** RI at 3.2 kHz

Fig. 4.22 Simulated transmittance spectra of the two types of AAT tunnels with a pair of M1 and M2 composed of four meta-atoms. The values of *d* for the four selected meta-atoms are 4.0, 6.5, 9.0, and 11.5 mm for M1; and 14.0, 16.5, 19.0, and 21.5 mm for M2

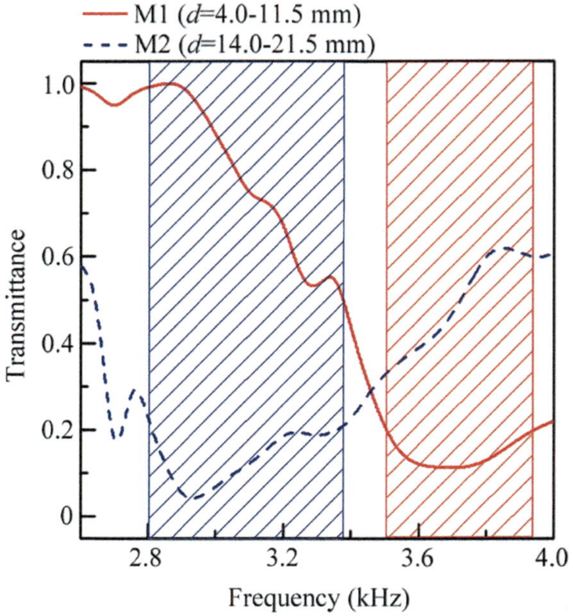

3.50–3.94 kHz (the red shaded region), the transmittance is lower than 0.2 for M1. Note that the optimized metasurface is the combination of M1 and M2. Therefore, the transmittance of the optimized AAT tunnel for RI is lower than 0.2 in the range 2.8–3.8 kHz (Fig. 4.20), which is almost the same as the superposition of the ranges 2.80–3.38 kHz for M2 and 3.50–3.94 kHz for M1.

However, in addition to the minimum value of 4.0 mm in Fig. 4.4a, the selected values of *d* for the hook-like meta-atoms are in the range 13.9–20.3 mm, and the AAT effect exists in the range 2.90–3.40 kHz. Thus, we deduce that the broadband characteristic of the optimized AAT tunnel is attributed to the hook-like meta-atoms with the parameter *d* within the range of 4.0–11.5 mm. Based on the optimized AAT tunnel, we can further optimize the AAT window by using the phase distribution of the metasurface in Fig. 4.19.

Figure 4.23 shows the transmittance spectra of the optimized AAT window for LI and RI. It is found that the AAT effect exists in the range 2.78–3.90 kHz (the grey shaded region). Compared with the results in Fig. 4.15a, the bandwidth of the optimized AAT window becomes much broader, and the fractional bandwidth reaches 0.36.

Figure 4.24a, b show the pressure distributions through the improved AAT window for LI and RI, respectively. The field distributions for LI and RI are similar to those in Fig. 4.16a, b, respectively, indicating that the mechanism of the improved AAT window is the same as that in Fig. 4.16a, b.

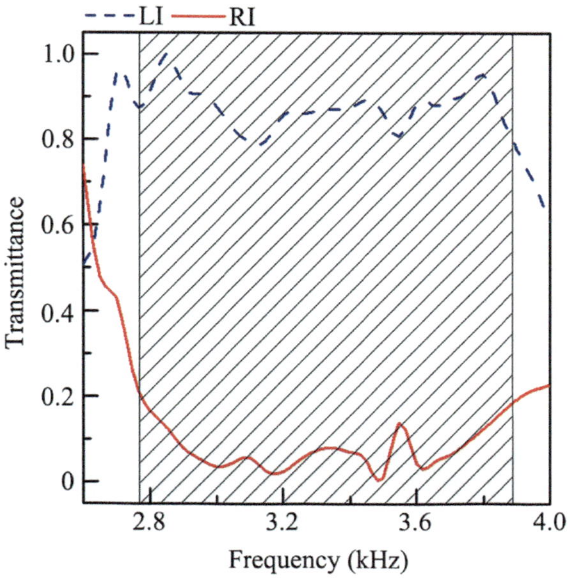

Fig. 4.23 Simulated transmittance spectra of the optimized AAT window for LI and RI

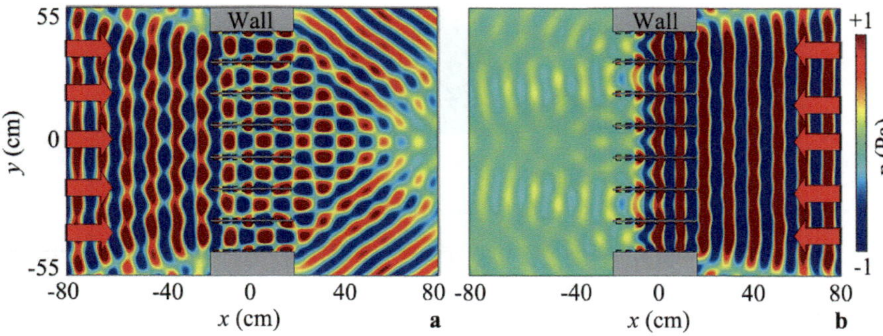

Fig. 4.24 Simulated pressure distributions caused by the optimized AAT window for **a** LI and **b** RI at 3.2 kHz

4.3 Acoustic Asymmetric Transmission by an Open Tunnel with Two Pairs of Triangle Cavities

4.3.1 Design and Performance of AAT Tunnel

As schematically shown in Fig. 4.25, the AAT system consists of an open tunnel attached with two pairs of triangle cavities on both sides [2]. The characteristics of the AAT tunnel are simulated using the COMSOL Multiphysics software, and the structural parameters of the AAT tunnel are $h_1 = 77.5$ mm, $l_1 = 155$ mm, $h_2 = 20$ mm, $l_2 = 80$ mm, $d = 30$ mm, and $w = 140$ mm. Additionally, the material parameters are as follows: density $\rho = 1180$ kg/m^3, longitudinal wave velocity $c_l = 2730$ m/s, and transversal wave velocity $c_t = 1430$ m/s for perspex; and $\rho = 1.21$ kg/m^3 and $c = 343$ m/s for air at 293 K. Throughout this work, the normal incidence of sound from the left and right sides of the tunnel are defined as LI and RI, respectively.

Figure 4.26 shows the simulated transmittance spectra of the AAT tunnel for LI and RI. The transmittance spectra exhibit an obvious difference between LI and RI in the range 4.32–4.42 kHz. The transmittance for LI is very low, with a minimum value of 0.008 at 4.40 kHz. However, the transmittance for RI is larger than 0.6.

To demonstrate the AAT behavior, we simulate the intensity distributions through the AAT tunnel at 4.40 kHz for LI and RI, as shown in Fig. 4.27a, b, respectively. The

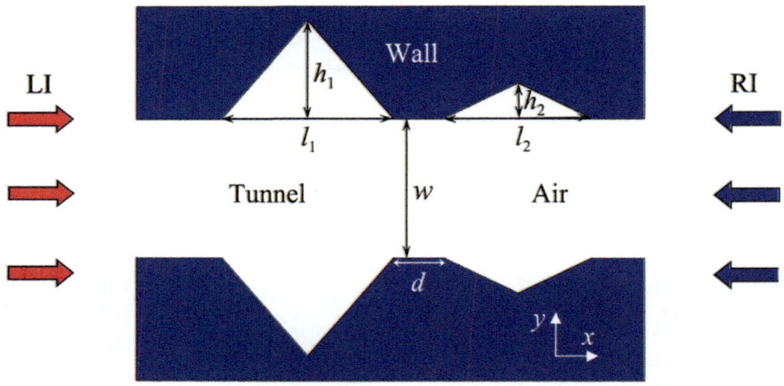

Fig. 4.25 Schematic of the AAT tunnel

Fig. 4.26 Simulated transmittance spectra of the AAT tunnel for LI and RI in Fig. 4.25

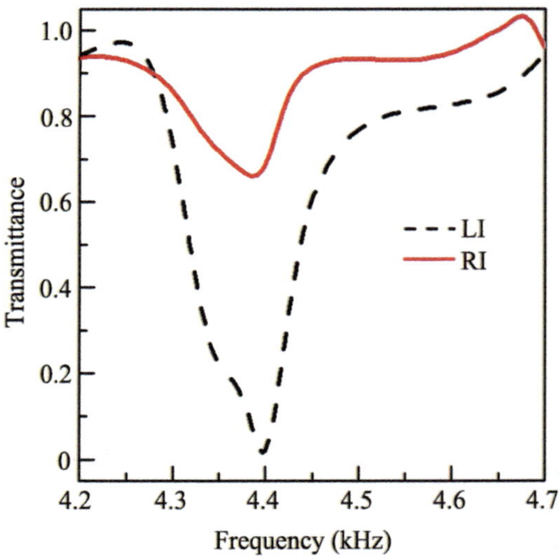

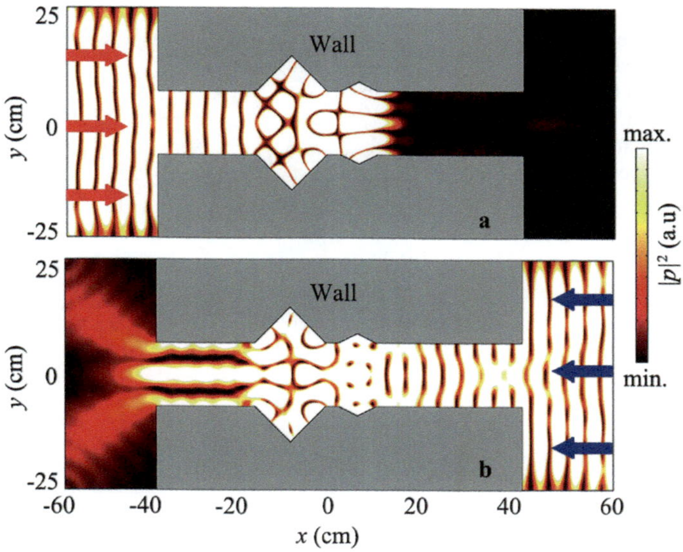

Fig. 4.27 Simulated intensity distributions caused by the AAT tunnel at 4.4 kHz for **a** LI and **b** RI

incident wave can transmit through the tunnel for RI (Fig. 4.27b), and the transmitted sound energy is divided into two beams. However, negligible transmission is observed for LI (Fig. 4.27a).

4.3.2 Physical Mechanism

To gain insight into the mechanism of AAT, we simulate the scattering and total pressure distributions caused by the triangle cavities at 4.40 kHz. For LI, the incident wave first reaches the position of a pair of large triangle cavities. Figure 4.28 presents the scattering pressure distribution caused by a single wall with a large triangle cavity, with the incident plane wave located on the left side. Note that three scattering waves, denoted as S1, S2, and S3, are excited by the large triangle cavity.

Figure 4.29 shows the simulated pressure amplitude distribution in the polar coordinate, where the direction of the incident wave is 180°, and the three peaks correspond to the three scattering waves in Fig. 4.28. Since the angle of S1 is greater than 90°, S1 can transmit through the triangle cavity. However, the angle of S3 is less than 90°, and the angle of S2 is close to 90°, so neither of them can pass through the triangle cavity.

To show the propagation path of S1 in the AAT tunnel, we simulate the total pressure distribution for S1 scattered by the triangle cavities, as shown in Fig. 4.30. There is an obvious standing wave between a pair of small triangle cavities, and the

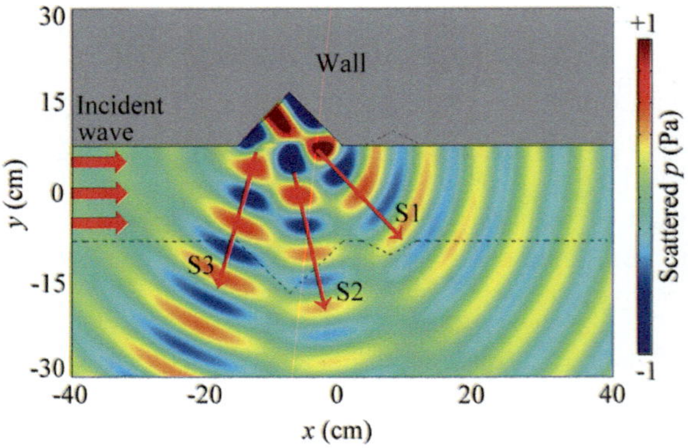

Fig. 4.28 Simulated scattering pressure distribution caused by a single wall with a large triangle cavity for LI at 4.4 kHz. S1, S2, and S3 indicate three types of scattering waves

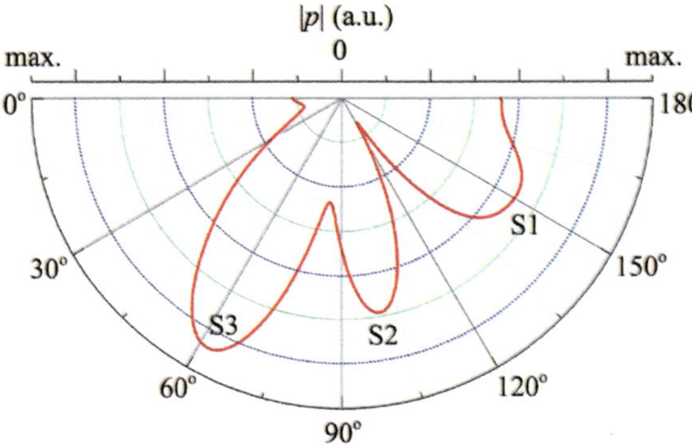

Fig. 4.29 Simulated scattering pressure distribution caused by a single wall with a large triangle cavity for LI at 4.4 kHz in the polar coordinate

sound energy in the right region is almost zero, which explains why the incident wave cannot pass through the tunnel for LI.

For RI, the incident wave first reaches the position of a pair of small triangle cavities. Figure 4.31 presents the total pressure distribution caused by a single wall with a

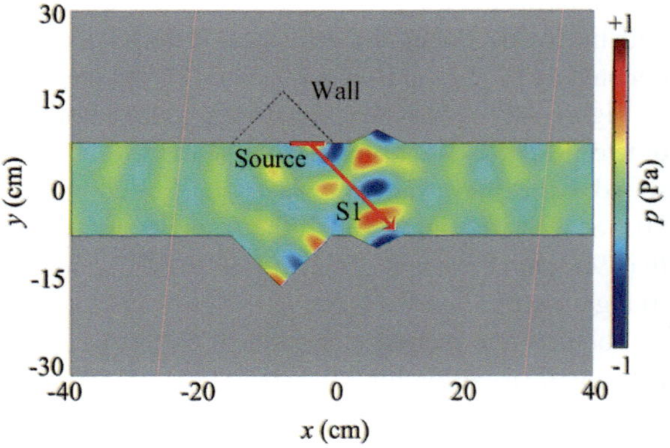

Fig. 4.30 Simulated total pressure distribution for S1 scattered by the triangle cavities at 4.4 kHz

small triangle cavity. The propagation direction of the incident acoustic wave remains unchanged with the small triangle cavity. Therefore, the acoustic wave reaches the large triangle cavity and is scattered into three components (S1, S2, and S3), of which S1 can finally pass through the tunnel. Based on the above analysis, we deduce that the AAT effect arises from asymmetric multiple scatterings caused by triangle cavities of different sizes, which is different from previous AM systems [5–7].

Based on the aforementioned simulations, we illustrate the acoustic propagation paths through the AAT tunnel for LI and RI, as shown in Fig. 4.32. As shown in Fig. 4.32a, the incident wave is first scattered by the large triangle cavities and obliquely reaches the small triangle cavity. After scattering by the small triangle cavity, the acoustic wave propagates along the vertical direction of the tunnel and cannot pass through the tunnel. However, for RI (Fig. 4.32b), the small triangle cavity has almost no influence on the incident wave. As a result, the acoustic wave is scattered by the large triangle cavities, and part of the incident wave's energy finally passes through the tunnel.

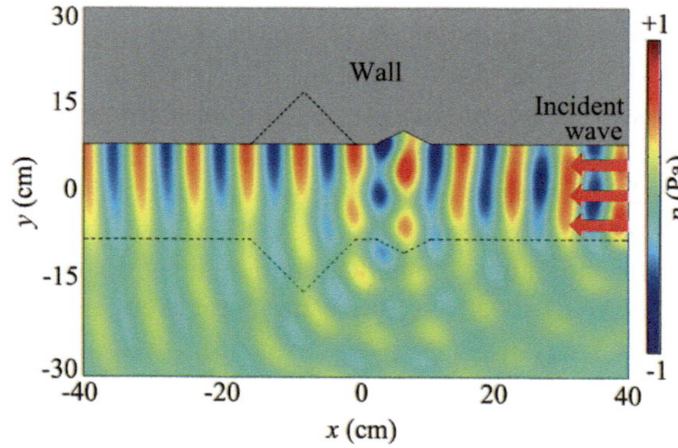

Fig. 4.31 Simulated total pressure distribution caused by a single wall with a small triangle cavity for RI at 4.4 kHz

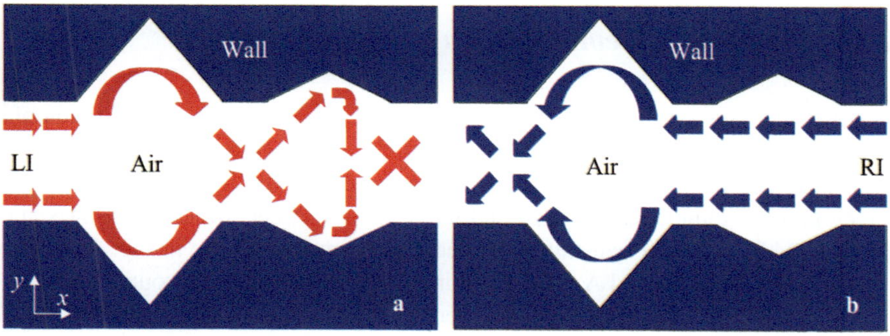

Fig. 4.32 Schematic of acoustic propagation paths inside the AAT tunnel for **a** LI and **b** RI

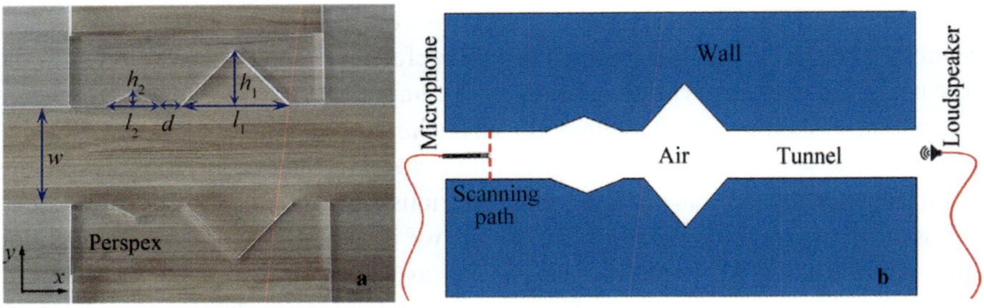

Fig. 4.33 **a** Photograph of the sample and **b** schematic of the experimental set-up

4.3.3 Experimental Measurement

To verify the performance of the AAT tunnel, we experimentally measure the transmittance spectra of the AAT tunnel. To accurately measure acoustic transmittance, we extend the AAT tunnel with waveguides on both sides to concentrate sound energy. Figure 4.33a shows the photograph of the sample, which is fabricated with perspex, and the parameters h_1, h_2, l_1, l_2, d, and w are the same as those in Fig. 4.25. Figure 4.33b shows the experimental set-up. The loudspeaker (25 × 25 mm^2) is placed on the right side of the waveguide, 1.0 m away from the AAT tunnel. A 0.25 inch-diameter Brüel & Kjær type-4961 microphone is used to detect the sound signal on the left side of the waveguide. Using the Pulse Labshop software, we can detect the pressure amplitude along the scanning line. Meanwhile, the transmittance spectra of the AAT tunnel with the same parameters are also simulated for comparison.

Figure 4.34 presents the measured and simulated transmittance spectra of the AAT tunnel for LI and RI in the range 4.20–4.70 kHz. The measured transmittance spectrum agrees with the simulated result, but there are some discrepancies between the two results, especially in the low-frequency range, due to the viscous loss of air.

4.3.4 Robustness Demonstration

Finally, we demonstrate the robustness of the AAT tunnel against the excitation of a cylindrical sound source. The parameters of the AAT tunnel are the same as those in Fig. 4.25. Figure 4.35 shows the transmittance spectra of the AAT tunnel excited by the cylindrical sound source for LI and RI. The AAT effect still exists in the range 4.32–4.42 kHz, and the transmittance spectra are similar to those in Fig. 4.26. This indicates that the designed AAT tunnel can be used for cylindrical sound sources.

Figure 4.36a, b show the simulated intensity distributions through the AAT tunnel at 4.4 kHz for LI and RI, respectively. The AAT effect persists under the excitation of the cylindrical sound source, with RI associated with high transmitted sound energy, while for LI, the transmitted sound energy is quite weak, similar to the results in Fig. 4.27.

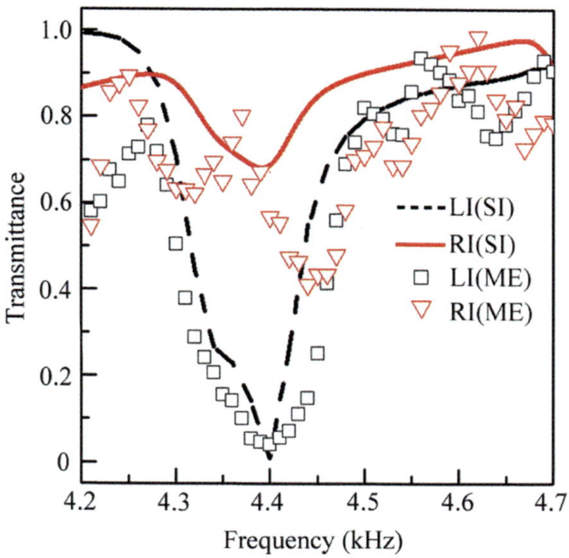

Fig. 4.34 Measured and simulated transmittance spectra of the AAT tunnel for LI and RI

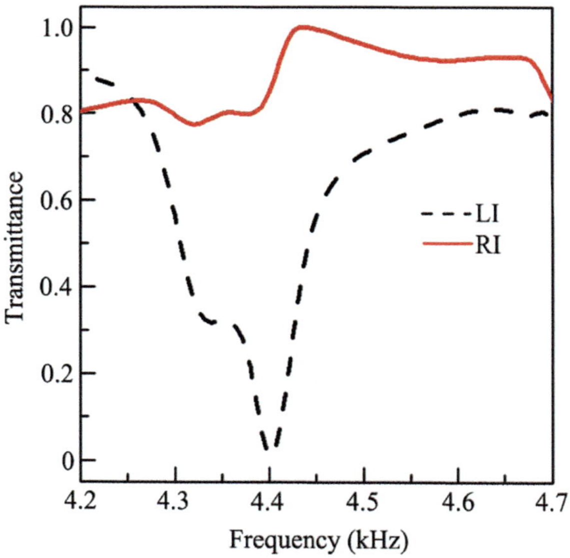

Fig. 4.35 Simulated transmittance spectra of the AAT tunnel under the excitation of a cylindrical acoustic source for LI and RI

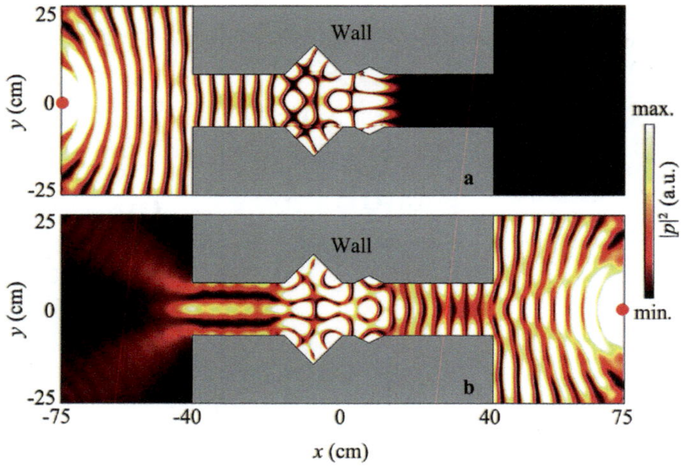

Fig. 4.36 Simulated intensity distributions caused by the AAT tunnel under the excitation of a cylindrical acoustic source at 4.4 kHz for **a** LI and **b** RI

4.4 Multi-Channel Acoustic Asymmetric Transmission by Mode-Conversion Meta-Atoms

4.4.1 Design and Performance of Mode-Conversion Meta-Atom

As schematically shown in Fig. 4.37, we propose a type of phased unit cell composed of different numbers (N) of V-shaped structures [3]. The structural parameters are selected as $l = 30$ mm, $h = 5.5$ mm, $t = 0.5$ mm, and $d = 0.25$ mm, and its opening angle is determined by N. Throughout this work, we use COMSOL Multiphysics software to simulate the performance of AAT. The phased unit cell is fabricated with epoxy resin to satisfy the sound-hard boundary. In the models, the material parameters are as follows: density $\rho = 1180$ kg/m^3 and sound velocity $c = 2720$ m/s for epoxy resin; and $\rho = 1.21$ kg/m^3 and $c = 343$ m/s for air.

Figure 4.38 shows the transmission spectrum and phase delays of the phased unit cells with different values of N at 7.1 kHz. As N increases, the phase delay gradually increases, and the transmission coefficient exceeds 0.7. Based on the mode-conversion theory in Eq. (2.52), we select two types of phased unit cells I and II (the blue open circles) with a phase difference of π to design a mode-conversion meta-atom. The selected values of N for phased unit cells I and II are 1 and 11, respectively, with transmission coefficients of 0.99 and 0.85. The two insets at the bottom of Fig. 4.38 show the pressure distributions caused by phased unit cells I and II, further demonstrating the phase difference between the two unit cells.

To realize the mode conversion between the zero-order and first-order waves, we calculate the theoretical continuous phase distribution based on Eq. (2.53), as shown

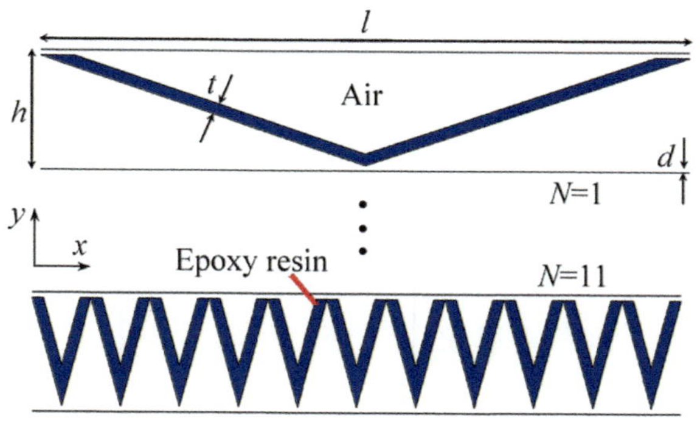

Fig. 4.37 Schematic of phased unit cells composed of different numbers (N) of V-shape structures. The total length l of each unit cell is the same, and the opening angle of the V-shaped structure is determined by N

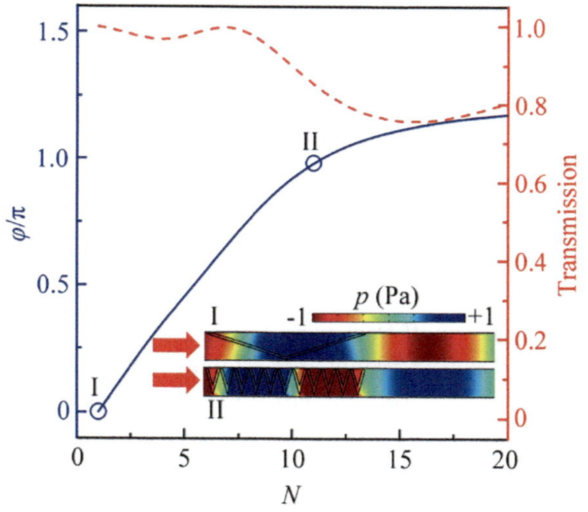

Fig. 4.38 Simulated transmission coefficients (the red dashed line) and phase delays (the blue solid line) of the phased unit cells with different values of N at 7.1 kHz. The two insets show the pressure distributions caused by phased unit cells I and II

in Fig. 4.39. Based on the theoretical phase distribution, we design a type of mode-conversion meta-atom using three pairs of phased unit cells I and II, with a total width of 33 mm, satisfying the propagation of the first-order wave at 7.1 kHz.

As shown in Fig. 4.40a, the zero-order wave can transmit through the phased unit cells and convert to the first-order wave. Similarly, the first-order wave can also convert to the zero-order wave, indicating that the proposed phased unit cells can realize mode conversion between the zero-order and first-order waves. Here, it is noted that the first-order wave in Fig. 4.40a is represented by two plane sound beams with an angle of $\pm 45°$ with respect to the x-axis. This is because the first-order wave in the waveguide with a width of 33 mm is formed as the zigzag propagation of the plane wave with an angle of 45°. Additionally, to realize the characteristic of AAT, we introduce a step waveguide placed on the right side of the phased unit cells (Fig. 4.40b), where the width $h_R = 21$ mm is smaller than half the wavelength of sound at 7.1 kHz. Thus, the first-order wave cannot pass through the step waveguide

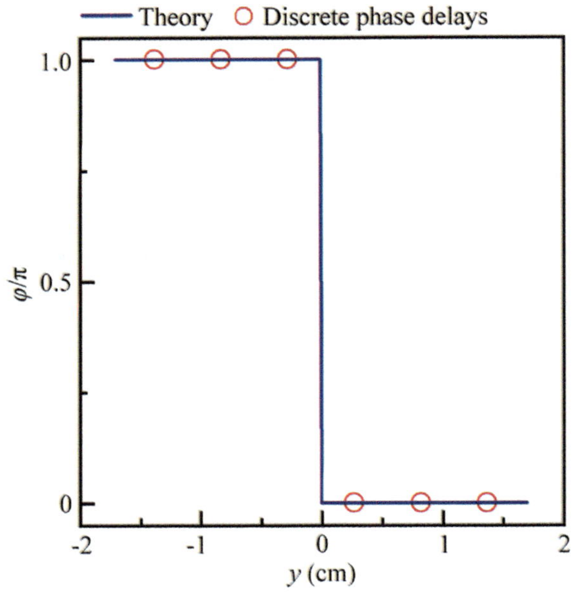

Fig. 4.39 Theoretical continuous phase distribution and discrete phase delays of the mode-conversion meta-atom

due to the cutoff frequency. The parameters of the mode-conversion meta-atom are selected as $l_1 = 50$ mm, $l_2 = 12$ mm, $h_1 = 6$ mm, $h_L = 33$ mm, and $t = 0.5$ mm.

The proposed meta-atom exhibits the characteristic of AAT based on mode conversion. To clearly demonstrate this, we present the propagation paths in the phased unit cell for the incidence of the zero-order wave. The abbreviations LI and RI refer to sound incidence from the left and right sides of the mode-conversion meta-atom, respectively. As shown in Fig. 4.41, the zero-order wave can transmit through the meta-atom and convert to the first-order wave for RI. However, due to the cutoff frequency of the first-order wave, it cannot pass through the step waveguide for LI. Figure 4.42 shows the simulated pressure distributions caused by the meta-atom

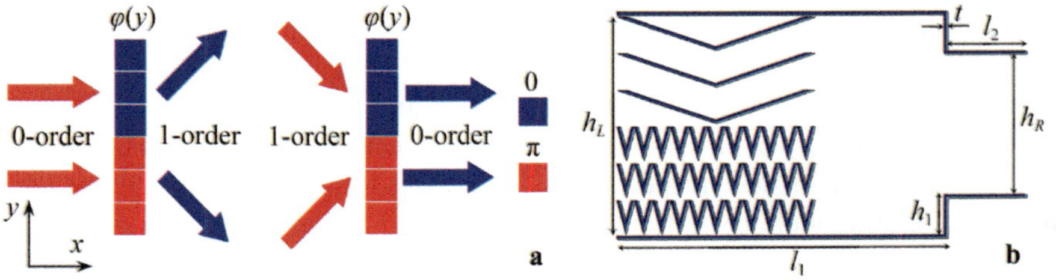

Fig. 4.40 **a** Schematic of mode conversion of sound caused by a meta-atom composed of six phased unit cells (I and II). The red and blue arrows represent the incident and transmitted acoustic waves, respectively. **b** Schematic of the mode-conversion meta-atom

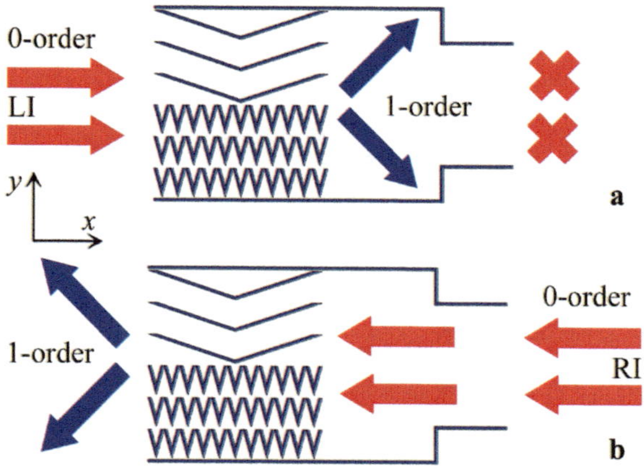

Fig. 4.41 Schematic of acoustic propagation paths through the mode-conversion meta-atom under the excitation of the zero-order wave for **a** LI and **b** RI

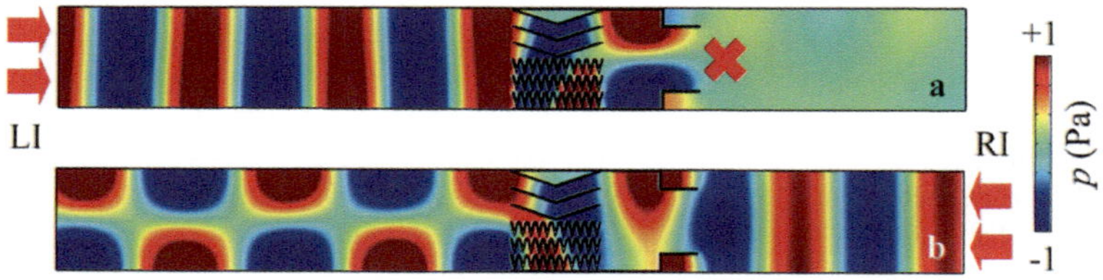

Fig. 4.42 Simulated pressure distributions caused by the mode-conversion meta-atom under the excitation of the zero-order wave for **a** LI and **b** RI at 7.1 kHz. The red arrows represent the incident zero-order wave

under the excitation of the zero-order wave for LI and RI. The simulated pressure distributions agree well with the propagation paths shown in Fig. 4.41, demonstrating high performance of AAT based on mode conversion.

Figure 4.43 shows the transmission spectra of the mode-conversion meta-atom under the excitation of the zero-order wave. In the black shaded region, the transmission coefficients are greater than 0.65 for RI, but lower than 0.4 for LI, showing the characteristic of AAT. Thus, the bandwidth of AAT for the mode-conversion meta-atom is 4.2 kHz, and the fractional bandwidth is 0.5, which is much wider than those of previous AAT systems with phase modulation. To explain this, we separately simulate the phase spectra of phased unit cells I and II and display the phase difference spectrum in Fig. 4.44. Note that in the working band of AAT (the black shaded region), the phase difference ranges from 0.82π to 1.52π, which is around the value of π. Therefore, the broadband characteristic of the mode-conversion meta-atom arises from the stable phase difference of both unit cells I and II around 7.1 kHz.

Moreover, when the mode-conversion meta-atom is excited by the first-order wave (Fig. 4.45), the characteristics of mode conversion and AAT caused by the mode-conversion meta-atom are opposite to those of the zero-order wave (Fig. 4.41).

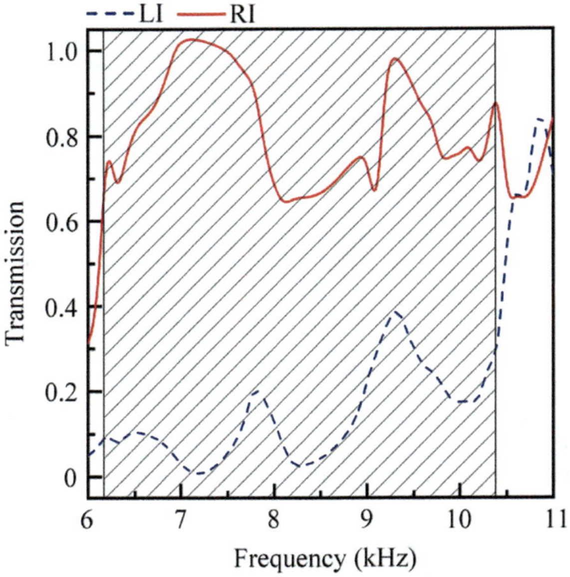

Fig. 4.43 Simulated transmission spectra of the mode-conversion meta-atom under the excitation of the zero-order wave for LI and RI

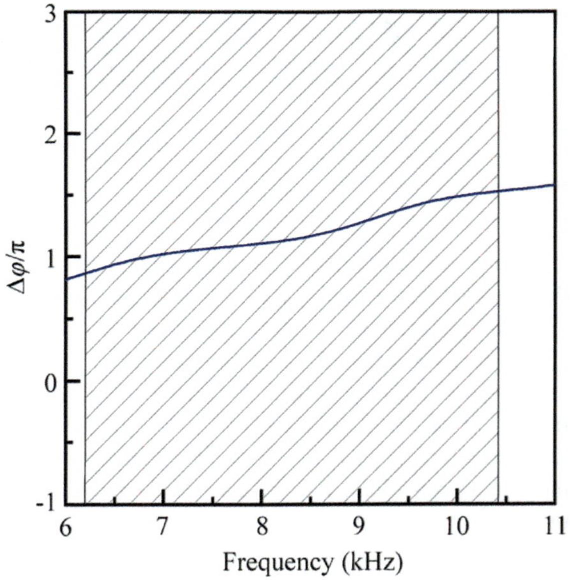

Fig. 4.44 Simulated phase-difference spectrum of phased unit cells I and II. The black shaded regions (6.2–10.4 kHz) represents the working band of AAT

Figure 4.46 shows the simulated pressure distributions under the excitation of the first-order wave for LI and RI, which match well with the corresponding propagation paths, demonstrating the characteristic of reversed AAT under the excitation of the first-order wave.

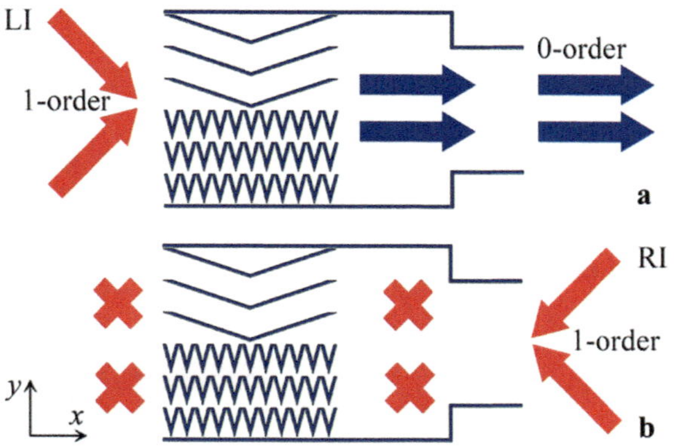

Fig. 4.45 Schematic of acoustic propagation paths through the mode-conversion meta-atom under the excitation of the first-order wave for **a** LI and **b** RI

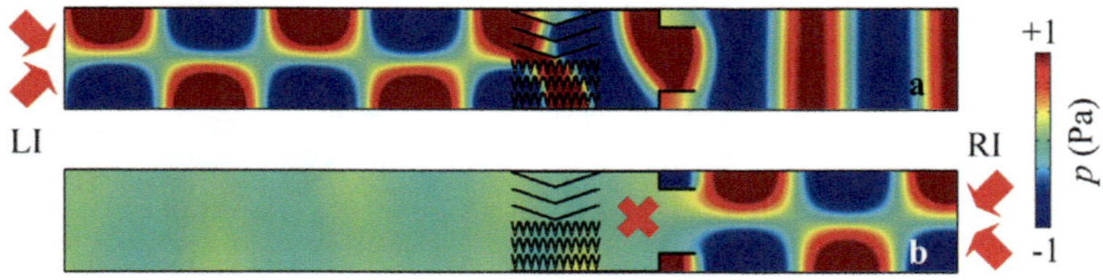

Fig. 4.46 Simulated pressure distributions caused by the mode-conversion meta-atom under the excitation of the first-order wave for **a** LI and **b** RI at 7.1 kHz. The red arrows represent the incident first-order wave

4.4.2 Design and Performance of AAT Lens

Based on the mode-conversion meta-atom, we design a type of multi-channel AAT lens [3]. Compared with previous AAT systems with phase manipulation, the designed AAT lenses have the advantages of a single-layer planar structure, multi-channel AAT, and broad bandwidth. Here, we use twelve mode-conversion meta-atoms to design a multi-channel AAT lens, as shown in Fig. 4.47a, where the twelve mode-conversion meta-atoms are placed close together. The acoustic propagation paths through the AAT lens under the excitation of the zero-order wave are shown in Fig. 4.47b, c. Similar to those in Fig. 4.41, the first-order wave is obtained and passes through the lens as two symmetric sound beams with refraction angles of ±45° for RI. However, for LI, the incident zero-order wave cannot pass through the step waveguide in each meta-atom.

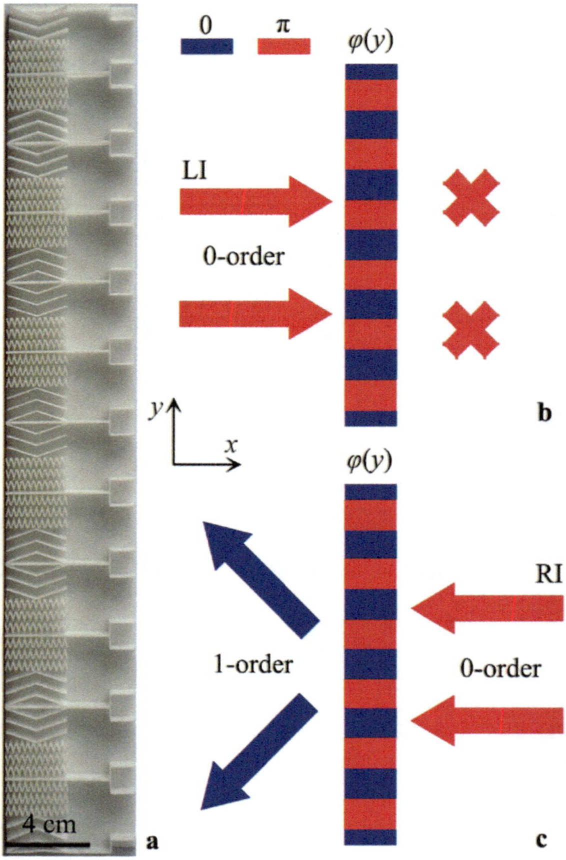

Fig. 4.47 **a** Photograph of the multi-channel AAT lens composed of twelve mode-conversion meta-atoms. Schematic of acoustic propagation paths through the AAT lens under the excitation of the zero-order wave for **b** LI and **c** RI

4.4.3 Experimental Measurement

To experimentally verify the performance, we conduct an experiment to measure the pressure distribution and transmission spectrum of the AAT lens. As shown in Fig. 4.48, the experiment is carried out in a planar waveguide system composed of two parallel plates with a size of 2 m × 2 m, where the anechoic environment is realized by placing wedge-shaped sound-absorbing foams at the boundaries of the planar waveguide. The incident plane wave is generated from a speaker array driven by a power amplifier. In the measurement, we scan the sound pressure using a microphone (Brüel & Kjær type-4961) moved by a set of 2D motorized linear stages (Newport: MIN300CC and ILS250CC). Another microphone is fixed towards the speaker array as a reference. The measured sound signals are recorded by the Brüel & Kjær 3160-A-022 module and analyzed using the PULSE Labshop software.

Figure 4.49 shows the simulated pressure distributions caused by the AAT lens for LI and RI. We can see that under the excitation of the zero-order wave, the sound

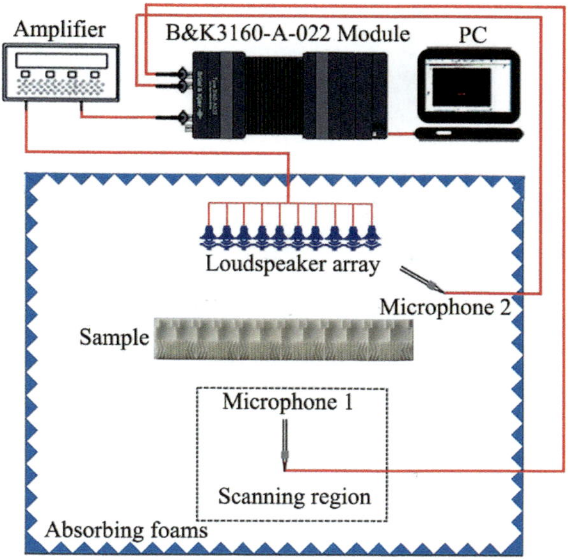

Fig. 4.48 Schematic of the experimental set-up for the AAT lens in Fig. 4.47a

energy can pass through the lens as two symmetric acoustic beams for RI, but cannot reach the right region for LI, showing a typical characteristic of AAT. The four insets at the bottom of Fig. 4.49 show the pressure distributions in the red open squares R1 and R2. The measured results match well with the simulated ones. Therefore, by designing the acoustic lens composed of mode-conversion meta-atoms, the AAT effect can be observed.

Figure 4.50 shows the measured and simulated transmittance spectra of the AAT lens for LI and RI. In the range of 6.2–9.0 kHz (the black shaded region), the transmittance spectra for LI and RI exhibit an obvious characteristic of AAT. Thus, the bandwidth of the AAT lens is 2.80 kHz, and the fractional bandwidth reaches about 0.37. Additionally, both types of results are consistent with each other, experimentally demonstrating the broadband characteristic of the AAT lens.

4.4.4 Performance Modulation

In addition to the aforementioned AAT effect of the lens, we observe multi-channel AAT in the same lens under the excitation of the first-order wave. As shown in Fig. 4.51, the acoustic propagation paths through the AAT lens under the excitation of the first-order wave are the same as those in Fig. 4.45. The incident first-order wave is converted into the zero-order wave and transmitted through the lens for LI, but cannot pass through the lens for RI. Compared with the results in Fig. 4.47, the AAT effect is reversed under the excitation of the first-order wave.

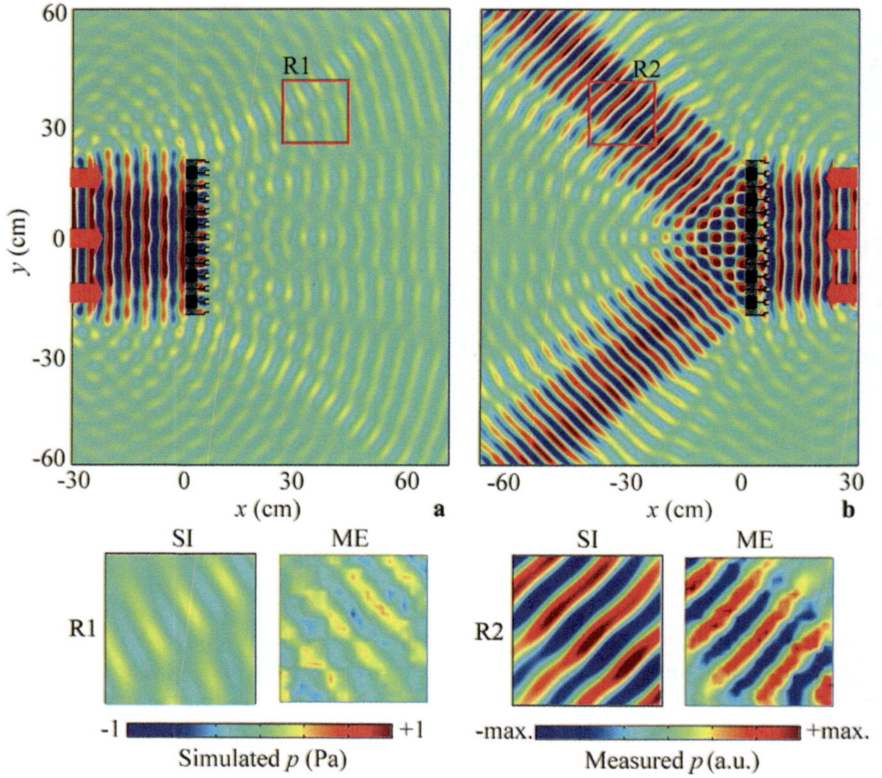

Fig. 4.49 Simulated pressure distributions through the AAT lens under the excitation of the zero-order wave for **a** LI and **b** RI at 7.1 kHz. The red arrows represent the incident zero-order wave, and the four insets at the bottom show the measured and simulated intensity distributions in red open squares R1 and R2

Figure 4.52 shows the simulated pressure distributions through the AAT lens under the excitation of the first-order wave for LI and RI. For RI, the first-order wave is blocked by the step waveguide of each mode-conversion meta-atom and cannot reach the left side. However, for LI, the first-order wave is converted into the zero-order wave and can pass through the AAT lens. The characteristic of AAT is opposite to that in Fig. 4.49, indicating that the designed AAT lens can reverse the AAT effect under the excitation of the first-order wave. The measured pressure distributions in red open squares R3 and R4 are presented at the bottom of Fig. 4.52, and the measured results agree with the simulated ones. Therefore, the proposed multi-channel AAT lens can realize the characteristics of AAT and reversed AAT under the excitation of the zero-order and first-order waves, respectively.

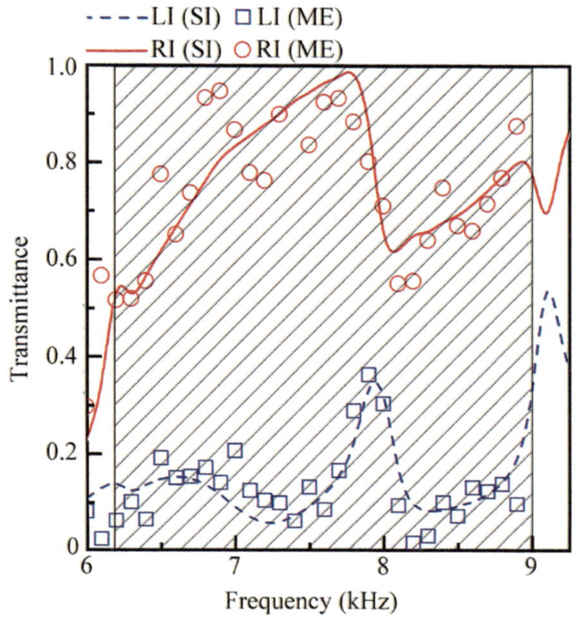

Fig. 4.50 Measured and simulated transmittance spectra of the AAT lens under the excitation of the zero-order wave for LI and RI

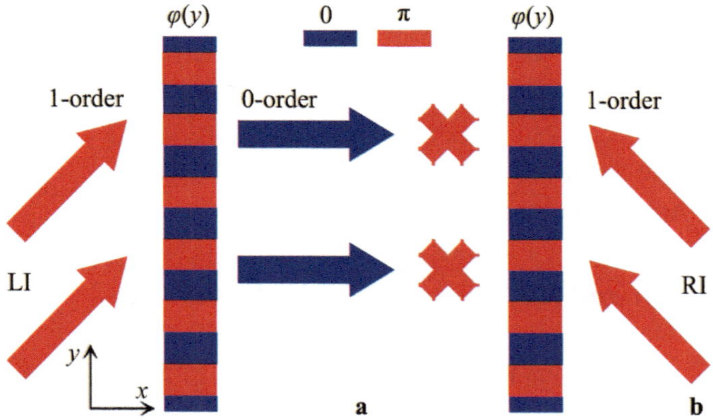

Fig. 4.51 Schematic of acoustic propagation paths through the AAT lens under the excitation of the first-order wave for **a** LI and **b** RI

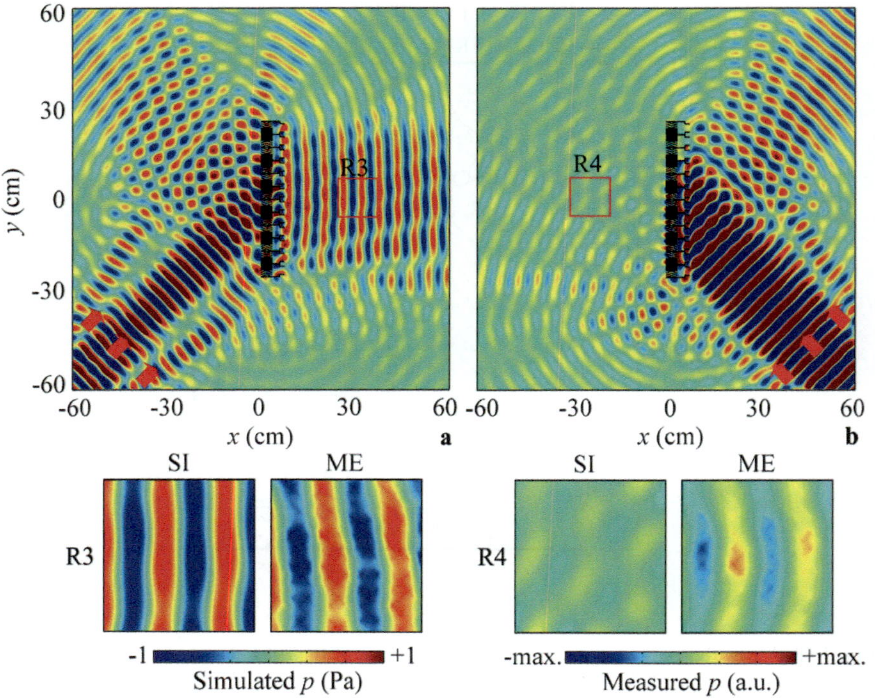

Fig. 4.52 Simulated pressure distributions through the AAT lens under the excitation of the first-order wave for **a** LI and **b** RI at 7.1 kHz. The red arrows represent the incident first-order wave, and the four insets at the bottom show the measured and simulated intensity distributions in red open squares R3 and R4

4.5 Tunable Acoustic Asymmetric Transmission by Phase-Controlled Prism

4.5.1 Design and Performance of AAT Prism

As schematically shown in Fig. 4.53, we design a type of AAT prism composed of 31 phased unit cells with a width of l [4]. The shape of the prism is an isosceles right triangle with $\alpha = 45°$ and a height of H. The abbreviations LI and RI represent the normal incidence of acoustic waves from the left and right sides of the AAT prism, respectively. Figure 4.53a, b show the acoustic propagation paths through the AAT prism for LI and RI, respectively, where the prism is placed in region II. For LI, the acoustic wave can transmit through the prism with a refraction angle θ_{t1}, but the transmitted sound energy is confined to the two surfaces on the right side of the prism with $\theta_{t1} = -90°$. Moreover, in the range $-90° < \theta_{t1} \leq \alpha - 90°$, the acoustic wave can transmit through the prism but cannot reach region I.

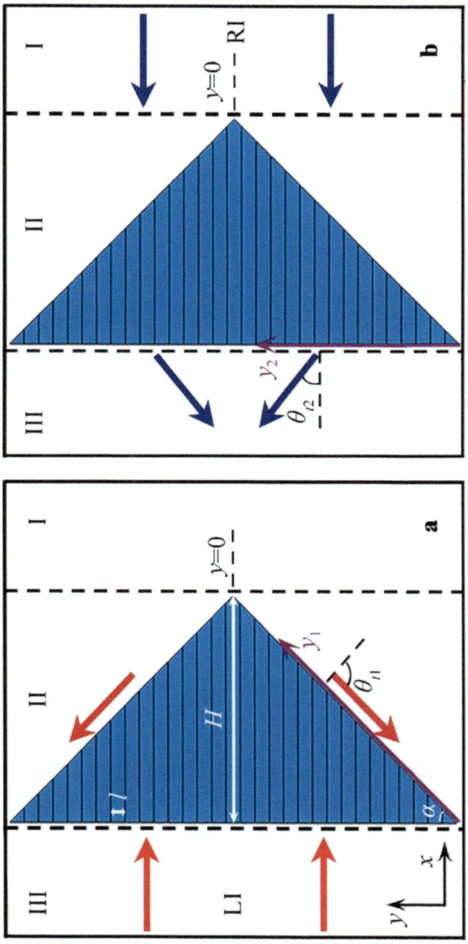

Fig. 4.53 Schematic of acoustic propagation paths through the AAT prism with different phase distributions for **a** LI and **b** RI

Here, we only study the lower half of the prism due to its structural symmetry. For LI, based on Eq. (2.41), the phase distribution φ_1 of the prism in the y_1 direction is written as:

$$\varphi_1 = \sqrt{2}\sin\theta_{t1}k(H + x). \tag{4.1}$$

where $k = 2\pi f/c$ is the wave vector in air, and c and f are the acoustic velocity and the frequency, respectively. Based on Eq. (4.1), the phase delays between two adjacent unit cells ($\Delta\varphi_1$) can be expressed as:

$$\Delta\varphi_1 = \sqrt{2}\sin\theta_{t1}kl. \tag{4.2}$$

For RI, the acoustic wave can pass through the prism and reach region III with a refraction angle θ_{t2} (Fig. 4.53b), and the left surface of the prism is defined as the y_2 axis. Based on Eq. (4.2), the phase delays between two adjacent unit cells ($\Delta\varphi_2$) can be written as:

$$\Delta\varphi_2 = \sqrt{2}\sin\theta_{t1}kl - kl + 2\pi m, \tag{4.3}$$

where $2\pi m$ is selected to keep the phase difference between two adjacent unit cells in the range of 2π, satisfying the generalized Snell's law. The phase distribution φ_2 of the prism in the y_2 direction can be written as:

$$\varphi_2 = \sqrt{2}\sin\theta_{t1}k(H + x) - kx + 2\pi m\frac{H + x}{l}. \tag{4.4}$$

Based on Eq. (2.41), the relationship between the refraction angles θ_{t2} and θ_{t1} is given as:

$$\sin\theta_{t2} = \sqrt{2}\sin\theta_{t1} - 1 + \frac{2\pi m}{kl}. \tag{4.5}$$

The aforementioned results show that the phase distributions of the output surfaces in the prism are different for LI and RI, as are the refraction angles θ_{t1} and θ_{t2}, indicating the feasibility of designing the AAT prism.

To design the AAT prism, we propose a meta-atom (with a length l = 16 mm) composed of two arrays of Helmholtz resonators and a central straight channel with a tunable width of h_1 immersed in air. As shown in Fig. 4.54a, the cavity width h is determined by the channel width h_1, and the other parameters are selected as d = 1.0 mm, $h = (l - 4d - h_1)/2$, w_1 = 1.5 mm, and $w = (l - 4d)/2$. The meta-atom is fabricated with epoxy resin to satisfy the sound-hard boundary condition. In this work, the COMSOL Multiphysics software is used to simulate the AAT characteristics of the prism. The material parameters are as follows: density ρ = 1180 kg/m^3, longitudinal wave velocity c_l = 2720 m/s, and transversal wave velocity c_t = 1460 m/s for epoxy resin; ρ = 1.21 kg/m^3 and c = 343 m/s for air.

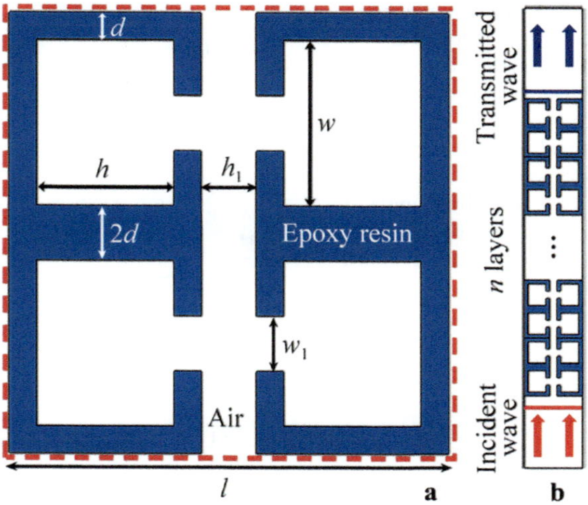

Fig. 4.54 Schematic of **a** the meta-atom and **b** the phased unit cell composed of *n*-layer meta-atoms

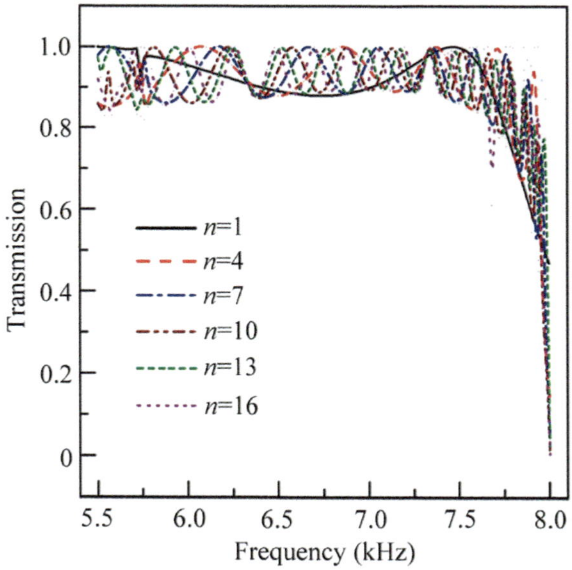

Fig. 4.55 Simulated transmittance spectra of the phased unit cells with different numbers (*n*) of meta-atoms

As shown in Fig. 4.54b, we design a phased unit cell composed of *n*-layer meta-atoms with $h_1 = 5.2$ mm, where the incident acoustic wave is located on the bottom-side. Figure 4.55 shows the transmittance spectra with different numbers of meta-atoms, where *n* is selected as 1, 4, 7, 10, 13, and 16. As *n* increases, the transmittance spectra have similar shapes, and the transmission coefficients are larger than 0.85 in the range of 5.5–7.7 kHz, indicating a broadband characteristic of the phased unit cell.

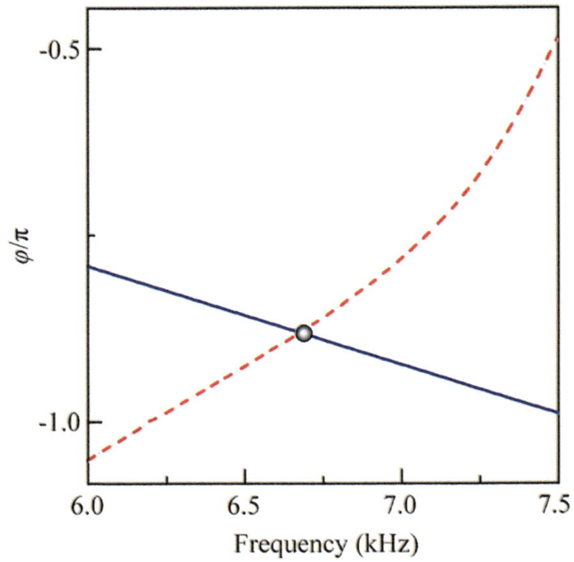

Fig. 4.56 Theoretically calculated phase delay spectrum (the blue solid line) between two adjacent unit cells based on Eq. (4.2) and simulated phase delay spectrum (the red dashed line) caused by a single meta-atom. The frequency of the black point is 6.66 kHz

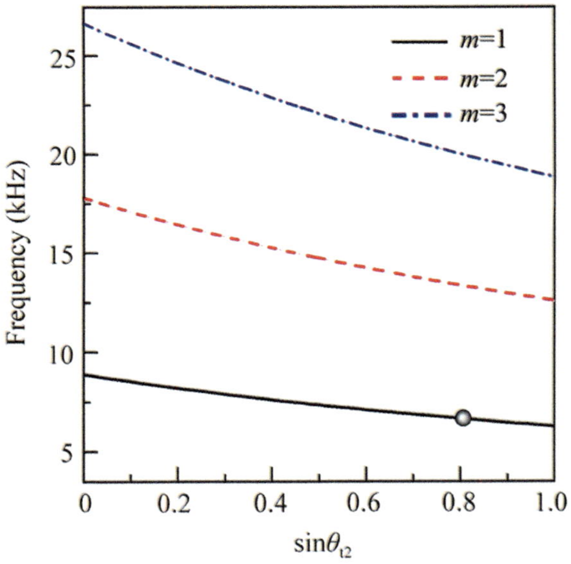

Fig. 4.57 The values of $\sin\theta_{t2}$ at different frequencies for $m = 1, 2$, and 3. The frequency of the black point is 6.66 kHz

We design an AAT prism based on the phased unit cells with different numbers of meta-atoms, where the phase gradient of the prism in the y_1 direction is selected as $\mathrm{d}\varphi_1/k\mathrm{d}y_1 = -1.0$, and thus θ_{t1} can be calculated as $-90°$. To obtain the desired phase delay of the meta-atom, we theoretically calculate the phase delay spectrum based on Eq. (4.2), as shown in Fig. 4.56. Additionally, we simulate the phase delay spectrum of a single meta-atom. Note that the theoretical phase delay is the same as that of the meta-atom at 6.66 kHz (the black point in Fig. 4.56).

Based on Eq. (4.5), we calculate the parameter $\sin\theta_{t2}$ at different frequencies for $m = 1, 2$, and 3 (Fig. 4.57), and the parameters m and θ_{t2} are obtained as 1 and 54° at 6.66 kHz (the black point), respectively. Therefore, we realize the AAT prism with $\theta_{t1} = -90°$ and $\theta_{t2} = 54°$ for LI and RI, respectively.

Figure 4.58 shows the photograph of the designed AAT prism composed of the 31 phased unit cells, where the phased unit cell at the center of the prism contains sixteen meta-atoms, and the numbers of the meta-atoms in the phased unit cells on both sides decrease gradually. The parameters of the meta-atom are the same as those in Fig. 4.54a. Based on this, we design the AAT prism with $\alpha = 45°$ and $H = 25.6$ cm, and the other parameters are selected as $f = 6.66$ kHz, $m = 1$, $c = 343$ m/s, $\theta_{t1} = -90°$, and $\theta_{t2} = 54°$.

Based on Eqs. (4.1) and (4.4), we theoretically calculate the continuous phase distributions in the y_1 and y_2 directions of the AAT prism at 6.66 kHz, which are

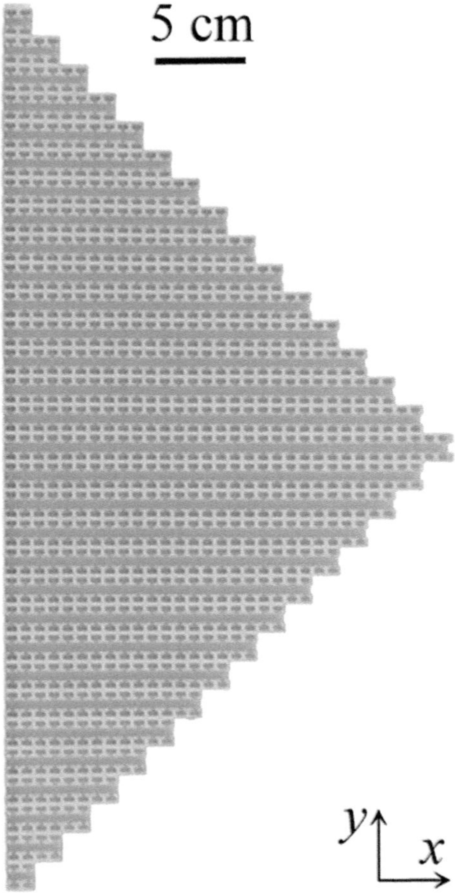

Fig. 4.58 Photograph of the AAT prism

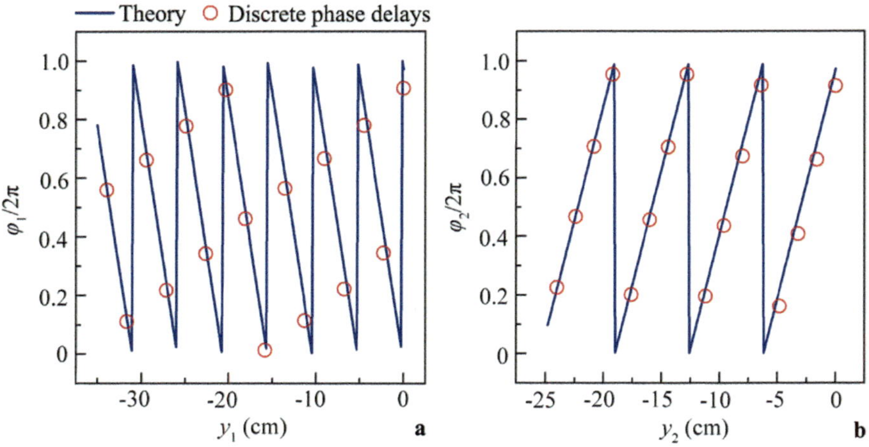

Fig. 4.59 Theoretical continuous phase distributions and sixteen discrete phase delays in the **a** y_1 and **b** y_2 directions for the lower half of the AAT prism at 6.66 kHz

shown in Fig. 4.59a, b, respectively. Additionally, we select sixteen discrete phase delays of the phased unit cells in the y_1 and y_2 directions for the lower half of the prism. Note that the phase distributions in the y_1 and y_2 directions of the AAT prism are different, and the discrete phase delays agree well with the theoretical ones in both directions.

Figure 4.60a, b show the simulated intensity distributions caused by the AAT prism for LI and RI, respectively. As shown in Fig. 4.60a, the transmitted sound energy is localized on both surfaces on the right side and cannot reach region I. However, for RI, the acoustic wave can transmit through the prism, and reach region III, showing an obvious AAT phenomenon. The simulated refraction angles for LI and RI match the theoretical values ($\theta_{t1} = -90°$ and $\theta_{t2} = 54°$, the white open arrows). Therefore, the designed prism with asymmetric phase distribution can achieve high-performance AAT.

Furthermore, we simulate the intensity distributions caused by the prism for LI and RI at three other selected frequencies within the working band of AAT, as shown in Fig. 4.61. As shown in Fig. 4.61a, the transmitted sound energy cannot pass through the prism at 6.2 kHz for LI. This is because the transmitted acoustic wave is converted into the evanescent surface wave. Additionally, as the incident frequency increases, the transmitted sound energy is not localized on the output surface for LI (Fig. 4.61c, e), where the refraction angles are within the range $-\pi/2 < \theta_{t1} < -\pi/4$. Therefore, the transmitted sound energy still cannot reach region I. For RI (Fig. 4.61b, d, and f), the acoustic wave can transmit through the prism and reach region III, and the refraction angle θ_{t2} increases gradually with the increase in the incident frequency. Therefore, by tuning the incident frequency, the AAT effect still exists, but the refraction angles for LI and RI change simultaneously. The refraction angles for LI and RI match well with the theoretical ones (the white open arrows) calculated by Eq. (2.41).

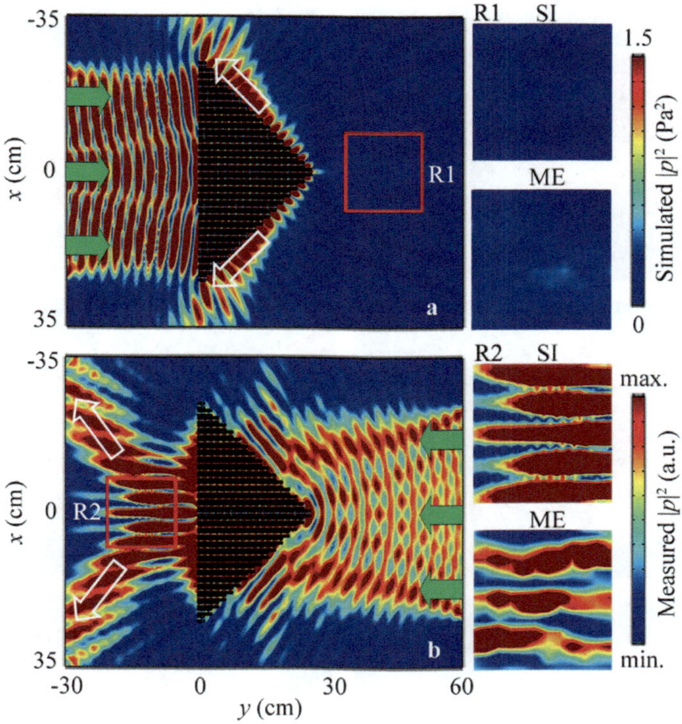

Fig. 4.60 Measured and simulated intensity distributions caused by the AAT prism for **a** LI and **b** RI at 6.66 kHz. The four insets at the right side show the measured and simulated intensity distributions in red open squares R1 and R2. The white open and the green arrows represent the theoretical refraction angles and the incident acoustic waves, respectively

4.5.2 Experimental Measurement

To verify the AAT performance, we measure the intensity distributions for LI and RI in red open squares R1 and R2 (with a size of 16 × 16 cm^2), where the experimental set-up is shown in Fig. 4.62. The experimental set-up and the sample are placed in a planar waveguide system composed of two parallel plates (dimension 2 m × 2 m × 1 cm). Wedge-shaped foams are placed at the boundaries of the planar waveguide to realize an anechoic environment. The incident plane wave is generated from a speaker array driven by the power amplifier. The pressure distribution is scanned by a microphone (Brüel & Kjær type-4961) moved by a set of 2D motorized linear stages (Newport: MIN300CC and ILS250CC). Another microphone is fixed towards the speaker array as a reference. The data is recorded by the Brüel & Kjær 3160-A-022 module and analyzed using the software of PULSE Labshop. As shown in the insets on the right side of Fig. 4.60, for both LI and RI, the measured and simulated field distributions agree well with each other, experimentally demonstrating the performance of the AAT prism.

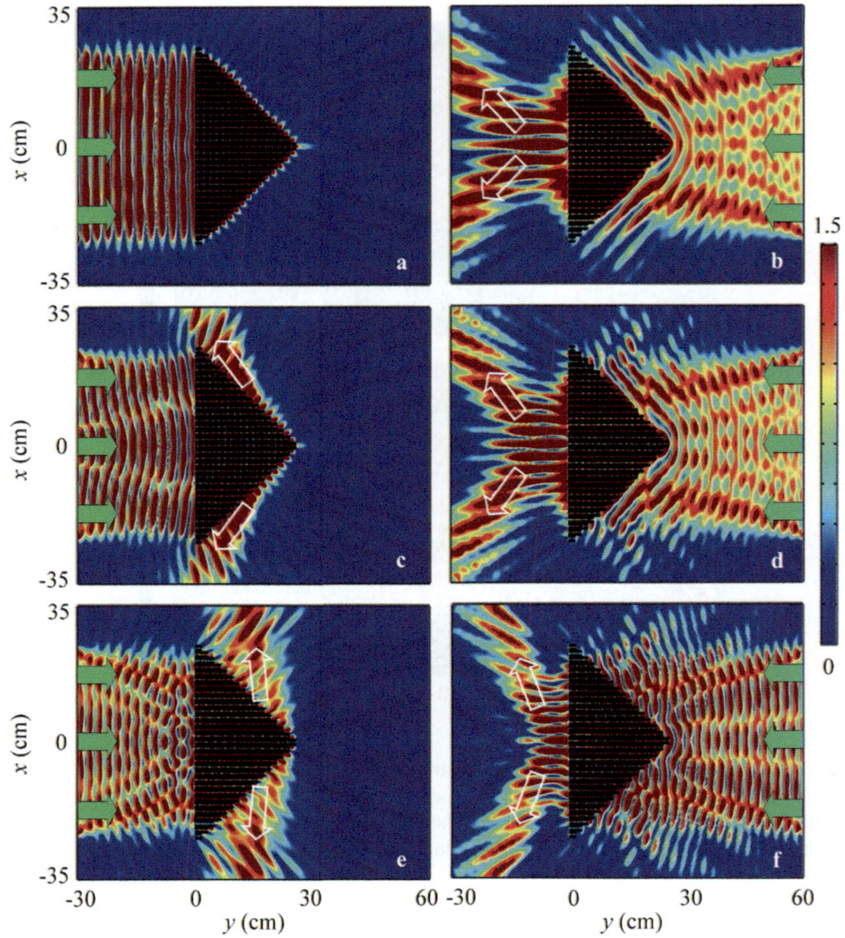

Fig. 4.61 Simulated intensity distributions caused by the AAT prism at **a**, **b** 6.2 kHz, **c**, **d** 6.7 kHz, and **e**, **f** 7.2 kHz for LI and RI

To present the bandwidth of the AAT prism, we simulate the transmittance spectra of the prism for LI and RI. Here, we define the transmittance as P/P_0, where P_0 and P are the integration of the acoustic intensity in the same regions without and with the prism, respectively, and the selected integration regions are regions I and III for LI and RI, respectively. As shown in Fig. 4.63a, the transmittance is lower than −8 dB for LI in the black shaded region, while the transmittance is greater than −4 dB for RI, where the transmittance spectra show a remarkable AAT behavior. Additionally, we experimentally measure the transmittance spectra for LI and RI, which agree with the simulated results in the black shaded region. However, there is a slight difference in the transmittance spectra between the two results. This is because the measured transmittance spectra are obtained by measuring and integrating the intensities in

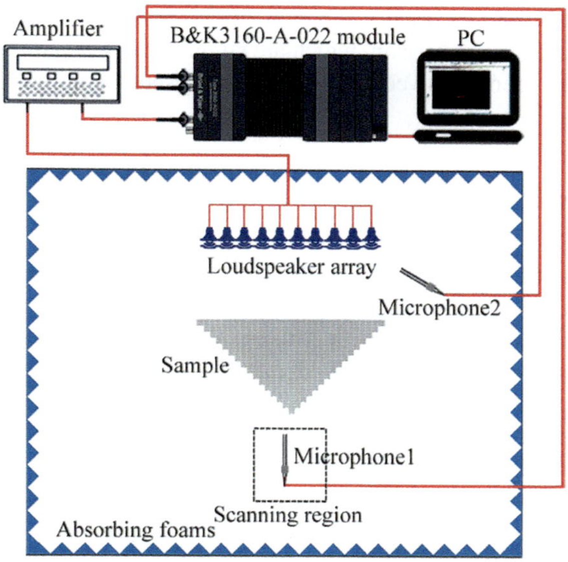

Fig. 4.62 Schematic of the experimental set-up for the AAT prism in Fig. 4.58

several scanning lines, and the corresponding intensity distributions are generally nonuniform.

To further evaluate the AAT performance, we provide the contrast ratios (R_c) of the AAT prism based on Eq. (3.1). As shown in Fig. 4.63b, the simulated R_c is greater than 0.7 in the black shaded region, indicating that the bandwidth of the AAT prism

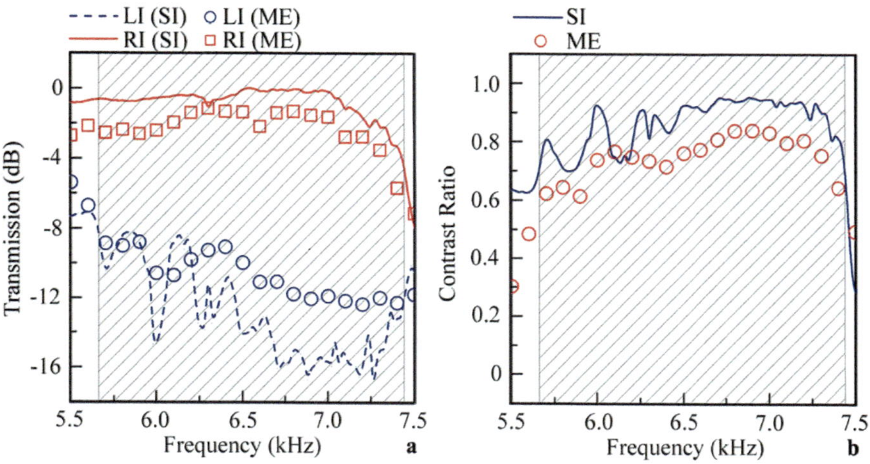

Fig. 4.63 **a** Measured and simulated transmittance spectra of the prism for LI and RI, and **b** the corresponding contrast ratios. The black shaded regions (5.66–7.44 kHz) represent the working band of AAT

is about 1.78 kHz, and its fractional bandwidth reaches about 0.26. The measured results are plotted for comparison, which are slightly lower than the simulated ones due to a small difference between the transmittance spectra in Fig. 4.63a.

4.5.3 Parameter Analysis

Next, we discuss the influences of the parameter w_2 of the meta-atom on the performance of the AAT prism. As shown in Fig. 4.64, we simulate the phase delays and transmittances of the meta-atoms with different values of w_1 at 6.66 kHz. Note that as w_1 increases, the phase delay gradually decreases, and the transmittances are greater than 0.8 in the range $0.8 \le w_1 \le 2.5$ mm. Therefore, by tuning the parameter w_1, the AAT performance can be controlled.

Figure 4.65 shows the intensity distributions caused by the prism with three different values of w_1 (the black points I, II and III in Fig. 4.64) for LI and RI. We can see that as w_1 increases, the AAT effect still exists. For LI (Fig. 4.65a, c, and e), as w_1 increases, the refraction angle θ_{t1} gradually decreases, but the transmitted sound energy cannot reach region I. In the case of RI (Fig. 4.65b, d, and f), the acoustic wave can transmit through the prism and reach region III, with the refraction angle θ_{t2} gradually decreasing as w_1 increases. The refraction angles agree well with the theoretical values. Therefore, the sound refraction for LI and RI can be controlled by the parameter w_1.

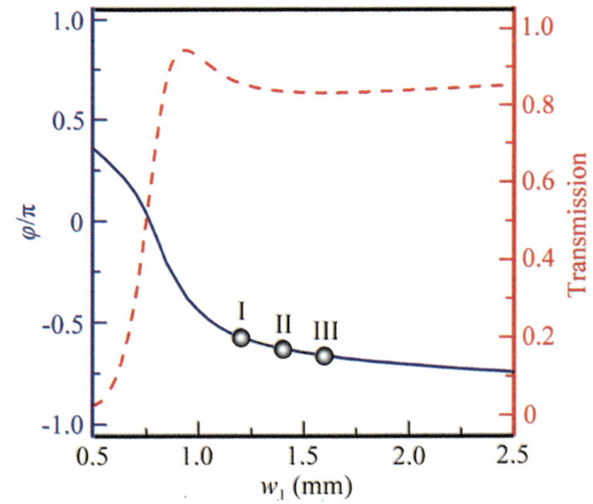

Fig. 4.64 Simulated phase delays (the blue solid line) and transmittances (the red dashed line) of the meta-atoms with different values of w_1 at 6.66 kHz. The black points I, II, and III correspond to $w_1 = 1.2$ mm, 1.4 mm, and 1.6 mm, respectively

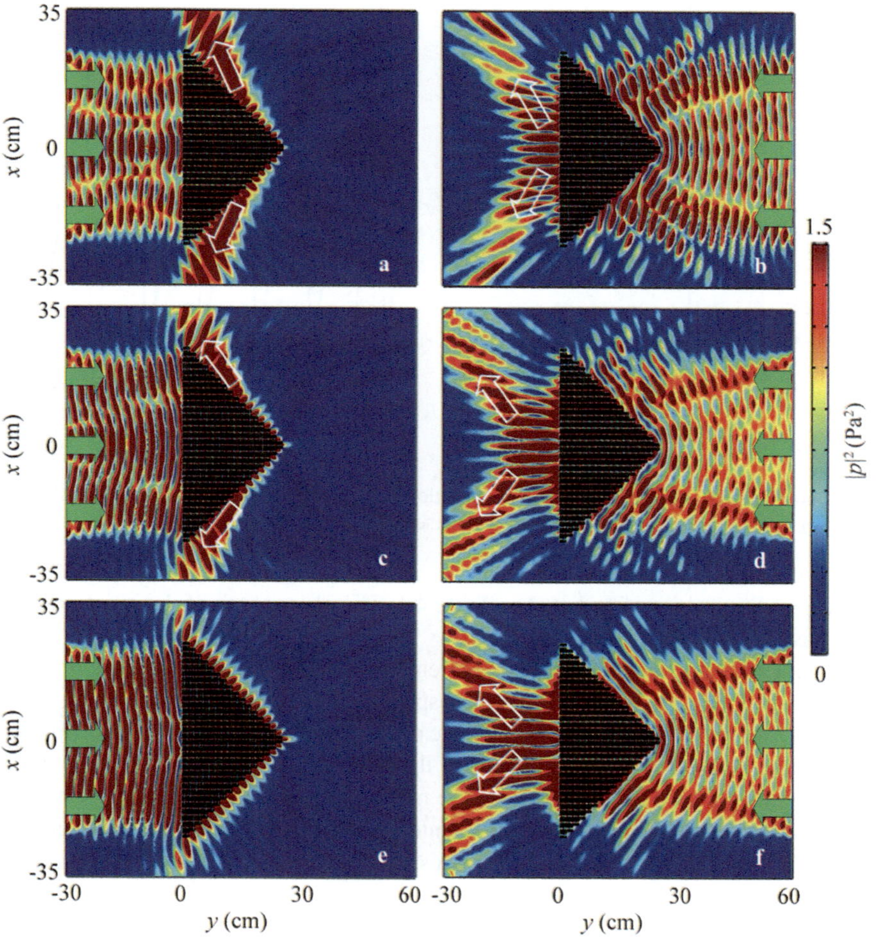

Fig. 4.65 Simulated intensity distributions caused by the AAT prism with **a**, **b** $w_1 = 1.2$ mm, **c**, **d** 1.4 mm, and **e**, **f** 1.6 mm for LI and TI at 6.66 kHz

4.5.4 *Design and Performance of Reversed AAT Prism*

Finally, we design a reversed AAT prism with characteristics opposite to those of the previously designed AAT prisms. In the reversed AAT prism, the acoustic wave can transmit through the prism for LI but can not pass through for RI. To achieve this performance, we select $d\varphi_2/kdx_2 = -1.1$, which converts the transmitted wave for RI into an evanescent wave with high performance [8].

Based on Eqs. (4.1), (4.4) and (4.5), we obtain $\theta_{t1} = -4°$ and $m = 0$, and calculate the phase distributions of the reversed AAT prism in the y_1 and y_2 directions, as shown in Fig. 4.66a, b, respectively. The sixteen red open circles are the discrete phase delays of the phased unit cells for the lower half of the reversed AAT prism, with the

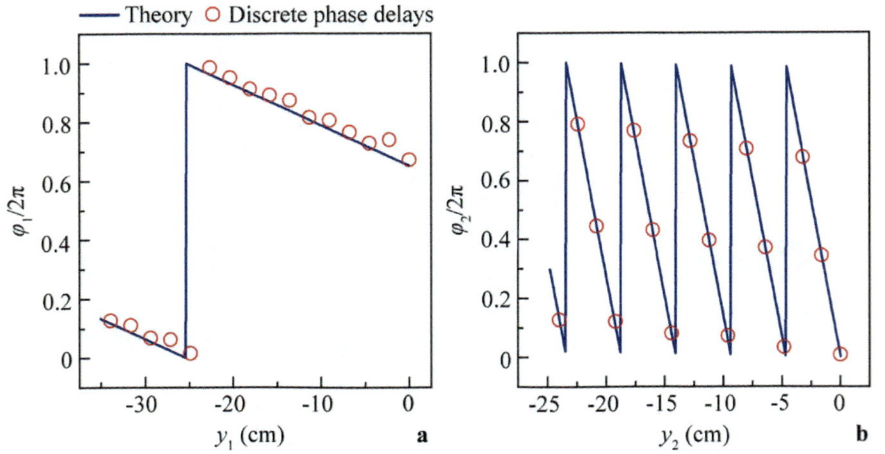

Fig. 4.66 Theoretical continuous phase distributions and sixteen discrete phase delays in the **a** y_1 and **b** y_2 directions for the lower half of the reversed AAT prism at 6.66 kHz

corresponding values of h_1 listed in Table 4.1. The photograph of the reversed AAT prism is shown in Fig. 4.67.

Figure 4.68a, b show the simulated intensity distributions caused by the reversed AAT prism for LI and RI at 6.66 kHz, respectively. Unlike the results in Fig. 4.60, the acoustic wave can transmit through the prism and reach region I for LI, while the transmitted sound energy is localized on the left surface of the prism, and can not reach region III for RI. Furthermore, the measured acoustic intensity distributions in red open squares R3 and R4 are presented in the right region, and the measured distributions agree with the simulated results. Therefore, the AAT effect can be switched by tuning the phase distributions of the prism.

Table 4.1 Selected sixteen discrete phase delays of the phased unit cells for the lower half of the reversed AAT prism with $d\varphi_2/kdx_2 = -1.1$

n	1	2	3	4	5	6	7	8	9	10	11	12	13	14	15	16
h_1 (mm)	10.0	4.6	7.2	5.2	6.8	8.9	6.7	8.1	6.6	7.6	8.9	5.9	8.3	7.2	4.9	8.8
Phase delay (rad)	0.79	4.97	2.79	0.75	4.82	2.71	0.49	4.61	2.48	0.45	4.43	2.32	0.19	4.26	2.16	0.03

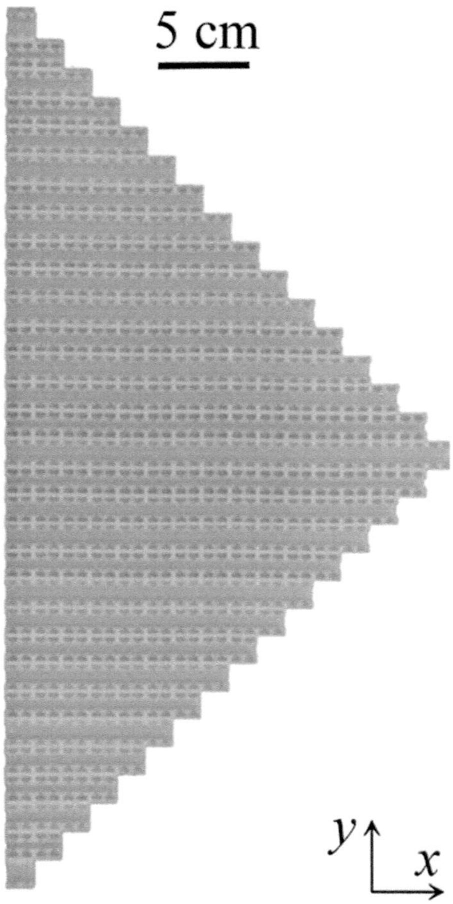

Fig. 4.67 Photograph of the reversed AAT prism

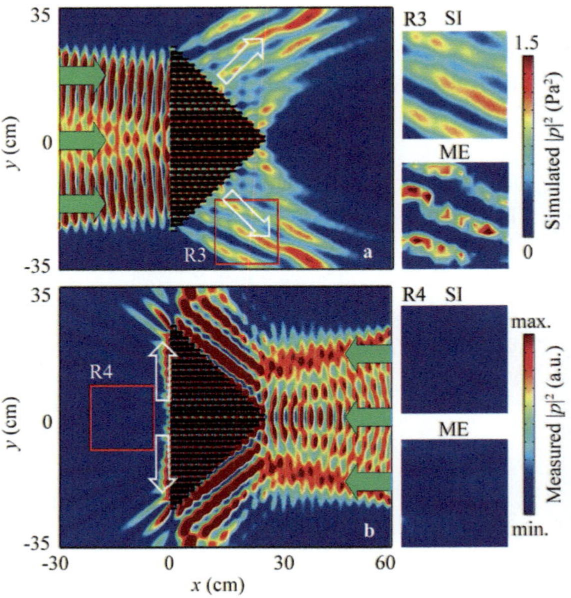

Fig. 4.68 Measured and simulated intensity distributions caused by the reversed AAT prism for **a** LI and **b** RI at 6.66 kHz. The insets on the right side show the measured and simulated intensity distributions in red open squares R3 and R4

4.6 Conclusion

This chapter explores the design and experimental validation of airborne AAT systems using various metamaterial structures, including ultrathin metasurfaces, triangle cavities, mode-conversion meta-atoms, and phased unit cells. The AAT effect is achieved through asymmetric sound manipulation, such as multiple reflections and ABRs, or by exploiting the cutoff frequency of higher-order waves. The chapter presents a broadband AAT tunnel/window, a multi-channel AAT lens, and a tunable AAT prism, all demonstrating high performance and broad bandwidths. The physical mechanisms, including the role of phase gradients and mode conversion, are discussed in detail. Experimental measurements confirm the simulated results, and the systems show robustness against different sound sources. The chapter concludes with the potential applications of these AAT systems in architectural acoustics, noise control, and acoustic functional components.

References

1. Ge Y, Sun HX, Yuan SQ, Lai Y (2018) Broadband unidirectional and omnidirectional bidirectional acoustic insulation through an open window structure with a metasurface of ultrathin hooklike meta-atoms. Appl Phys Lett 112(24):243502. https://doi.org/10.1063/1.5025812
2. Ge Y, Sun HX, Yuan SQ, Xia JP (2017) Asymmetric acoustic transmission in an open channel based on multiple scattering mechanism. Appl Phys A 123(5):328. https://doi.org/10.1007/s00339-017-0956-8

3. Chen JH, Qian J, Guan YJ, Ge Y, Yuan SQ, Sun HX, Lai Y, Liu XJ (2021) Broadband bidirectional and multi-channel unidirectional acoustic insulation by mode-conversion phased units. Front Mater 8:766491. https://doi.org/10.3389/fmats.2021.766491
4. Wang Y, Xia JP, Sun HX, Yuan SQ, Ge Y, Si QR, Guan YJ, Liu XJ (2019) Multifunctional asymmetric sound manipulations by a passive phased array prism. Phys Rev Appl 12(2):024033. https://doi.org/10.1103/PhysRevApplied.12.024033
5. Zhu YF, Zou XY, Liang B, Cheng JC (2015) Acoustic one-way open tunnel by using metasurface. Appl Phys Lett 107(11):113501. https://doi.org/10.1063/1.4930300
6. Zhu YF, Gu ZM, Liang B, Yang J, Yang J, Yin LL, Cheng JC (2016) Asymmetric sound transmission in a passive non-blocking structure with multiple ports. Appl Phys Lett 109(10):103504. https://doi.org/10.1063/1.4962435
7. Zhang HL, Zhu YF, Liang B, Yang J, Cheng JC (2017) Sound insulation in a hollow pipe with subwavelength thickness. Sci Rep 7:44106. https://doi.org/10.1038/srep44106
8. Yu NF, Genevet P, Kats MA, Aieta F, Tetienne JP, Capasso F, Gaburro Z (2011) Light propagation with phase discontinuities: generalized laws of reflection and refraction. Science 334(6054):333–337. https://doi.org/10.1126/science.1210713

Chapter 5
Acoustic Focusing in Ring-Shaped Metamaterials

5.1 Introduction

AF is a special phenomenon where sound energy is focused onto a specific region. Traditionally, AF can be achieved using curved lenses or reflectors, which are difficult to integrate. However, with the development of acoustic metamaterials, it has become feasible to design high-performance AF lenses. In recent years, ring-shaped metamaterials have attracted significant attention due to their rich eigenmodes and resonant cavity structures, which can be used to achieve perfect AF with high intensity using weak sound energy. This occurs because when the incident frequency of the acoustic wave matches the eigenfrequency of the ring-shaped structure, the eigenmodes of the ring are excited, forming resonance inside it. As a result, the sound energy excited by structural resonance propagates along a specific path within the ring, creating a perfect focus with high intensity. This can enable long-distance and high-precision underwater detection for sonar systems. However, achieving the perfect AF effect in ring-shaped metamaterials remains a challenge.

In this chapter, we design four types of AF systems using ring-shaped metamaterials and discuss their potential applications. First, we propose an AF system composed of a brass circular ring immersed in water [1]. The sound energy is focused on a perfect point at the center of the ring, resulting from the eigenmode of the ring structure. Additionally, we discuss the potential applications of the AF effect in black box detectors in the sea and medical ultrasound treatment. Second, we theoretically study and experimentally demonstrate a multi-focus AF system consisting of a brass circular ring immersed in water [2]. The multi-focus AF, with one or more focal points, can be achieved at the center of the ring due to its rich Mie-resonance modes. Beyond that, we discuss the potential application of the multi-focus AF effect in acoustic encryption communication in detail. Third, we propose an AF system composed of dual-layer circular rings immersed in air [3]. The sound energy is focused on a perfect point without diffraction, resulting from the eigenmode of the dual-layer structure. We investigate the potential applications of the AF

H. Sun et al., *Acoustic Metamaterials*, Microfluidics and Nanophotonics: Science and Engineering 2, https://doi.org/10.1007/978-981-96-8524-0_5

effect in acoustic beam splitters and directional transmitters with energy enhancement. Finally, we design a ring-shaped AF system consisting of eight multiple-cavity unit cells with rich artificial Mie-resonances [4]. The AF effect arises from the compound monopole Mie resonance (MMR) of the system based on out-phase resonance. The maximum intensity of the focus can reach 9000 Pa^2, and the AF system exhibits strong robustness against the incident direction and the shape of the ring-shaped structure. The designed ring-shaped AF systems have great potential in underwater signal detection, medical ultrasound treatment, acoustic communication, and acoustic energy harvesting.

5.2 Acoustic Focusing in Metal Ring-Shaped Structures

5.2.1 Design and Performance of AF System

As schematically shown in Fig. 5.1, the designed AF system consists of a brass circular ring immersed in water [1], where the parameters R and r represent the outer and inside radii of the circular ring, respectively. Throughout this work, the COMSOL Multiphysics software is utilized to simulate the characteristics of AF. In the model, the acoustic-structure boundary is applied to the outer and inside boundaries of the circular ring, and the material parameters are as follows: density $\rho = 8400$ kg/m^3, longitudinal wave velocity $c_l = 4400$ m/s and transversal wave velocity $c_s = 2200$ m/s for brass; $\rho = 998$ kg/m^3 and $c = 1483$ m/s for water at 296 K.

Next, we simulate the pressure amplitude distributions without and with the circular ring at 181.7 kHz. The parameters R and r are selected as 30.0 and 25.5 mm, respectively, and a cylindrical sound source is located at (−40 mm, 0). As shown in Fig. 5.2a, the sound energy attenuates rapidly along the propagation distance.

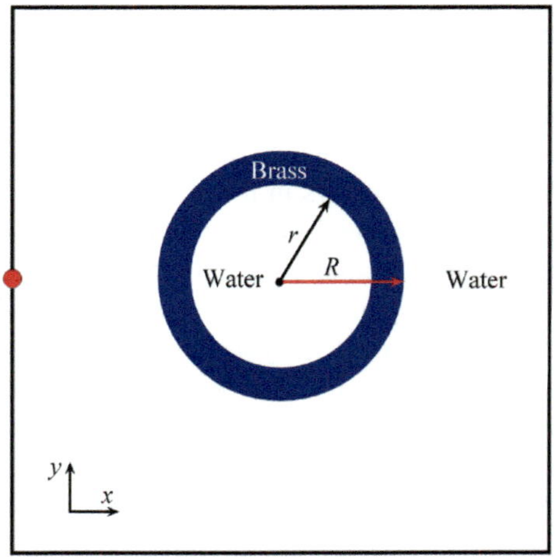

Fig. 5.1 Schematic of the AF system

However, with the circular ring (Fig. 5.2b), the sound energy is focused at a perfect point at the center of the ring structure. Additionally, the AF effect is also achieved at 152.7 and 123.7 kHz, as shown in Fig. 5.2c, d, respectively. By comparing the results in Fig. 5.2b–d, the focus becomes smaller as the frequency increases, and the difference between the two adjacent frequencies is constant (29.0 kHz). Therefore, we can predict the frequency of AF for the circular ring.

Next, we simulate the deformation of the displacement distribution in the circular ring at 181.7 kHz. As shown in Fig. 5.3, the eigenmode is excited in the circular ring at 181.7 kHz, and the vibration directions of the eigenmode agree with the radial direction of the circular ring. It is therefore deduced that the propagation directions of the acoustic waves are restructured by the eigenmode, and all sound energy is finally focused on a point at the center of the ring structure. This is fundamentally different from previous AF lenses based on negative refraction [5–8].

Figure 5.4 show the phase distributions without and with the circular ring at 181.7 kHz, corresponding to Fig. 5.2a, b, respectively. As shown in Fig. 5.4a, the excited phase distributions are a series of undisturbed curved lines without the circular ring. However, in Fig. 5.4b, the phase distributions are converted into a series of concentric circles with the same phase values due to the excitation of the eigenmode, and are eventually focused on a point at the centre of the circular ring.

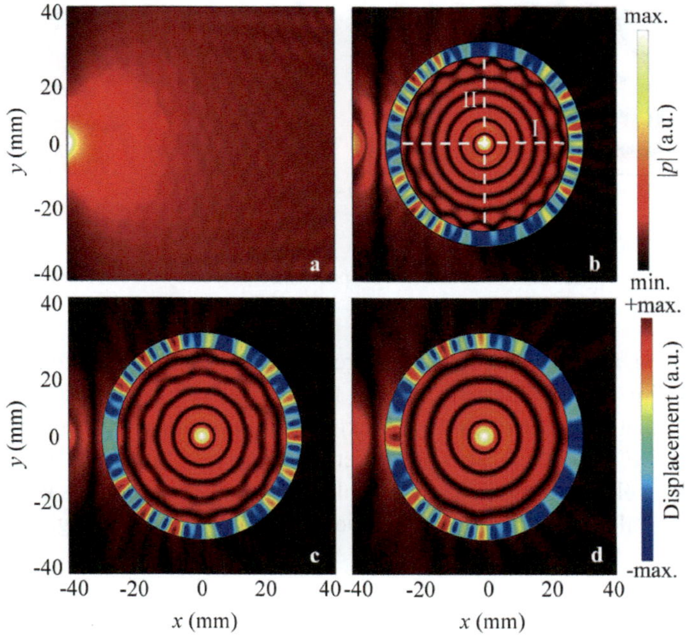

Fig. 5.2 **a** Simulated pressure amplitude distribution in free space at 181.7 kHz, and pressure amplitude and displacement distributions caused by the circular ring under the excitation of a cylindrical sound source at **b** 181.7 kHz, **c** 152.7 kHz, and **d** 123.7 kHz

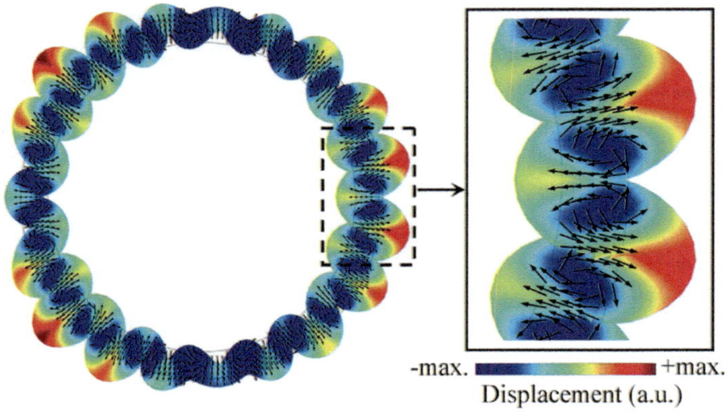

Fig. 5.3 Simulated deformation of displacement distribution caused by the circular ring at 181.7 kHz. The black arrows represent the vibration directions of the displacement

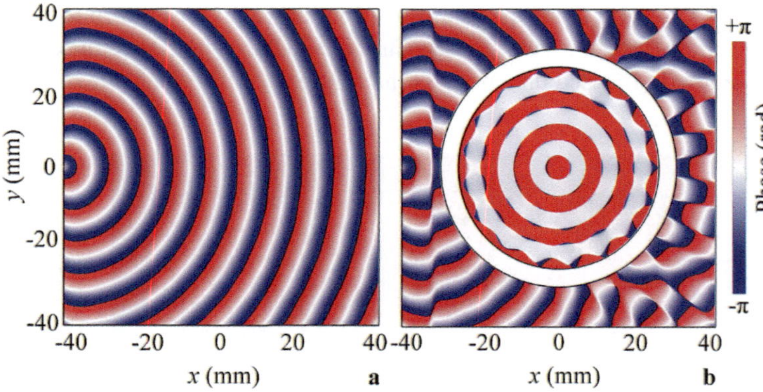

Fig. 5.4 Simulated phase distributions **a** in free space and **b** caused by the circular ring at 181.7 kHz

Figure 5.5a, b present the simulated pressure amplitude distributions along lines I and II in Fig. 5.2b, with those without the circular ring plotted for comparison. We can see that high-performance AF exists in both x and y directions. Compared to previous AF lenses, the focus at the center of the circular ring is a perfect point rather than an elliptical spot, and the circular ring-shaped structure is much simpler. This indicates that perfect AF can be achieved using the circular ring.

5.2.2 Parameter Analysis

To investigate the relationship between the size of the circular ring and the performance of AF, we simulate the pressure amplitude distributions with the circular ring

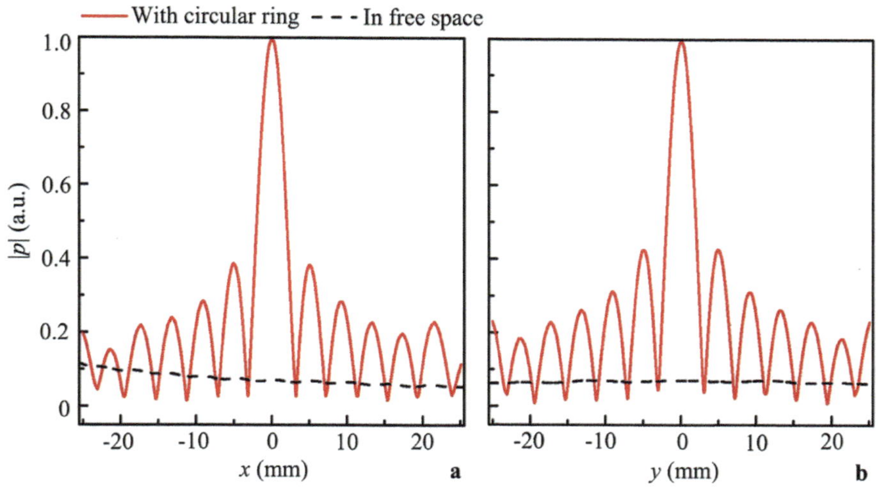

Fig. 5.5 Simulated pressure amplitude distributions along lines **a** I and **b** II in Fig. 5.2b

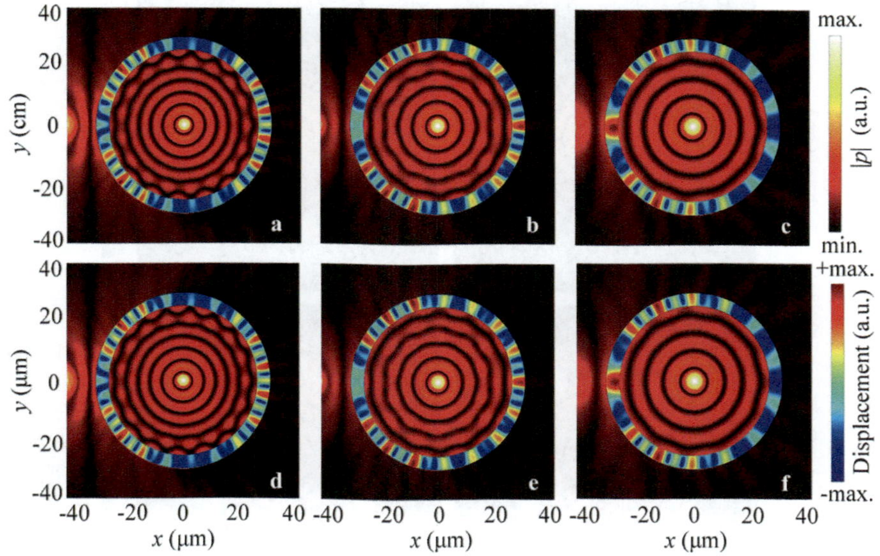

Fig. 5.6 Simulated pressure amplitude and displacement distributions caused by the circular rings with $R = 30.0$ cm and $r = 25.5$ cm at **a** 18.17 kHz, **b** 15.27 kHz, and **c** 12.37 kHz; and with $R = 30.0$ μm and $r = 25.5$ μm at **d** 181.7 MHz, **e** 152.7 MHz, and **f** 123.7 MHz

by scaling both parameters R and r by a factor of n, while keeping the material parameters constant. Compared to Fig. 5.2b–d, the pressure amplitude and displacement distributions remain almost the same, and the frequencies of AF are reduced by a factor of 10 in Fig. 5.6a–c, as the size of the circular ring structure is magnified by 10 times. Additionally, compared to Fig. 5.6a–c, the size of the circular ring is reduced by 10^4 times, and the frequencies of AF are magnified by 10^4 times in Fig. 5.6d–f, indicating that the frequency of AF is inversely proportional to the size of the circular ring.

5.2.3 Potential Application

The proposed ring-shaped structure may have substantial practical significance in detecting black boxes of crashed planes in the sea. It is known that the traditional detection systems are affected greatly by the temperature (T), the salinity (S), and the depth (h) in the sea. This is because the density (ρ) [9] and acoustic velocity (c) [10] of sea water are determined by these parameters.

Here, we design a black box detector using the brass circular ring in different environments. Figure 5.7 shows the excited AF effect in the circular ring in different

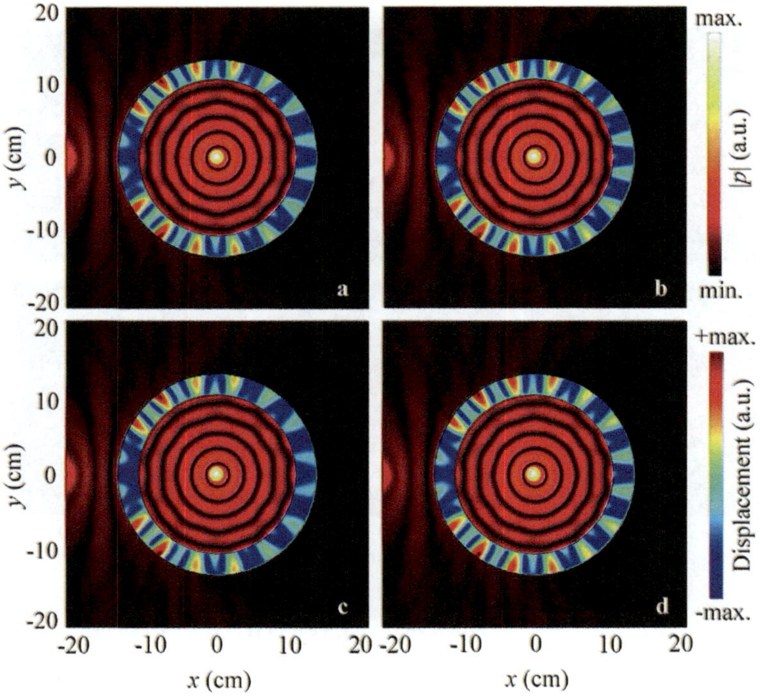

Fig. 5.7 Simulated pressure amplitude and displacement distributions caused by the circular ring in different environments with **a** $T = 293$ K, $S = 32$ g/kg, $h = 1000$ m, $\rho = 1023.7$ kg/m^3, and $c = 1530.2$ m/s; **b** $T = 302$ K, $S = 32$ g/kg, $h = 1000$ m, $\rho = 1021.7$ kg/m^3, and $c = 1550.4$ m/s; **c** $T = 293$ K, $S = 25$ g/kg, $h = 1000$ m, $\rho = 1018.4$ kg/m^3, and $c = 1510.1$ m/s; **d** $T = 293$ K, $S = 32$ g/kg, $h = 3000$ m, $\rho = 1023.7$ kg/m^3, and $c = 1564.3$ m/s

environments at 37.50 kHz, which is the special incident frequency of the black box. In the simulations, the parameters ρ and c of the sea water outside the ring structure vary due to changes in T, S, and h, while the parameters of the seawater inside the ring remain the same as in Fig. 5.2. The parameters R and r are selected as 13.17 and 10.37 cm, respectively. As shown in Fig. 5.7b–d, despite changes in T, S and h, the AF effect is realized at the same position and frequency. This demonstrates that the excited AF caused by the circular ring is unaffected by the external environment and can be used as a black box detector in complex and variable conditions.

Another potential application of the AF effect is in medical ultrasound [11, 12]. It is known that the high-intensity focused ultrasound (HIFU) technique has been applied in several types of tumor treatments, but healthy human tissues and cells may be damaged during HIFU treatment. Here, we propose a method for selectively killing cancerous tissues and cells based on the fact that the AF effect is closely related to the size and shape of ring-shaped structures. To verify the feasibility and effectiveness, we design circular, square, and elliptical ring structures immersed in water, as shown in Fig. 5.8. The acoustic wave is selectively focused in the three types of ring-shaped structures at 209.7, 95.1, and 105.9 MHz, as shown in Fig. 5.9a–c, respectively. Meanwhile, the other two ring-shaped structures are largely unaffected, indicating that the AF effect caused by ring-shaped structures has great potential in medical ultrasound treatment.

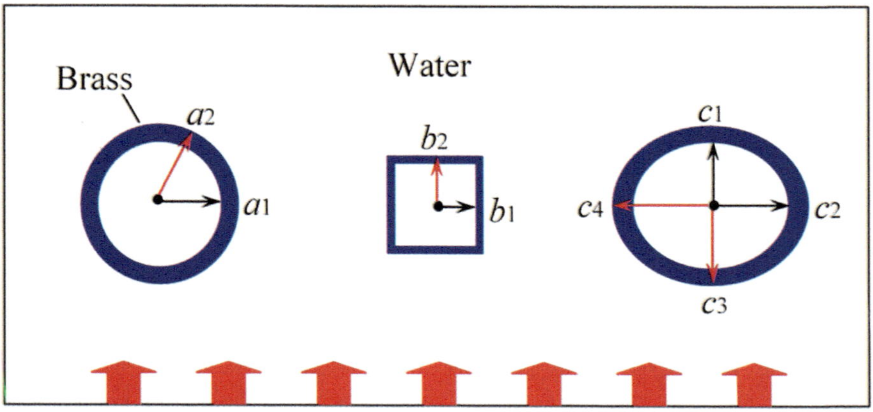

Fig. 5.8 Schematic of three ring-shaped structures ($a_1 = 15.0$ μm and $a_2 = 20.0$ μm for the circular ring; $b_1 = 10.0$ μm and $b_2 = 12.5$ μm for the square ring; $c_1 = 15.0$ μm, $c_2 = 20.0$ μm, $c_3 = 20.0$ μm, and $c_4 = 25.0$ μm for the elliptical ring)

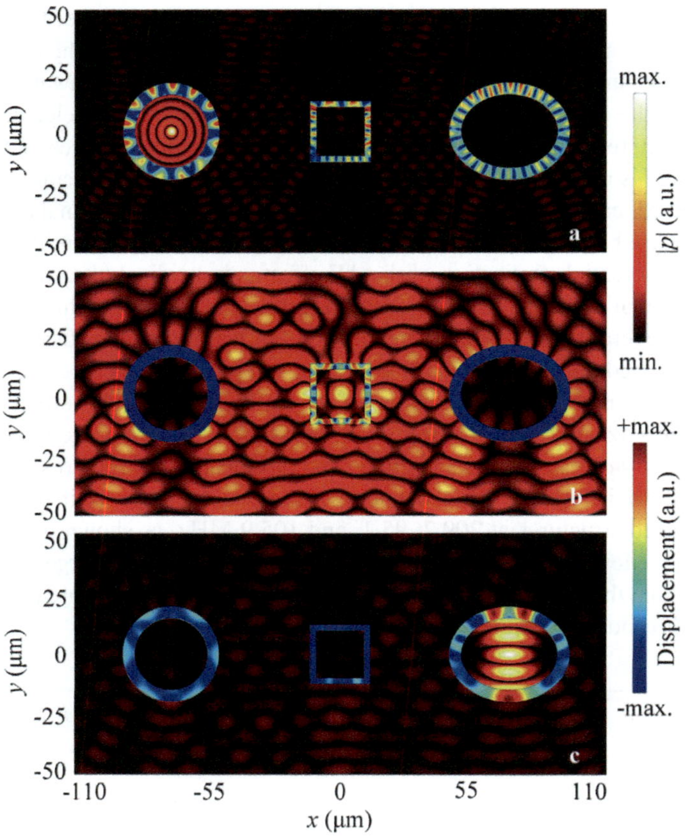

Fig. 5.9 Simulated pressure amplitude and displacement distributions caused by the ring-shaped structures at **a** 209.7 MHz, **b** 95.1 MHz, and **c** 105.9 MHz

5.3 Multi-focus Acoustic Focusing in Metal Ring-Shaped Structures

5.3.1 Design and Performance of Multi-focus AF System

As schematically shown in Fig. 5.10, the multi-focus AF lens consists of a brass circular ring immersed in water [2], with the incident plane wave placed at the left boundary. The inner ($r = a$) and outer ($R = b$) radii of the circular ring are 10 and 15 mm, respectively. Throughout this work, we study the characteristics of multi-focus AF numerically, theoretically, and experimentally. The COMSOL Multiphysics software is used to simulate the multi-focus AF effect. The material parameters are as follows: density $\rho = 8400$ kg/m^3 and sound speed $c = 4400$ m/s for brass; $\rho = 998$ kg/m^3 and $c = 1483$ m/s for water at 296 K.

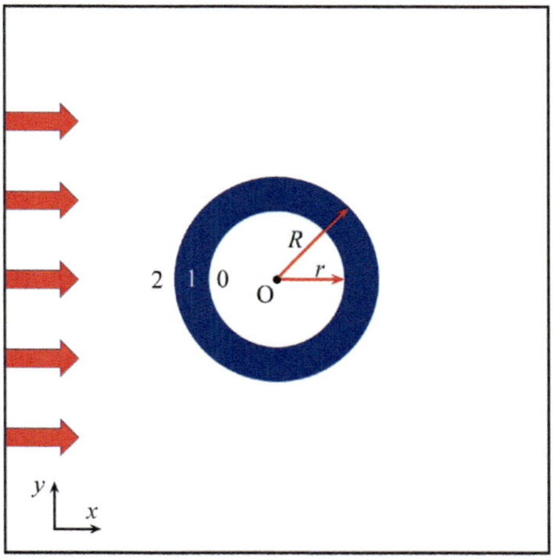

Fig. 5.10 Schematic of the multi-focus AF lens

Next, we theoretically analyze the AF effect based on the acoustic scattering theory. Here, we consider the incident plane wave as $p_{in} = P_0 e^{i(k_0 r \cos\varphi - \omega t)}$, where P_0 is the initial amplitude, i indicates the imaginary component, k_0 is the wave number, and ω is the angular frequency. The pressure can be expressed in terms of a spherical Bessel function $J_{\text{n}}(kr)$ and a Hankel function of the first kind $H_{\text{n}}^{(1)}(kr)$. Thus, the incident plane wave in the cylindrical coordinate can be written as [13]:

$$p_{in} = P_0 \sum_{n=-\infty}^{+\infty} i^n J_n(k_0 r) e^{in\theta}, \tag{5.1}$$

and the acoustic pressures p_0, p_1, and p_2 in regions 0, 1, and 2 can be expressed as [13]:

$$p_0 = P_0 \sum_{n=-\infty}^{+\infty} \left[i^n J_n(k_1 r) + B_{0n} H_n^{(1)}(k_0 r) \right] \mathrm{e}^{in\theta}, \tag{5.2}$$

$$p_1 = P_0 \sum_{n=-\infty}^{+\infty} \left[A_{1n} J_n(k_1 r) + B_{1n} H_n^{(1)}(k_1 r) \right] \mathrm{e}^{in\theta}, \tag{5.3}$$

$$p_2 = P_0 \sum_{n=-\infty}^{+\infty} A_{2n} J_n(k_2 r) \mathrm{e}^{in\theta}, \tag{5.4}$$

where k_0, k_1, and k_2 are the wave numbers, and A_{1n}, A_{2n}, B_{0n}, and B_{1n} are the expansion coefficients corresponding to the spherical Bessel and Hankel functions in regions 0, 1, and 2 (Fig. 5.10), respectively.

The radial velocities v_{0r}, v_{1r}, and v_{2r} in regions 0, 1, and 2 are obtained by the functional relation $v_r = (i/\omega\rho)(\partial p/\partial r)$ as follows:

$$v_{0r} = P_0 \sum_{n=-\infty}^{+\infty} \frac{i^n}{\rho_0 c_0} [i^n J_n'(k_0 r) + B_{0n} H_n^{(1)\prime}(k_0 r)] e^{in\theta}, \tag{5.5}$$

$$v_{1r} = P_0 \sum_{n=-\infty}^{+\infty} \frac{i^n}{\rho_1 c_1} \left[A_{1n} J_n'(k_1 r) + B_{1n} H_n^{(1)\prime}(k_1 r)\right] e^{in\theta}, \tag{5.6}$$

$$v_{2r} = P_0 \sum_{n=-\infty}^{+\infty} \frac{i^n}{\rho_2 c_2} A_{2n} J_n'(k_2 r) e^{in\theta}, \tag{5.7}$$

where ρ_0, ρ_1, and ρ_2 are the densities, and c_0, c_1, and c_2 are the sound velocities in regions 0, 1, and 2, respectively. By using the orthogonality of the function $\exp(in\theta)$ and the continuous scalar acoustic pressure and radial velocity at the interfaces $r = a$ and b, we obtain:

$$\begin{bmatrix} H_n^{(1)}(k_0 b) & -J_n(k_1 b) & -H_n^{(1)}(k_1 b) & 0 \\ 0 & J_n(k_1 a) & H_n^{(1)}(k_1 a) & J_n(k_2 a) \\ \frac{H_n^{(1)\prime}(k_0 b)}{\rho_0 c_0} & -\frac{J_n'(k_1 b)}{\rho_1 c_1} & -\frac{H_n^{(1)\prime}(k_1 b)}{\rho_1 c_1} & 0 \\ 0 & -\frac{J_n'(k_1 a)}{\rho_1 c_1} & -\frac{H_n^{(1)\prime}(k_1 a)}{\rho_1 c_1} & -\frac{J_n'(k_2 a)}{\rho_2 c_2} \end{bmatrix} \begin{bmatrix} B_{0n} \\ A_{1n} \\ B_{1n} \\ A_{2n} \end{bmatrix} = \begin{bmatrix} -i^n J_n(k_0 b) \\ 0 \\ -\frac{i^n J_n'(k_0 b)}{\rho_0 c_0} \\ 0 \end{bmatrix}. \tag{5.8}$$

Here, we rewrite Eq. (5.8) as $\mathbf{A}\boldsymbol{x} = \mathbf{b}$, and theoretically calculate the eigenfrequencies of the Mie resonance modes by solving the eigenvalue equation $\det(\mathbf{A}) = 0$. In the equation $\det(\mathbf{A}) = 0$, the zero-order ($n = 0$), first-order ($n = 1$), second-order ($n = 2$) and third-order ($n = 3$) Bessel and Hankel functions correspond to the monopole, dipole, quadrupole, and octupole Mie resonance modes, respectively. As shown in Fig. 5.11a–d, we theoretically calculate the monopole, dipole, quadrupole, and octupole Mie resonance modes in the brass circular ring at 538.36, 574.55, 610.04, and 644.90 kHz, respectively, where one, two, four, and six focuses are obtained at the center of the brass circular ring.

Moreover, we theoretically calculate the pressure amplitude distributions caused by the circular ring under the excitation of the incident plane wave. The coefficients B_{0n}, A_{1n}, B_{1n} and A_{2n} of the Bessel and Hankel functions at each eigenfrequency can be obtained by solving Eq. (5.8). Figure 5.11e–h show the theoretical pressure distributions at 538.36, 574.55, 610.04, and 644.90 kHz, respectively. We can see that the incident waves are focused on one, two, four, and six focal points at the center of the circular ring, consistent with the results in Fig. 5.11a–d. Therefore, we deduce that the multi-focus AF effect arises from the Mie resonance modes of the circular ring. The simulated pressure distributions caused by the circular ring are shown in Fig. 5.11i–l, corresponding to Fig. 5.11e–h, respectively. The simulated results agree well with the theoretical ones.

In addition to the sixth eigenfrequencies of the Mie resonance modes, we find that each Mie-resonance mode also exists at a series of other eigenfrequencies. As shown in Fig. 5.12, we theoretically calculate and numerically simulate the pressure distributions at the fourth and second eigenfrequencies of the monopole and dipole

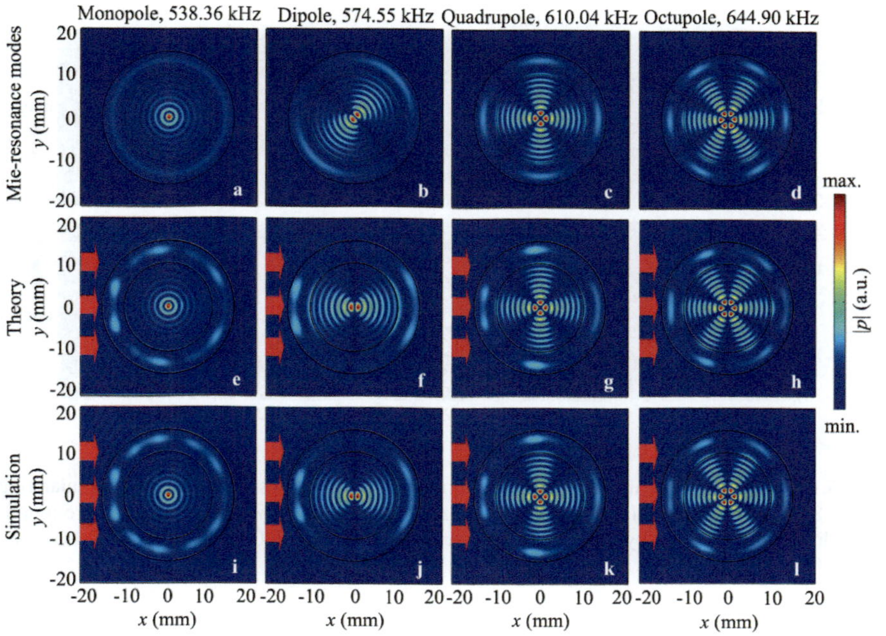

Fig. 5.11 Theoretical pressure amplitude eigenfunctions of the circular ring for the **a** monopole, **b** dipole, **c** quadrupole, and **d** octupole Mie resonance modes. Theoretically calculated pressure amplitude distributions caused by the circular ring at **e** 538.36 kHz, **f** 574.55 kHz, **g** 610.04 kHz, and **h** 644.90 kHz. Simulated pressure amplitude distributions caused by the circular ring at **i** 538.36 kHz, **j** 574.55 kHz, **k** 610.04 kHz, and **l** 644.90 kHz. The red arrows represent the incident plane wave

modes. Note that the acoustic waves are also focused on one and two focuses at the center of the circular ring, and the eigenfrequencies and the pressure distributions in the theoretical results agree with those in the simulated results. Therefore, we deduce that the multi-focus AF effect can be excited at a series of eigenfrequencies.

5.3.2 Experimental Measurement

To verify the AF effect in the circular ring, we experimentally measure the pressure distributions by a Schlieren imaging system [14], which is advantageous for detecting acoustic field distributions in a limited space. Figure 5.13a shows a photograph of the sample, which is a brass cylinder with a length of 50.0 mm, an inner radius of 9.9 mm and an outer radius of 15.0 mm. Figure 5.13b presents the schematic of the Schlieren imaging system, where the brass cylinder is placed between an ultrasonic transducer (500 kHz central frequency and 1.0 inch diameter) and an absorber immersed in a water tank. The system consists of a He–Ne Laser, convex lenses L_0 ($f_0 = 10$ mm), L_1 ($f_1 = 200$ mm), L_2 ($f_2 = 500$ mm), and

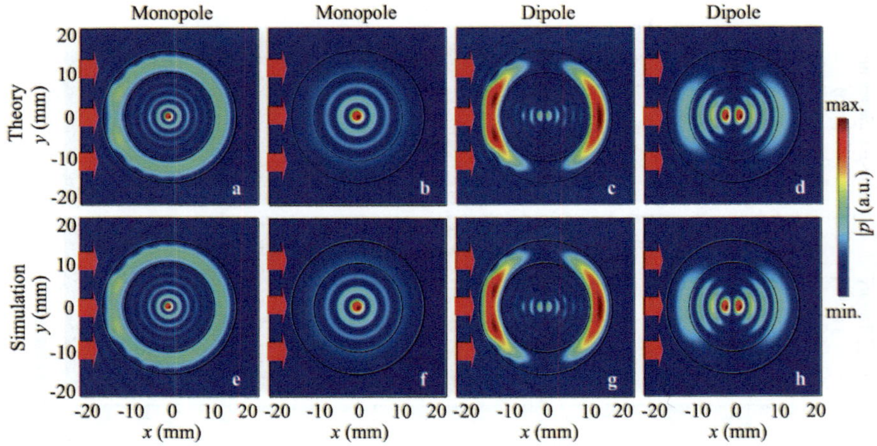

Fig. 5.12 Theoretically calculated pressure amplitude distributions caused by the circular ring at **a** 386.46 kHz, **b** 240.11 kHz, **c** 419.92 kHz, and **d** 276.06 kHz. Simulated pressure amplitude distributions caused by the circular ring at **e** 386.46 kHz, **f** 240.11 kHz, **g** 419.92 kHz, and **h** 276.06 kHz

L_3 ($f_3 = 300$ mm), and an Intensification Charge-Coupled Device (ICCD) camera. The system has three focal planes: the object plane, the Schlieren filtering plane, and the image plane. In the measurement, the laser beam with a wavelength of 632.8 nm is converted into a parallel laser beam through lenses L_0 and L_1, and then incident on the object plane. The parallel laser beam is modulated by the pressure distribution inside the cylinder. The lens L_2 is the transform lens, and an aperture is placed on the Schlieren filtering plane to avoid zero-order diffracted light. The lens L_3 is the imaging lens, and the ICCD camera captures the sound images generated from the brass cylinder.

Figure 5.14 shows the pressure distributions of the seventh and eighth monopole modes, where the parameters in the theoretical calculations are the same as those of the sample. The acoustic pressure distribution inside the brass cylinder is clearly

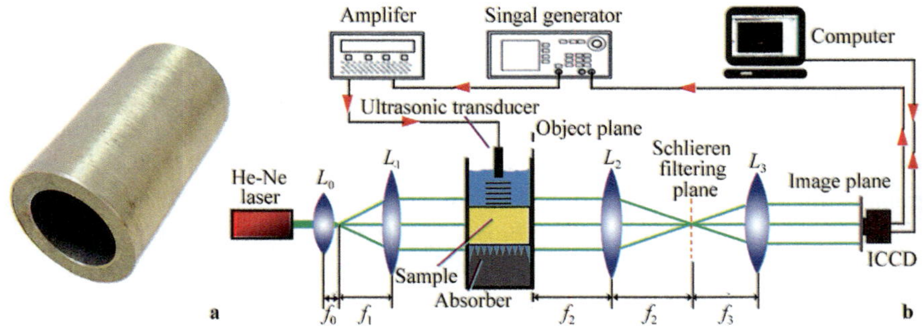

Fig. 5.13 **a** Photograph of the sample and **b** schematic of the experimental set-up

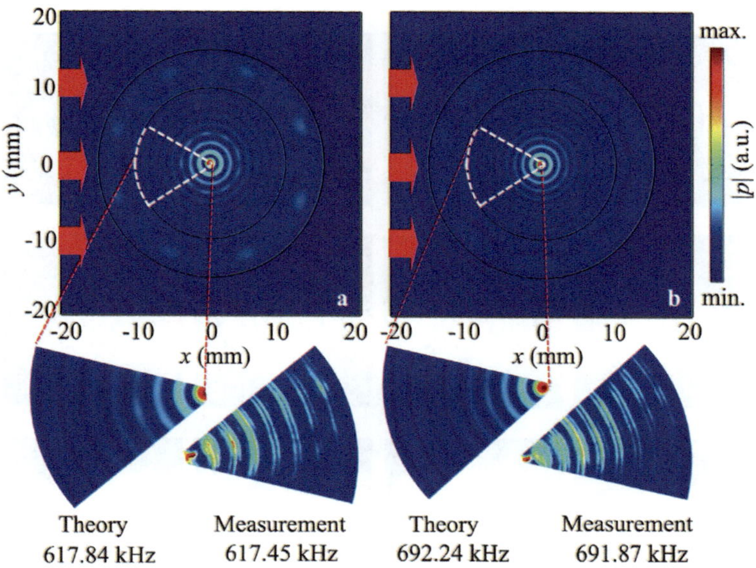

Fig. 5.14 Theoretically calculated and experimentally measured pressure amplitude distributions caused by the circular ring for the **a** seventh and **b** eighth monopole modes

measured by the Schlieren imaging system, and the sound energy is focused on a perfect point in the experiment. The eigenfrequencies and the measured pressure distributions agree with the theoretical results, although there exists a slight difference in pressure distributions (different color) between them. This is because measured results are obtained by integrating the sound energy along the length of the brass cylinder, while the theoretical results are calculated in a 2D cross-section of the cylinder.

5.3.3 Potential Application

The multi-focus focusing effect has potential applications in acoustic encryption communication. Here, we take the word "ONE" as an example. Based on the American standard code for information interchange (ASCII), the letters O, N, and E are expressed as the quaternary arrays 1321, 1320, and 1221, respectively. We use the eigenfrequencies of the second monopole, dipole, quadrupole, and octupole Mie-resonance modes of the circular ring to represent the numbers 0, 1, 2, and 3, respectively, with these eigenfrequencies are in the range of 240–344 kHz. As shown in Fig. 5.15a–d, we represent the letter O using the multi-focus AF in the circular ring at 276.06, 343.88, 310.50, and 276.06 kHz. The letters N and E are represented in Fig. 5.15e–h and i–l, respectively. Therefore, we can achieve acoustic encryption communication using the multi-focus focusing effect.

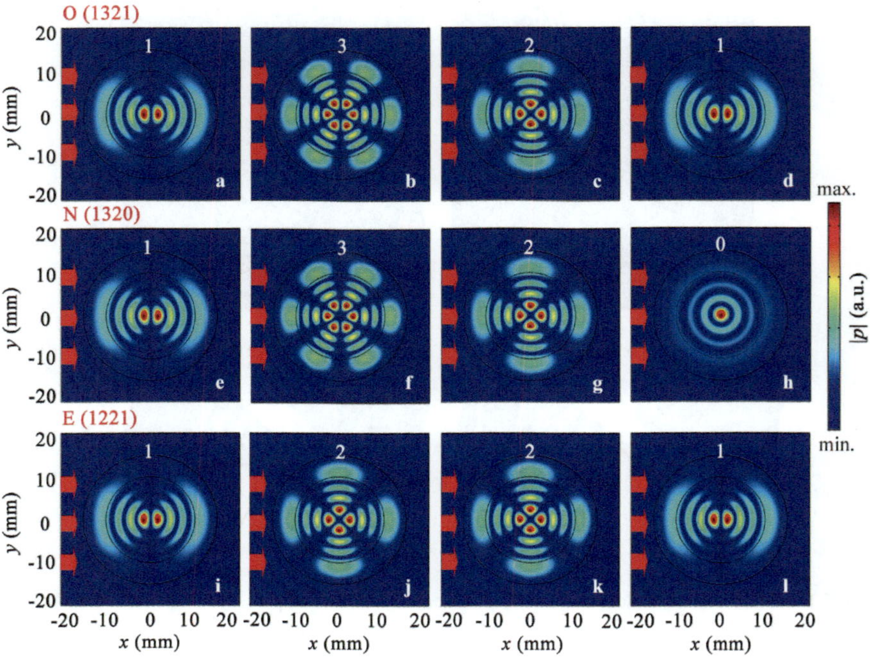

Fig. 5.15 Simulated pressure amplitude distributions caused by the circular ring at **a** 276.06 kHz, **b** 343.88 kHz, **c** 310.50 kHz, and **d** 276.06 kHz for letter O; at **e** 276.06 kHz, **f** 343.88 kHz, **g** 310.50 kHz, and **h** 240.11 kHz for letter N; at **i** 276.06 kHz, **j** 310.50 kHz, **k** 310.50 kHz, and **l** 276.06 kHz for letter E

5.4 Acoustic Focusing in Dual-Layer Ring-Shaped Structures

5.4.1 *Design and Performance of Dual-Layer AF System*

As schematically shown in Fig. 5.16a, the designed AF lens is a dual-layer ring-shaped structure composed of brass cylinders immersed in air [3], with its cross-section shown in Fig. 5.16b. The incident cylindrical source is located on the left side of the model. The radii of the inner (R_1) and outer (R_2) layers of the AF lens are 13.0 and 22.5 mm, respectively, and the diameters of the cylinders in the inner (r_1) and outer (r_2) layers are 2.0 and 3.0 mm, respectively, with the angle of $\theta = 9°$ between two adjacent cylinders in both layers. Throughout this work, the characteristics of the AF lens are simulated using the COMSOL Multiphysics software, and the material parameters are as follows: density $\rho = 8400$ kg/m^3, longitudinal wave velocity $c_l = 4400$ m/s, and transversal wave velocity $c_s = 2200$ m/s for brass; $\rho = 1.21$ kg/m^3 and $c = 344$ m/s for air. In the models, the acoustic-structure boundary is applied to the boundaries of each brass cylinder.

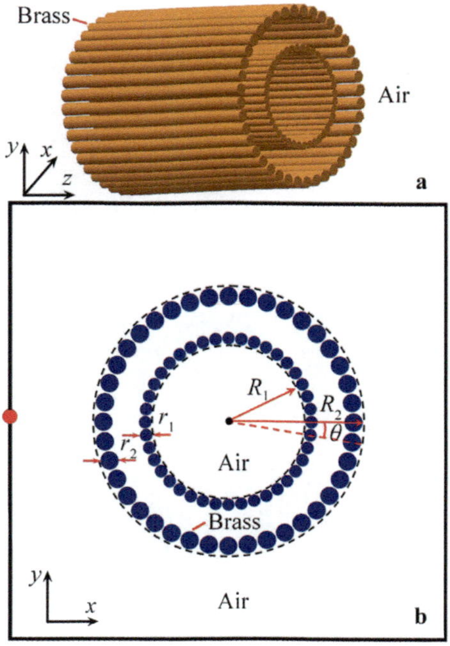

Fig. 5.16 **a** Schematic of the AF lens, and **b** cross-section of the lens

Figure 5.17a, b show the pressure amplitude distributions without and with the AF lens at 42.63 kHz, respectively. The incident cylindrical source is located at (−40 mm, 0), and the initial pressure amplitude $|p_0|$ is 1.0 Pa. As shown in Fig. 5.17a, without the AF lens, the cylindrical wave attenuates rapidly with increasing propagation distance. However, with the AF lens, the sound energy is focused on a perfect point without diffraction at the center of the dual-layer structure, as shown in Fig. 5.17b. Moreover, the AF effect also exists at a series of other frequencies, such as 29.50 and 55.32 kHz, as shown in Fig. 5.17c, d, respectively. The AF effect exhibits high performance at these frequencies as well. By comparing the results in Fig. 5.17b–d, the size of the focus becomes smaller as the frequency increases due to the wavelength of sound.

Figure 5.18a, b show the pressure amplitude distributions caused by lenses with single-layer and three-layer ring-shaped structures, respectively, where the other parameters are the same as those in Fig. 5.17b. As shown in Fig. 5.18a, b, the AF effect also exists. Compared to the results in Fig. 5.17b, the AF effect is much weaker in the single-layer lens, but is almost the same in the three-layer lens. The maximum values of $|p/p_0|$ at the focus are 0.43, 1.19, and 1.22 in Figs. 5.18a, 5.17b and 5.18b, respectively. Therefore, we design the AF lens with a two-layer ring-shaped structure.

Figure 5.19a, b present the simulated pressure amplitude distributions caused by the circular ring along lines I and II in Fig. 5.17b, with the corresponding results in free space plotted for comparison. Note that the AF effect exists in both the x and y directions, and the widths of the focus in both directions are $\lambda/2$. Additionally, compared to the results in free space, the amplitude at the focus is enhanced by 12

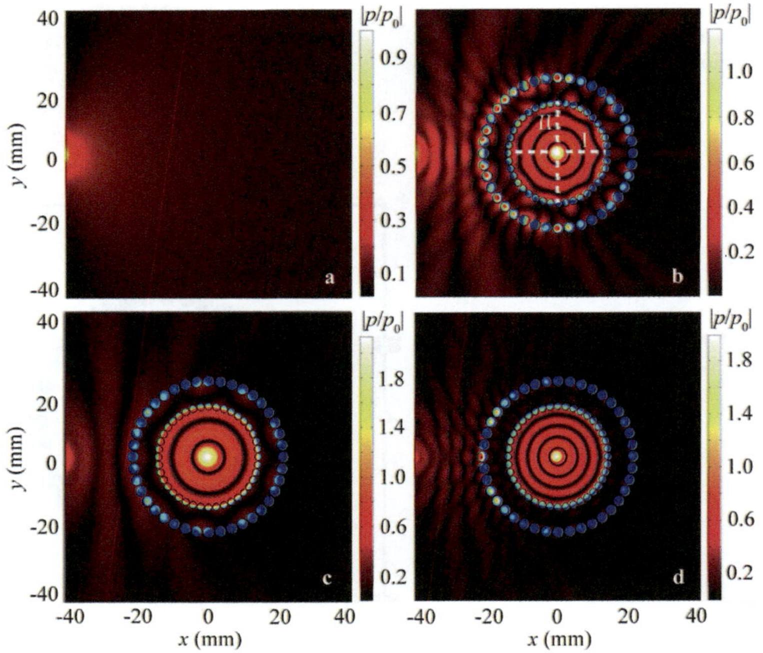

Fig. 5.17 Simulated pressure amplitude distribution ($|p/p_0|$) **a** in free space at 42.63 kHz, and caused by the AF lens under the excitation of a cylindrical sound source at **b** 42.63 kHz, **c** 29.50 kHz, and **d** 55.32 kHz

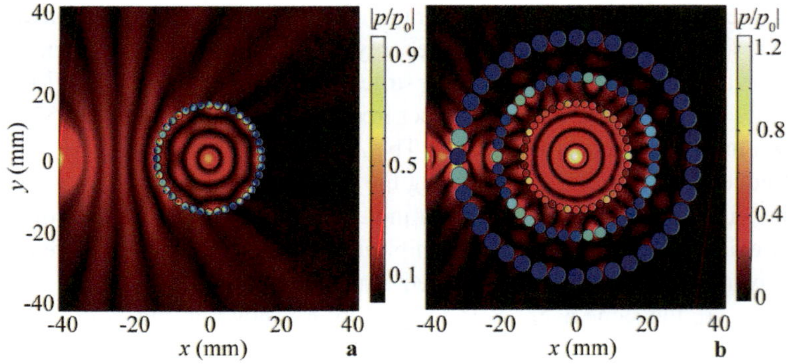

Fig. 5.18 Simulated pressure amplitude distributions caused by the AF lenses with **a** single-layer ring-shaped structure at 42.51 kHz and with **b** three-layer ring-shaped structure at 42.59 kHz

times with the AF lens, indicating high-performance AF caused by the dual-layer ring-shaped structure.

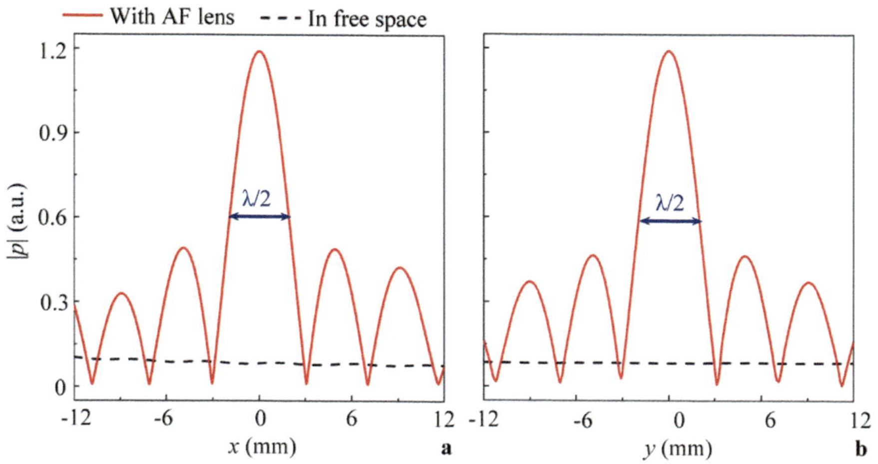

Fig. 5.19 Simulated pressure amplitude distributions along lines **a** I and **b** II in Fig. 5.17b

5.4.2 Equivalent Model

To verify the performance of the AF lens, we design its EM model composed of a circular ring structure in air, as shown in Fig. 5.20. The parameters R_1 and R_2 are the same as those in Fig. 5.16b. Figure 5.21 shows the absolute value of the velocity $|v_{\text{eff}}|$ and the real part of the density ρ_{eff} of the EM model, calculated based on Eqs. (2.15)–(2.18). Figure 5.22 presents the simulated pressure amplitude distribution caused by the EM at 42.63 kHz, and the amplitude distribution inside the EM is similar to that in Fig. 5.17b, confirming the validity of the designed AF lens.

Fig. 5.20 Schematic of the EM model for the AF lens

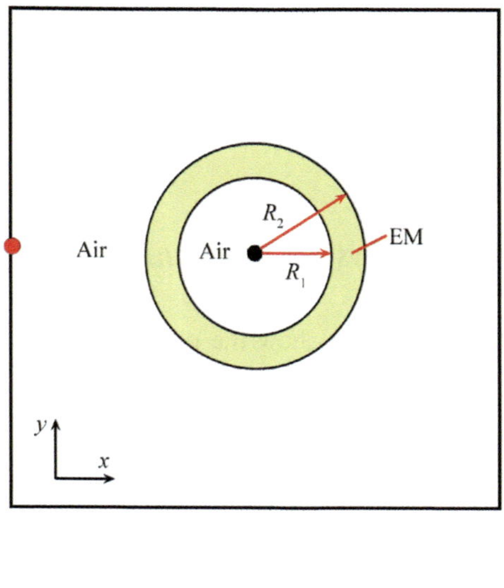

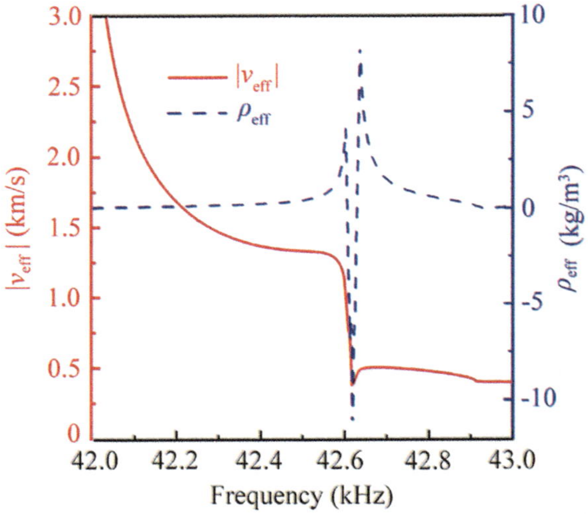

Fig. 5.21 Theoretically calculated real part of ρ_{eff} and $|v_{\text{eff}}|$ spectra of the EM

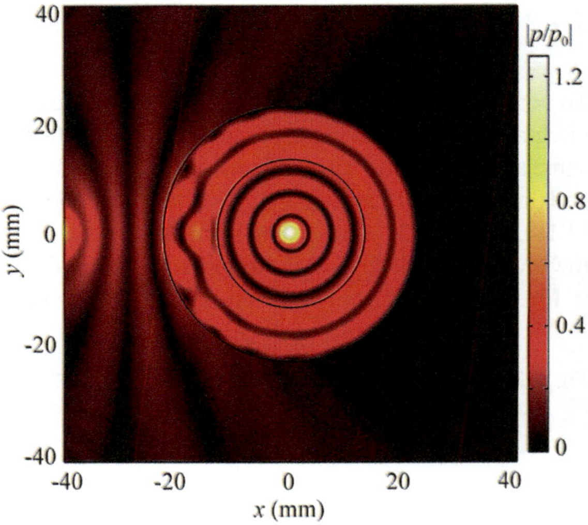

Fig. 5.22 Simulated pressure amplitude distribution caused by the EM at 42.63 kHz

5.4.3 *Physical Mechanism*

Figure 5.23a–c show the phase distributions corresponding to Figs. 5.17a, b, and 5.22, respectively. We can see that the phase distributions are a series of undisturbed circular lines in free space (Fig. 5.23a). However, in Fig. 5.23b, c, the phase distributions are reconstructed inside the AF lens, transforming into a series of concentric circles with the same phase values, eventually focusing on a point at the centre of the AF lens.

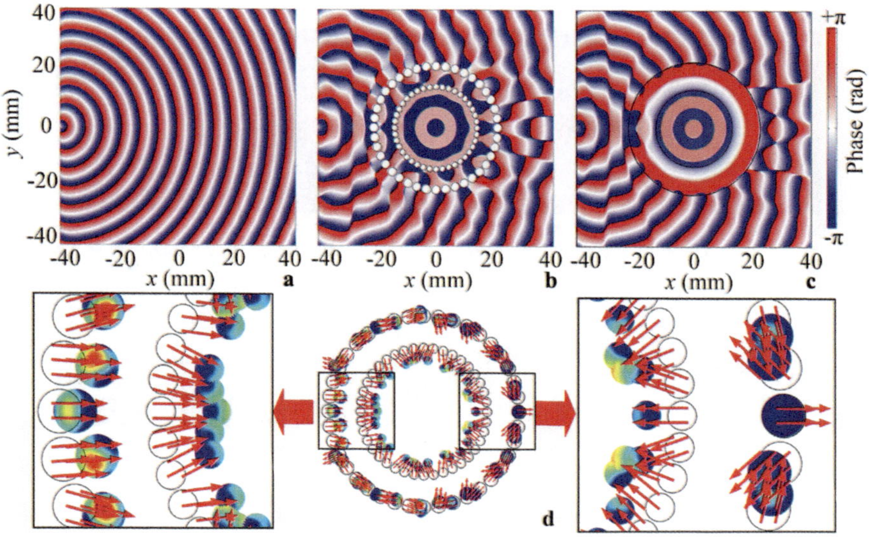

Fig. 5.23 Simulated phase distributions **a** in free space and caused by **b** the AF lens and **c** the EM at 42.63 kHz, and **d** simulated deformation of displacement distribution caused by the dual-layer ring-shaped structure. The red arrows represent the vibration directions of the displacement

To provide insight into the mechanism of the AF effect, we simulate the deformation of the total displacement distributions in the dual-layer ring-shaped structure in Fig. 5.17b, as shown in Fig. 5.23d. We can see that the Mie-resonance mode [15] is excited in the AF lens at the eigenfrequency (43.62 kHz), and the vibration directions point toward the focal spot, especially in the inside ring-shaped structure. It is deduced that the propagation directions of the cylindrical wave are restructured by the Mie-resonance mode [16], and the cylindrical waves are finally focused on a perfect point at the center of the AF lens.

5.4.4 Parameter Analysis

Next, we simulate the pressure amplitude distributions by separately scaling the sizes of the outer (R_2 and r_2) and inner (R_1 and r_1) ring-shaped structures by a factor of n. As shown in Fig. 5.24, the frequencies of AF remain almost the same with different sizes of the outer ring-shaped structure, but are inversely proportional to the size of the inner ring-shaped structure. This indicates that the frequencies of AF are closely related to the size of the inner ring-shaped structure.

In addition, the AF effect also exists in ring-shaped structures of different shapes. To verify this, we design four types of dual-layer ring-shaped structures with different shapes, which are presented in Fig. 5.25. As shown in Fig. 5.26, the cylindrical sound waves are focused at the center of the four types of dual-layer ring-shaped structures

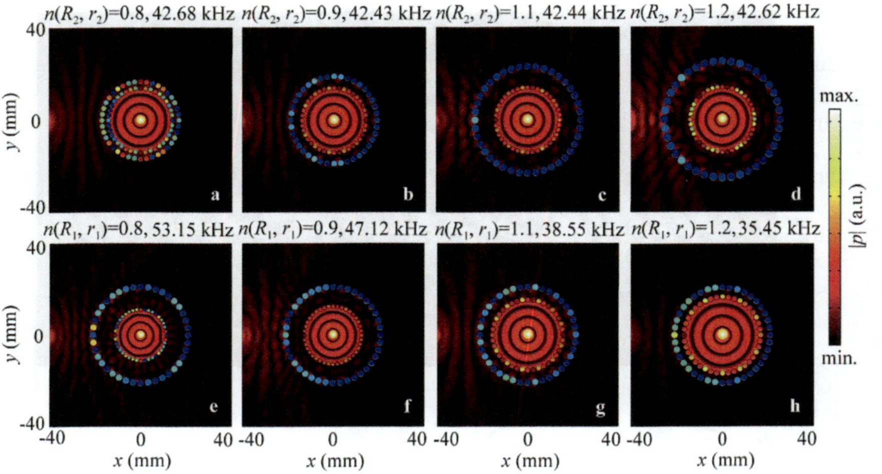

Fig. 5.24 Simulated pressure amplitude distributions caused by the AF lens with the parameters **a–d** R_2 and r_2 and **e–h** R_1 and r_1 separately scaled by a factor of n

at 17.19, 42.50, 37.42, and 29.55 kHz, and the incident cylindrical sound sources can be located at arbitrary positions outside the dual-layer ring-shaped structures.

5.4.5 Potential Application

The AF effect caused by the dual-layer ring-shaped structures has potential applications in beam splitters. It is known that the sound energy is divided into several beams and significantly reduced in beam splitters. Here, we propose an acoustic beam splitter with enhanced sound energy using a dual-layer square ring-shaped structure, as shown in Fig. 5.25a. The incident cylindrical wave is located at the center of the dual-layer ring-shaped structure, and the initial amplitude $|p_0|$ is also 1.0 Pa. Additionally, the pressure amplitude distribution in free space is simulated for comparison (Fig. 5.27a), where the white circular lines represent the contour lines of $|p/p_0|$.

As shown in Fig. 5.27b, the output sound energy caused by the dual-layer square ring-shaped structure is divided into four sound beams. Compared to the result in Fig. 5.27a, the pressure amplitude at the position of the sound source is enhanced by 40 times, and the amplitude of each beam through the dual-layer structure is about 15 times in Fig. 5.27b. To clearly exhibit the characteristic of the output sound beam, we show the pressure amplitude distribution outside the dual-layer ring-shaped structure with a different range of the color bar (the right inset in Fig. 5.27b), and the pressure amplitudes of the output sound beam are significantly larger than those in free space (Fig. 5.27a).

Another potential application of the proposed dual-layer ring-shaped structure is in the design of an acoustic directional transmitter. Here, we consider the dual-layer

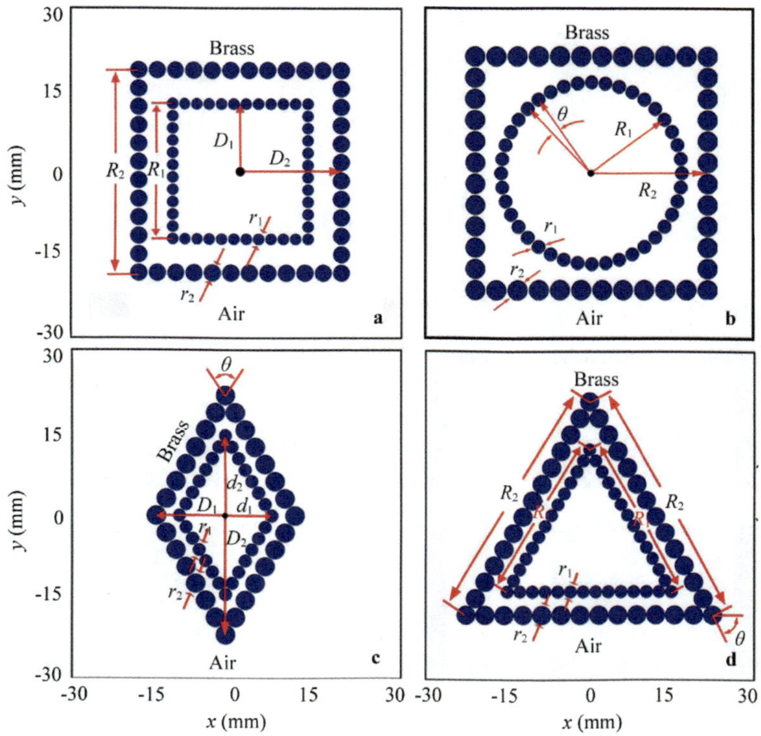

Fig. 5.25 Schematic of dual-layer **a** square (R_1 = 24.00 mm, R_2 = 36.00 mm, D_1 = 12.00 mm, D_2 = 18.00 mm, r_1 = 2.00 mm and r_2 = 3.00 mm), **b** square and circular ($\theta = 9°$, R_1 = 14.00 mm, R_2 = 18.00 mm, D_1 = 12.00 mm, D_2 = 18.00 mm, r_1 = 2.00 mm and r_2 = 3.00 mm), **c** rhomboic ($\theta = 60°$, d_1 = 12.50 mm, d_2 = 7.22 mm, D_1 = 3.61 mm, D_2 = 18.75 mm, r_1 = 2.00 mm and r_2 = 3.00 mm), and **d** triangular ($\theta = 60°$, R_1 = 26.27 mm, R_2 = 39.40 mm, r_1 = 2.00 mm and r_2 = 3.00 mm) ring-shaped structures

square ring-shaped structure (Fig. 5.27b) surrounded by a square ring-shaped structure with an opening on the upper boundary, as schematically shown in Fig. 5.28. Figure 5.29 shows the pressure amplitude distribution caused by the designed acoustic directional transmitter, where the output sound energy is concentrated in the direction of the opening. Compared to the result in Fig. 5.27b, the amplitude of the output acoustic wave is magnified by 1.5 times in Fig. 5.29, as shown in right inset.

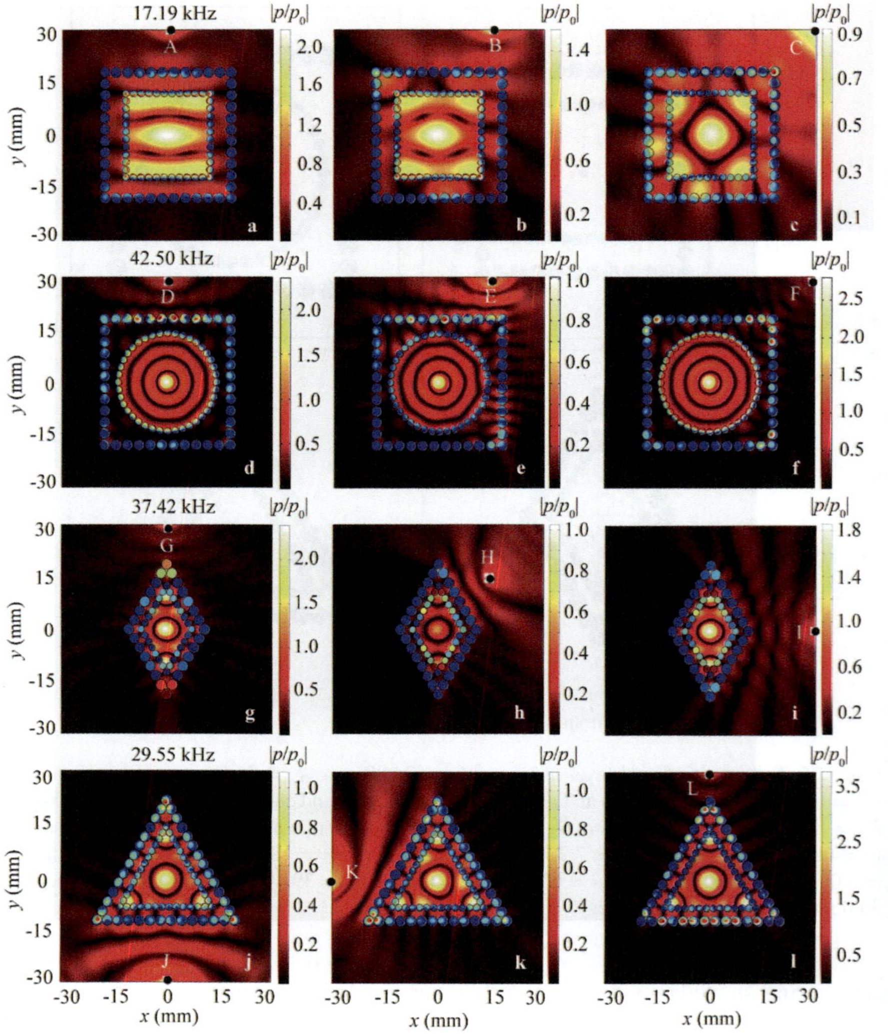

Fig. 5.26 Simulated pressure amplitude distributions caused by the dual-layer **a–c** square, **d–f** square and circular, **g–i** rhomboid, and **j–l** triangular ring-shaped structures with different incident positions of the cylindrical sound source

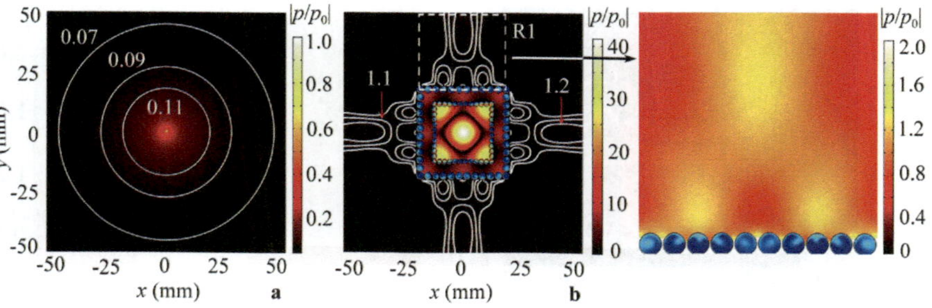

Fig. 5.27 Simulated pressure amplitude distributions **a** in free space and **b** caused by the dual-layer ring-shaped structure at 17.19 kHz. The inset on the right side shows the simulated pressure amplitude distribution in white open square R1 in (**b**). The white circular lines represent the contour lines of amplitude

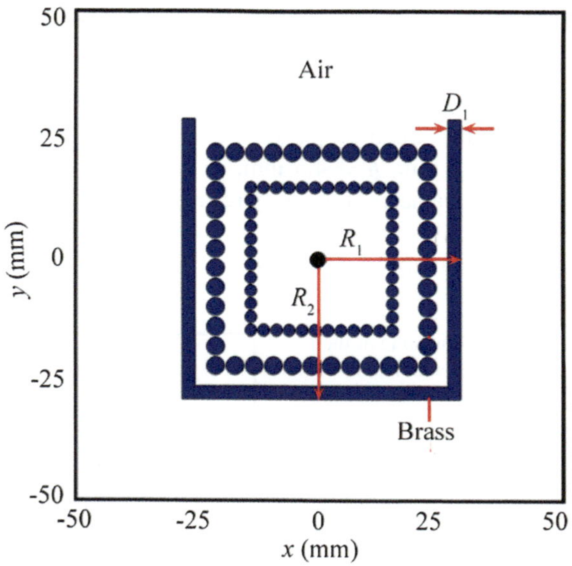

Fig. 5.28 Schematic of the acoustic directional transmitter ($R_1 = 21.50$ mm, $R_2 = 23.50$ mm, and $D_1 = 2.00$ mm)

5.5 Acoustic Focusing in Ring-Shaped Mie Resonance-Based Structures

5.5.1 Design and Performance of Multiple-Cavity Unit Cell

As schematically shown in Fig. 5.30, we propose a circular multiple-cavity unit cell consisting of a central cavity surrounded by eight interconnected same identical cavities [4]. These cavities are filled with air and constructed from epoxy resin to satisfy the sound-hard boundary condition. The solid frame has a thickness t, the channels have a width w, and the open width of the cavities is h. The outer and inner

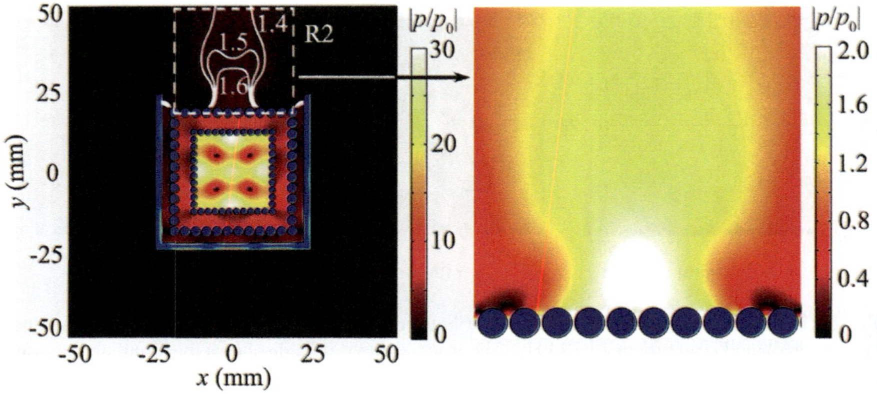

Fig. 5.29 Simulated pressure amplitude distributions caused by the acoustic directional transmitter at 17.19 kHz. The inset on the right side shows the simulated pressure amplitude distribution in white open square R2. The white circular lines represent the contour lines of amplitude

radii are R and r, respectively. We use COMSOL Multiphysics, a FE software, to simulate the sound propagation characteristics. The selected structural parameters are t = 2.0 mm, h = 4.5 mm, w = 4.6 mm, r = 3.0 cm, and R = 5.0 cm. The material parameters for epoxy resin are density $\rho = 1050$ kg/m^3, Young's modulus E = 5.08 GPa, and Poisson's ratio $\nu = 0.35$. For air, density $\rho = 1.21$ kg/m^3 and acoustic velocity $c = 343$ m/s.

Fig. 5.30 Schematic of the multiple-cavity unit cell

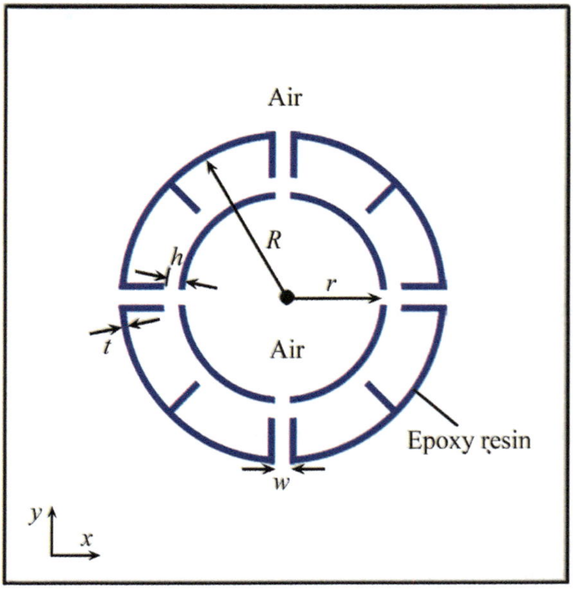

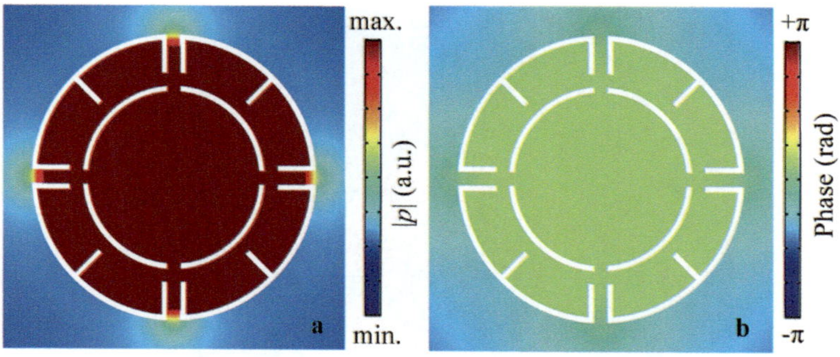

Fig. 5.31 Simulated **a** pressure amplitude and **b** phase eigenfunctions of the MMR mode for the multiple-cavity unit cell

Further simulations about the eigenmodes of the multiple-cavity unit cell reveal typical characteristics of Mie resonances. These resonances are closely related to the multiple-cavity structure of the unit cell. Figure 5.31a, b show the pressure and phase eigenfunctions of the MMR mode at 634 Hz. The MMR mode concentrates sound energy within the multiple-cavity unit cell (Fig. 5.31a), and all cavities exhibit in-phase characteristics (Fig. 5.31b). Additionally, the diameter of the multiple-cavity unit cell is about 0.185λ.

As shown in Fig. 5.32, the MMR mode of the multiple-cavity unit cell can be excited by an incident plane wave at 634 Hz, with an initial amplitude of 1.0 Pa. The sound energy is primarily concentrated in the unit cell due to the MMR mode, and the intensity at the center is about 30 times greater than that of the incident wave. Figure 5.33a shows the simulated intensities along lines I and II in Fig. 5.32. The intensities along lines I and II exceed 26 Pa^2, reaching about 32 Pa^2 at the center of the unit cell. Figure 5.33b displays the intensity spectrum at the center of the unit cell. The intensity exceeds 10 Pa^2 in the range of 590–680 Hz, resulting in a fractional bandwidth of 0.14.

Additionally, we simulate the dipole and quadrupole Mie resonance modes of the multiple-cavity unit cell, as shown in Fig. 5.34. Unlike the MMR mode, the pressure and phase eigenfunctions are not uniformly distributed within the unit cell for these modes. The rich Mie resonances of the multiple-cavity unit cell provide the foundation for designing various acoustic devices [15–20].

5.5.2 Design and Performance of AF System

Next, we design an AF system composed of eight multiple-cavity unit cells arranged in a square shape, as shown in Fig. 5.35. The distance between adjacent unit cells is $d = 14.0$ cm. Figure 5.36a, b show the simulated intensity and phase eigenfunctions of the MMR mode of the AF system at 716 Hz. The MMR mode concentrates sound energy in the central region of the AF system, and the phase in this region is

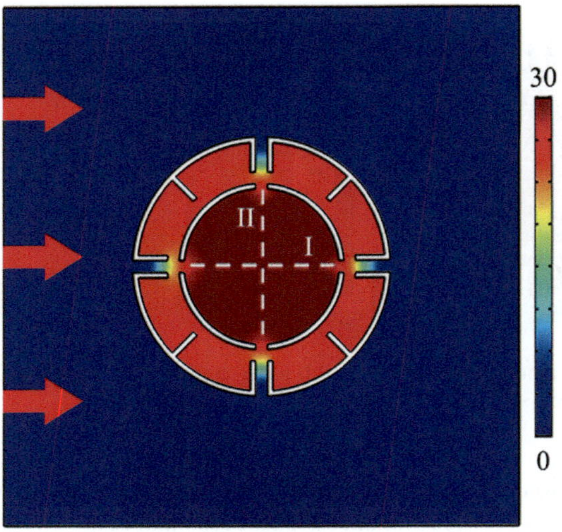

Fig. 5.32 Simulated intensity distributions caused by the multiple-cavity unit cell at 634 Hz. The red arrows represent the incident acoustic wave

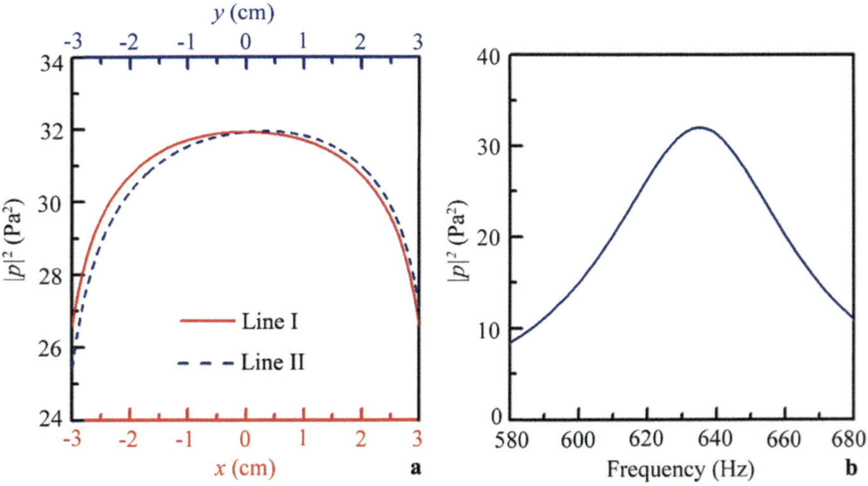

Fig. 5.33 **a** Simulated acoustic intensity distributions along lines I and II in Fig. 5.32. **b** Simulated acoustic intensity spectrum at the center of the multiple-cavity unit cell

uniform (Fig. 5.36b), similar to the characteristics of the MMR mode in the multiple-cavity unit cell (Fig. 5.31). The MMR mode also exists in the eight unit cells, with higher intensities in the four unit cells at the midpoints compared to those at the corners, due to their proximity to the center. Furthermore, the dipole (948 Hz) and quadrupole (1020 Hz) Mie resonance modes of the AF system are simulated, as shown in Fig. 5.37. These modes exhibit characteristics similar to those in Fig. 5.34.

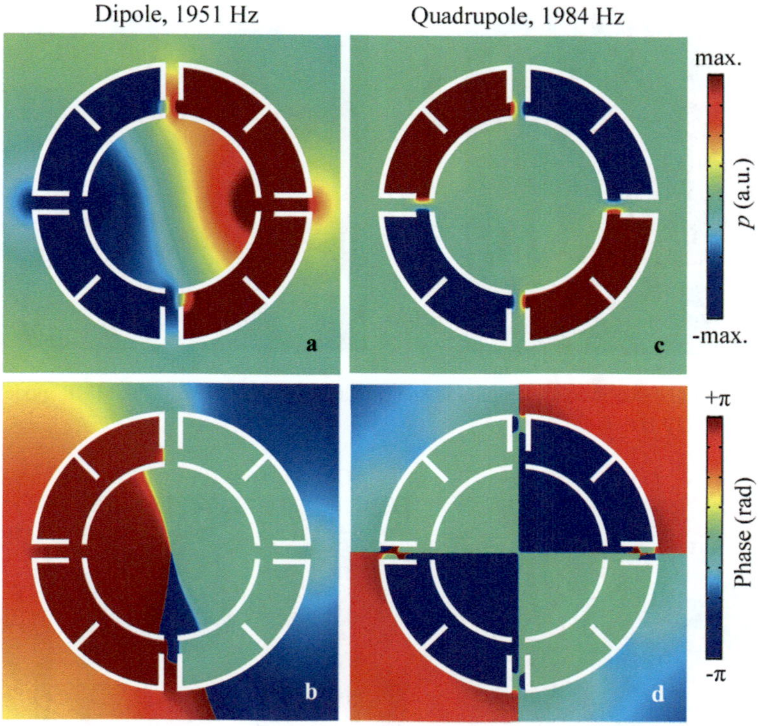

Fig. 5.34 Simulated pressure and phase eigenfunctions of the **a**, **b** dipole and **c**, **d** quadrupole Mie resonance modes of the multiple-cavity unit cell

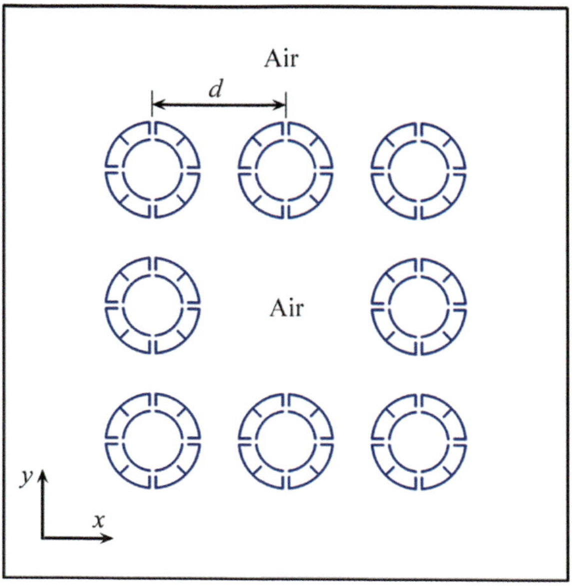

Fig. 5.35 Schematic of the AF system composed of eight multiple-cavity unit cells

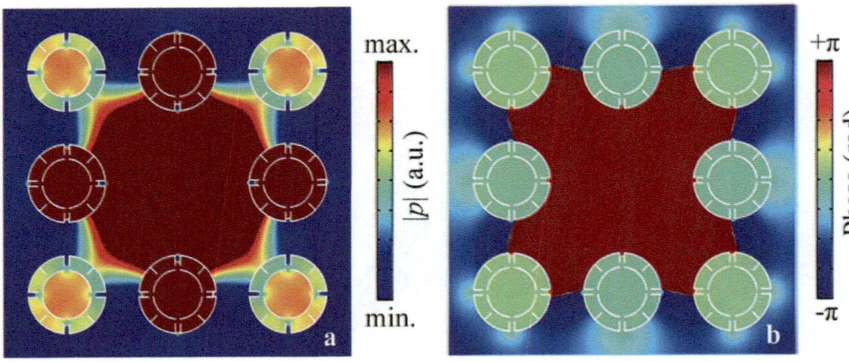

Fig. 5.36 Simulated **a** pressure amplitude and **b** phase eigenfunctions of the MMR mode for the AF system at 716 Hz

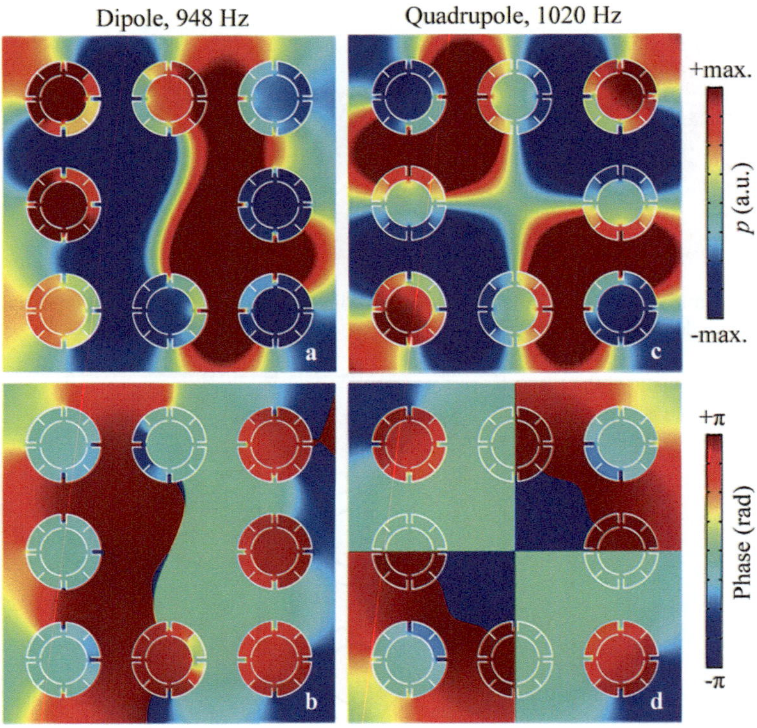

Fig. 5.37 Simulated pressure and phase eigenfunctions of the **a, b** dipole and **c, d** quadrupole Mie resonance modes for the AF system

Similar to the MMR of the multiple-cavity unit cell, the MMR of the AF system can also be excited by the incident plane wave at 716 Hz, as shown in Fig. 5.38a. The sound energy is focused in the central region of the AF system, with an intensity about 60 times greater than that of the incident wave. The length of the system is only about 0.79λ, demonstrating its high-performance focusing capability and sub-wavelength size. The high intensity in the central region makes the AF system suitable for acoustic energy harvesting in low-energy environments. The inset on the right side of Fig. 5.38a shows the intensity distribution within the AF system. All acoustic velocity directions point toward the center of the AF system due to the MMR mode, causing sound energy from the external region and the eight unit cells to be absorbed into the central region. Thus, the AF effect is realized based on the MMR mode of the system. The performance of AF can be further enhanced by coupling the MMR modes of the unit cells. For comparison, we simulate the MMR mode and corresponding AF in a system composed of eight solid cylinders (Fig. 5.38b), with the same configuration and cylinder size as in Fig. 5.38a. However, the acoustic velocity direction mainly aligns with the incident direction, and the AF effect is not obvious. The intensity at the center of the system is significantly lower than in Fig. 5.38a because the MMR mode arises from the square structure without involving the coupling of the MMR modes of the eight unit cells (the inset at the right side in Fig. 5.38b).

Figure 5.39 shows the simulated intensities along lines III and IV in Fig. 5.38, with corresponding results in free space for comparison. Three peaks are observed along lines III and IV within the multiple-cavity unit cells, indicating that the AF effect in the central region is enhanced by the coupling of the MMR modes of the unit cells. The maximum intensities in the unit cells exceed 120 Pa^2 and 80 Pa^2 along lines III and IV, respectively, while those for the solid cylinders and free space are close to zero. This further confirms that the AF effect in Fig. 5.38b is due to the square structure and does not involve the MMR mode of the cylinders. For the AF system composed of eight multiple-cavity unit cells, the intensity at the center reaches about 60 Pa^2, and the FLHM of the focus is only 16 cm (about λ/3) (Fig. 5.39a, b). Therefore, subwavelength AF can be achieved in low-energy environments using the MMR mode of the system composed of multiple-cavity unit cells.

5.5.3 Physical Mechanism

Figure 5.40 shows the intensity spectra at points A, B, and C in Fig. 5.38a. Intensity peaks are observed simultaneously at all three points at 716 Hz (f_4). Additionally, at 542 Hz (f_1), 608 Hz (f_2), and 619 Hz (f_3), intensity peaks are present at points B and/or C, but are very weak at point A. These frequencies differ from the MMR mode frequency of a single multiple-cavity unit cell (634 Hz), indicating that the MMR mode at these frequencies arises from the mutual coupling of the unit cells. However, the MMR mode of the AF system only exists at f_4, suggesting a different coupling mechanism at f_4 compared to f_1, f_2, and f_3. Figure 5.41a–c show the intensity

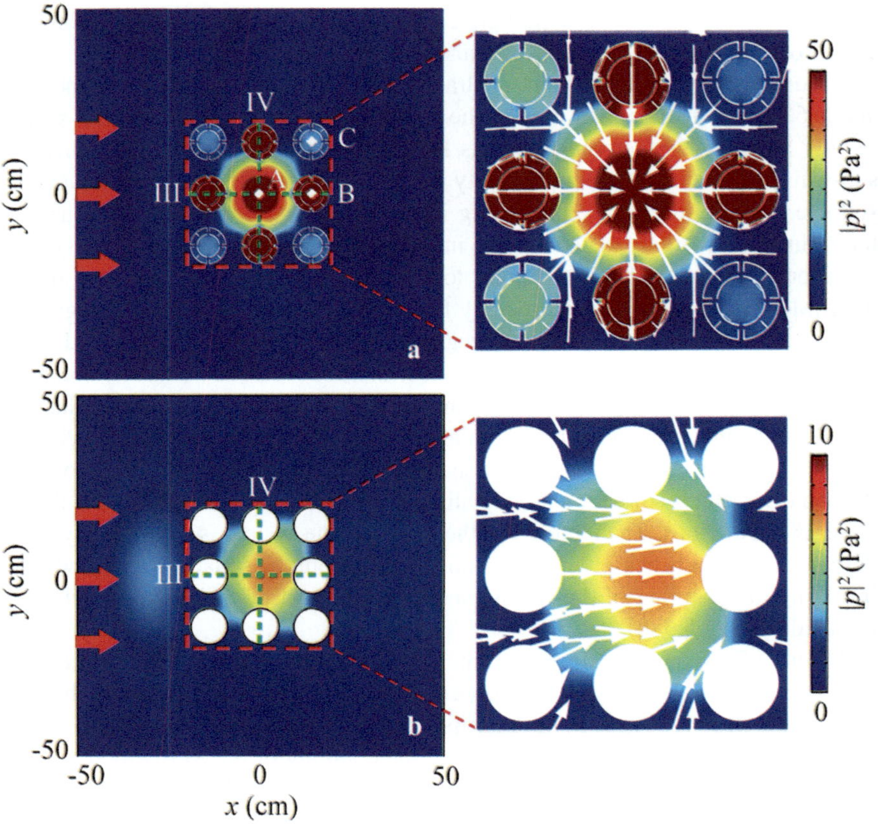

Fig. 5.38 Simulated intensity distributions caused by the two types of AF systems composed of eight **a** multiple-cavity unit cells and **b** solid cylinders at 716 Hz. The insets on the right side show the simulated intensity distributions inside both systems, with the white arrows representing the acoustic velocity directions

distributions caused by the AF system at f_1, f_2, and f_3, respectively. The MMR mode exists in the eight unit cells at these frequencies, with sound energy primarily concentrated within the unit cells. However, the MMR mode and corresponding AF effect of the system are not excited at these frequencies.

To understand the mechanism of the AF effect, we simulate the phase distribution of the AF system at 716 Hz (f_4), as shown in Fig. 5.42. Similar to Fig. 5.36b, the phase distributions are nearly uniform in each unit cell, characteristic of the MMR mode. However, the phase at the center of the system is almost opposite to that in the unit cells. Figure 5.43 shows the simulated phase spectra at points A (φ_A) and B (φ_B) in Fig. 5.42. The phase difference between points A and B ($|\varphi_A - \varphi_B|$) is π at 716 Hz, indicating an out-of-phase relationship between the center region of the system and the eight unit cells. This suggests that the MMR mode arises from an out-of-phase resonance caused by the mutual coupling between the system and the

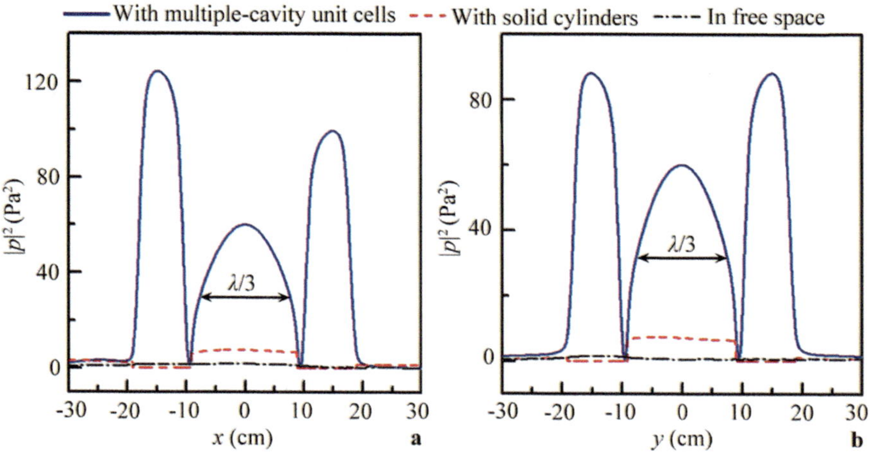

Fig. 5.39 Simulated acoustic intensity distributions along lines III and IV in Fig. 5.38a, b, with corresponding results in free space

Fig. 5.40 Simulated intensity spectra at points A, B, and C in Fig. 5.38a

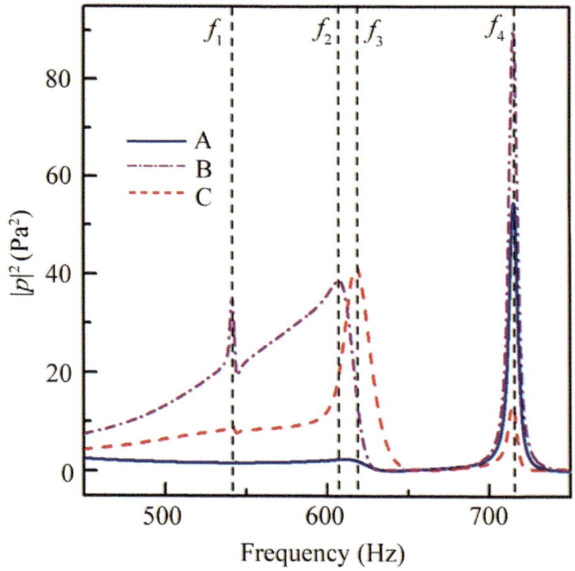

unit cells. Figure 5.44a–c show the phase distributions at 542 Hz (f_1), 608 Hz (f_2), and 619 Hz (f_3), respectively. The phase distributions vary in each unit cell, and the phase differences do not exhibit the out-of-phase characteristic observed at f_4. Therefore, the MMR mode of the AF system does not exist at these frequencies.

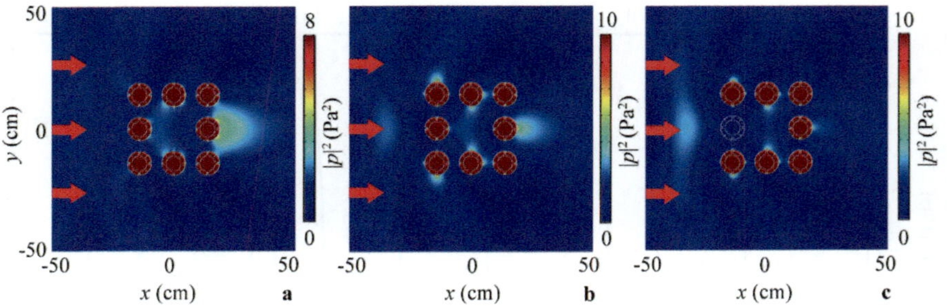

Fig. 5.41 Simulated intensity distributions caused by the AF system at **a** 542 Hz (f_1), **b** 608 Hz (f_2), and **c** 619 Hz (f_3)

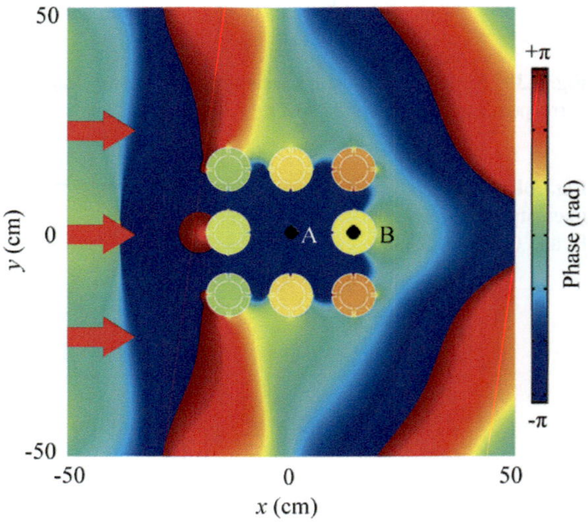

Fig. 5.42 Simulated phase distribution caused by the AF system at 716 Hz (f_4)

5.5.4 Parameter Analysis

We investigate the control of the AF effect by tuning the inner radius r and the distance d. Figure 5.45a, b show the eigenfrequencies of the MMR mode for systems with different values of r/r_0 and d/d_0, where $r_0 = 3.0$ cm and $d_0 = 14.0$ cm. The eigenfrequencies of the MMR mode increase with r/r_0 but decrease with d/d_0, indicating that these parameters can tune the eigenfrequencies of the MMR mode. Figure 5.45c, d show the intensity at the center of the AF system for different values of r/r_0 and d/d_0. The intensity varies with these parameters, reaching 9000 Pa^2 and 110 Pa^2 for $r/r_0 = 0.73$ and $d/d_0 = 0.93$, respectively. This demonstrates that sound energy in the central region can be controlled by adjusting r and d.

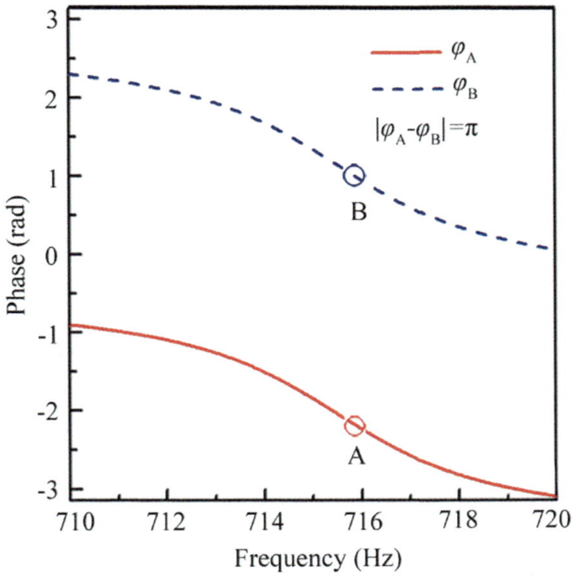

Fig. 5.43 Simulated phase spectra at points A and B in Fig. 5.38a

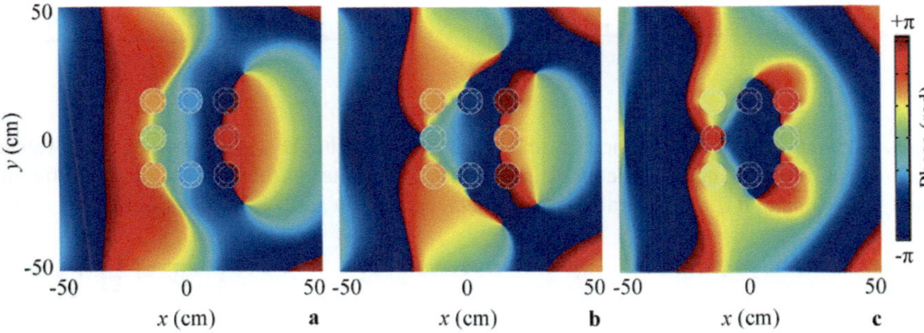

Fig. 5.44 Simulated phase distributions caused by the AF system at **a** 542 Hz (f_1), **b** 608 Hz (f_2), and **c** 619 Hz (f_3)

5.5.5 Robustness Demonstration

The AF effect based on the MMR mode exhibits strong robustness. As shown in Fig. 5.46, the AF effect remains effective at different incident angles, with intensities at the center of the system of approximately 40, 30, and 20 Pa^2 for incident angles of 15°, 30°, and 45°, respectively. The AF effect can be excited at various incident angles, demonstrating its robustness. The decrease in sound intensity in the central region is due to the reduction in sound energy captured by the system surface as the incident angle increases.

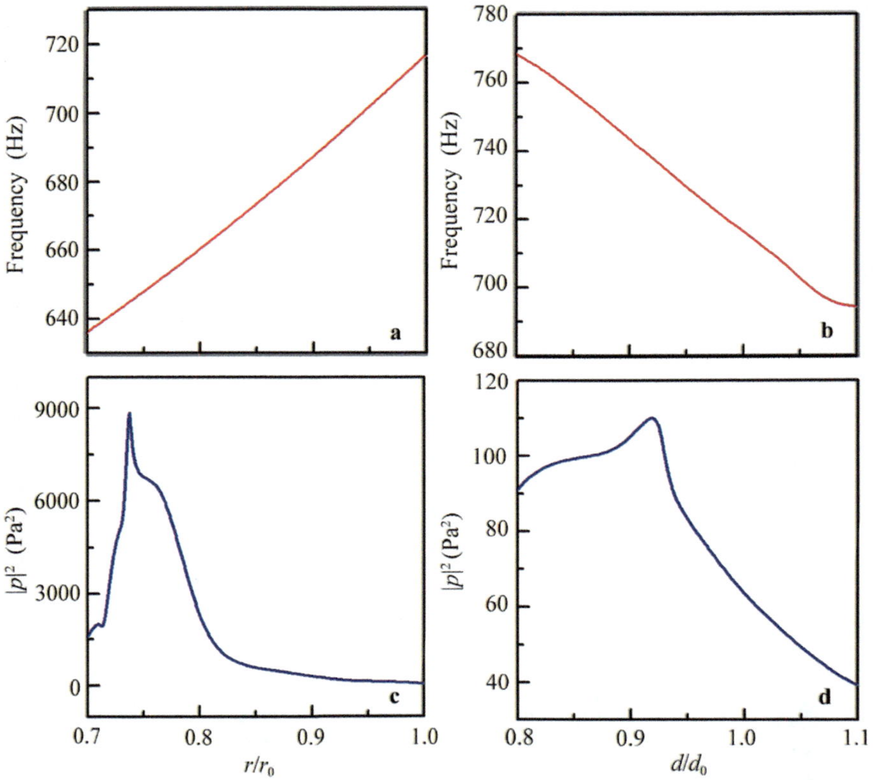

Fig. 5.45 Simulated eigenfrequencies of the MMR mode for the AF system with different values of **a** r/r_0 and **b** d/d_0. Simulated acoustic intensities at the center of the AF system with different values of **c** r/r_0 and **d** d/d_0. The values of r_0 and d_0 are 3.0 cm and 14.0 cm, respectively

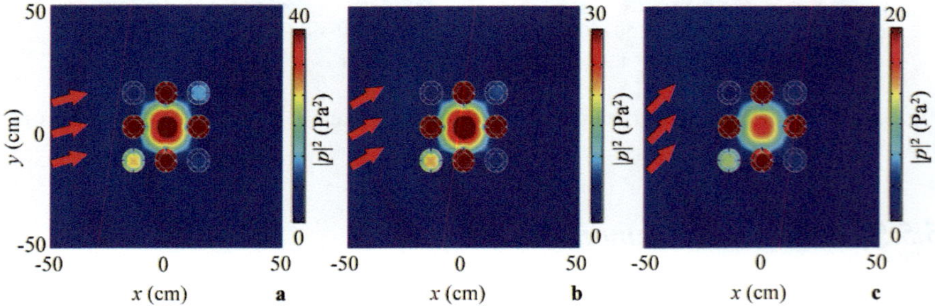

Fig. 5.46 Simulated intensity distributions caused by the system under the plane wave excitation at angles of **a** 15°, **b** 30°, and **c** 45° at 716 Hz

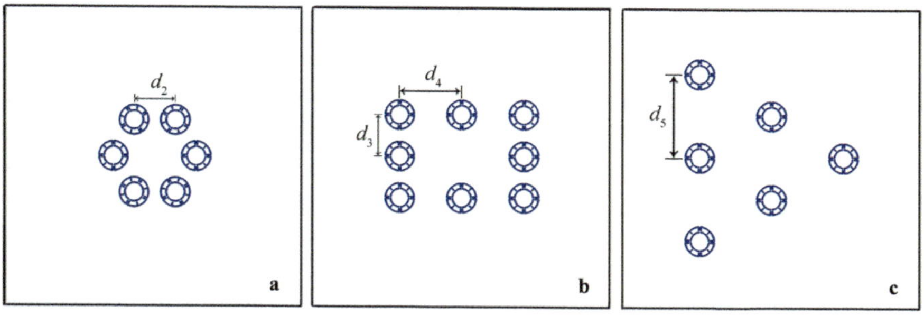

Fig. 5.47 Schematic of the **a** hexagon, **b** rectangle, and **c** triangle ring-shaped systems

The proposed AF effect also exists in systems with different shapes. We design three ring-shaped systems using different numbers of unit cells: hexagon, rectangle, and triangle ring-shaped systems, as shown in Fig. 5.47a–c. The distances between adjacent unit cells are $d_2 = 14$ cm, $d_3 = 14$ cm, $d_4 = 21$ cm, and $d_5 = 28$ cm, with the unit cell parameters matching those in Fig. 5.30.

Similar to the results shown in Fig. 5.38a, the AF effect is also obtained in the central region of the three ring-shaped systems with different shapes (Fig. 5.48). The intensities at the center reach about 100, 20, and 60 Pa^2 in Fig. 5.48a–c, respectively. This indicates that the AF effect based on the MMR mode can be achieved in various ring-shaped systems. Such a flexible AF system can be applied to a wide range of specific scenarios involving ring-shaped AF systems. These results demonstrate the strong robustness of the AF effect based on the MMR mode.

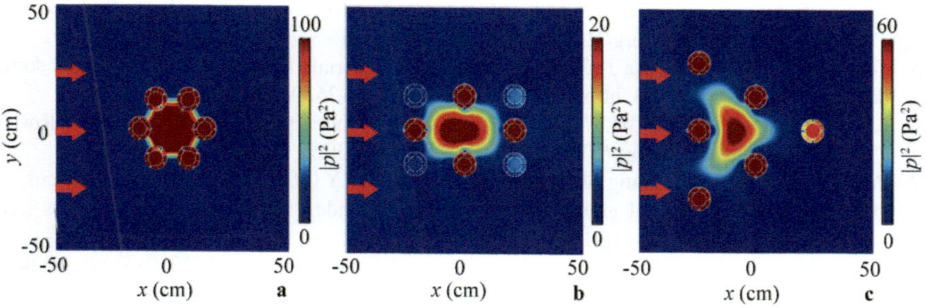

Fig. 5.48 Simulated intensity distributions caused by **a** hexagon (738 Hz), **b** rectangle (754 Hz), and **c** triangle (713 Hz) ring-shaped systems

5.6 Conclusion

This chapter investigates the phenomenon of AF using ring-shaped metamaterials, both in water and air. The AF effect is achieved through the excitation of eigenmodes in the ring structures, leading to perfect focusing of sound energy at specific points. The chapter discusses single-layer and multi-layer ring structures, demonstrating multi-focus AF and its potential applications in underwater detection and medical ultrasound. The physical mechanisms, including Mie-resonance modes and the role of ring geometry, are analyzed. Experimental measurements using Schlieren imaging demonstrate the theoretical predictions. The chapter also explores the potential applications of AF in acoustic encryption communication, beam splitters, and directional transmitters, highlighting the versatility and high performance of ring-shaped metamaterials in sound manipulation.

References

1. Xia JP, Sun HX (2015) Acoustic focusing by metal circular ring structure. Appl Phys Lett 106(6):063505. https://doi.org/10.1063/1.4908117
2. Xia JP, Sun HX, Cheng Q, Xu Z, Chen H, Yuan SQ, Zhang SY, Ge Y, Guan YJ (2016) Theoretical and experimental verification of acoustic focusing in metal cylinder structure. Appl Phys Express 9(5):057301. https://doi.org/10.7567/APEX.9.057301
3. Guan YJ, Sun HX, Liu SS, Yuan SQ, Xia JP, Ge Y (2016) Acoustic focusing through two layer annuluses in air. Chin Phys B 25(10):104302. https://doi.org/10.1088/1674-1056/25/10/104302
4. Gao WT, Xia JP, Sun HX, Yuan SQ, Ge Y, Liu XJ (2019) Acoustic energy harvesting for low-frequency airborne sound based on compound Mie resonances. Appl Phys Express 12(4):044002. https://doi.org/10.7567/1882-0786/ab07e5
5. Torrent D, Sánchez-Dehesa J (2007) Acoustic metamaterials for new two-dimensional sonic devices. New J Phys 9:323. https://doi.org/10.1088/1367-2630/9/9/323
6. Lin SCS, Huang TJ, Sun JH, Wu TT (2009) Gradient-index phononic crystals. Phys Rev B 79(9):094302. https://doi.org/10.1103/PhysRevB.79.094302
7. Peng SS, He ZJ, Jia H, Zhang AQ, Qiu CY, Ke MZ, Liu ZY (2010) Acoustic far-field focusing effect for two-dimensional graded negative refractive-index sonic crystals. Appl Phys Lett 96(26):263502. https://doi.org/10.1063/1.3457447
8. Martin TP, Nicholas M, Orris GJ, Cai LW, Torrent D, Sánchez-Dehesa J (2010) Sonic gradient index lens for aqueous applications. Appl Phys Lett 97(11):113503. https://doi.org/10.1063/1.3489373
9. Millero FJ, Huang F (2009) The density of seawater as a function of salinity (5 to 70 g kg^{-1}) and temperature (273.15 to 363.15 K). Ocean Sci 5(2):91–100. https://doi.org/10.5194/os-5-91-2009
10. Mackenzie KV (1981) Nine-term equation for sound speed in the oceans. J Acoust Soc Am 70(3):807–812. https://doi.org/10.1121/1.386920
11. Laude V, Gérard D, Khelfaoui N, Jerez-Hanckes CF, Benchabane S, Khelif A (2008) Subwavelength focusing of surface acoustic waves generated by an annular interdigital transducer. Appl Phys Lett 92(9):094104. https://doi.org/10.1063/1.2891055
12. Chen YC, Nawaz AA, Zhao YH, Huang PH, McCoy JP, Levine SJ, Wang L, Huang TJ (2014) Standing surface acoustic wave (SSAW)-based microfluidic cytometer. Lab Chip 14(5):916–923. https://doi.org/10.1039/c3lc51139a

13. Cheng Y, Liu XJ (2009) Extraordinary resonant scattering in imperfect acoustic cloak. Chin Phys Lett 26(1):014301. https://doi.org/10.1088/0256-307X/26/1/014301
14. Unverzagt C, Olfert S, Henning B (2010) A new method of spatial filtering for Schlieren visualization of ultrasound wave fields. Phys Procedia 3(1):935–942. https://doi.org/10.1016/j.phpro.2010.01.120
15. Cheng Y, Zhou C, Yuan BG, Wu DJ, Wei Q, Liu XJ (2015) Ultra-sparse metasurface for high reflection of low-frequency sound based on artificial Mie resonances. Nat Mater 14(10):1013–1019. https://doi.org/10.1038/NMAT4393
16. Xia JP, Sun HX, Yuan SQ, Zhang SY (2015) Extraordinary acoustic transmission based on source pattern enhancement and reconstruction by metal cylinder structure. Appl Phys Express 8(10):104301. https://doi.org/10.7567/APEX.8.104301
17. Zhou C, Yuan BG, Cheng Y, Liu XJ (2016) Precise rainbow trapping for low-frequency acoustic waves with micro Mie resonance-based structures. Appl Phys Lett 108(6):063501. https://doi.org/10.1063/1.4941664
18. Zhang J, Cheng Y, Liu XJ (2017) Extraordinary acoustic transmission at low frequency by a tunable acoustic impedance metasurface based on coupled Mie resonators. Appl Phys Lett 110(23):233502. https://doi.org/10.1063/1.4985154
19. Lu GX, Ding EL, Wang YY, Peng XY, Cui J, Liu XZ, Liu XJ (2017) Realization of acoustic wave directivity at low frequencies with a subwavelength Mie resonant structure. Appl Phys Lett 110(12):123507. https://doi.org/10.1063/1.4979105
20. Landi M, Zhao JJ, Prather WE, Wu Y, Zhang LK (2018) Acoustic Purcell effect for enhanced emission. Phys Rev Lett 120(11):114301. https://doi.org/10.1103/PhysRevLett.120.114301

Chapter 6
Acoustic Focusing by Phase-Controlled Metamaterials

6.1 Introduction

Phase-controlled metamaterials are designed and fabricated using artificial microstructure technology, enabling precise shaping of acoustic wavefronts. By tuning structural parameters of phased unit cells and adjusting their output phases, incident acoustic waves experience different phase delays as they pass through the phase-controlled metamaterials. By precisely arranging these phased unit cells, sound energy can be focused at specific spatial positions. This technology has significant application potential in medical ultrasound diagnosis and imaging, non-destructive testing, architectural design, ocean energy collection, military applications, and more. However, previously demonstrated AF lenses are typically composed of a series of phased unit cells with varying parameters, which increases fabrication complexity and requires high-precision machining. Additionally, most phased unit cells are specific resonant structures with fixed bandwidths, limiting the AF lenses to narrow frequency bands and restricting their practical applications. Therefore, designing broadband AF lenses composed of phased meta-atoms with identical sizes remains a significant challenge.

In this chapter, we numerically design and experimentally demonstrate three types of AF lenses using phase-controlled metamaterials based on the generalized Snell's law. First, we propose a broadband AF lens composed of six types of phased unit cells, each consisting of different numbers of cavity meta-atoms [1]. The phase delay mechanism differs from other phased unit cells by tuning structural parameters. The fractional bandwidth of this AF lens is 0.24. Additionally, we discuss the design and performance of binary-phase AF lenses and reflected AF lenses in detail. Second, we design an ultra-broadband AF lens with phase control, where the phased unit cells consist of different numbers of V-shaped meta-atoms [2]. The measured fractional bandwidths of the AF lenses composed of eight and two types of phased unit cells are 1.12 and 0.76, respectively, demonstrating ultra-broad bandwidth and high robustness. Finally, we propose a flexible AF lens with phase control, constructed from

H. Sun et al., *Acoustic Metamaterials*, Microfluidics and Nanophotonics: Science and Engineering 2, https://doi.org/10.1007/978-981-96-8524-0_6

metafiber bundles [3]. Each metafiber, made of eight circular and narrow rectangular cavities, exhibits flexible characteristics. The fractional bandwidth of the AF effect is 0.2. Furthermore, using the flexibility of the metafiber, we design AF lenses with different focal lengths and incorporate rigid scatterers within the lens structure. We also discuss the applications of metafiber bundles in direction converters, waveform converters, and AF lenses for cylindrical sound sources. The designed AF lenses have potential applications in acoustic communication, acoustic energy harvesting, non-destructive detection, and medical ultrasound.

6.2 Acoustic Focusing by Cavity Meta-Atoms with Phase Control

6.2.1 Design and Performance of Cavity Meta-Atom

Figure 6.1a schematically illustrates a cavity meta-atom composed of a square cavity and two symmetric inclined channels [1]. The length and width of the cavity meta-atom are l and d, respectively. The length of the solid structure, the length of the square cavity, and the width and inclined angle of both channels are h, a, w, and θ, respectively. The solid part of the cavity meta-atom is made of epoxy resin to satisfy the sound-hard boundary condition. Figure 6.1b shows an EM model of the cavity meta-atom. Throughout this work, we use the COMSOL Multiphysics software for numerical simulations. The material parameters are as follows: $\rho = 1180$ kg/m^3, $c_l = 2720$ m/s and $c_t = 1460$ m/s for epoxy resin; $\rho = 1.21$ kg/m^3 and $c = 343$ m/s for air. The parameters l, d, h, a, w and θ are set to 20 mm, 7 mm, 12 mm, 5 mm, 2 mm, and 58°, respectively.

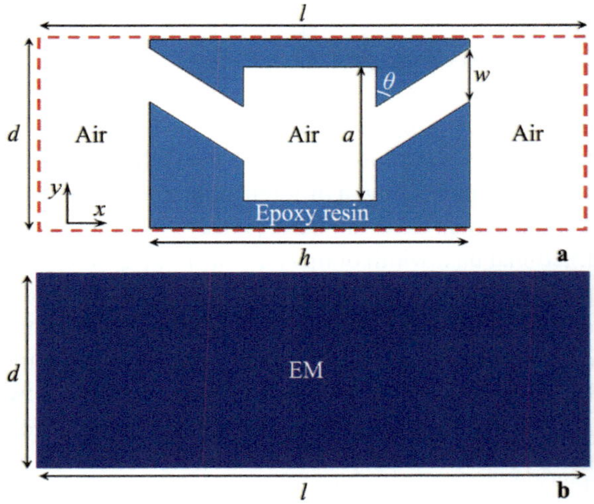

Fig. 6.1 **a** Schematic of the cavity meta-atom and **b** its EM model

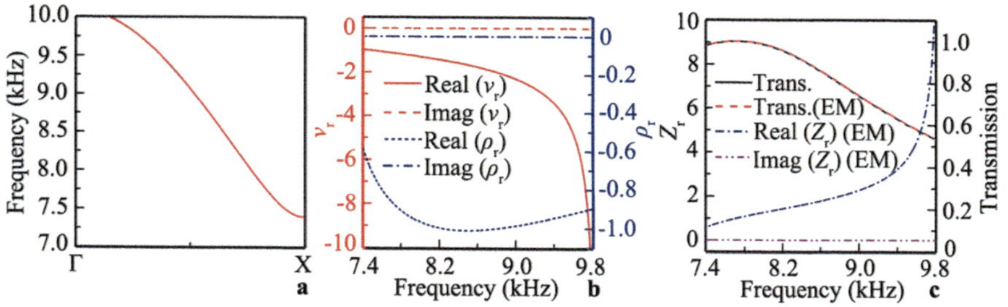

Fig. 6.2 **a** Simulated dispersion relationship of the cavity meta-atom in the ΓX direction. **b** Simulated effective acoustic velocity, density, and **c** acoustic impedance of the EM, and simulated transmission spectra of the cavity meta-atom and its EM

Figure 6.2a presents the dispersion relation of the cavity meta-atom in the ΓX direction. Above 7.4 kHz, the ΓX direction is a pass band, indicating that acoustic waves can propagate through the cavity meta-atom in the x direction. The relative effective velocity v_r, density ρ_r, and impedance Z_r of the EM, normalized by those of air, can be calculated using Eqs. (2.15)–(2.18). Figure 6.2b shows the effective acoustic velocity v_r and density ρ_r. In the range of 7.4–9.8 kHz, the imaginary parts of v_r and ρ_r are negligible, indicating no resonance-induced loss in the cavity meta-atom. Figure 6.2c presents the effective acoustic impedance Z_r and the transmission spectra. The black solid and red dashed lines represent the transmission spectra of the cavity meta-atom and its EM model, respectively. Both spectra agree well, with a transmission coefficient greater than 0.7 in the range of 7.4–9.1 kHz. This is because the real part of Z_r of the EM is approximately 1.0, and the imaginary part is close to zero. Thus, the acoustic impedance of the EM matches well with air, and the high transmission of the cavity meta-atom is attributed to weak viscous loss in air.

6.2.2 Design and Performance of Phased Unit Cell

As shown in Fig. 6.3, we design a phased unit cell composed of n-layer cavity meta-atoms in air, with the incident acoustic wave located on the left side of the phased unit cell. The distance between adjacent meta-atoms (l–h) is 8 mm. Figure 6.4 shows the transmission spectra of phased unit cells with different numbers of cavity meta-atoms, where n is set to 2, 3, 4, 5, and 6. The transmission spectra exhibit similar shapes as n increases, and the transmission coefficients exceed 0.7 in the range of 7.4–9.1 kHz. Thus, the designed phased unit cells have a broad working bandwidth.

Figure 6.5 shows the phase delays and transmission coefficients of phased unit cells with different values of n at 7.95 kHz. As n increases, the phase delay gradually increases, and the transmission coefficient approaches 1.0. The phase delays for the six blue open circles are about $\pi/3$, $2\pi/3$, π, $4\pi/3$, π, $5\pi/3$, and 2π, corresponding to phased unit cells with $n = 1$, 2, 3, 4, 5, and 6, respectively. Thus, the phase delays of the phased unit cells can cover the full 2π range using the selected six cavity

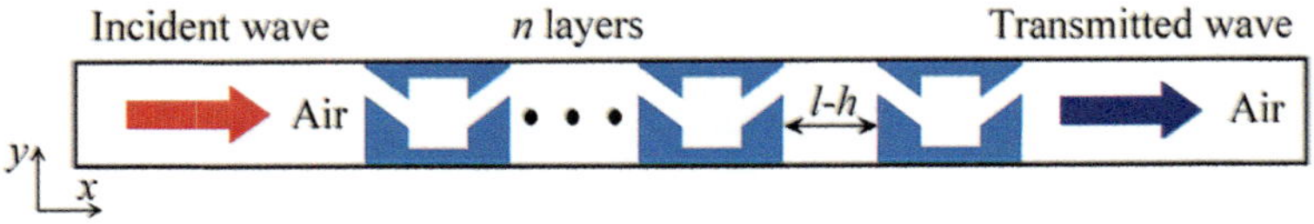

Fig. 6.3 Schematic of the phased unit cell composed of n-layer cavity meta-atoms

Fig. 6.4 Simulated transmission spectra of phased unit cells with different numbers of cavity meta-atoms

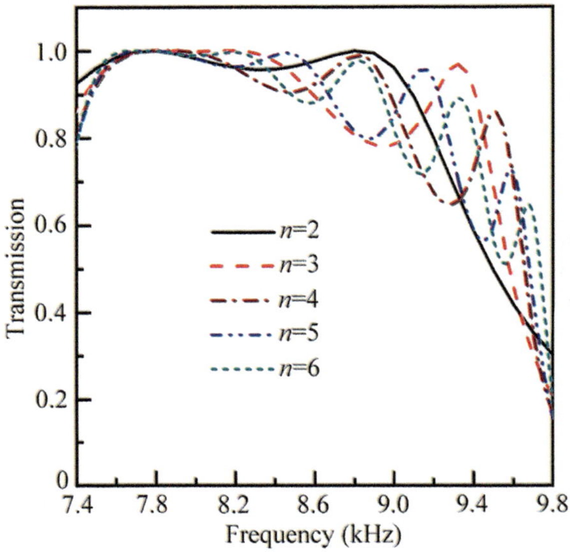

meta-atoms. Figure 6.6 shows the simulated pressure distributions caused by phased unit cells with $n = 1, 2, 3, 4, 5$, and 6. The phase delays are equally spaced from 0 to 2π in steps of $\pi/3$. Therefore, the phased unit cells can arbitrarily control the acoustic wavefront by tuning the number of cavity meta-atoms. This design concept differs from other phased unit cells that adjust structural parameters [4–8] and offers advantages in ease of fabrication and flexible operation.

6.2.3 Design and Performance of AF Lens

Considering an acoustic wave with normal incidence in the x direction and phase distributions in the y direction, the refraction angle θ_t can be determined using Eq. (2.41). To design an AF lens with a focal length of e, the propagation paths of sound through the AF lens are shown in Fig. 6.7, where the refraction angle θ_t is expressed as

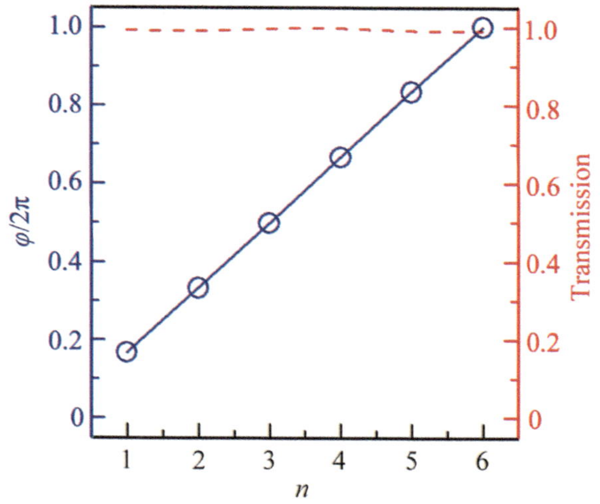

Fig. 6.5 Simulated phase delays (the blue solid line) and transmission coefficients (the red dash line) of phased unit cells with different values of n at 7.95 kHz

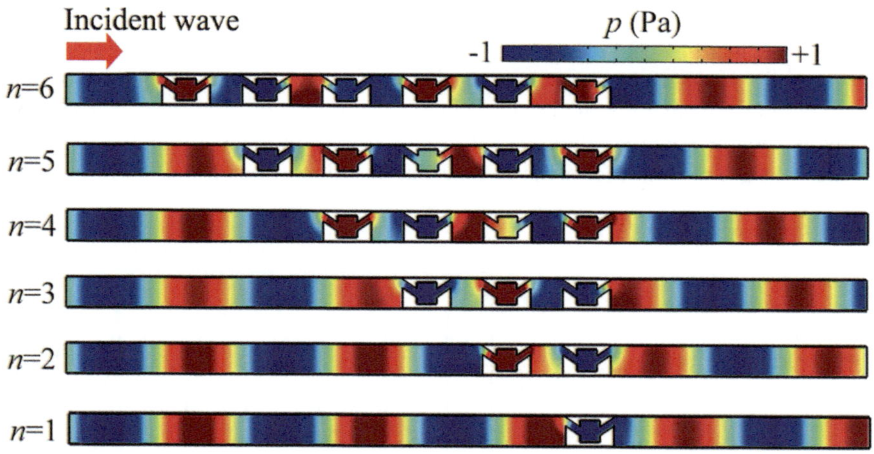

Fig. 6.6 Simulated pressure distributions caused by phased unit cells with different values of n at 7.95 kHz

$$\sin\theta_t = -\frac{y}{\sqrt{e^2 + y^2}}. \tag{6.1}$$

Therefore, the phase distribution of the AF lens in the y direction satisfies [6]:

$$\varphi(y) = -k\sqrt{y^2 + b^2}. \tag{6.2}$$

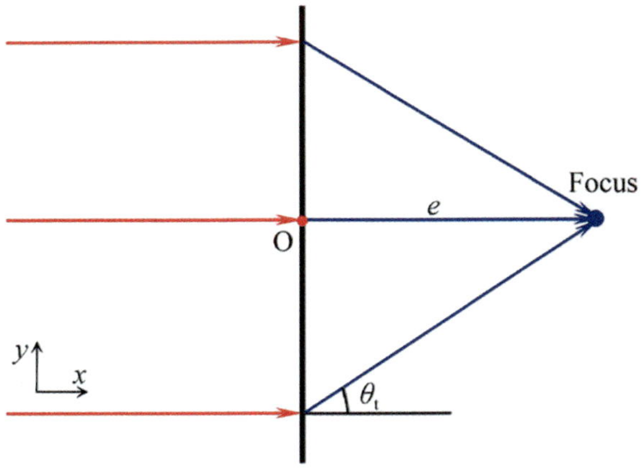

Fig. 6.7 Schematic of acoustic propagation paths through the AF lens

In the AF lens, the parameters are set to $e = 30$ cm, $f = 7.95$ kHz, and $c = 343$ m/s. The theoretical continuous phase distribution in the y direction is shown in Fig. 6.8. Based on the phase delays of the six types of phased unit cells, we select 81 discrete phase delays in the y direction to design the AF lens, as shown in Fig. 6.9 [1].

Figure 6.10 shows the intensity distribution caused by the AF lens with and without viscous loss in air. As shown in Fig. 6.10a, the acoustic wave transmits through the lens, and the focus is located at (26 cm, 0) on the right side of the lens. Thus, the focal length is 26 cm, close to the theoretical value of 30 cm. Additionally, as shown in Fig. 6.10b, the AF effect with viscous loss is almost the same as that in Fig. 6.10a. Figure 6.11a, b present the intensity distributions along lines I and II in Fig. 6.10,

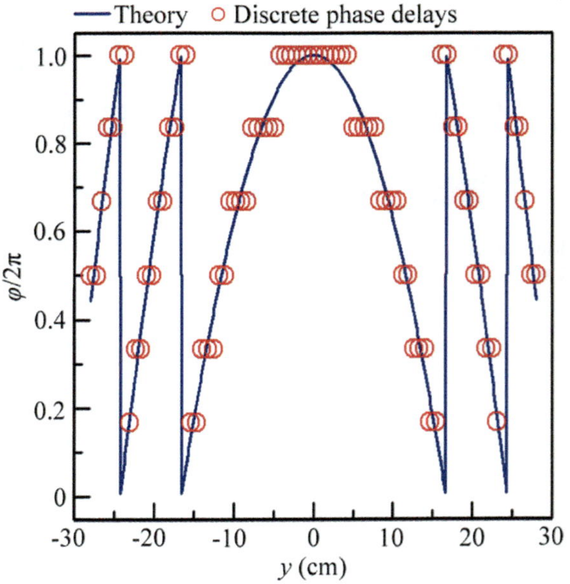

Fig. 6.8 Theoretical continuous phase distribution and discrete phase delays of the AF lens with the six types of phased unit cells

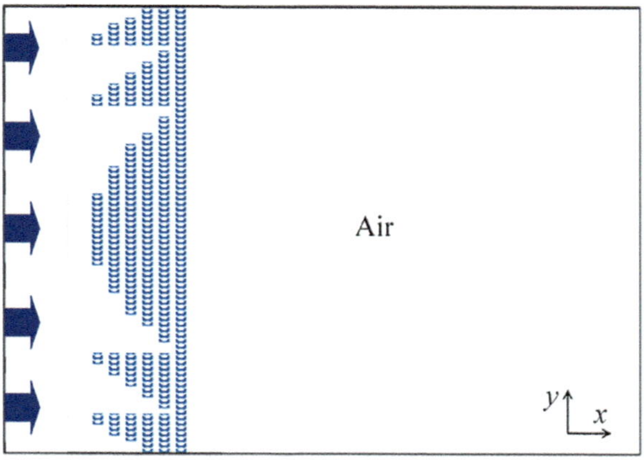

Fig. 6.9 Schematic of the AF lens with the six types of phased unit cells

respectively, with results in free space shown for comparison. Compared to free space, the intensity at the focus increases by about 10 times for the AF lens without viscous loss. Moreover, the intensity at the focus decreases by only 4.6% with viscous loss. Thus, the influence of viscous loss on focusing performance is negligible, and the lens exhibits high-performance AF.

The designed AF lens has a broad working bandwidth. Figure 6.12 shows the intensity distributions caused by the AF lens at different frequencies, with the lens parameters identical to those in Fig. 6.10a. The AF effect is evident at these frequencies, and the working bandwidth reaches 2.1 kHz. This results from the similar phase distributions of the lens and the high transmission of the phased unit cells within the AF working band. To demonstrate the broadband characteristic of the AF lens, we calculate the Q factor, defined as $Q = f_0/\Delta f$ [9], where f_0 and Δf are the center frequency and the bandwidth, respectively.

Figure 6.13 shows the maximum intensity spectrum at the focus. The parameters f_0 and Δf are about 8.4 kHz and 2.1 kHz, respectively, indicating a Q factor of about 4.0. Thus, the proposed AF lens has the advantage of broad bandwidth.

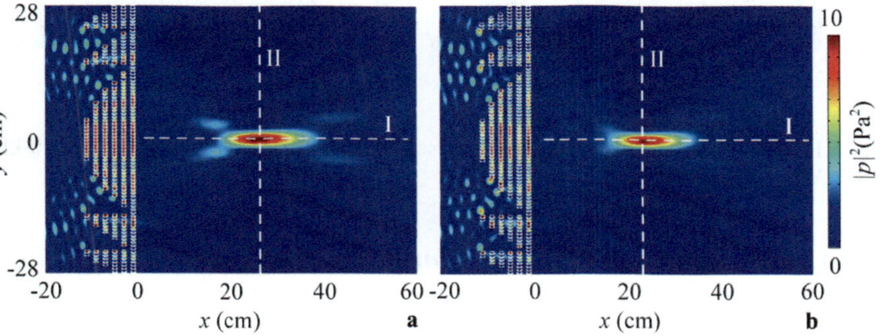

Fig. 6.10 Simulated intensity distributions caused by the AF lens **a** without and **b** with viscous loss of air at 7.95 kHz

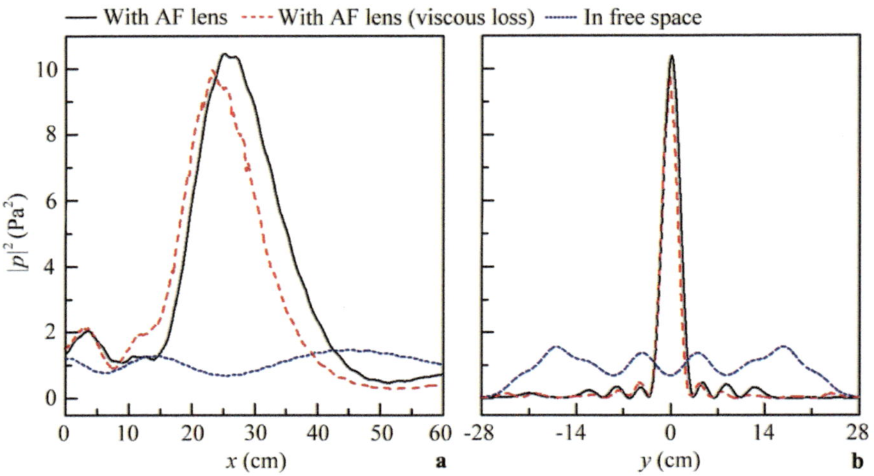

Fig. 6.11 Simulated intensity distributions along lines **a** I and **b** II in Fig. 6.10a, b

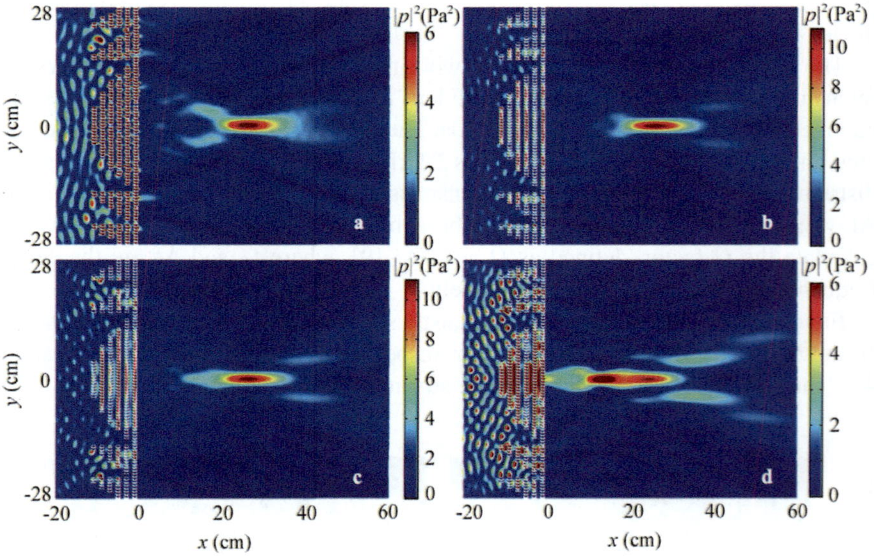

Fig. 6.12 Simulated intensity distributions caused by the AF lens at **a** 7.56 kHz, **b** 8.50 kHz, **c** 9.00 kHz, and **d** 9.66 kHz

6.2.4 Experimental Measurement

Figure 6.14a shows a sample photograph. The sample consists of a base plate and cavity meta-atoms fabricated using epoxy resin via 3D printing. The base plate measures 567 mm in length and 120 mm in width, and the sample parameters match

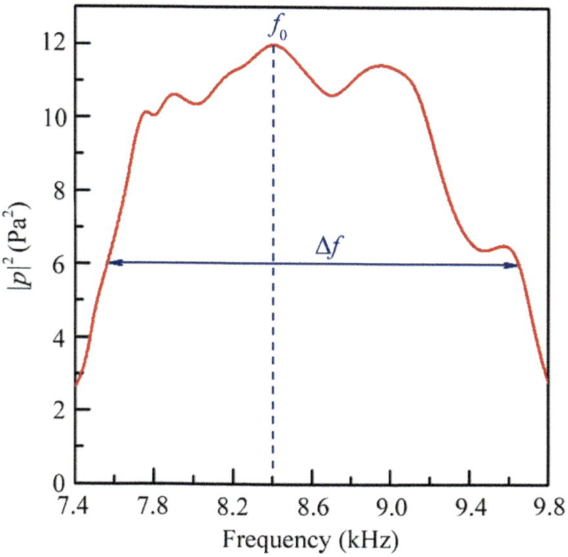

Fig. 6.13 Simulated maximum intensity spectrum at the focus

those in Fig. 6.10a. Figure 6.14b illustrates the schematic of the sample assembly. The base plate features a series of raised cylinders on its surface, and the cavity structure is fixed to the base plate using four of these raised cylinders. Each cavity meta-atom has four holes on its bottom surface, matching the size of the raised cylinders, as shown in Fig. 6.14b. This design allows the cavity meta-atoms to be securely attached to the base plate, forming the AF lens. After the experiment, the cavity meta-atoms can be easily separated from the base plate and reused for other acoustic devices. Compared to other phased unit cells with varying geometric parameters, the cavity meta-atoms, being of the same size, are easier to fabricate and assemble without the need to identify different geometric parameters.

Figure 6.15 shows the experimental set-up. The experiment is conducted in a planar waveguide composed of two parallel plates measuring 2 m × 2 m × 1 cm. Wedge-shaped absorbing foams at the boundaries of the planar waveguide create an anechoic environment. The incident plane wave is generated by a speaker array driven by a power amplifier. Acoustic pressure distributions are scanned using a microphone (Brüel & Kjær type-4961) moved by a set of 2D motorized linear stages (Newport: MIN300CC and ILS250CC). Another microphone is fixed toward the speaker array as a reference. The sound signal is recorded by the Brüel & Kjær 3160-A-022 module and analyzed using the PULSE Labshop software.

Figure 6.16a shows the simulated intensity distribution caused by the AF lens. The upper insets display the measured and simulated intensity distributions at the focus, which agree well with each other. Additionally, the intensity distributions along lines III and IV in Fig. 6.16a are presented in Fig. 6.16b. We observe that the intensity distributions agree very well between the measurement and the simulation.

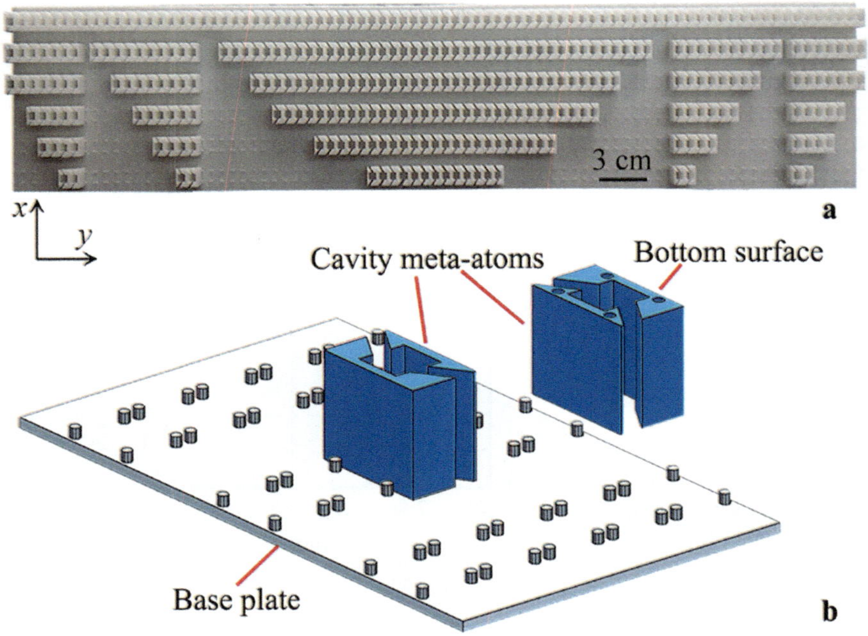

Fig. 6.14 **a** Photograph of the sample and **b** schematic of the sample assembly

Fig. 6.15 Schematic of the experimental set-up for the AF lens in Fig. 6.14a

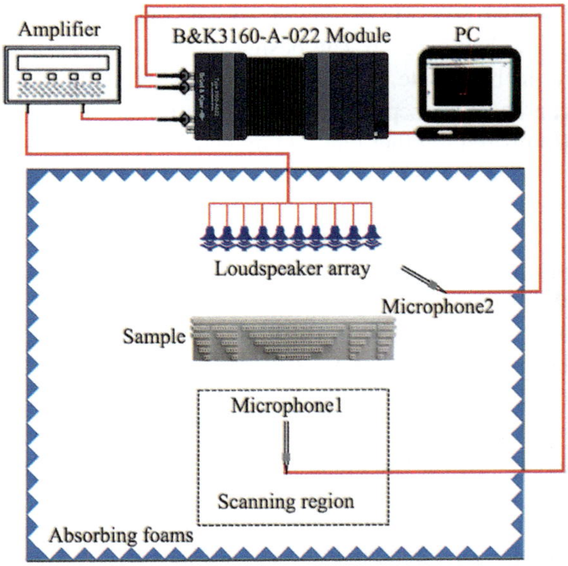

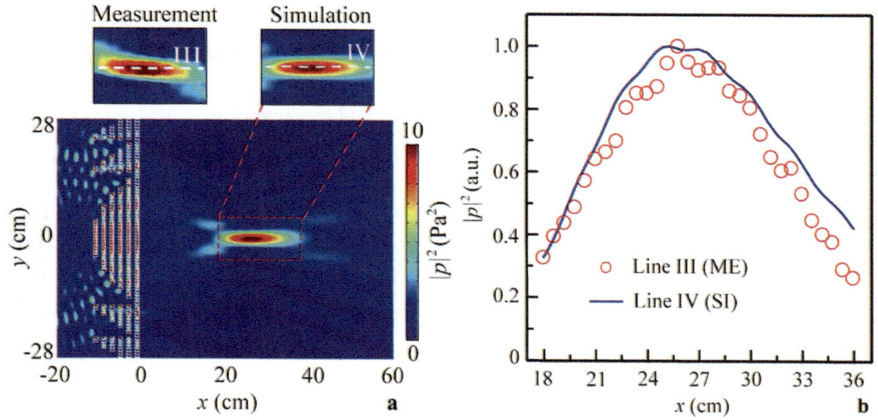

Fig. 6.16 **a** Measured and simulated intensity distributions caused by the AF lens at 7.95 kHz, and **b** the intensity distributions along lines III and IV in (**a**)

The aforementioned results demonstrate the performance of the AF lens with the six types of phased unit cells.

6.2.5 Performance Modulation

Figure 6.17a, b show the theoretical continuous phase distributions and discrete phase delays of the AF lenses with focal lengths of 10 cm and 50 cm, respectively. The parameters of the six types of phased unit cells are the same as those in Fig. 6.5. Based on the discrete phase delays in Fig. 6.17a, b, we design AF lenses with focal lengths of 10 and 50 cm, as shown in Fig. 6.18a, b, respectively. The AF effect exists for both lenses, with focal lengths of 7 and 44 cm, respectively, close to their theoretical values. This demonstrates the ability to design AF lenses with different focal lengths.

6.2.6 Design and Performance of Binary-Phase AF Lens

To further simplify the AF lens structure, we design an AF lens using two types of phased unit cells with a phase difference of π, while keeping the other parameters unchanged. We select the phased unit cells with $n = 1$ and 4 for this design. Figure 6.19 shows the theoretical continuous phase distributions and discrete phase delays of the AF lens with the two types of phased unit cells. The corresponding AF lens is schematically shown in Fig. 6.20.

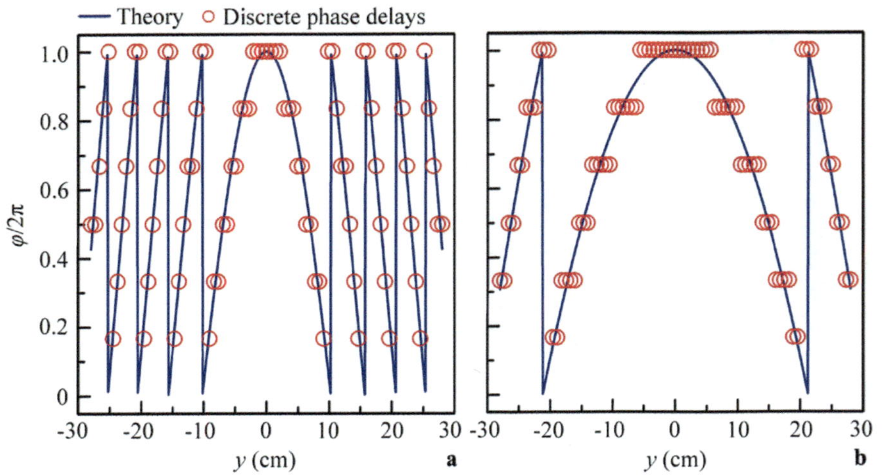

Fig. 6.17 Theoretical continuous phase distribution and discrete phase delays of the AF lenses with focal lengths of **a** 10 cm and **b** 50 cm at 7.95 kHz

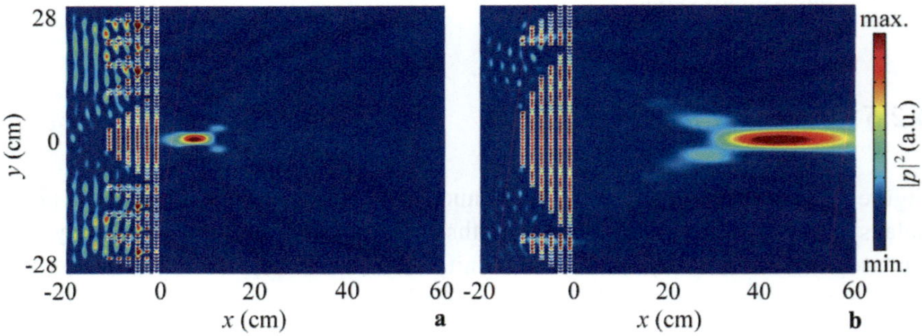

Fig. 6.18 Simulated intensity distributions caused by the AF lenses with focal lengths of **a** 10 cm and **b** 50 cm at 7.95 kHz

Figure 6.21 shows the intensity distributions caused by the AF lens. A clear focus is obtained at the position of (25 cm, 0). Figure 6.22 presents the longitudinal and transverse intensity distributions through the focus with and without the AF lens. The intensity at the focus caused by the AF lens is about 8 times greater than that in free space, indicating that the AF lens with the two types of phased unit cells exhibits high performance. Compared to the AF lens with six types of unit cells, the intensity at the focus decreases significantly in Fig. 6.22. However, the proposed AF lens, composed of only two types of phased unit cells, offers advantages in ease of fabrication and high robustness.

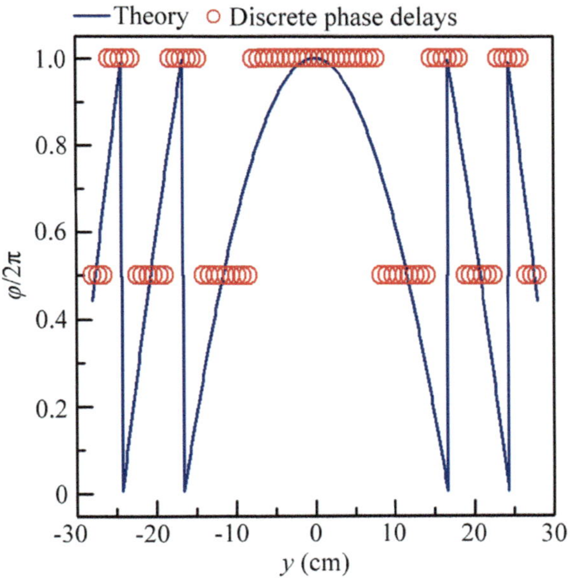

Fig. 6.19 Theoretical continuous phase distribution and discrete phase delays of the AF lens with the two types of phased unit cells ($n = 1$ and 4)

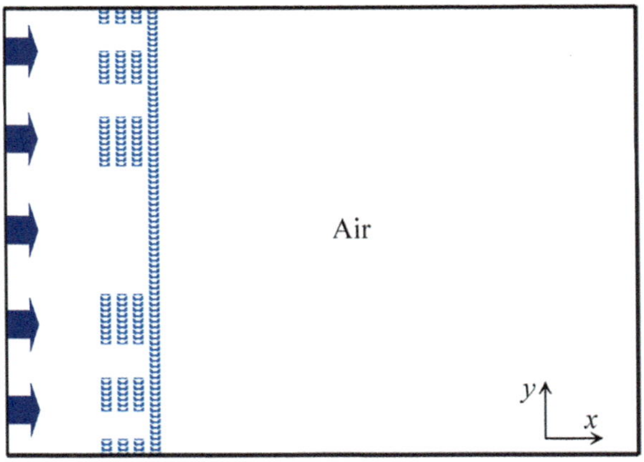

Fig. 6.20 Schematic of the AF lens with the two types of phased unit cells

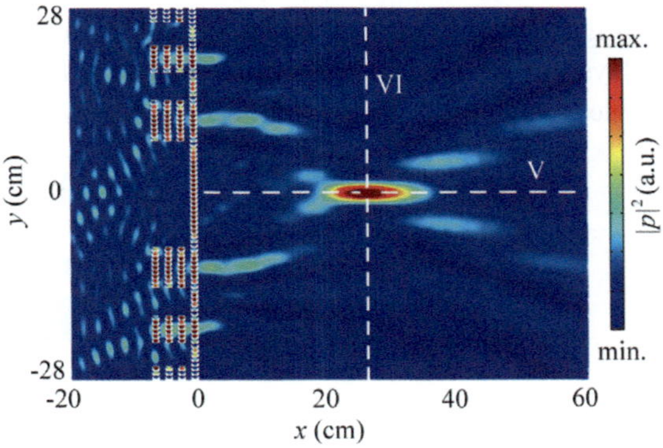

Fig. 6.21 Simulated intensity distributions caused by the AF lens with the two types of phased unit cells at 7.95 kHz

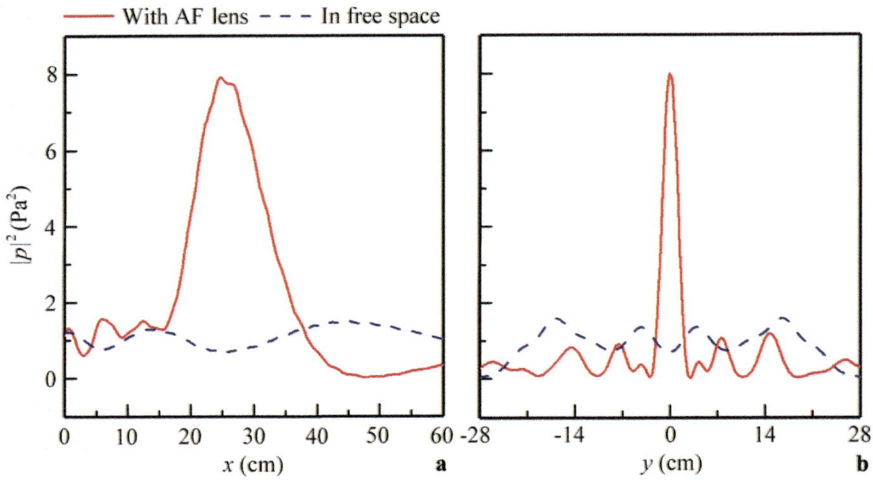

Fig. 6.22 Simulated intensity distributions along lines **a** V and **b** VI with and without the AF lens in Fig. 6.21

6.2.7 Design and Performance of Reflected AF Lens

Finally, we design a reflected AF lens based on cavity meta-atoms. Figure 6.23 shows the reflected unit cell composed of *n*-layer cavity meta-atoms. The left side of the unit cell is set as a sound-hard boundary, and the incident acoustic wave is located on the right side.

Figure 6.24 shows the phase delays of the reflected unit cells with different values of *n*. The phase delays for $n = 1$, 2, and 3 are about 0.54π, 1.24π, and 2π, respectively. As *n* increases, the reflected phase delay gradually increases, covering the full 2π range with only three cavity meta-atoms. This is because the propagation path of sound in the reflected unit cell is about twice that in the transmitted unit cell. Figure 6.25 shows the pressure distributions caused by the reflected unit cells with $n = 1$, 2, and 3. As *n* increases, the reflected phase delay increases. Based on the three types of reflected unit cells, we design a reflected AF lens.

Figure 6.26 shows the theoretical continuous phase distribution and discrete phase delays of the reflected AF lens, with parameters *e*, *f*, and *c* matching those of the AF lens in Fig. 6.8. Figure 6.27 schematically illustrates the reflected AF lens. The

Fig. 6.23 Schematic of the reflected unit cell composed of *n*-layer cavity meta-atoms

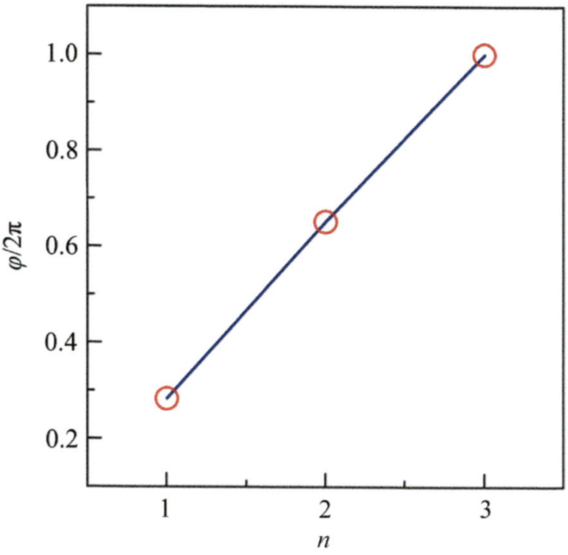

Fig. 6.24 Simulated phase delays of the reflected unit cells with different values of n at 7.95 kHz

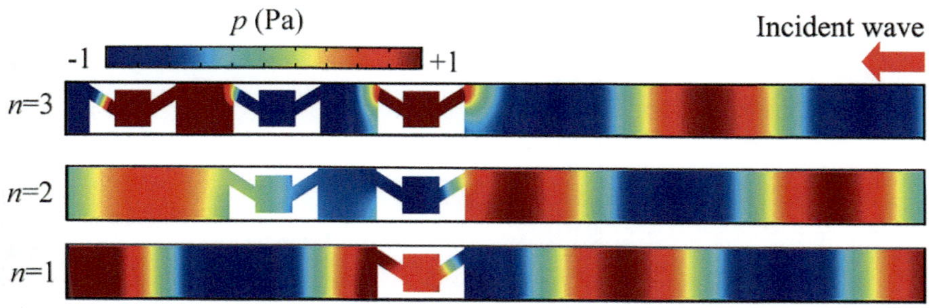

Fig. 6.25 Simulated reflected pressure distributions caused by the reflected unit cells with different values of n at 7.95 kHz

reflected intensity distribution caused by the AF lens is shown in Fig. 6.28. The acoustic wave is reflected by the AF lens, and the reflected sound energy is focused on the right side of the lens.

Figure 6.29 presents the longitudinal and transverse intensity distributions through the focus with and without the AF lens, with field distributions in free space shown for comparison. Compared to free space, the intensity at the focus with the AF lens is increased by 11 times, which is slightly larger than that in Fig. 6.11. This is due to the reflection coefficient of the unit cells being about 1.0, owing to the sound-hard boundary. Therefore, the reflected AF lens achieves high-performance AF.

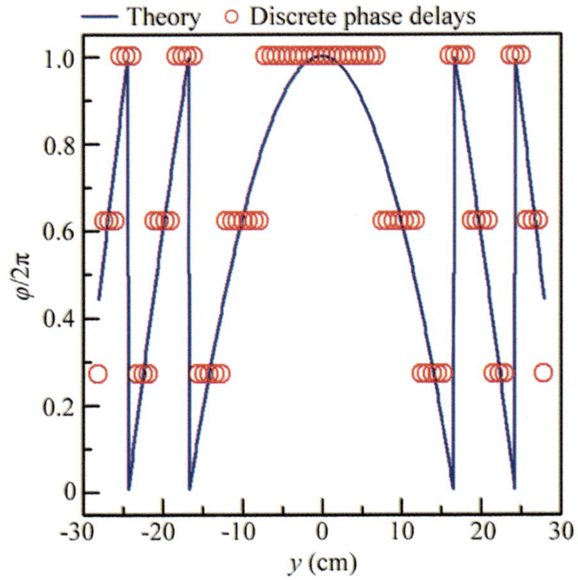

Fig. 6.26 Theoretical continuous phase distribution and discrete phase delays of the reflected AF lens composed of the three types of reflected unit cells

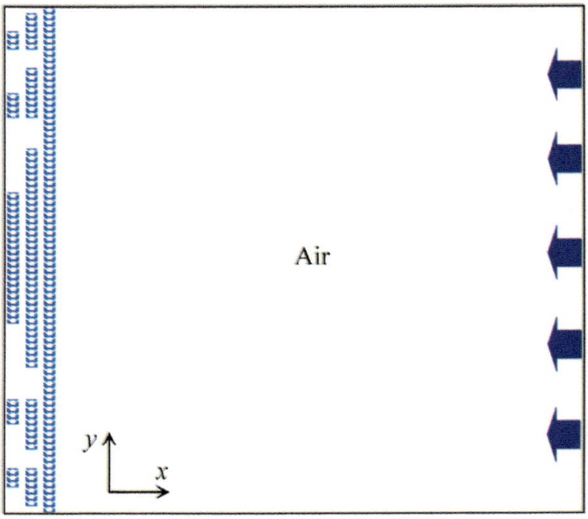

Fig. 6.27 Schematic of the reflected AF lens

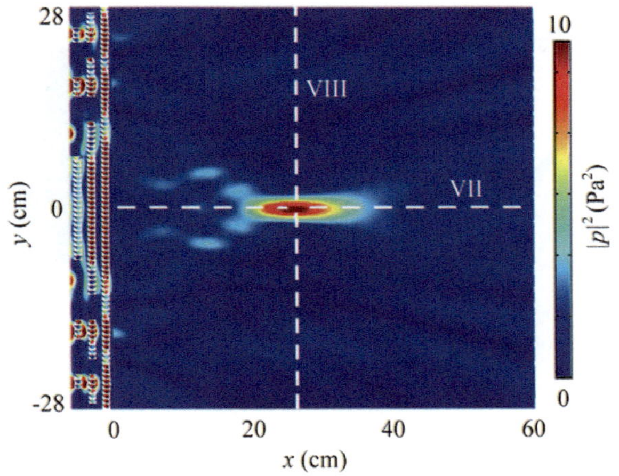

Fig. 6.28 Simulated reflected intensity distributions caused by the reflected AF lens at 7.95 kHz

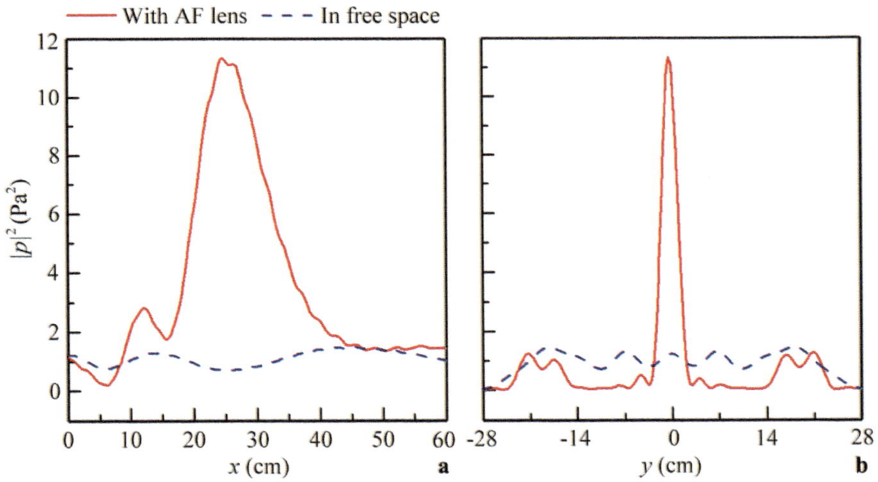

Fig. 6.29 Simulated reflected intensity distributions along lines **a** VII and **b** VIII with and without the reflected AF lens in Fig. 6.28

6.3 Ultra-Broadband Acoustic Focusing by V-Shaped Meta-Atoms with Phase Control

6.3.1 Design and Performance of V-Shaped Meta-Atom

As shown in Fig. 6.30a, the proposed V-shaped meta-atom consists of two symmetric inclined slides made of epoxy resin [2]. The structure of the V-shaped meta-atom is determined by the following parameters: $l = 14.6$ mm, $h = 9.0$ mm, $\theta = 78°$, $w =$

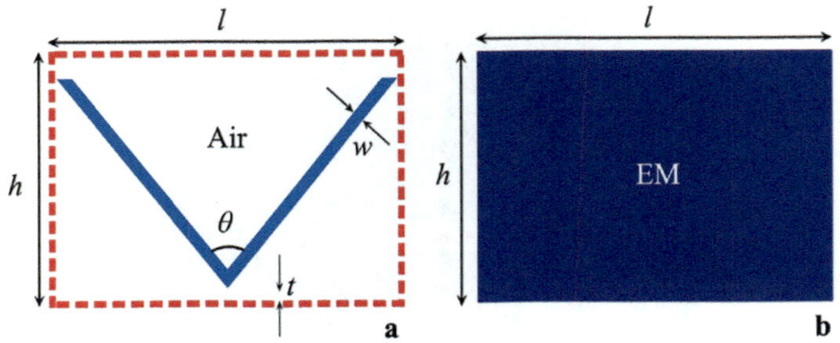

Fig. 6.30 **a** Schematic of the V-shaped meta-atom and **b** its EM model

0.5 mm, and $t = 0.5$ mm. Throughout this work, we use the COMSOL Multiphysics software to simulate the characteristics of AF. The V-shaped meta-atom is made of epoxy resin to satisfy the sound-hard boundary condition. The material parameters used in the models are as follows: density $\rho = 1.21$ kg/m^3 and sound velocity $c =$ 343 m/s for air; $\rho = 1180$ kg/m^3 and $c = 2720$ m/s for epoxy resin. Figure 6.30b shows an EM model of the meta-atom, which is used to simulate its effective parameters based on Eqs. (2.15)–(2.18).

Figure 6.31 shows the dispersion relationship of the V-shaped meta-atom. We can see that the ΓX direction exhibits a pass band below 13.0 kHz, indicating that acoustic waves can transmit through the V-shaped meta-atom in the x direction. As shown in Fig. 6.32a, the real part of the effective acoustic impedance Z_r is below 4 for frequencies under 13.0 kHz, with most values around 1.0 below 9.0 kHz. Additionally, the imaginary part of Z_r is close to zero, demonstrating high performance of the V-shaped meta-atom. We also simulate the relative refractive index n_r of the meta-atom and compare it with the EM model, as shown in Fig. 6.32b. Both results agree well below 13.0 kHz.

6.3.2 Design and Performance of Phased Unit Cell

By varying the number (N) of V-shaped meta-atoms, we design a phased unit cell with different phase delays and high transmission, which is shown in Fig. 6.33. In this design, adjacent meta-atoms are placed close to each other. The incident sound source is located on the left side of the phased unit cell, and the upper and lower boundaries (the red solid lines) are set as periodic boundary conditions. Figure 6.34a shows the relative refractive index n_r of the phased unit cells with different numbers of meta-atoms. As N increases, the refractive index spectra exhibit a gentle and similar trend, with nearly equal differences between adjacent curves. Additionally, the transmission spectra of the phased unit cells have similar shapes, with transmission coefficients greater than 0.7 below 13.0 kHz, as shown in Fig. 6.34b. This demonstrates the feasibility of designing an AF lens with an ultra-broadband characteristic.

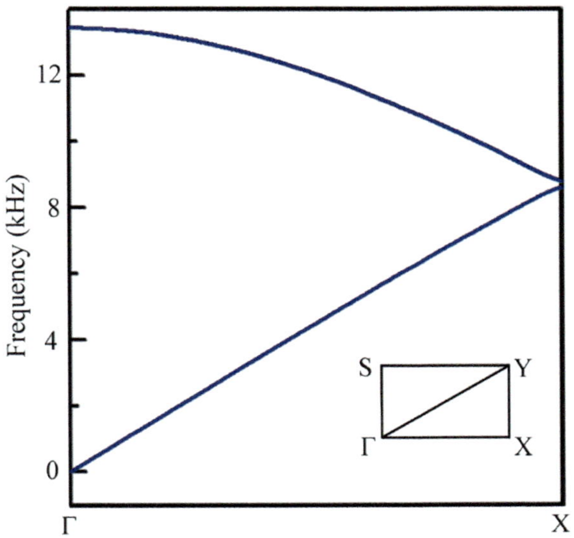

Fig. 6.31 Simulated dispersion relationship of the V-shaped meta-atom

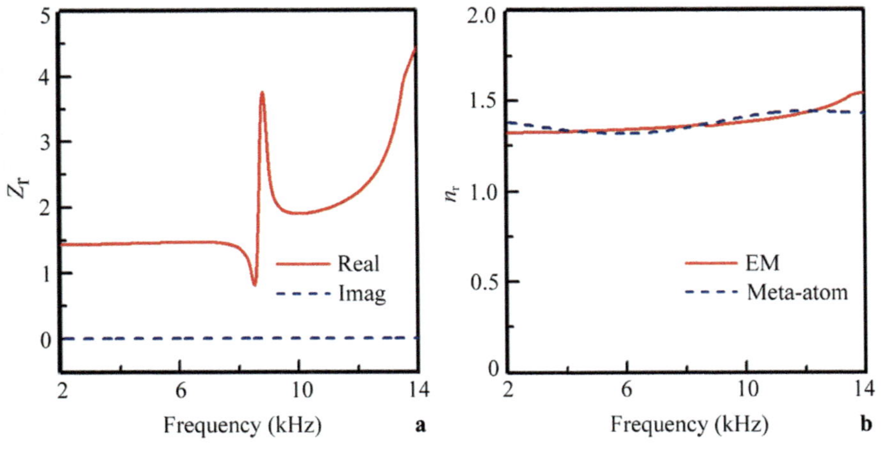

Fig. 6.32 **a** Simulated effective acoustic impedance of the EM and **b** relative refractive index spectra of the meta-atom and the EM

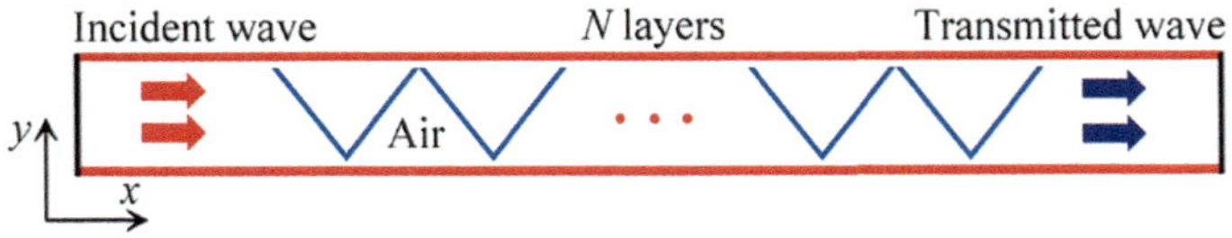

Fig. 6.33 Schematic of the phased unit cell composed of N-layer V-shaped meta-atoms

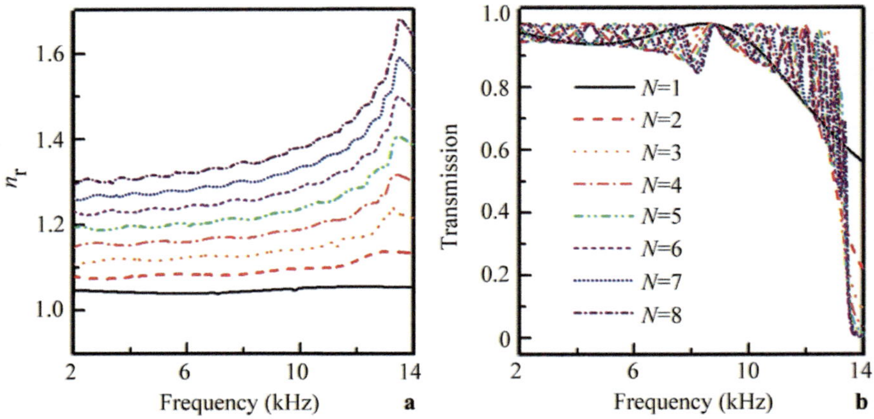

Fig. 6.34 **a** Simulated relative refractive index n_r and **b** transmission spectra of the phased unit cells with different values of N

Next, we simulate the transmission coefficients and phase delays of the phased unit cells with different values of N at 8.58 kHz, as shown in Fig. 6.35. For N ranging from 1 to 8, the transmission coefficients exceed 0.9, and the phase delays increase gradually, covering the entire 2π range with a step of $\pi/4$. For comparison, we theoretically calculate the phase delays of the phased unit cells based on the relative refractive index n_r of the EM at 8.58 kHz and the wave number k in air. The theoretical phase delays are consistent with the simulated results. To further illustrate this, we simulate the pressure distributions caused by phased unit cells with different values of N. Figure 6.36 presents the simulated pressure distributions, showing that as N increases, the phase delays gradually increase and cover the entire 2π range.

Fig. 6.35 Simulated transmission coefficients and phase delays of the phased unit cells with different values of N at 8.58 kHz

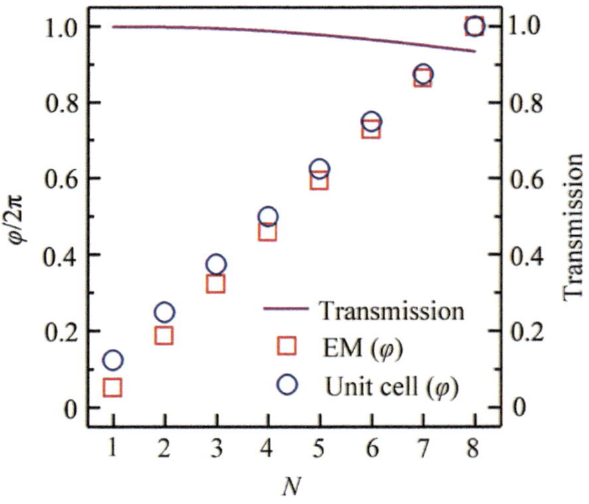

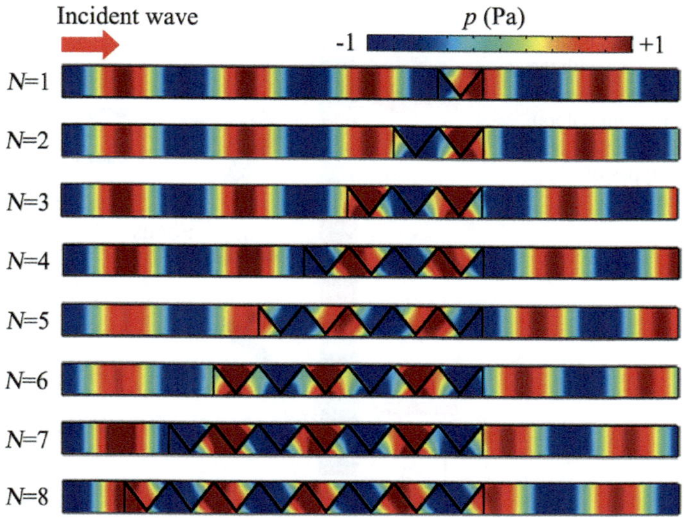

Fig. 6.36 Simulated pressure distributions caused by the phased unit cells with different values of N at 8.58 kHz

6.3.3 Design and Performance of Ultra-Broadband AF Lens

Based on Eq. (6.2), we calculate the theoretical and discrete phase distributions of the AF lens using the parameters $b = 30$ cm, $f = 8.58$ kHz, and $c = 343$ m/s. The acoustic propagation paths through the AF lens with the eight types of phased unit cells and the sample photograph of are shown in Fig. 6.37a, b, respectively. Figure 6.38 shows the phase distribution of the AF lens with the eight types of phased unit cells, where the blue solid lines represent the theoretical continuous phase distribution and the red open circles represent the discrete phase delays.

Next, we simulate the intensity distribution caused by the AF lens, with the initial amplitude of the incident wave set to 1.0 Pa. As shown in Fig. 6.39, the incident wave passes through the AF lens and focuses on a point (29.2 cm, 0) on the right side, close to the theoretical focal length of 30 cm. We measure the intensity distribution in the focusing region (red open square R1), and the experimental set-up is schematically shown in Fig. 6.40. The sample is placed in a planar waveguide composed of two parallel plates (2 m × 2 m × 1 cm), with wedge-shaped absorbing foams placed at four boundaries to create an anechoic environment. The incident wave is generated by a speaker array driven by a power amplifier. Two microphones (Brüel & Kjær type-4961) are used to measure sound pressure. Microphone 1 is moved flexibly by a set of 2D motorized linear stages to measure sound amplitudes, while microphone 2 is fixed toward the speaker array to detect reference signals. The measured data is recorded and analyzed using the Brüel & Kjær 3160-A-022 module.

As shown in the insets on the right side of Fig. 6.39, the measured intensity distribution of the focus matches well with the simulated result, demonstrating the

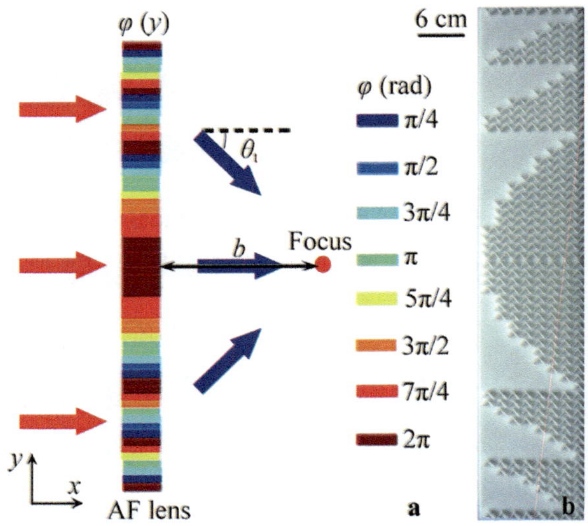

Fig. 6.37 **a** Schematic of acoustic propagation paths through the AF lens with the eight types of phased unit cells (shown in eight colors). **b** Photograph of the AF lens with the eight types of phased unit cells

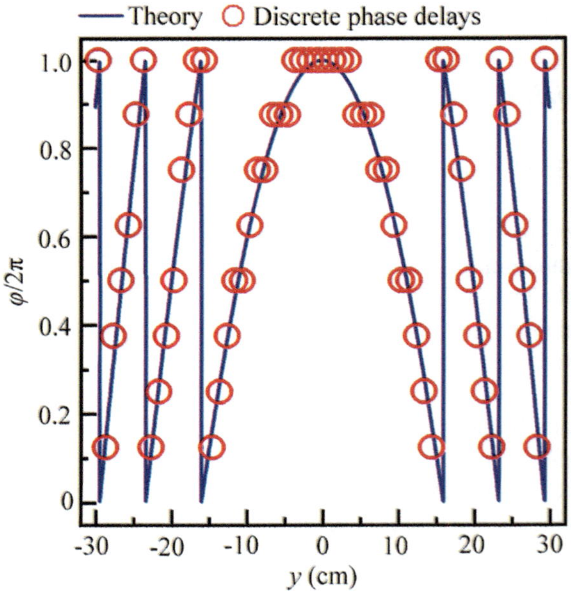

Fig. 6.38 Theoretical continuous phase distribution and discrete phase delays of the AF lens with the eight types of phased unit cells

feasibility of the AF lens. To quantify the focusing performance, we display the normalized intensity distributions along lines I and II for both simulated and measured results in Fig. 6.39, as shown in Fig. 6.41. The corresponding result in free space is also plotted for comparison. The maximum value of $|p|^2$ is enhanced by 10 times compared to free space, indicating high performance of the AF lens. The measured intensity distributions in both directions agree well with the simulated ones. To show

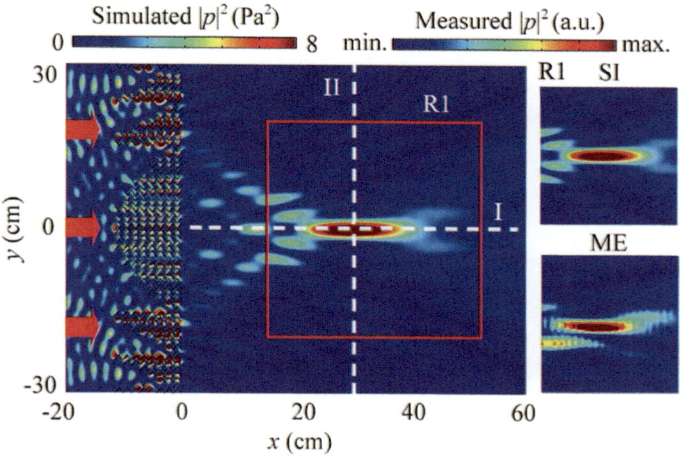

Fig. 6.39 Measured and simulated intensity distributions caused by the AF lens at 8.58 kHz. The red arrows represent the incident acoustic wave

Fig. 6.40 Schematic of the experimental set-up for the AF lens in Fig, 6.37b

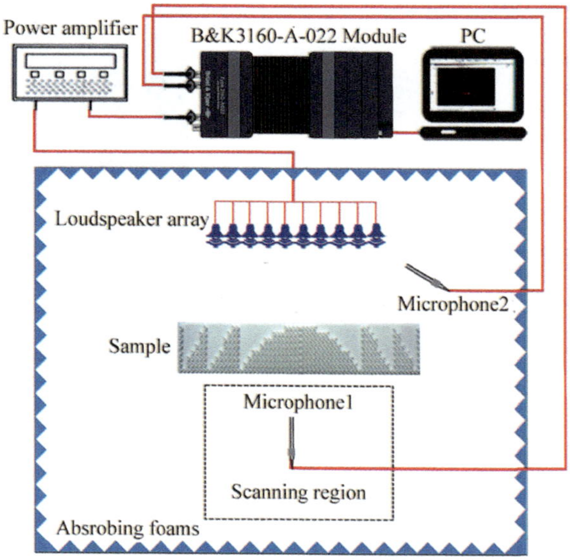

the working bandwidth of the lens, we simulate the intensity spectrum along line II in Fig. 6.39, as shown in Fig. 6.42. In the range of 3.75–13.25 kHz, the maximum value of $|p|^2$ at the focus exceeds 4.0 Pa^2. Based on this, the working bandwidth of the AF lens reaches 9.5 kHz, with a fractional bandwidth of about 1.12, demonstrating an ultra-broadband characteristic of the lens.

In addition to the aforementioned results, we further experimentally verify the ultra-broadband characteristic of the AF lens. Figure 6.43 shows the measured and

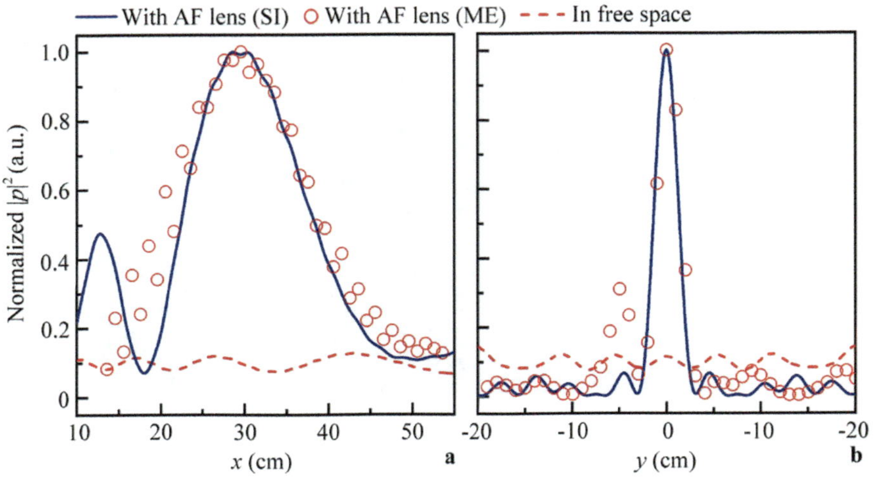

Fig. 6.41 Measured and simulated intensity distributions along lines **a** I and **b** II in Fig. 6.39

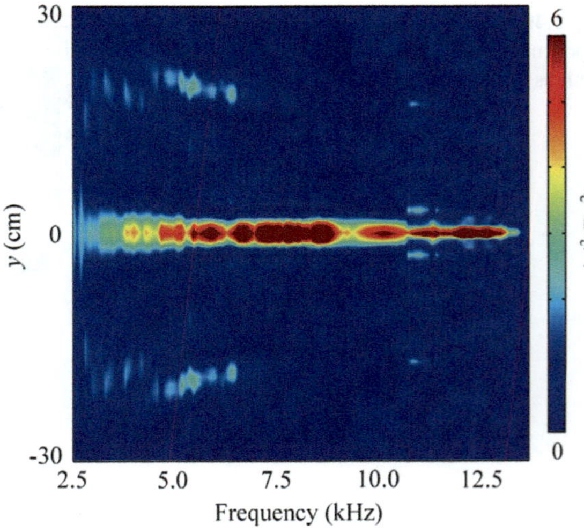

Fig. 6.42 Simulated intensity spectrum along line II in Fig. 6.39

simulated intensity distributions caused by the AF lens at different frequencies. The measured results and their corresponding simulations in red open squares R2–R5 are plotted on the upper and bottom sides. The measurements and simulations agree well with each other, as the refractive index spectra of the eight types of unit cells exhibit a gentle and similar trend below 13.0 kHz (Fig. 6.34a). Thus, the ultra-broadband characteristic of the AF lens is experimentally demonstrated.

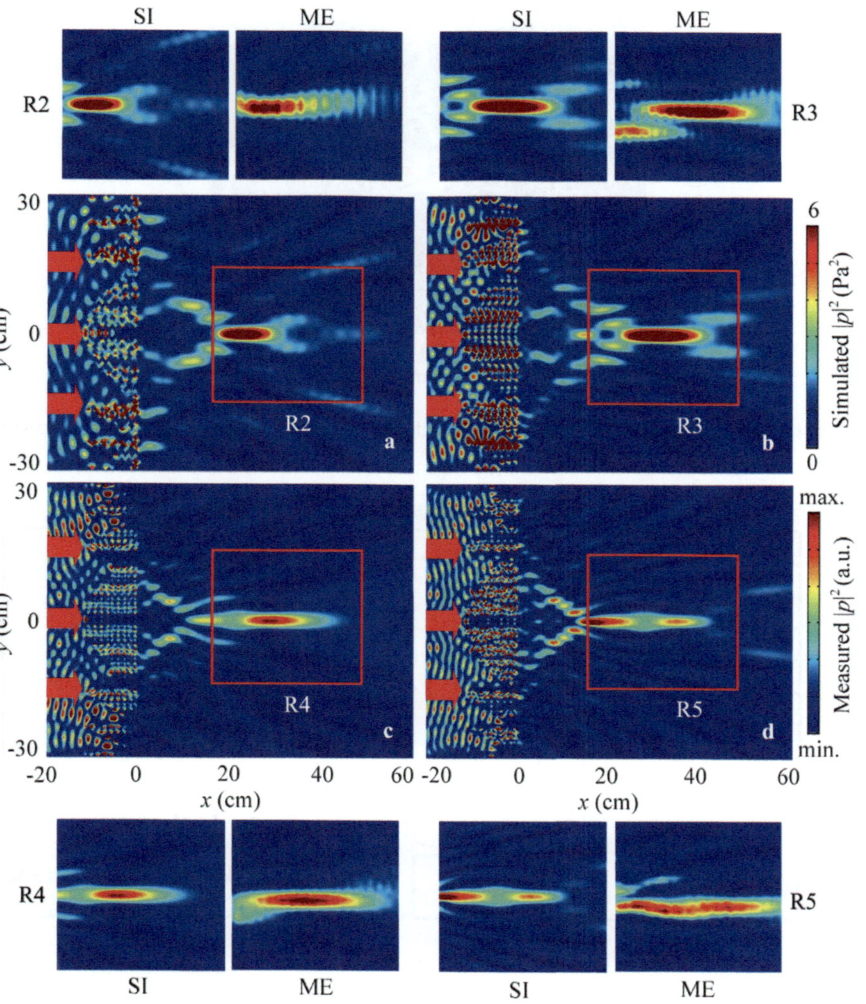

Fig. 6.43 Measured and simulated intensity distributions caused by the AF lens at **a** 6.7 kHz, **b** 8.7 kHz, **c** 10.0 kHz and **d** 11.5 kHz. The eight insets show measured and simulated intensity distributions in red open squares R2–R5

6.3.4 Parameter Analysis

We also discuss the influence of viscous loss on the AF effect. As shown in Fig. 6.44, introducing the viscous loss of air has almost no effect on the AF performance compared to the case without viscous loss (Fig. 6.39). To further illustrate this, we simulate the intensity distributions along the focus with and without viscous loss, as shown in Fig. 6.45. The maximum intensity decreases by only 4% with viscous loss, indicating that the viscous loss of the AF lens is negligible.

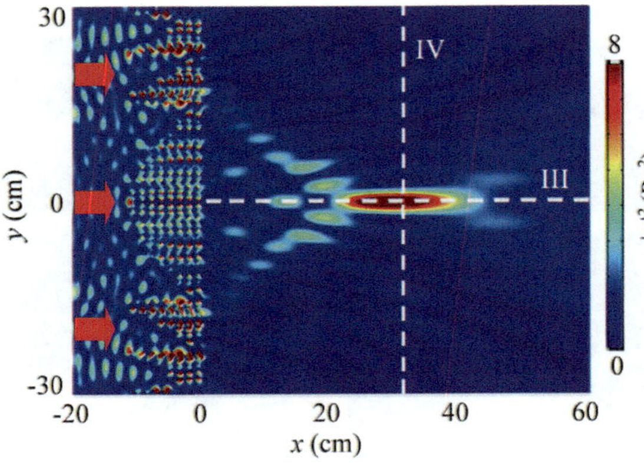

Fig. 6.44 Simulated intensity distributions caused by the AF lens with viscous loss of air at 8.58 kHz

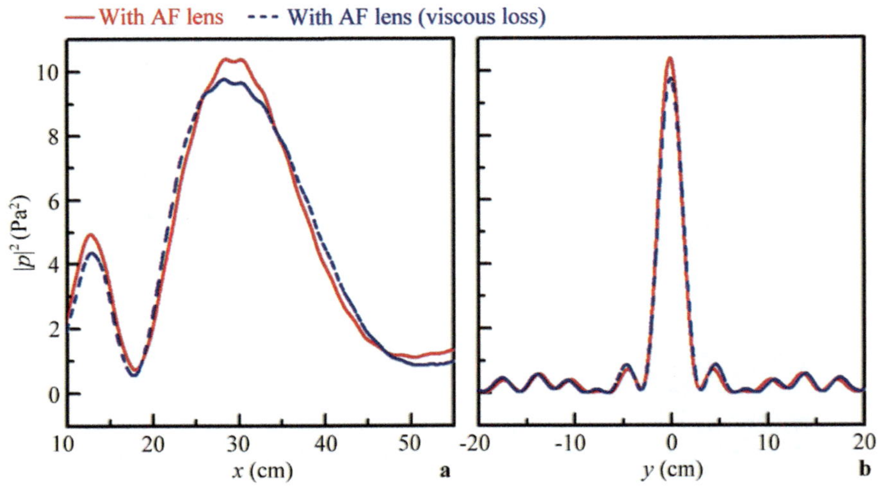

Fig. 6.45 Simulated intensity distributions along lines **a** I and III and **b** lines II and IV in Fig. 6.39 and 6.44

6.3.5 Design and Performance of Binary-Phase AF Lens

To demonstrate the robustness of the AF lens, we design a lens composed of two types of phased unit cells (Fig. 6.46a) with a phase difference of π, corresponding to $N = 4$ and 8. The sample photograph is shown in Fig. 6.46b. The phase distribution of the AF lens is displayed in Fig. 6.47. Figure 6.48 shows the simulated intensity distribution caused by the AF lens at 8.58 kHz. The AF effect is still present, and the maximum intensity at the focus is close to that in Fig. 6.39.

Figure 6.49 presents the simulated intensity spectrum along the white dashed line in Fig. 6.48. The working bandwidth of the AF lens is 5.4 kHz, with a fractional

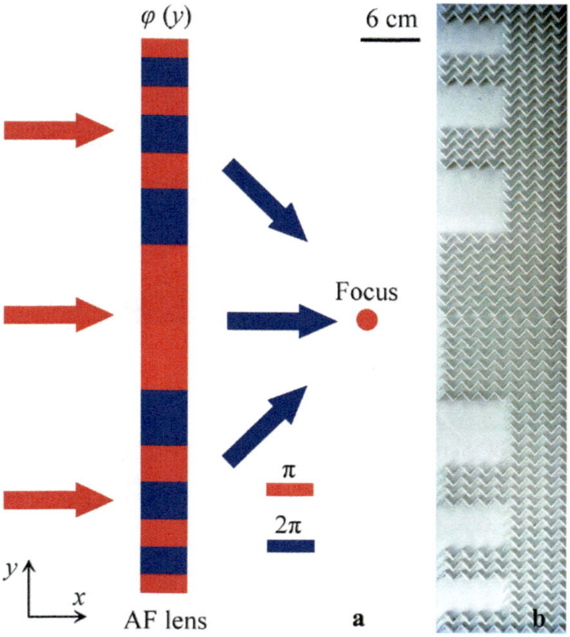

Fig. 6.46 **a** Schematic of acoustic propagation paths through the AF lens with two types of phased unit cells ($N = 4$ and 8). **b** Photograph of the AF lens

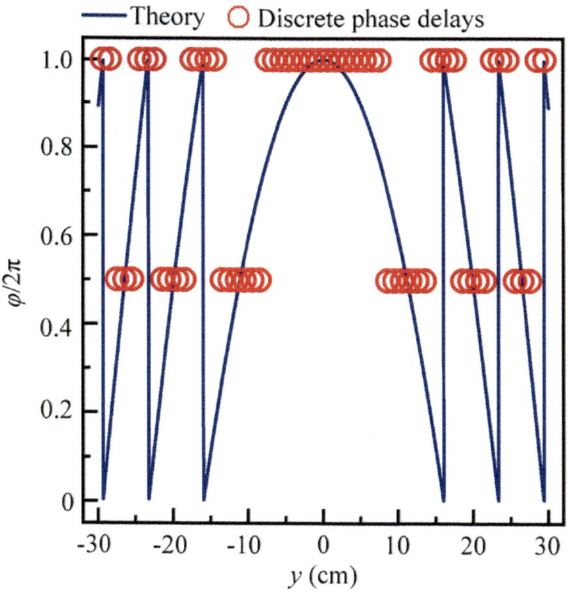

Fig. 6.47 Theoretical continuous phase distribution and discrete phase delays of the AF lens with two types of phased unit cells

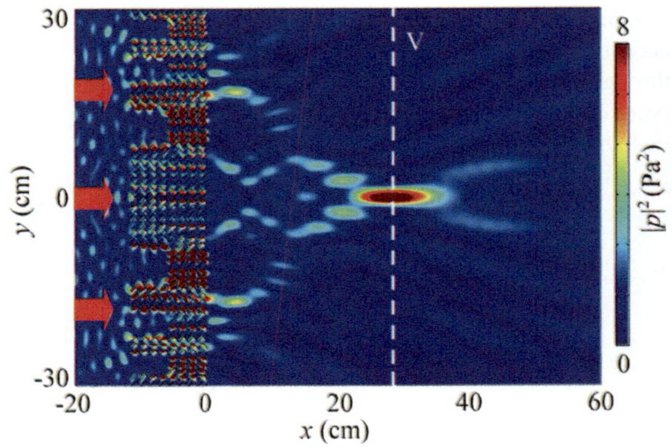

Fig. 6.48 Simulated distribution of acoustic intensity field caused by the AF lens at 8.58 kHz

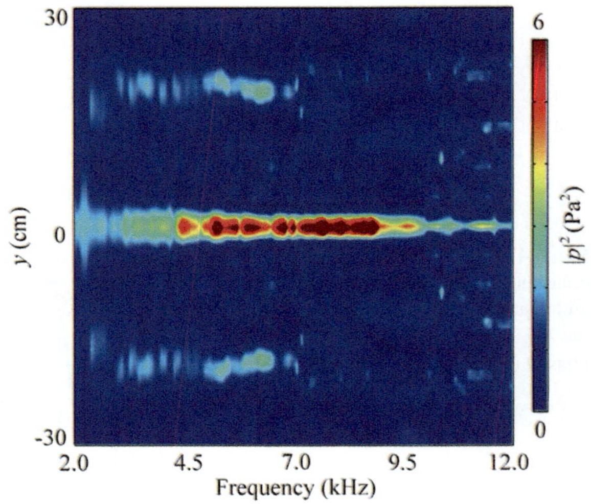

Fig. 6.49 Simulated intensity spectrum along line V in Fig. 6.48

bandwidth of 0.76, demonstrating the high robustness of the ultra-broadband AF lens. The reduction in bandwidth is attributed to the larger difference between the theoretical and discrete phase distributions (Fig. 6.47). Furthermore, the maximum intensities at the focus exceed 2.0 Pa^2 in the range of 2–12 kHz, further confirming the ultra-broadband characteristic of the lens.

Finally, we experimentally verify the robustness of the ultra-broadband feature. As shown in Fig. 6.50, the lens maintains high AF performance at 5.3, 6.3, 7.3, and 8.3 kHz, with the maximum values of $|p|^2$ at the focus exceeding 4.0 Pa^2. The measured and simulated results in red open squares R6–R9 match well with each other. Therefore, the ultra-broadband characteristic of the AF lens composed of two types of phased unit cells is experimentally confirmed, demonstrating its high robustness.

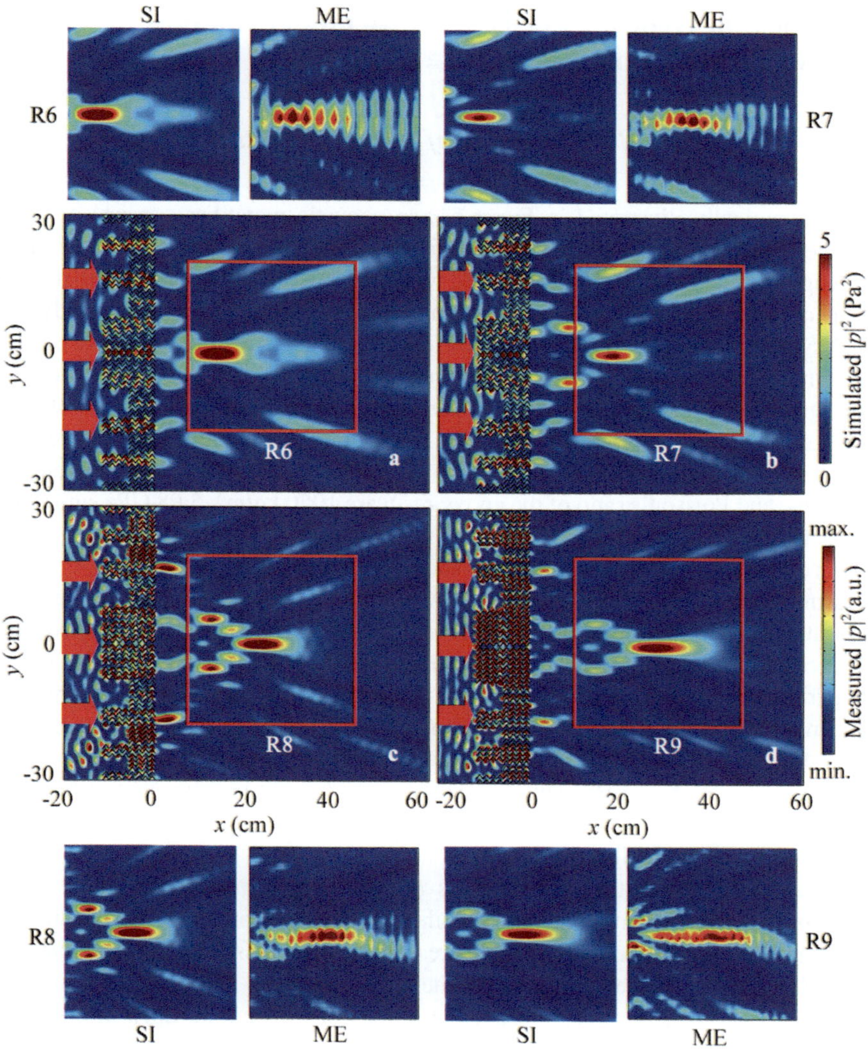

Fig. 6.50 Measured and simulated intensity distributions caused by the AF lens composed of two types of phased unit cells at **a** 5.3 kHz, **b** 6.3 kHz, **c** 7.3 kHz, and **d** 8.3 kHz. The eight insets show measured and simulated intensity distributions in red open squares R6–R9

6.4 Flexible Acoustic Focusing by Metafiber Bundles with Phase Control

6.4.1 Design and Performance of Acoustic Metafiber

As schematically illustrated in Fig. 6.51, an acoustic metafiber consists of eight unit cells [3]. Each unit cell contains a circular cavity with radius r and a narrow rectangular cavity with length $l - 2r$ and width d. The cavities are filled with air and made of epoxy resin to satisfy the sound-hard boundary condition. In this study, the COMSOL Multiphysics software is used to simulate the focusing characteristics. In the simulations, the acoustic-structure boundary condition is applied to the outer and inner boundaries of the metafiber. The parameters of the metafiber are $l = 30$ mm, $d = 1.5$ mm, and $r = 5$ mm. Additionally, the material parameters of air are density $\rho = 1.21$ kg/m^3 and acoustic velocity $c = 344$ m/s. The mass density, Young's modulus, and Poisson's ratio of the epoxy resin are 1050 kg/m^3, 5.08 GPa, and 0.35, respectively.

An example of a metafiber unit cell is shown in Fig. 6.52a, along with the simulated pressure amplitude distribution of mode A (the blue point) in band II (the red solid line) at $k = 0$. The amplitude distributions of mode A on both sides of the unit cell are identical. Similar characteristics can also be obtained in band III [10]. In this work, the performance of the metafiber in band II is studied in detail. Figure 6.52b shows the transmission spectrum of an acoustic metafiber. The transmission coefficient exceeds 0.5 in the range of 7.38–8.75 kHz (the black shaded region) and approaches 1.0 in the range of 7.50–8.21 kHz. This frequency range closely matches that of band II. Thus, the high transmission and broad bandwidth characteristics are attributed to the eigenmode of the metafiber.

Figure 6.53 shows the phase delays φ and the transmission coefficients of metafibers with different values of r, while keeping other parameters constant. As shown in Fig. 6.53, as the parameter r increases, the transmission coefficient remains above 0.8, and the phase delay φ spans the entire 2π range. To clearly demonstrate the phase delays of the metafibers, eight values of r (the blue open circles) are

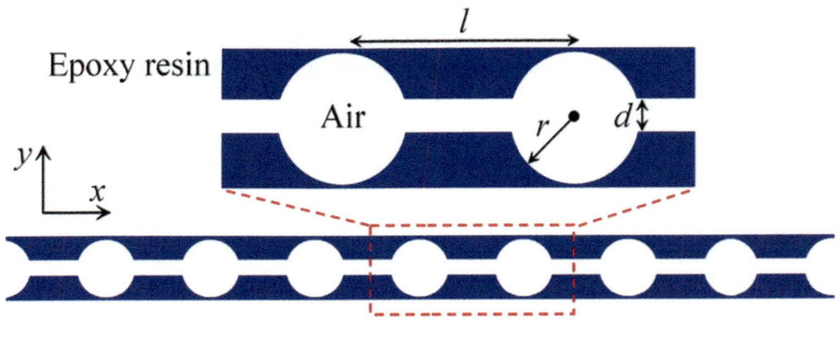

Fig. 6.51 Schematic of the acoustic metafiber

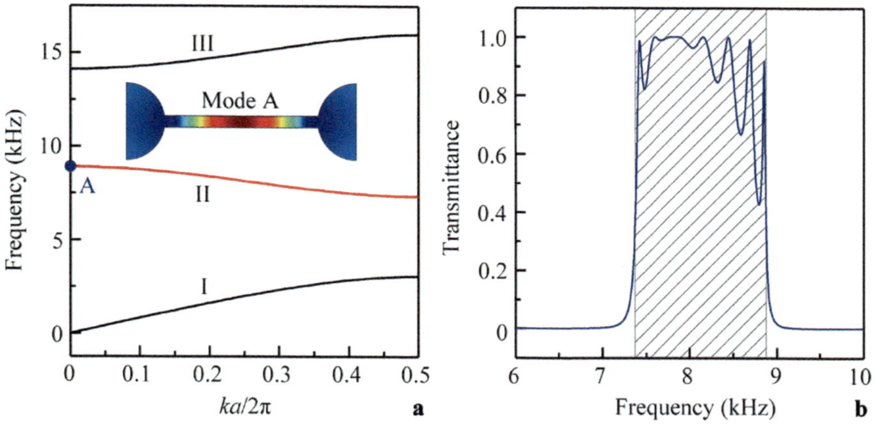

Fig. 6.52 **a** Simulated pressure amplitude eigenfunction of a metafiber unit cell at $k = 0$ and its dispersion relationship. The eigenfrequency of mode A is 8893 Hz. **b** Simulated transmission spectrum of the acoustic metafiber

selected to achieve eight equally spaced phase delays from 0 to 2π. The values of r for the eight blue open circles are 5.547, 5.419, 5.277, 5.113, 4.916, 4.669, 4.382, and 3.660 mm, corresponding to phase delays of $\pi/4$, $\pi/2$, $3\pi/4$, π, $5\pi/4$, $3\pi/2$, $7\pi/4$, and 2π, respectively. Figure 6.54 presents the pressure distributions through the metafibers with different values of r, where the incident acoustic wave is located on the left side. The results clearly show that the transmitted waveform matches the incident waveform, and the phase delay is equally spaced from 0 to 2π in steps of $\pi/4$. Therefore, acoustic propagation paths can be arbitrarily manipulated by designing desired phase distributions of metafiber bundles.

Fig. 6.53 Simulated phase delays (the blue solid line) and transmission coefficients (the red dashed line) for acoustic metafibers with different values of r at 8.0 kHz

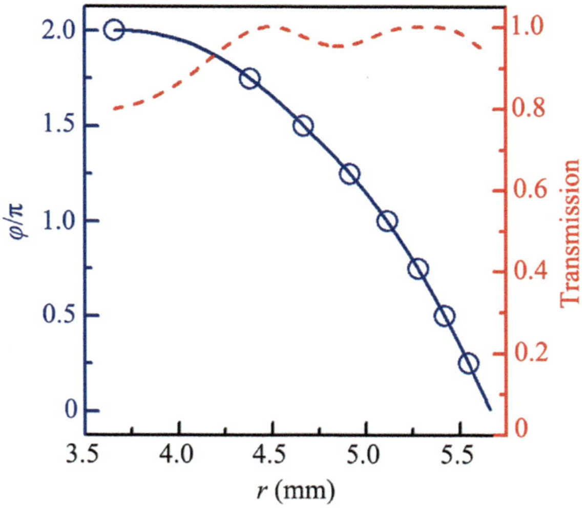

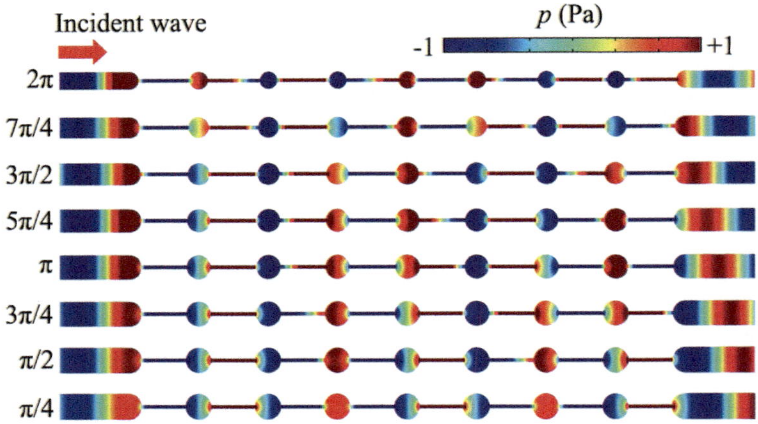

Fig. 6.54 Simulated pressure distributions caused by the eight acoustic metafibers with different values of r (the blue open circles in Fig. 6.53) at 8.0 kHz

To demonstrate the flexibility of the acoustic metafiber, the metafiber structure is bent at different angles θ at the center point. Figure 6.55 shows the transmission coefficients and phase delays of the acoustic metafibers with different values of θ. It is evident that the phase delay and transmission coefficients remain nearly unchanged for different values of θ, with the transmission coefficient close to 1.0. Thus, a flexible acoustic metafiber with high transmission and phase manipulation can be achieved.

Fig. 6.55 Schematic of the acoustic metafiber with bending angle θ. Simulated phase delays (the blue solid line) and transmission coefficients (the red dashed line) for acoustic metafibers with different values of θ at 8.0 kHz

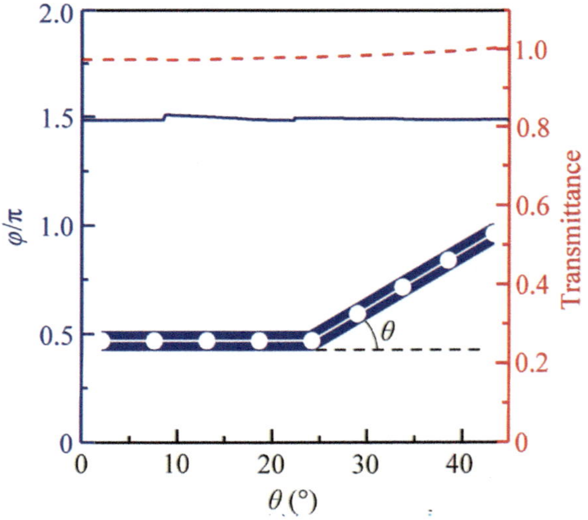

6.4.2 Design and Performance of AF Lens

Based on Eq. (6.2), an AF lens can be designed using acoustic metafibers with different values of r. The parameters are selected as $b = 30$ cm, $f = 8.0$ kHz, and $c_0 = 343$ m/s. The theoretical continuous phase distribution of the AF lens in the y direction is shown in Fig. 6.56. Based on the theoretical curve, 31 discrete phase delays are selected to design the AF lens with metafibers, with a distance of 12 mm between adjacent metafibers.

Figure 6.57 shows the intensity distribution through the metafiber bundles with different bending angles. As shown in Fig. 6.57a, the acoustic wave can transmit through the metafiber bundle and focus on the right side, indicating that the AF effect can be achieved with the metafiber bundle. The center position of the focus is located at (52.6 cm, 0), and the focal length is 28.6 cm, which is close to the theoretical value of 30.0 cm. Moreover, due to the flexible nature of the metafiber bundle, the focus position can be controlled by bending the metafiber bundle without altering its phase distribution. As examples, the metafiber bundle is bent at 15°, 30°, and 45° at the center point, as shown in Fig. 6.57b–d, respectively. The AF effect with high performance exists at different bending angles. However, as the bending angle increases, the focus position gradually shifts to the upper left, resulting from the change in the output surface position of the metafiber bundle.

To further demonstrate the performance of the AF lens at different bending angles, the intensity distributions along lines I and II in Fig. 6.57 are simulated, and the corresponding results are displayed in Fig. 6.58. Without bending the metafiber bundle (0°), the intensity at the center of the focus reaches about 9.0 Pa^2, indicating

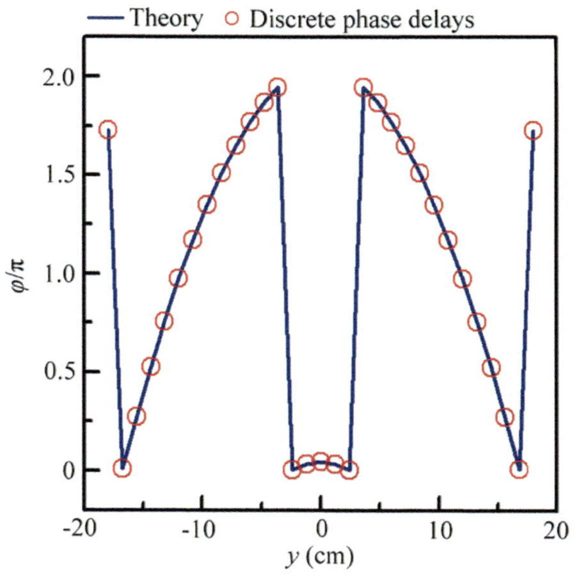

Fig. 6.56 Theoretical continuous phase distribution and discrete phase delays of the AF lens composed of 31 metafibers at 8.0 kHz

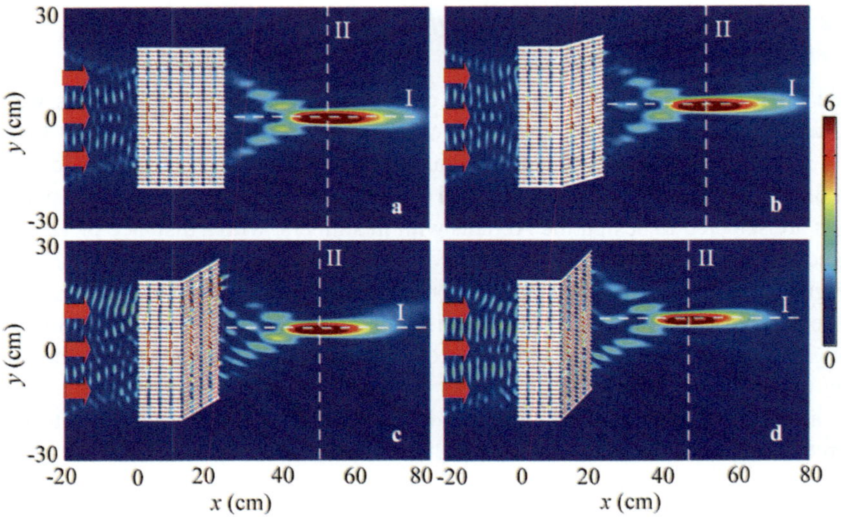

Fig. 6.57 Simulated intensity distributions caused by the AF lenses with bending angles of **a** 0°, **b** 15°, **c** 30°, and **d** 45° at 8.0 kHz

high focusing performance in both directions. As the bending angle increases, the focus position gradually shifts to the upper left, and the intensity at the center of the focus remains above 8.0 Pa^2, which is at least eight times greater than that in free space. Additionally, as shown in Fig. 6.58b, the FLHM can reach 0.66λ. These results demonstrate that the metafiber bundles can achieve the AF effect with high performance, and the focus position can be controlled by bending the metafiber bundle due to its flexible nature.

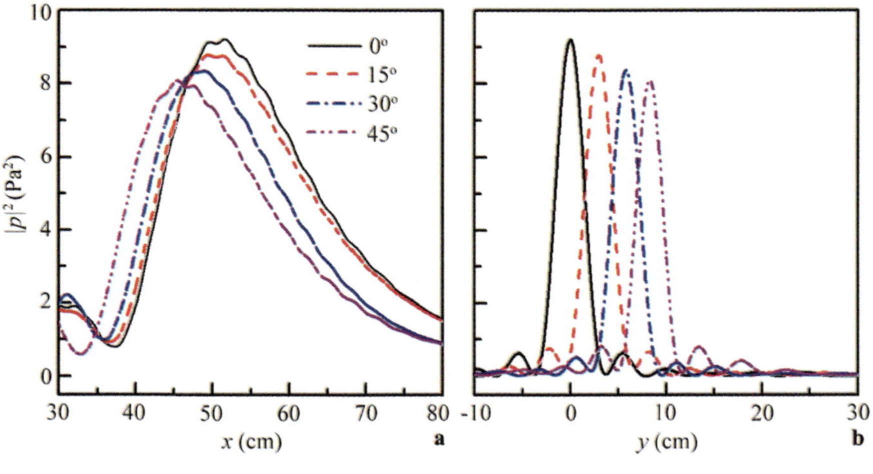

Fig. 6.58 Simulated intensity distributions along lines **a** I and **b** II in Fig. 6.57a–d

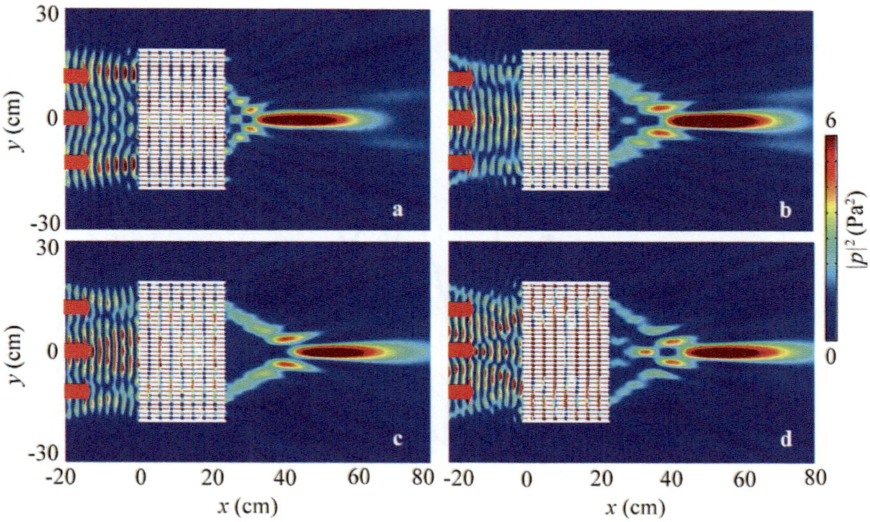

Fig. 6.59 Simulated intensity distribution caused by the AF lens at **a** 7.6 kHz, **b** 7.9 kHz, **c** 8.2 kHz, and **d** 8.6 kHz

Owing to the broadband characteristics of the metafibers, the proposed AF lens has a broad working bandwidth. Figure 6.59 shows the intensity distributions caused by the metafiber bundle at 7.6, 7.9, 8.2, and 8.6 kHz, with the lens parameters being the same as those in Fig. 6.57a. As shown in Fig. 6.59, the proposed lens exhibits high focusing performance at these frequencies, with intensities at the center of the focus exceeding 4.0 Pa^2. Additionally, the focus position shifts to the right as the frequency increases, which can be used to manipulate the focal length of the lens. Figure 6.60 shows the intensity spectrum at the center of the focus. In the range of 7.15–8.65 kHz (the black shaded region), the intensities at the center of the focus exceed 3.0 Pa^2. Based on this, the working bandwidth of the lens is 1.5 kHz, resulting in a fractional bandwidth of 0.2. Thus, the proposed AF lens has a broad bandwidth, and the focal length can be controlled by the incident frequency.

6.4.3 Experimental Measurement

Figure 6.61 shows a photograph of the sample composed of 31 acoustic metafibers with a bending angle of 30°, fabricated using epoxy resin via 3D printing. The parameters of the sample are the same as those in Fig. 6.57c. The experimental set-up is shown in Fig. 6.62, where the experiment is conducted in a planar waveguide composed of two parallel plates (dimensions 2 m × 2 m × 1 cm). Wedge-shaped sound-absorbing foams at the boundaries of the planar waveguide create an anechoic environment. The incident plane wave is generated by a speaker array driven by a

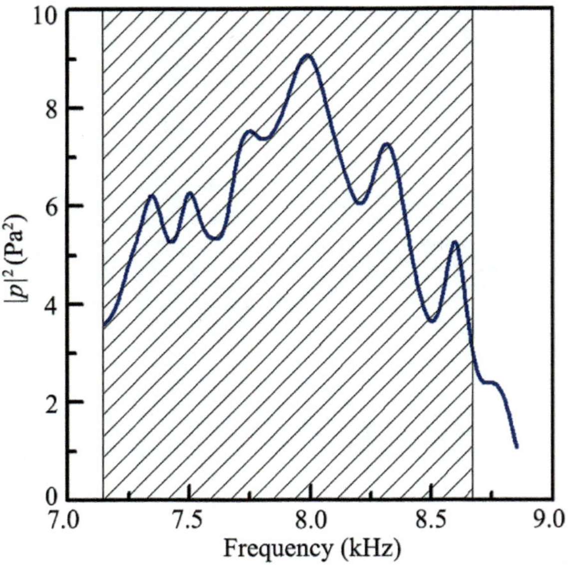

Fig. 6.60 Simulated intensity spectrum at the center of the focus

power amplifier. The acoustic pressure distribution is scanned using a microphone (Brüel & Kjær type-4961) moved by a set of 2D motorized linear stages (Newport: MIN300CC and ILS250CC). Another microphone is fixed toward the speaker array to measure the reference acoustic signal. The data is recorded by the Brüel & Kjær 3160-A-022 module and analyzed using PULSE Labshop software.

Figure 6.63 shows the simulated intensity distribution caused by the metafiber bundle, with the insets on the right sides presenting the simulated and measured intensity distributions at the focus. The experimental results agree well with the simulated ones. Figure 6.64 shows the intensity distributions along lines III and IV in Fig. 6.63. The intensity distributions from the measured and simulated results are in excellent agreement. These results demonstrate that the AF lens composed of metafiber bundles with a certain bending angle exhibits high focusing performance.

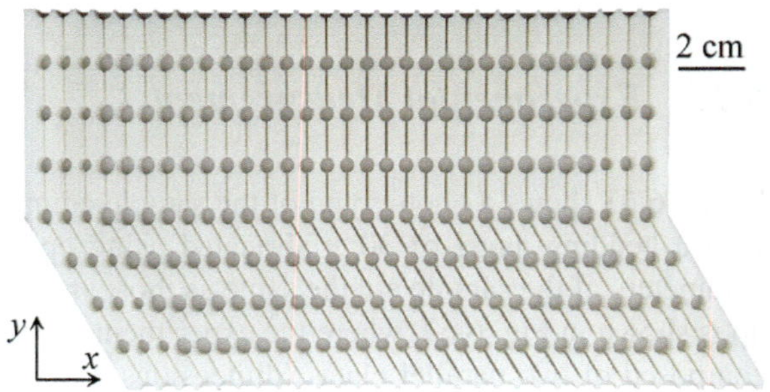

Fig. 6.61 Photograph of the AF lens with a bending angle of 30°

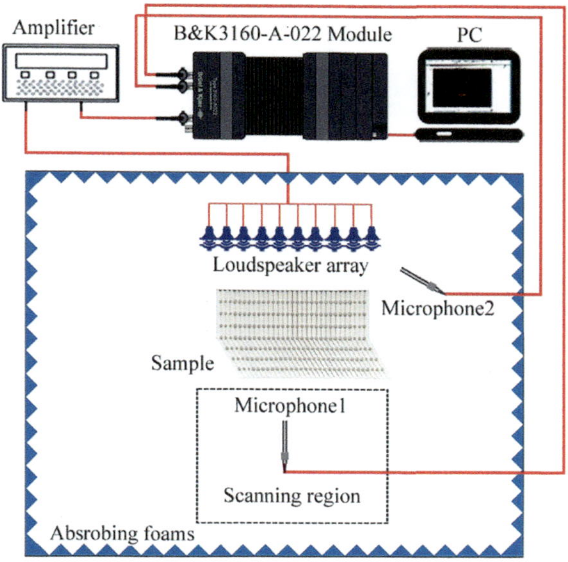

Fig. 6.62 Schematic of the experimental set-up for the AF lens in Fig. 6.61

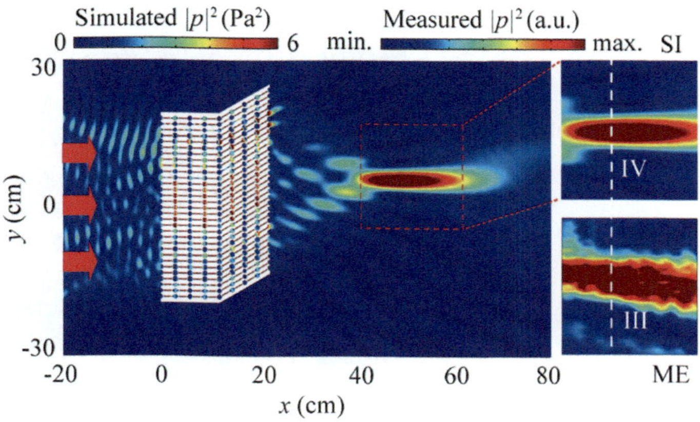

Fig. 6.63 Measured and simulated intensity distributions caused by the sample at 8.0 kHz

6.4.4 Performance Modulation

Based on its flexible characteristics, the designed AF lens can bypass rigid scatterers placed inside the metafiber bundle. As shown in Fig. 6.65a, b, a diamond and a square hidden region are designed inside the AF lens by symmetrically bending the metafiber bundle at 30° and 45°, respectively. The other parameters of the AF lens are the same as those in Fig. 6.57a. Rigid scatterers are placed in the two hidden regions of the AF lens. The rigid scatterers have no effect on the acoustic waves inside

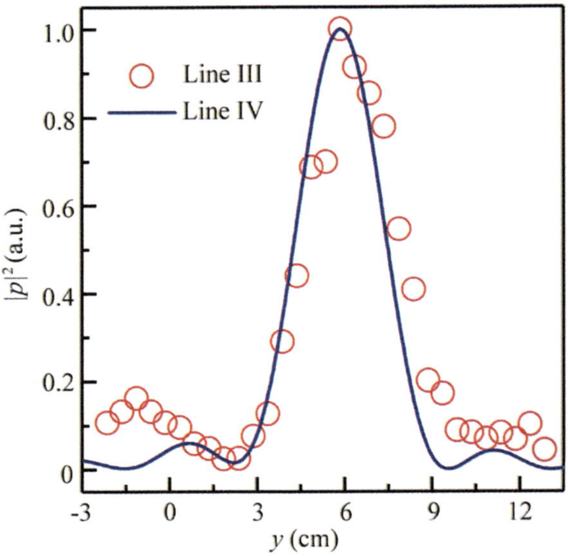

Fig. 6.64 Measured and simulated intensity distributions along lines III and IV in Fig. 6.63

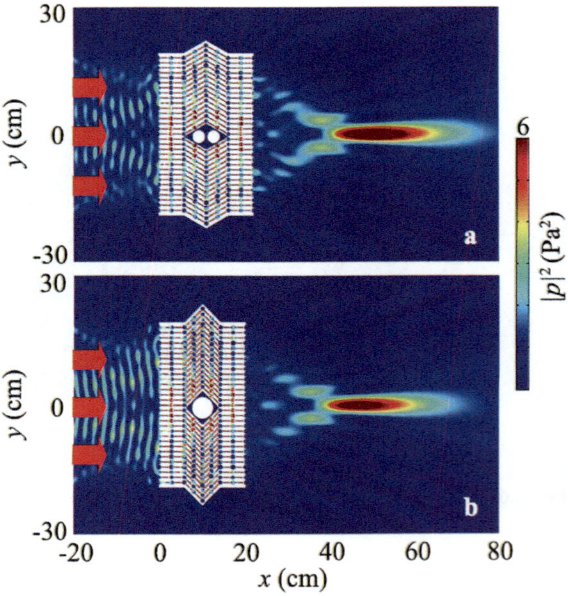

Fig. 6.65 Simulated intensity distribution caused by the AF lens at 8.0 kHz, with the positions of the second and sixth circular cavities in the metafiber bundles symmetrically bent at **a** 30° and **b** 45°

the metafiber bundles, and the AF effect is still achieved on the right side of the metafiber bundles, with the intensity at the center of the focus nearly identical to that in Fig. 6.57a. Based on this, the AF lens with different bending angles and positions can bypass rigid scatterers of various shapes and positions, further expanding the potential applications of the AF lens.

Furthermore, we can change the propagation direction of sound using metafiber bundles, even for the focusing direction. As shown in the blue region of Fig. 6.66, a right-angled direction converter is designed using 31 metafibers with different parameters. Here, w_i ($i = 1, 2, \ldots, 31$) represents the length of the ith metafiber, e is the distance between adjacent metafibers, and the length difference $w_{i+1} - w_i = 2e$. To achieve a perfect change in the propagation direction of sound, the parameters of each metafiber are adjusted to ensure a continuous phase distribution between both sides of the direction converter. The phase delays of the metafibers primarily depend on the radius of the circular cavity r, the width of the rectangular cavity d, the length of the rectangular cavity l, and the number of unit cells N in the metafibers. The lengths of the first and last metafibers are $w_1 = 3.394$ mm and $w_{31} = 54.306$ mm, respectively, and the distance between adjacent metafibers is $e = 8.485$ mm. The parameters l, N, r, and d for the 31 metafibers in the direction converter are shown in Fig. 6.67a, b. Based on these parameters, as shown in Fig. 6.67c, the differences in phase delays $\Delta\varphi$ on both sides of these metafibers are less than 0.02π, and most of the transmission coefficients exceed 0.9, approaching 1.0.

As shown in Fig. 6.68a, the incident plane wave is located on the left side of the direction converter and passes through it with high transmission. Notably, the propagation direction of the output wave is perpendicular to that of the incident wave, and the transmitted acoustic waveform matches the incident one. Additionally, as shown in Fig. 6.68b, the phase distribution on both surfaces of the direction converter is

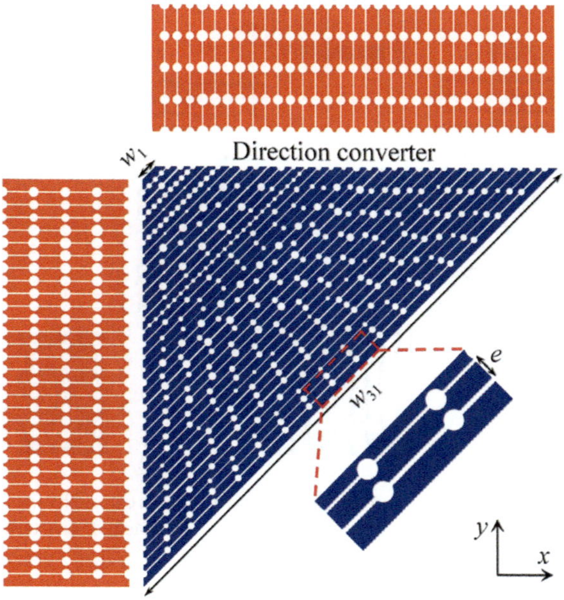

Fig. 6.66 Schematic of the right-angled direction converter

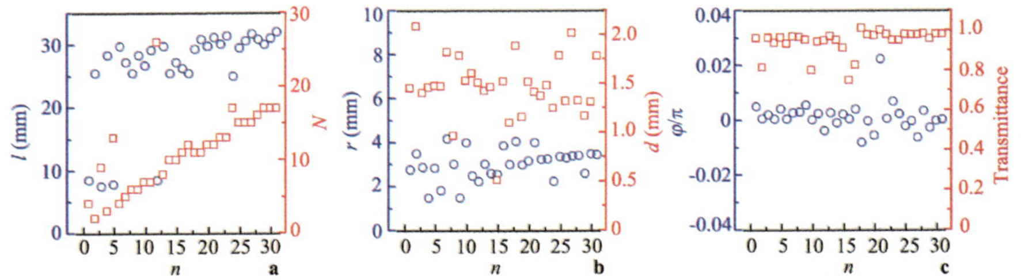

Fig. 6.67 Parameters **a** l and N, **b** r and d, and **c** the phase differences and transmission coefficients of the 31 metafibers in the right-angled direction converter

continuous, with the phase difference close to zero. Therefore, the direction converter enables the change in the propagation direction of sound.

By simultaneously incorporating the AF lens (the orange regions in Fig. 6.66), the direction of AF can also be controlled using the direction converter. Figure 6.68c shows the intensity distribution caused by an attached structure composed of the AF lens and the direction converter, where the numbers of metafibers in both systems are the same. We can see that the transmitted sound energy is focused on the upper side of the attached structure, and the focusing direction is perpendicular to the incident one. These results demonstrate that the AF lens composed of metafiber bundles offers the advantages of a flexible structure, broad bandwidth, and convenient control.

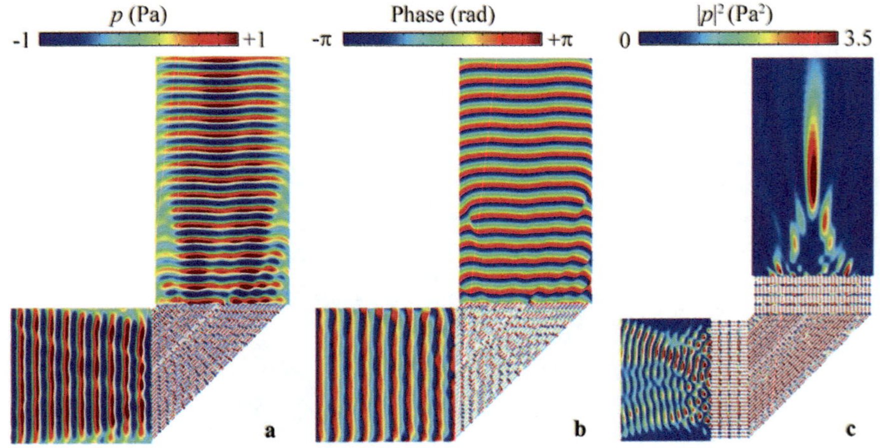

Fig. 6.68 Simulated **a** acoustic pressure and **b** phase distributions caused by the direction converter at 8.0 kHz. **c** Simulated intensity distribution caused by the attached structure composed of the direction converter and the focusing lens at 8.0 kHz

6.4.5 *Design and Performance of AF Lens for Cylindrical Sound Source*

Based on the generalized Snell's law, a focusing lens composed of metafiber bundles can be designed for a cylindrical sound source. In this design, the discrete phase distributions of the metafiber bundles in the y direction satisfy:

$$\varphi_2(y) = -k\left(\sqrt{y^2 + l_1^2} + \sqrt{y^2 + l_2^2}\right), \tag{6.3}$$

where l_1 and l_2 are the distances between the cylindrical acoustic source and the metafiber bundles and the focal length, respectively. In the simulations, the parameters are selected as $l_1 = l_2 = 20$ cm, and the theoretical continuous phase distribution of the AF lens is shown in Fig. 6.69. Based on the theoretical phase curve, the metafiber bundles are designed using 31 discrete phase delays in the y direction, with a distance of 12 mm between adjacent metafibers.

Figure 6.70a shows the intensity distribution caused by the AF lens under the excitation of the cylindrical sound source. The cylindrical acoustic wave transmits through the metafiber bundle and focuses on the right side. The center position of the focus is located at (43.8 cm, 0), and the focal length is 19.8 cm, which is close to the theoretical value of 20 cm. This result demonstrates that the AF effect for the cylindrical sound source can also be achieved using the metafiber bundle. Additionally, as shown in Fig. 6.70b–d, as the bending angle increases, the AF effect is maintained, and the focus position gradually shifts to the upper left, consistent with the results in

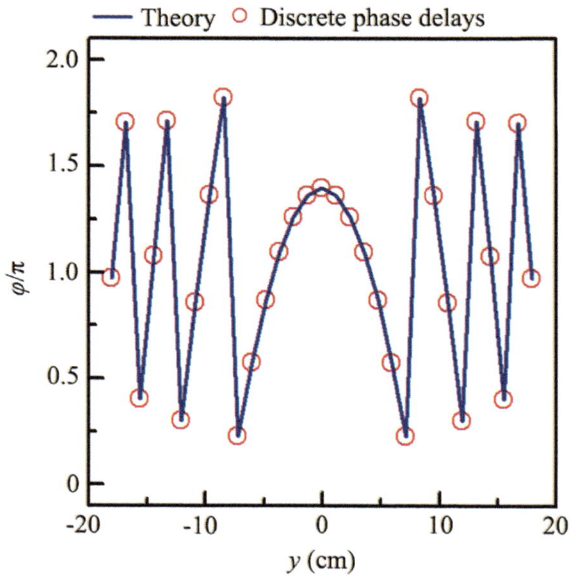

Fig. 6.69 Theoretical continuous phase distributions and discrete phase delays caused by the AF lens for the cylindrical sound source at 8.0 kHz

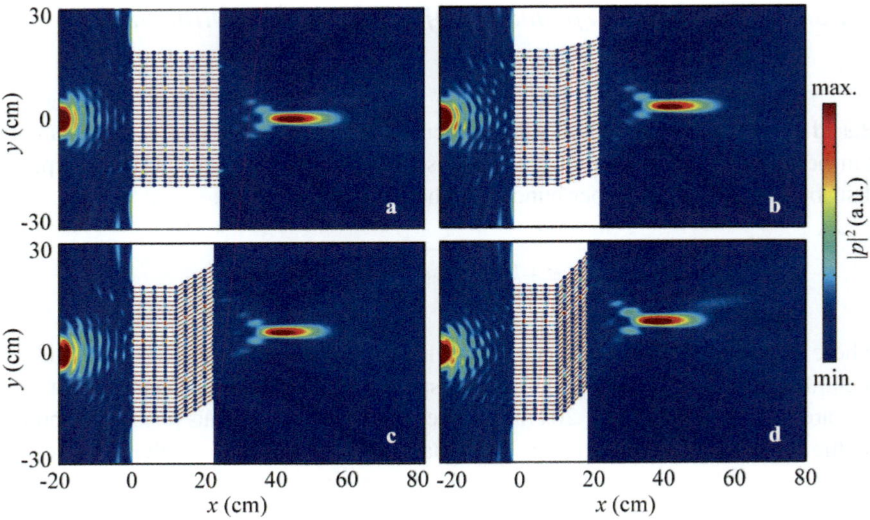

Fig. 6.70 Simulated pressure distributions through the AF lens with bending angles of **a** 0°, **b** 15°, **c** 30°, and **d** 45° under the excitation of the cylindrical sound source at 8.0 kHz

Fig. 6.57b–d. Based on these findings, we conclude that the metafiber bundles can be applied for wavefront conversion and AF for cylindrical sound sources.

6.5 Conclusion

This chapter explores AF techniques based on phase-controlled metamaterials, focusing on the design of three types of AF lenses. First, a broadband AF lens composed of cavity meta-atoms is proposed, achieving wideband focusing by adjusting the number and structural parameters of the cavity meta-atoms. Second, an ultra-broadband AF lens using V-shaped meta-atoms is designed, demonstrating high bandwidth and robustness. Finally, a flexible AF lens based on metafiber bundles is introduced, showcasing its focusing performance at various bending angles and its potential applications in acoustic communication, energy harvesting, and non-destructive testing. This chapter provides new design approaches for practical applications of AF technology.

References

1. Qian J, Xia JP, Sun HX, Yuan SQ, Ge Y, Yu XZ (2017) Broadband acoustic focusing by cavity structures with phase manipulations. J Appl Phys 122(24):244501. https://doi.org/10.1063/1.4998223

2. Zhang TC, Chen JH, Qian J, Ge Y, Yuan SQ, Sun HX, Liu XJ (2020) Observation of ultra-broadband acoustic focusing based on V-shaped meta-atoms. Adv Mater Technol 5(2):1900956. https://doi.org/10.1002/admt.201900956
3. Sun HX, Chen JH, Ge Y, Yuan SQ, Liu XJ (2018) Broadband and flexible acoustic focusing by metafiber bundles. J Phys D: Appl Phys 51(24):245102. https://doi.org/10.1088/1361-6463/aac1aa
4. Li Y, Liang B, Gu ZM, Zou XY, Cheng JC (2013) Reflected wavefront manipulation based on ultrathin planar acoustic metasurfaces. Sci Rep 3:2546. https://doi.org/10.1038/srep02546
5. Xie YB, Wang WQ, Chen HY, Konneker A, Popa BI, Cummer SA (2014) Wavefront modulation and subwavelength diffractive acoustics with an acoustic metasurface. Nat Commun 5:5553. https://doi.org/10.1038/ncomms6553
6. Li Y, Jiang X, Li RQ, Liang B, Zou XY, Yin LL, Cheng JC (2014) Experimental realization of full control of reflected waves with subwavelength acoustic metasurfaces. Phys Rev Appl 2(6):064002. https://doi.org/10.1103/PhysRevApplied.2.064002
7. Wang WQ, Xie YB, Konneker A, Popa BI, Cummer SA (2014) Design and demonstration of broadband thin planar diffractive acoustic lenses. Appl Phys Lett 105(10):101904. https://doi.org/10.1063/1.4895619
8. Xie YB, Konneker A, Popa BI, Cummer SA (2013) Tapered labyrinthine acoustic metamaterials for broadband impedance matching. Appl Phys Lett 103(20):201906. https://doi.org/10.1063/1.4831770
9. Harlow JH (2004) Electric Power Transformer Engineering. CRC Press
10. Xia JP, Sun HX, Yuan SQ (2017) Modulating sound with acoustic metafiber bundles. Sci Rep 7:8151. https://doi.org/10.1038/s41598-017-07232-6

Chapter 7
Acoustic Focusing by Thermoacoustic Phase-Controlled Metamaterials

7.1 Introduction

Thermoacoustic phase-controlled metamaterials are artificial structures capable of manipulating sound based on temperature gradients. By adjusting the temperature of air in each phased unit cell, different phase delays of sound can be achieved, enabling the creation of desired refractive index gradients and complex sound manipulations. In these metamaterials, acoustic waves propagate through a single medium—air—at varying temperatures. This approach effectively overcomes the significant impedance mismatch between lens materials (such as metal, epoxy resin, and rubber) and the ambient medium (air or water), thereby enabling various broadband sound effects. Utilizing temperature gradients, several acoustic devices have been realized, including an AF lens [1–4], a reduced acoustic cloak [5], an acoustic omnidirectional absorber [6], and an acoustic system with asymmetric transmission [7]. These achievements inspire us to develop broadband AF lenses using thermoacoustic phase-controlled metamaterials.

In this chapter, we design three types of AF lenses based on thermoacoustic phase-controlled metamaterials, guided by the generalized Snell's law, and discuss their potential applications. First, we propose a broadband self-healing AF lens composed of two symmetric thermoacoustic phased arrays (TPAs) of Airy beams. Each phased unit cell is designed using air at different temperatures [8]. The self-healing AF effect arises from the energy overlap of two symmetric Airy beams, resulting in minimal energy loss even when a scatterer is present in the focusing path. The bandwidths of the transmitted and reflected self-healing AF lenses are approximately 3.5 kHz and 4.8 kHz, respectively, demonstrating their broadband characteristics. We also discuss the feasibility of thermal insulation films within the lens. Second, we design a long-focus AF lens constructed from a TPA of Bessel-like beams [9]. The full length at half maximum (FLHM) of the focus is 16.8λ, and the fractional bandwidth is 0.77, showcasing its long-focus and broadband AF properties. We further investigate the long-focus AF lens for cylindrical acoustic sources and its design feasibility. Finally,

H. Sun et al., *Acoustic Metamaterials*, Microfluidics and Nanophotonics: Science and Engineering 2, https://doi.org/10.1007/978-981-96-8524-0_7

we propose a reflected long-focus AF lens consisting of a TPA of Bessel-like beams [10]. The focus length is about 2.0 m, and the fractional bandwidth is 0.29, also exhibiting long-focus and broadband AF characteristics. These AF lenses, based on TPAs, have potential applications in medical ultrasound, directional bird repelling, and acoustic energy harvesting.

7.2 Self-Healing Acoustic Focusing by Thermoacoustic Phased Arrays

7.2.1 Design and Performance of Thermoacoustic Phased Unit Cell

Assuming air behaves as an ideal fluid, the acoustic velocity and density of air are determined by the ambient temperature T, expressed as [3, 11]:

$$c = \sqrt{\gamma RT/M}, \tag{7.1}$$

$$\rho = p_0 M/RT, \tag{7.2}$$

where $\gamma = 1.4$ is the ratio of the molar heat capacities of air, $M = 28.97 \times 10^{-3}$ kg/mol is the molar mass of air, $R = 8.31$ J/(mol K) is the molar gas constant, and $p_0 =$ 101.325 kPa is the pressure at 273 K.

Figure 7.1 illustrates a thermoacoustic phased unit cell. The unit cell has a length l and width h, with the air inside at temperature T. The blue solid lines at the upper and lower sides represent rigid thermal insulation boundaries with thickness d, while the red dashed lines at the left and right sides are thermal insulation films that do not affect sound transmission. The incident acoustic wave is located at the left boundary, and the ambient air temperature (T_0) remains constant. Throughout this work, we use the COMSOL Multiphysics software to simulate the focusing characteristics. The unit cell parameters are $l = 8$ cm, $h = 1$ cm, $d = 0.05$ cm, and $T_0 = 300$ K. We account for the effect of temperature on the thermal parameters of air [3, 11], and the acoustic velocity and density of air are calculated using Eqs. (7.1) and (7.2), respectively.

Figure 7.2 shows the phase delays and transmission spectra of phased unit cells at different temperatures. As T increases, the phase delays cover the entire 2π range, and the transmission coefficient remains above 0.8. The temperatures corresponding to the eight blue open circles are 330, 410, 510, 650, 890, 1370, 2010, and 2820 K, with phase delays of 2π, $7\pi/4$, $3\pi/2$, $5\pi/4$, π, $3\pi/4$, $\pi/2$, and $\pi/4$, respectively.

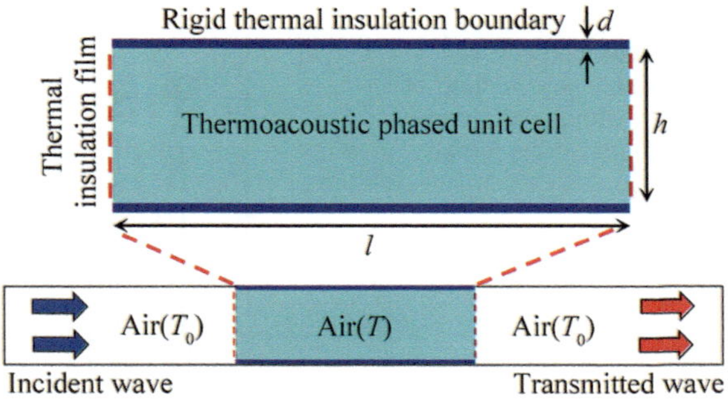

Fig. 7.1 Schematic of the thermoacoustic phased unit cell

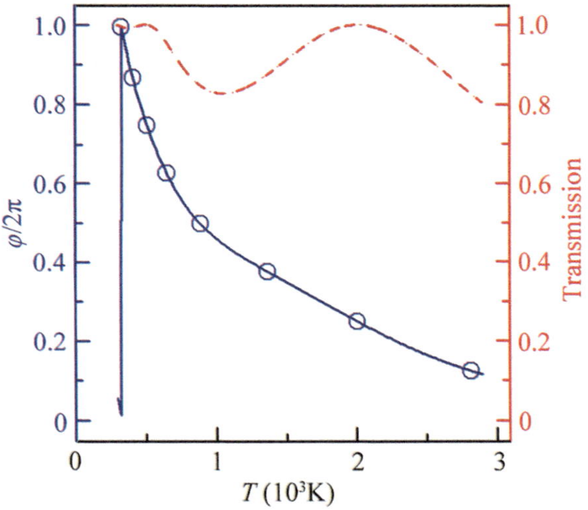

Fig. 7.2 Simulated transmission coefficients (the red dashed line) and phase delays (the blue solid line) of phased unit cells at different values of T at 5.0 kHz

7.2.2 *Design and Performance of Self-Healing AF Lens*

Based on Eq. (2.41), we can control the refraction angle θ_t of sound through the TPA by adjusting the phase gradient of $d\varphi(y)/dy$. To realize an Airy beam of sound, the phase distribution of $\varphi(y)$ of the TPA in the y direction can be expressed as [12, 13]:

$$\varphi(y) = -\arg[Ai(by)\exp(ay + ikby\sin\theta)], \tag{7.3}$$

where $Ai(by) = \frac{\int_0^\infty \cos(t^3/3+byt)\mathrm{d}t}{\pi}$ is the Airy function, k is the wave number, a is a positive value limiting the energy of the Airy beam, b is the transverse scale, and θ controls the bending direction of the Airy beam. In this work, we propose an AF lens composed of two TPAs of Airy beams symmetric about $y = 0$. The two symmetric Airy beams from the TPAs focus on the overlapping region, leveraging the self-bending characteristic of the Airy beam.

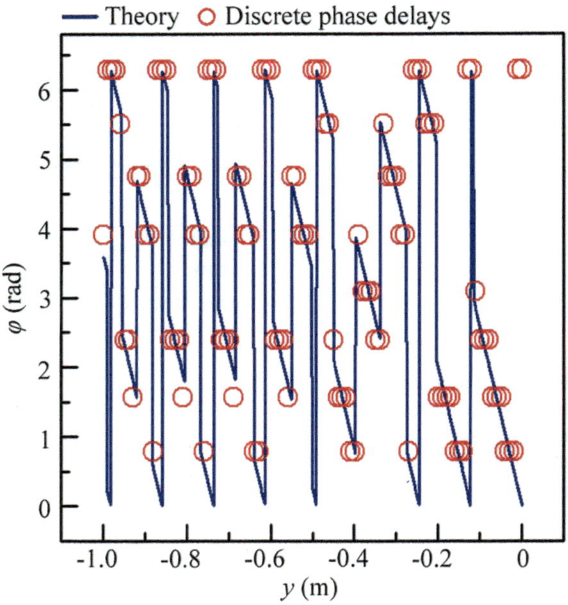

Fig. 7.3 Theoretical continuous phase distribution and discrete phase delays of the TPA with eight types of phased unit cells at 5.0 kHz

For the TPA of the Airy beam, the parameters are set as $a = 0.5$, $b = 20$, $\theta = 0.8°$, $f = 5.0$ kHz, and $c_0 = 343$ m/s. Based on Eq. (7.3), the theoretical continuous phase distribution in the y direction is calculated, as shown in Fig. 7.3. Using the eight types of phased unit cells with a step of $\pi/4$ from Fig. 7.2, we select 101 discrete phase delays in the y direction to design the TPA of the Airy beam.

Figure 7.4a, b show the intensity distributions of the Airy beam generated by the theoretical continuous phase distribution and the TPA, respectively. As shown in Fig. 7.4b, the transmitted sound energy forms an Airy beam with a clear self-bending characteristic, with energy concentrated in the main lobe. The Airy beam generated by the designed TPA closely matches the theoretical one in Fig. 7.4a. Figure 7.4c, d present the intensity distributions of the AF effect. The AF lens, composed of two symmetric TPAs from Fig. 7.4b, allows the acoustic wave to pass through and focus on the right side, demonstrating the AF effect achieved by two symmetric Airy beams. The results in Fig. 7.4c, d agree well, confirming the feasibility and effectiveness of the designed thermoacoustic AF lens.

Figure 7.5 shows the intensity distributions along lines I and II in Fig. 7.4d, with results in free space provided for comparison. The AF effect is evident in both horizontal and vertical directions. The focus is located at (133.8 cm, 0), with the intensity at the center of the focus about 15 times higher than in free space, indicating high focusing performance.

The proposed AF lens exhibits broad bandwidth. Figure 7.6 shows the intensity distributions caused by the AF lens at 4.3, 6.0, and 7.8 kHz, using the same parameters as in Fig. 7.4d. The AF lens maintains high focusing performance at these frequencies, with a bandwidth of 3.5 kHz. As the frequency increases, the focus size gradually

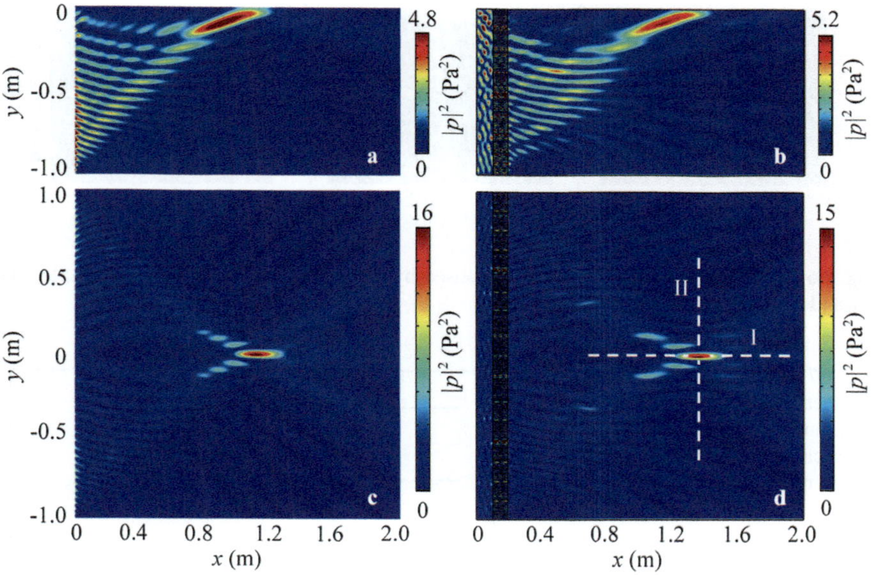

Fig. 7.4 Simulated intensity distributions caused by **a** the theoretical continuous phase distribution and **b** the TPA of the Airy beam, and those caused by **c** the theoretical continuous phase distribution and **d** the thermoacoustic AF lens at 5.0 kHz

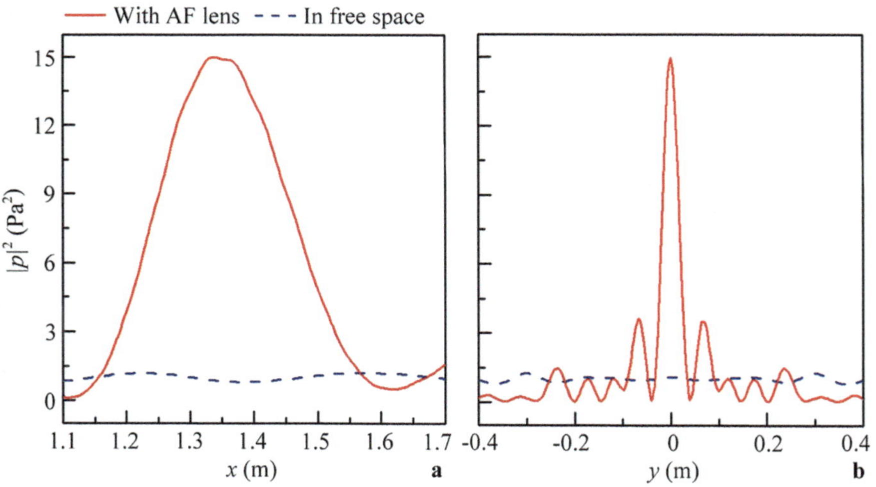

Fig. 7.5 Simulated intensity distributions along lines **a** I and **b** II in Fig. 7.4d

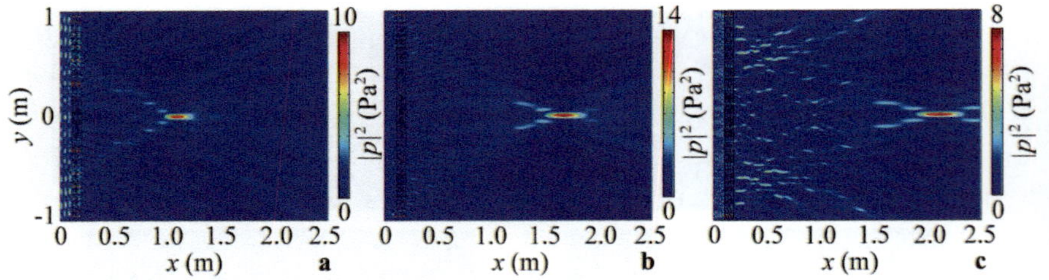

Fig. 7.6 Simulated intensity distributions caused by the AF lens at **a** 4.3 kHz, **b** 6.0 kHz, and **c** 7.8 kHz

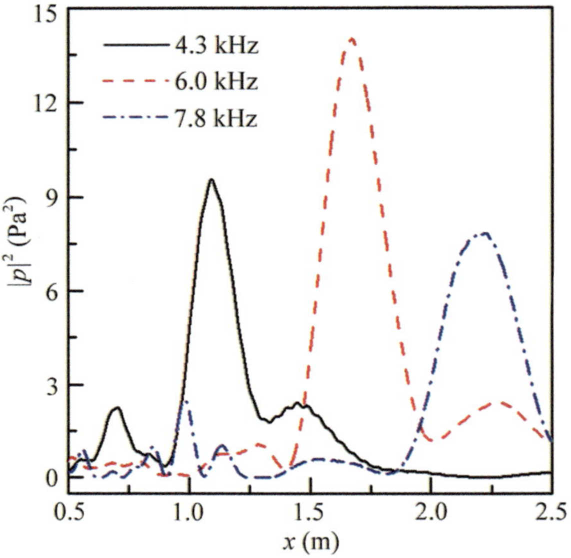

Fig. 7.7 Simulated transverse intensity distributions through the focus caused by the AF lens at different frequencies

increases, and the position shifts to the right, as shown in Fig. 7.7. This phenomenon results from the change in the propagation path of the Airy beam with increasing frequency. Thus, the proposed AF lens exhibits broadband characteristics, and the focal length can be controlled by the incident frequency.

7.2.3 Demonstration of Self-Healing Characteristic

Due to the self-healing property of the Airy beam, a scatterer placed in the propagation path of the AF has minimal impact on the AF effect. Figure 7.8 shows the intensity distributions caused by the AF lens with a rectangular scatterer placed horizontally and vertically in the propagation path. The scatterer is positioned at (70 cm, 0), with a length of 20 cm and a width of 12 cm. While the scatterer causes strong

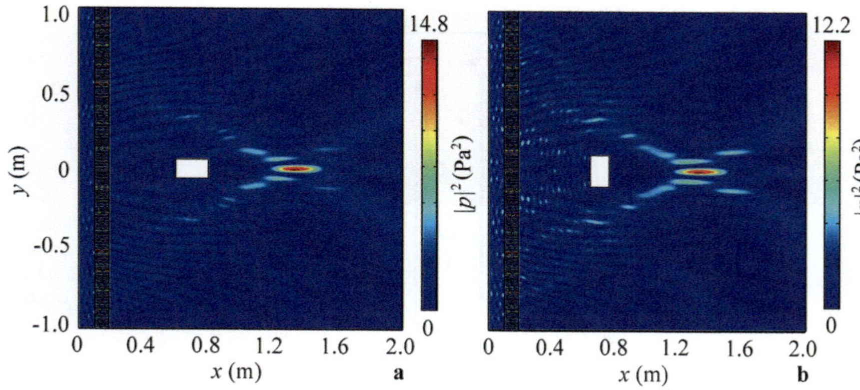

Fig. 7.8 Simulated intensity distributions caused by the AF lens with a rectangular scatterer placed **a** horizontally and **b** vertically in the propagation path

sound scattering in the near field, the AF effect remains nearly unchanged compared to Fig. 7.4d. This is due to the self-healing property of the Airy beam, allowing the transmitted sound energy to circumvent the scatterer and focus with minimal energy loss. However, the focusing performance is slightly affected by the size of the scatterer in the y direction, as seen in Fig. 7.8b.

To further verify the self-healing characteristic, we design a conventional AF lens using the eight types of phased unit cells from Fig. 7.2, with the same length and number of unit cells as in Fig. 7.4d. Figure 7.9 shows the theoretical continuous phase distribution and discrete phase delays of the conventional AF lens. Figure 7.10a–c show the intensity distributions caused by the conventional AF lens without a scatterer and with horizontal and vertical rectangular scatterers, respectively. While the AF effect exists in all cases, the intensity at the center of the focus decreases significantly with the scatterer present, unlike the proposed AF lens. This demonstrates that the scatterer has a greater impact on the focusing performance of the conventional AF lens.

Figures 7.11 and 7.12 show the longitudinal intensity distributions through the focus caused by the two types of AF lenses with the rectangular scatterer, normalized by the maximum value without a scatterer. With the horizontal scatterer, the normalized intensity at the center of the focus for the conventional AF lens is about 0.73, while the designed AF lens remains nearly unchanged (Fig. 7.11a). With the vertical scatterer, the normalized intensities for the designed and conventional AF lenses are about 0.82 and 0.60, respectively (Fig. 7.12). These results further confirm the superior self-healing characteristic of the designed AF lens.

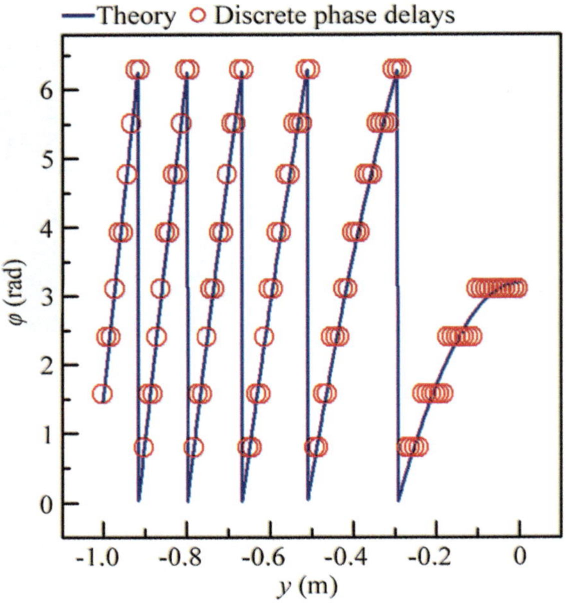

Fig. 7.9 Theoretical continuous phase distribution and discrete phase delays of the conventional AF lens

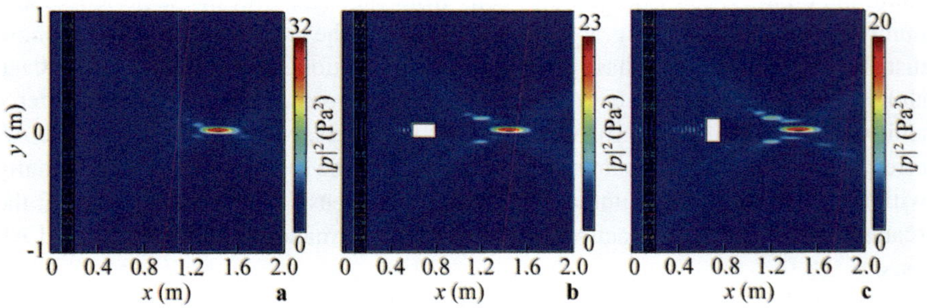

Fig. 7.10 Simulated intensity distributions caused by the AF lens **a** without a scatterer and with a rectangular scatterer placed **b** horizontally and **c** vertically in the propagation path at 5.0 kHz

7.2.4 Parameter Analysis

We next analyze the influence of the bending angle θ of the Airy beam on the focusing performance. Figure 7.13a, b show the theoretical continuous phase distributions and discrete phase delays of the AF lenses with different values of θ. Figure 7.14a, b show the intensity distributions caused by the AF lenses with different values of θ. The AF effect persists across different θ values, but as θ increases, the focus shifts to the right

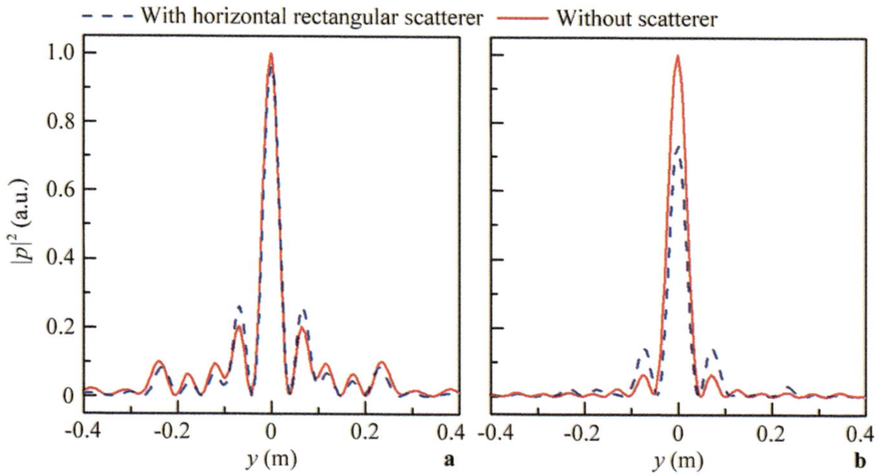

Fig. 7.11 Simulated longitudinal intensity distributions through the focus caused by the **a** designed and **b** conventional AF lenses with and without the horizontal rectangular scatterer at 5.0 kHz

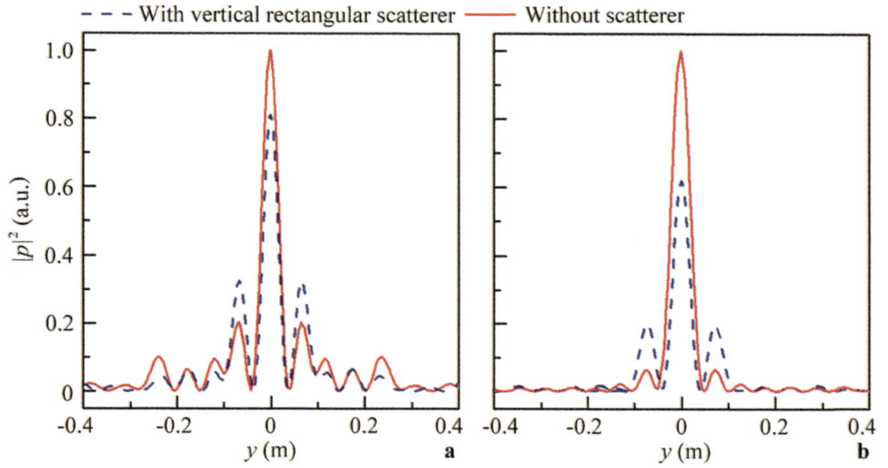

Fig. 7.12 Simulated longitudinal intensity distributions through the focus caused by the **a** designed and **b** conventional AF lenses with and without the vertical rectangular scatterer at 5.0 kHz

due to changes in the curvature of the Airy beam. Figure 7.15 shows the transverse intensity distributions through the focus for different θ values, confirming that the focus position shifts to the right as θ increases.

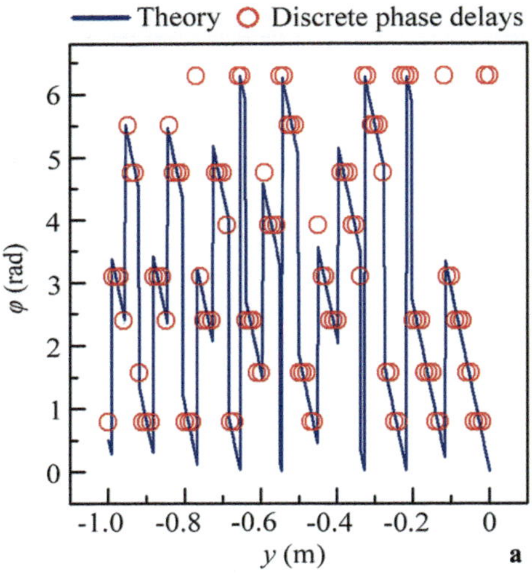

Fig. 7.13 Theoretical continuous phase distributions and discrete phase delays of the AF lenses with **a** $\theta = 0.9°$ and **b** $1.0°$ at 5.0 kHz

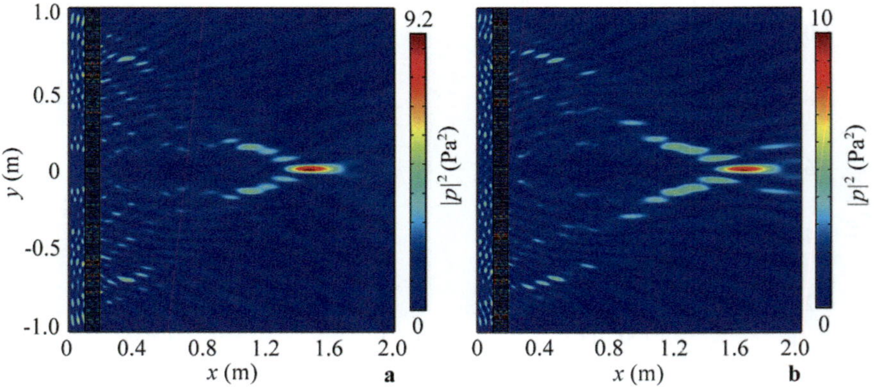

Fig. 7.14 Simulated intensity distributions caused by the AF lens with **a** $\theta = 0.9°$ and **b** $1.0°$ at 5.0 kHz

7.2.5 Feasibility Analysis

To verify the feasibility of the thermal insulation films in the AF lens, we select an aerogel film with a thickness of 0.05 cm as the thermal insulation material for the phased unit cells. The aerogel film has high temperature resistance and excellent heat insulation properties, effectively suppressing heat transfer. Additionally, its acoustic impedance is close to that of air, allowing sound energy to pass through with minimal loss. The material parameters of the aerogel [3] are as follows: density $\rho = 3.00$ kg/m^3, sound velocity $c = 333$ m/s, thermal conductivity $k = 0.03$ W/(m K), and

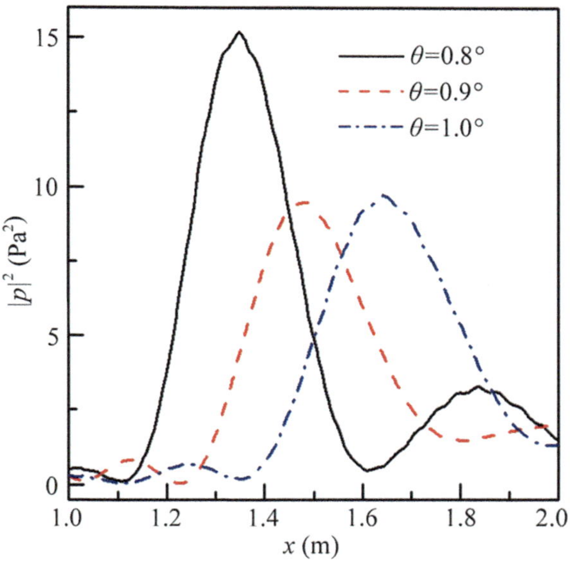

Fig. 7.15 Simulated transverse intensity distributions through the focus caused by the AF lenses with different values of θ at 5.0 kHz

thermal capacity $C_p = 0.05$ J/(kg K). The structure parameters and the number of phased unit cells remain the same as in Fig. 7.4d. Figure 7.16 shows the intensity distributions caused by the AF lens with aerogel films. Compared to Fig. 7.4d, the focusing performance remains nearly unchanged, confirming the feasibility of the proposed AF lens with thermal insulation films. However, experimental challenges remain, such as maintaining stable temperature distributions across the high and varying temperatures of each phased unit cell.

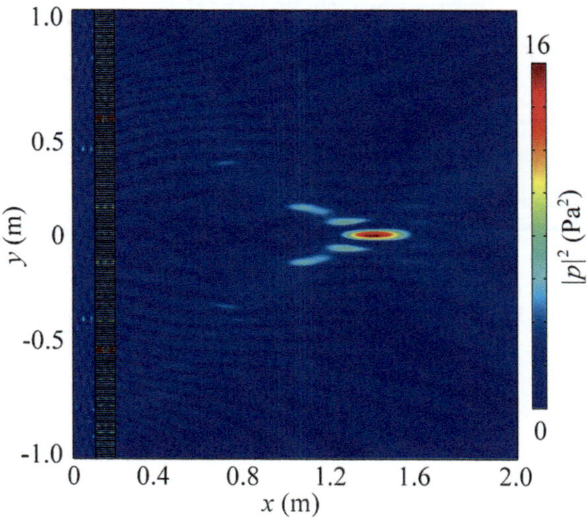

Fig. 7.16 Simulated intensity distributions caused by the AF lens with aerogel films at 5.0 kHz

7.2.6 Design and Performance of Reflected Phased Unit Cell

As shown in Fig. 7.17, we design a reflected unit cell for thermoacoustic phase control, consisting of three rigid thermal insulation boundaries (the blue solid lines) with thickness d and a thermal insulation film (the red dashed line) in air at temperature T. The length and width of the reflected unit cell are l and h, respectively. The acoustic wave is vertically incident from the right side and reaches the unit cell. At the left side of the unit cell, the acoustic wave is reflected by the rigid thermal insulation boundary and returns to the right side. The parameters of the reflected unit cell are $l = 6$ cm, $h = 1$ cm, and $d = 0.05$ cm.

Figure 7.18 shows the reflected phase delays of the unit cells at different air temperatures. As the temperature increases, the reflected phase delay covers the entire 2π range. However, compared to the results in Fig. 7.2, the reflected phase distribution changes significantly. This is because the propagation distance of sound in the reflected unit cell is about twice that in the transmitted unit cell shown in Fig. 7.1. The temperatures corresponding to the eight blue open circles are 400, 500, 630, 770, 900, 1030, 1230, and 1610 K, with reflected phase delays of 2π, $7\pi/4$, $3\pi/2$, $5\pi/4$, π, $3\pi/4$, $\pi/2$, and $\pi/4$, respectively.

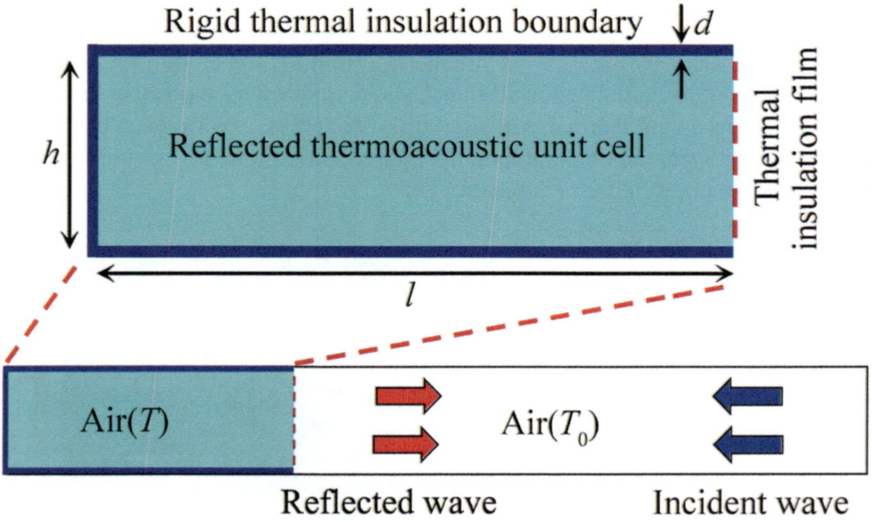

Fig. 7.17 Schematic of the reflected thermoacoustic unit cell

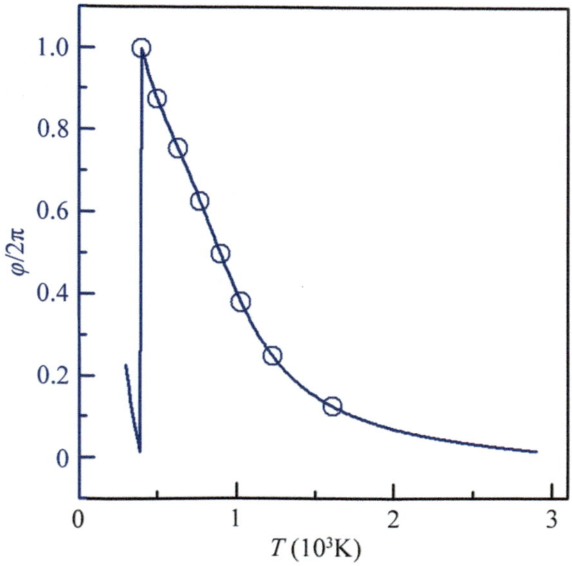

Fig. 7.18 Simulated reflected phase delays of the reflected unit cells ($l = 6$ cm) at different temperatures at 5.0 kHz

7.2.7 Design and Performance of Reflected Self-Healing AF Lens

Figure 7.19a shows the reflected intensity distribution caused by the reflected phased array of the Airy beam, where the phase distribution is the same as in Fig. 7.3. The reflected sound energy generated by the phased array forms an Airy beam, exhibiting a clear self-bending characteristic similar to the result in Fig. 7.4b. Figure 7.19b shows the intensity distribution caused by the reflected AF lens composed of two symmetric TPAs from Fig. 7.19a. The AF effect is achieved using two symmetric reflected Airy beams. Figure 7.20 shows the transverse and longitudinal intensity distributions along lines I and II in Fig. 7.19b, with results in free space provided for comparison. The focus is located at (133.8 cm, 0), and the intensity at the center of the focus is about 15 times higher than that in free space, demonstrating high performance of the reflected AF lens.

The reflected AF lens also exhibits broadband characteristics. Figure 7.21 shows the reflected intensity distribution caused by the reflected AF lens at different incident frequencies, using the same parameters as in Fig. 7.19b. The bandwidth of the reflected AF lens reaches up to 4.8 kHz. As the frequency increases, the size of the focus gradually increases, and the focus position shifts to the right, similar to the results in Fig. 7.6. Figure 7.22 shows the transverse intensity distributions through the focus at different frequencies, confirming the broadband focusing characteristics of the reflected AF lens. Thus, the reflected AF lens has a broad bandwidth, and the focal length can be controlled by adjusting the incident frequency.

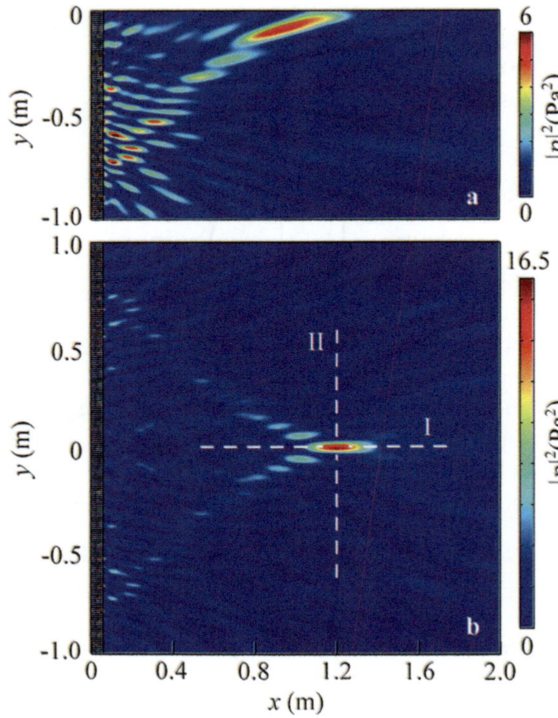

Fig. 7.19 Simulated reflected intensity distributions caused by **a** the reflected phased array of the Airy beam and **b** the reflected AF lens at 5.0 kHz

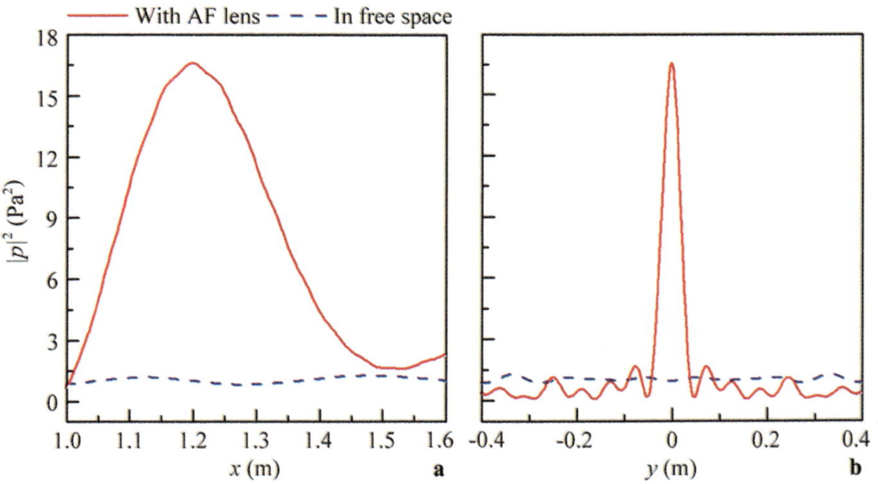

Fig. 7.20 Simulated transverse intensity distributions along lines **a** I and **b** II in Fig. 7.19 at 5.0 kHz

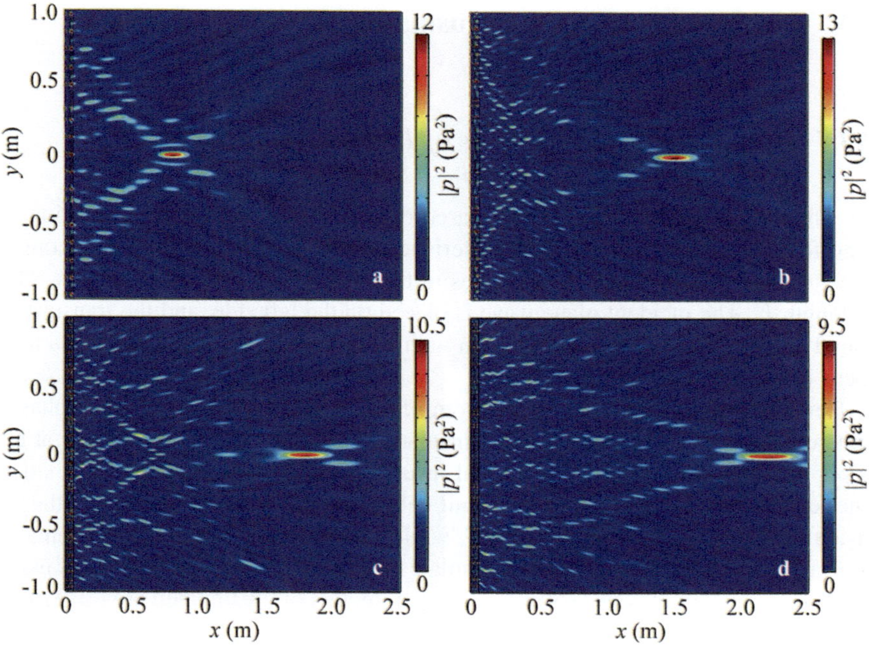

Fig. 7.21 Simulated reflected intensity distributions caused by the reflected AF lens at **a** 4.0 kHz, **b** 6.2 kHz, **c** 7.5 kHz, and **d** 8.8 kHz

Fig. 7.22 Simulated transverse intensity distributions through the focus caused by the reflected AF lens at different frequencies

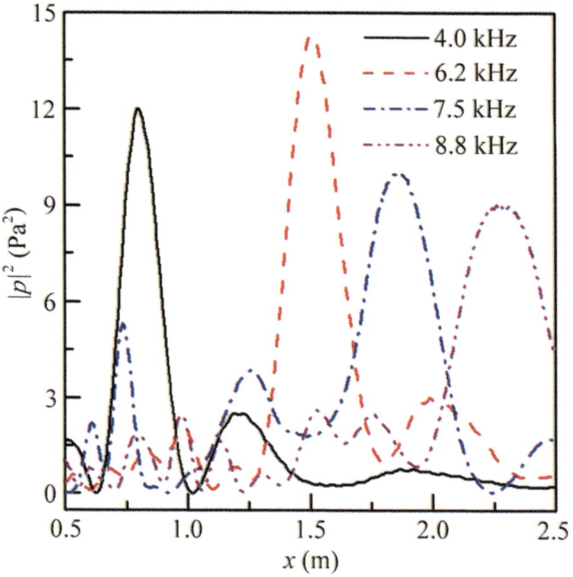

7.3 Long-Focus Acoustic Focusing by Thermoacoustic Phased Arrays

7.3.1 Design and Performance of Long-Focus AF Lens

The long-focus AF lens is designed based on the Bessel-like beam, leveraging its long focus and non-diffraction characteristics. As shown in Fig. 7.23, an acoustic wave with normal incidence propagates in the x direction and impinges on a TPA of length H. The incident plane wave is placed on the left side, and the transmitted sound energy forms a Bessel-like beam with a base angle β, converging into a long focus (the black solid line).

Here, we use the phased unit cell from Fig. 7.1 to design the TPA. Each phased unit cell consists of two rigid thermal insulation boundaries with thickness d at the upper and lower sides and two thermal insulation films at the left and right sides. The length and width of the phased unit cell are l and h, respectively, and the air inside the unit cell is at temperature T, while the ambient temperature T_0 remains constant. Throughout this work, we simulate the long-focus AF characteristics using the COMSOL Multiphysics software. The parameters of the unit cell are $l = 12$ cm, $h = 1$ cm, $d = 0.05$ cm, and $T_0 = 300$ K. The acoustic velocity and density of air are calculated using Eqs. (7.1) and (7.2), respectively. Thermal convection is neglected in the simulations, and the heat flux at the external boundaries is continuous.

Figure 7.24a shows the phase delays and transmittance spectrum of the phased unit cells at different temperatures. The phase delays cover the entire 2π range for temperatures below 1000 K, and the transmittance exceeds 0.9. We select eight types of phased unit cells (the blue open circles) with temperatures of 480, 400, 340, 290, 260, 910, 720, and 580 K, corresponding to phase delays from $\pi/4$ to 2π in steps of $\pi/4$, to design the long-focus AF lens. The lens parameters are $H = 81$ cm, $\beta = \pi/20$, $f = 5.0$ kHz, and $c_0 = 343$ m/s. To realize a Bessel-like beam, the phase distribution $\varphi(y)$ of the TPA in the y direction satisfies [14]:

$$\varphi(y) = -k_0|y|\sin\beta, \tag{7.4}$$

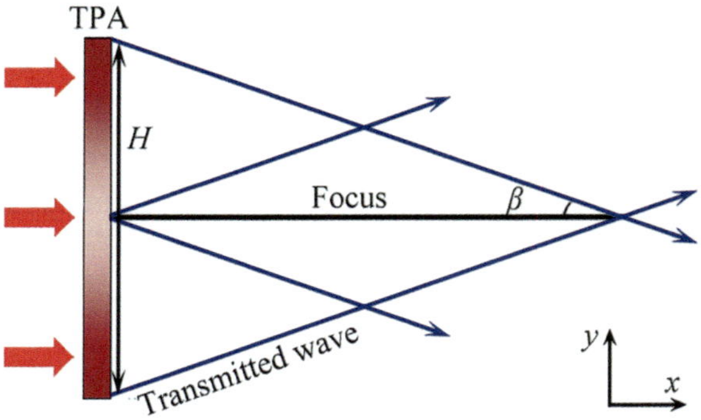

Fig. 7.23 Schematic of the AF lens based on acoustic Bessel-like beam

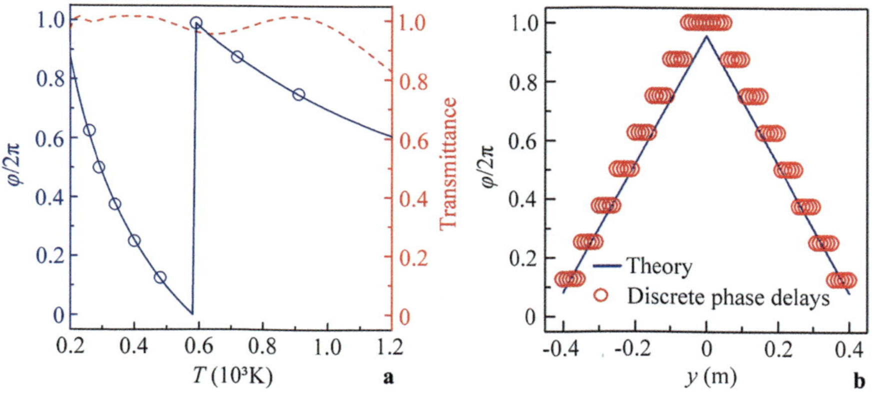

Fig. 7.24 **a** Simulated transmittance spectrum and phase delays of thermoacoustic phased unit cells at different temperatures. **b** Theoretical continuous phase distribution and 81 discrete phase delays of the long-focus AF lens at 5.0 kHz

where $k_0 = 2\pi f/c_0$ is the wave number, c_0 is the acoustic velocity of air at 300 K, f is the frequency, β is the base angle of the Bessel-like beam. Based on Eq. (7.4), we calculate the phase distribution $\varphi(y)$, as shown in Fig. 7.24b. Using the eight types of phased unit cells, we design the long-focus AF lens with 81 discrete phase delays from Fig. 7.24b.

Figure 7.25a, b show the intensity distributions caused by the theoretical continuous phase distribution and the TPA of the long-focus AF lens, respectively, with results in free space provided for comparison (Fig. 7.25c). The transmitted sound energy exhibits the characteristic of the Bessel-like beam, converging into a long focus with weak diffraction. This has potential applications in scenarios such as sound systems in large stadiums and bird-repelling equipment at airports. The long-focus AF characteristics from the TPA (Fig. 7.25b) agree with its theoretical results (Fig. 7.25a).

To further analyze the performance of the long-focus AF lens, we provide the intensity distributions through the focus along lines I and II in Fig. 7.25b, shown in Fig. 7.26a, b, respectively. The corresponding results in free space along lines III and IV in Fig. 7.25c are also displayed for comparison. The FLHM of the focus reaches about 16.8λ, making the focus size much longer than that of conventional AF lenses. The center of the focus is at (137.1 cm, 0), and its intensity is about 6.0 times higher than that in free space, demonstrating the high performance of the long-focus AF lens.

To demonstrate the working band of the long-focus AF lens, we simulate the intensity distributions caused by the long-focus AF lens at 4.2, 5.2, and 6.2 kHz, as shown in Fig. 7.27a–c, respectively. The lens parameters are the same as those in Fig. 7.25b. The lens exhibits high-performance long-focus AF at three frequencies.

Figure 7.28 shows the maximum intensity spectrum at the focus, with a bandwidth at half maximum (Δf) of about 4.18 kHz. The fractional bandwidth reaches 0.77, demonstrating the broadband characteristic of the long-focus AF lens. As the

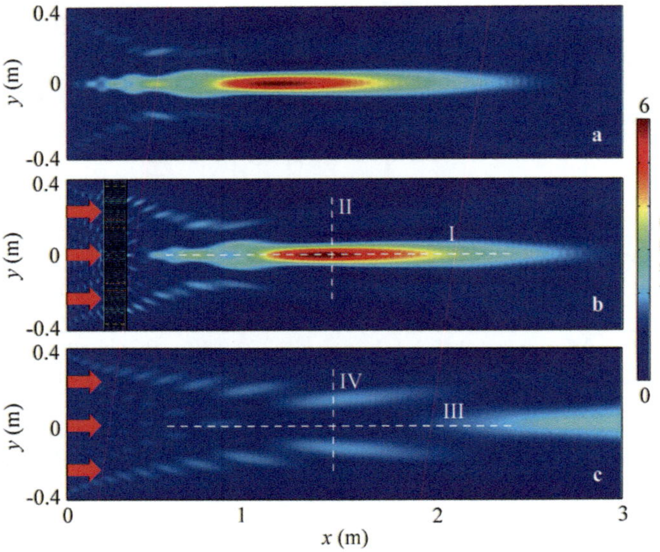

Fig. 7.25 Simulated intensity distributions caused by **a** the theoretical continuous phase distribution and **b** the TPA of the long-focus AF lens, and **c** in free space at 5.0 kHz. The red arrows in **b** and **c** represent incident acoustic waves

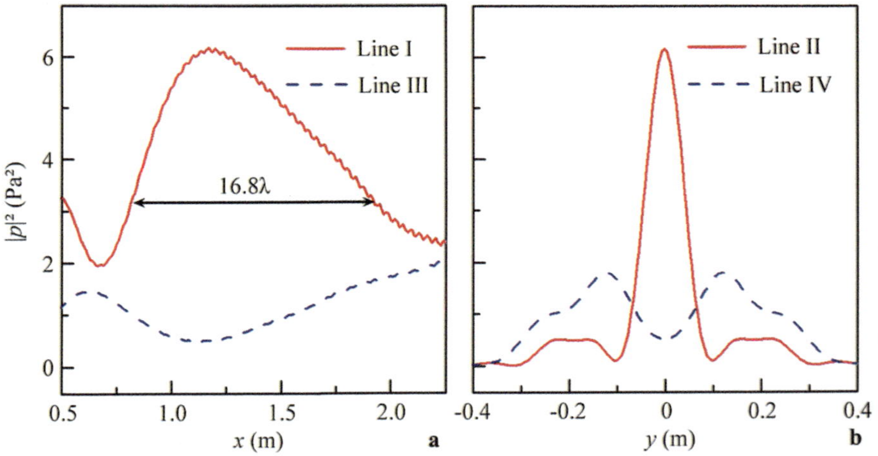

Fig. 7.26 Simulated intensity distributions along **a** lines I and III, and **b** lines II and IV in Fig. 7.25

frequency increases, the long focus shifts to the right. This is because the incident angle and phase distribution of the lens remain unchanged, causing the refraction angle θ_t to decrease with increasing frequency, as per Eq. (2.41). Thus, the long focus gradually moves to the right. These results confirm that the proposed long-focus AF lens has broad bandwidth and tunable focus.

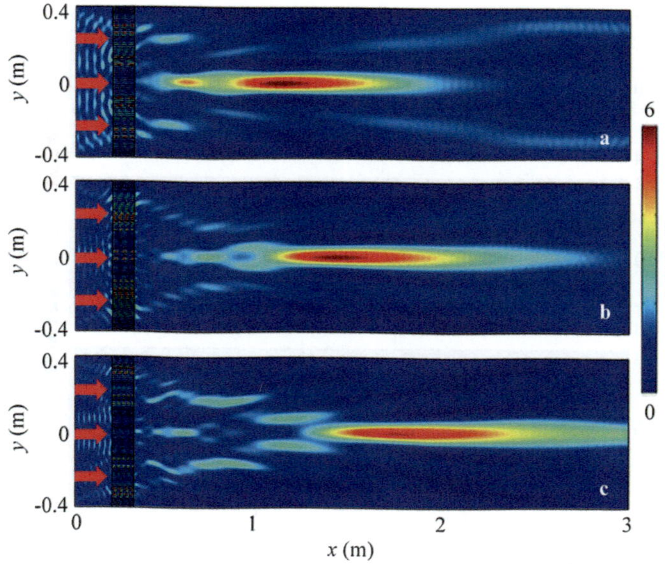

Fig. 7.27 Simulated reflected intensity distributions caused by the long-focus AF lens at **a** 4.2 kHz, **b** 5.2 kHz, and **c** 6.2 kHz

Fig. 7.28 Maximum intensity spectrum at the focus

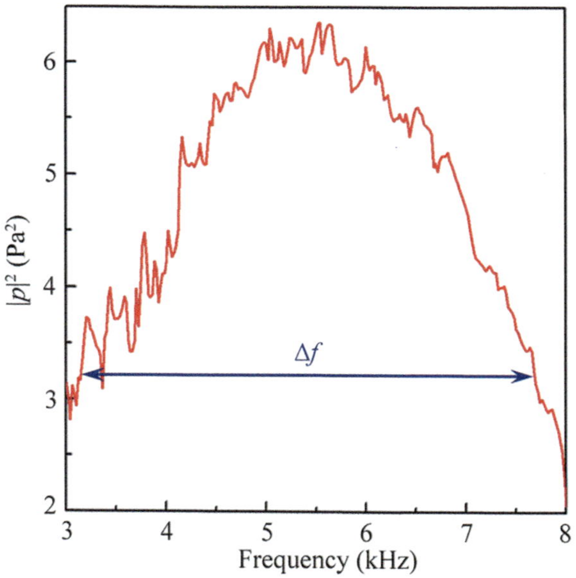

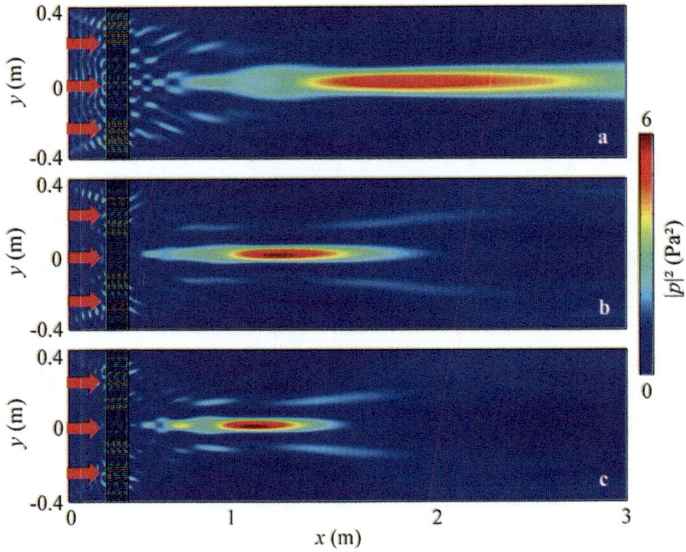

Fig. 7.29 Simulated reflected intensity distributions caused by the long-focus AF lenses with **a** $\beta = 6°$, **b** 12°, and **c** 15° at 5.0 kHz

7.3.2 Parameter Analysis

We also analyze the influence of the base angle β of the Bessel-like beam on the long-focus AF effect. Figure 7.29 shows the intensity distributions caused by the long-focus AF lenses with $\beta = 6°$, 12°, and 15°. The long-focus AF effect persists across different values of β, but the focus size decreases, and the focus shifts to the left as β increases. This is because the refraction angles of the sound beams increase with β, causing the interference region to move leftward and shrink. Figure 7.30 shows the transverse intensity distributions through the focus for different values of β, confirming that the focus shifts left and decreases in size as β increases, consistent with the results in Fig. 7.29.

7.3.3 Feasibility Analysis

To demonstrate the feasibility of the proposed long-focus AF lens, we design a thermoacoustic unit cell, as shown in Fig. 7.31. Two identical tungsten sheets with thickness d_1 are placed at the upper and lower sides of the unit cell. The center region (with length l_1 and temperature T) of each tungsten sheet raises the temperature of the air inside the unit cell. A vacuum layer with thickness d_2 provides thermal insulation between the tungsten sheets at different temperatures, and an aerogel film with thickness d_3 acts as a thermal insulation film on both sides of the unit

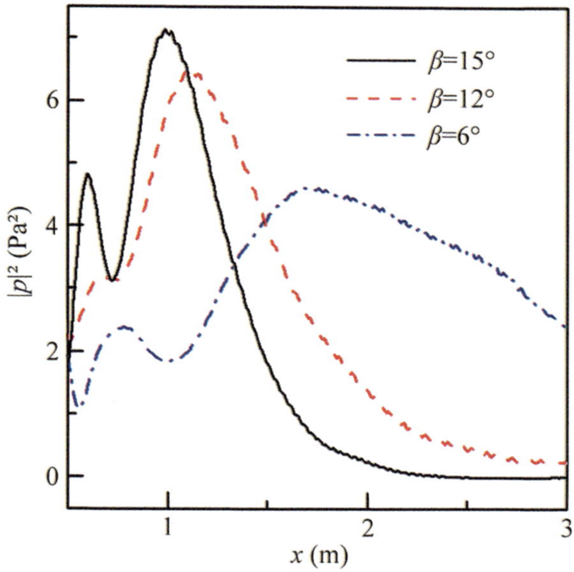

Fig. 7.30 Simulated reflected transverse intensity distributions through the focus caused by the long-focus AF lenses with different values of β at 5.0 kHz

cell. The aerogel film has high heat insulation, temperature resistance, and acoustic impedance matching with air, allowing sound waves to pass through while effectively hindering heat transfer. The length and width of the unit cells are the same as those in Fig. 7.24a, with additional parameters $l_1 = 4.0$ cm, $d_1 = 0.01$ cm, $d_2 = 0.08$ cm, and $d_3 = 0.10$ cm. Other parameters include: sound velocity $c = 5334$ m/s, density $\rho = 17{,}800$ kg/m^3, thermal capacity $C_p = 132$ J/(kg K), and thermal conductivity $K = 175$ W/(m K) for tungsten; $c = 333$ m/s, $\rho = 3$ kg/m^3, $C_p = 0.05$ J/(kg K), and $K = 0.03$ W/(m K) for aerogel [3, 11].

Figure 7.32 shows the simulated temperature distributions inside the two unit cells with $T_1 = 400$ K and $T_2 = 900$ K at $t = 8$ s, where t represents the heating time. The heat produced in the center region transfers to other regions [15], resulting in a relatively uniform temperature distribution in the unit cell at $t = 8$ s. To design the long-focus AF lens, we simulate the phase delays and transmittances of the unit cells at different temperatures at $t = 8$ s, as shown in Fig. 7.33. The phase delay covers the entire 2π range for temperatures below 1000 K, and the transmittances exceed 0.9. We select eight unit cells with temperatures of 520, 430, 360, 300, 250, 980, 790, and 640 K, corresponding to phase delays from $\pi/4$ to 2π in steps of $\pi/4$, to design the long-focus AF lens. Figure 7.34 shows the simulated intensity distributions caused by the long-focus AF lens with these eight unit cells. Similar to Fig. 7.25b, the transmitted intensity distributions exhibit a long focus with high performance at $t =$ 8 s. We also analyze the influence of heating time on the long-focus AF performance. Figure 7.35 shows the intensity distributions caused by the long-focus AF lens at $t = 2$ and 28 s. The long-focus AF effect persists across different heating times, confirming the design feasibility of the lens.

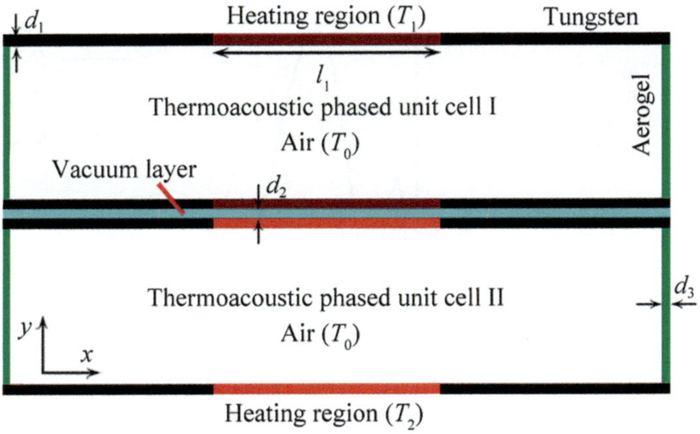

Fig. 7.31 Schematic of thermoacoustic phased unit cells I and II with temperatures of T_1 and T_2 for the heating region, respectively

Fig. 7.32 Simulated temperature distributions in thermoacoustic phased unit cells I and II with $T_1 = 400$ K and $T_2 = 900$ K in the heating region at $t = 8$ s

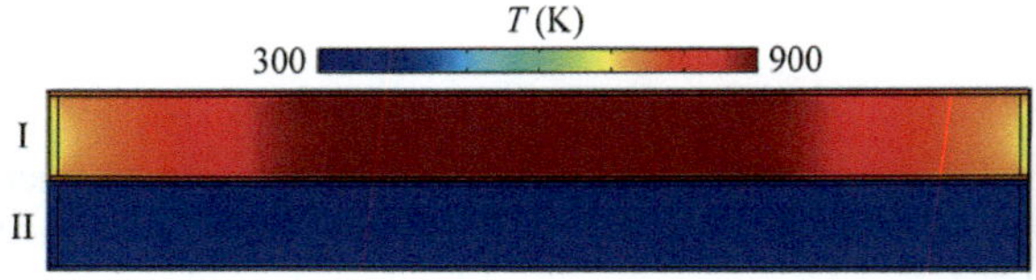

Fig. 7.33 Simulated transmittances and phase delays of thermoacoustic unit cells at different temperatures in the heating region at $t = 8$ s

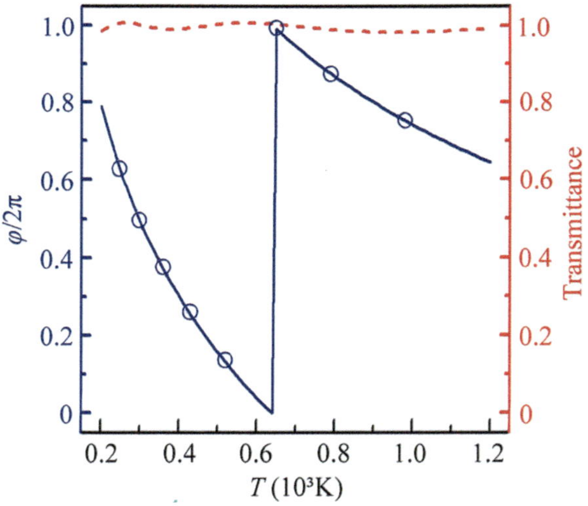

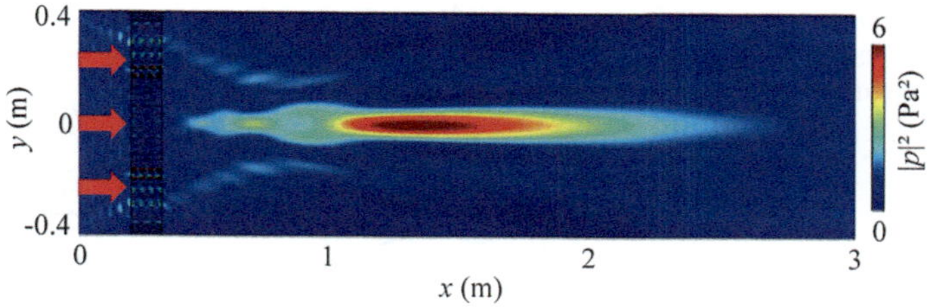

Fig. 7.34 Simulated intensity distributions caused by the long-focus AF lens with the eight types of unit cells at 5.0 kHz

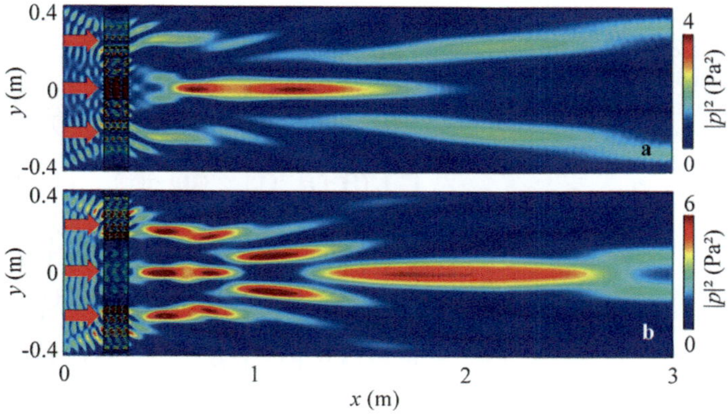

Fig. 7.35 Simulated intensity distributions caused by the long-focus AF lens at **a** $t = 2$ s and **b** 28 s at 5.0 kHz

7.3.4 *Design and Performance of Long-Focus AF Lens for Cylindrical Sound Source*

Next, we design a long-focus AF lens for a cylindrical sound source. As shown in Fig. 7.36, the cylindrical sound source is placed at (0, 0), and the transmitted sound energy forms a long focus, similar to Fig. 7.23. The distance between the cylindrical sound source and the lens is $L = 0.3$ m, and the other lens parameters are the same as in Fig. 7.25b. The phase distribution $\varphi(y)$ of the TPA satisfies [16]:

$$\varphi(y) = -k_0\left(|y| \sin\beta + \sqrt{y^2 + L^2}\right). \tag{7.5}$$

Based on Eq. (7.5), we calculate the theoretical continuous phase distribution of the long-focus AF lens for the cylindrical sound source, as shown in Fig. 7.37. The long-focus AF lens is designed using 81 discrete phase delays based on the selected eight unit cells.

Figure 7.38a shows the intensity distribution caused by the long-focus AF lens under the excitation of the cylindrical sound source, with the corresponding result

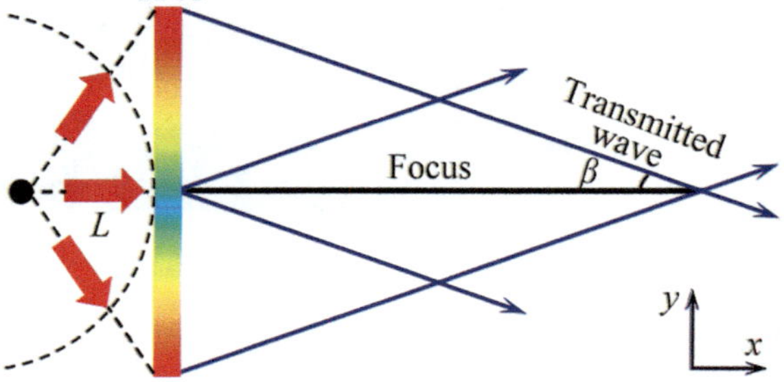

Fig. 7.36 Schematic of the long-focus AF lens for the cylindrical sound source

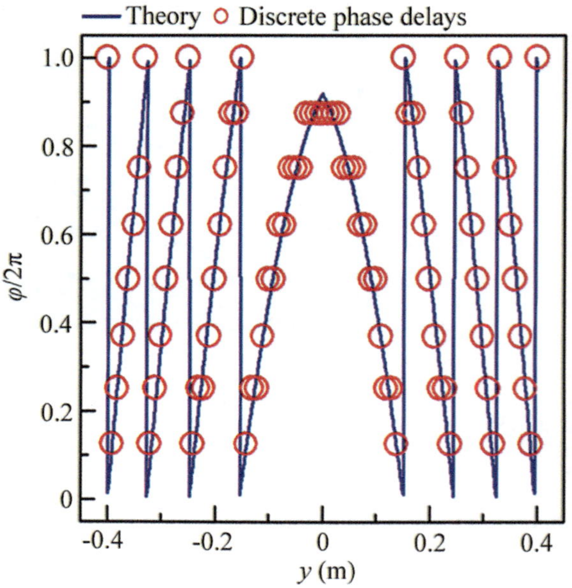

Fig. 7.37 Theoretical continuous phase distributions and 81 discrete phase delays of the long-focus AF lens for the cylindrical sound source

in free space provided for comparison (Fig. 7.38b). The cylindrical acoustic wave passes through the AF lens and converges into a long focus, exhibiting long-focus AF characteristics similar to those in Fig. 7.25b. We also simulate the intensity distributions through the focus along lines V and VI and in free space along lines VII and VIII in Fig. 7.38, as shown in Fig. 7.39. The FLHM of the long focus reaches 24.3λ, with very weak diffraction. The center of the focus is at (147 cm, 0), and its intensity is about 8.0 times larger than that in free space, demonstrating the high performance of the long-focus AF lens for the cylindrical sound source.

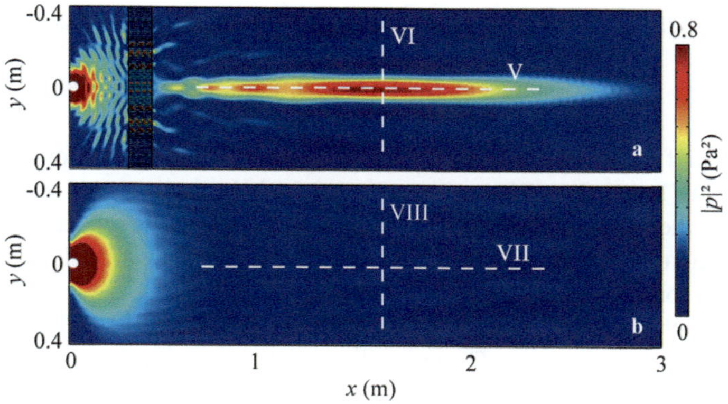

Fig. 7.38 Simulated reflected intensity distributions **a** caused by the long-focus AF lens and **b** in free space under the excitation of the cylindrical sound source (the white points) at 5.0 kHz

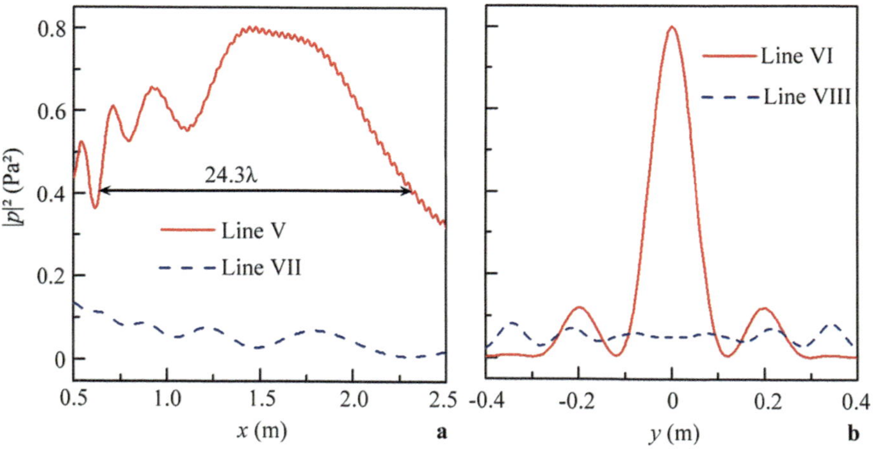

Fig. 7.39 Simulated reflected intensity distributions along **a** lines V and VII and **b** lines VI and VIII in Fig. 7.38

7.4 Reflected Long-Focus Acoustic Focusing by Thermoacoustic Phased Arrays

7.4.1 Design and Performance of Reflected Long-Focus AF Lens

Similar to the design of the long-focus AF lens discussed earlier, we design a reflected long-focus AF lens composed of a TPA of the Bessel-like beam, as shown in Fig. 7.40. The incident acoustic wave (the black arrows) is placed on the right side, and the reflected sound energy forms a Bessel-like beam, focusing on a long focus AB. The relationship between the length of the focus D and the base angle β is given by:

$$D = h/\tan\beta, \tag{7.6}$$

where h is half the length of the reflected AF lens. The parameters for the reflected long-focus AF lens are $y = 81$ cm, $\beta = \pi/20$, $f = 5.0$ kHz, and $c_0 = 343$ m/s. As shown in Fig. 7.41, the theoretical continuous phase distribution of the reflected Bessel-like beam is calculated using Eq. (7.4), and 81 discrete phase delays are selected to design the reflected long-focus AF lens.

To realize the reflected long-focus AF lens, we design a reflected unit cell with thermoacoustic phase control. The unit cell consists of three rigid thermal insulation

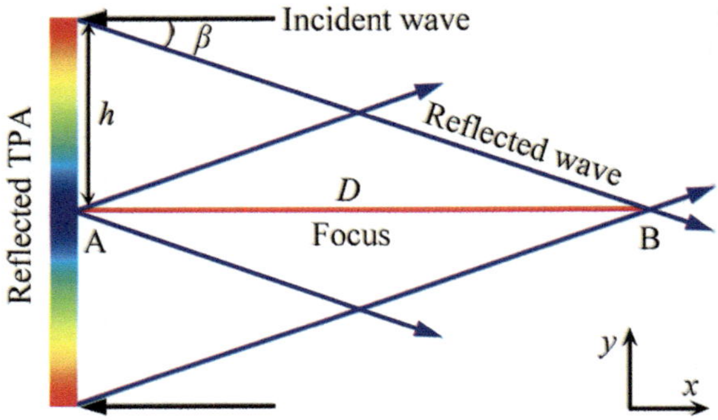

Fig. 7.40 Schematic of the reflected long-focus AF lens

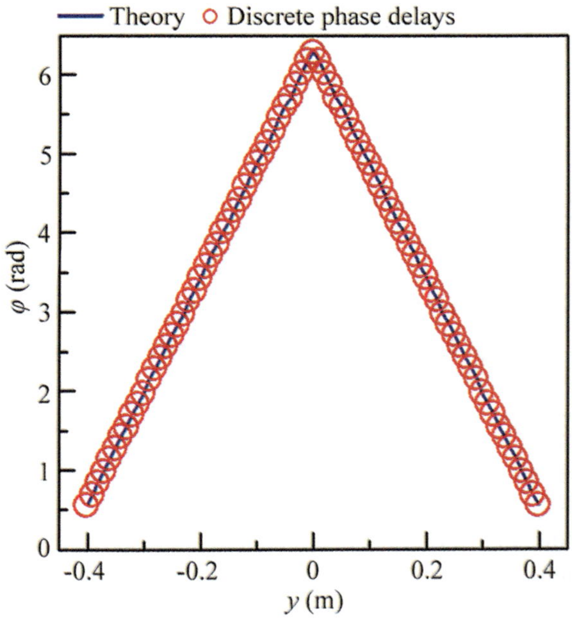

Fig. 7.41 Theoretical continuous phase distribution and 81 discrete phase delays of the reflected long-focus AF lens at 5.0 kHz

boundaries (the blue solid lines) and a thermal insulation film (the red dashed line) in air at temperature T, as shown in Fig. 7.17. The thickness of the rigid thermal insulation boundary is d, and the length and width of the unit cell are l and h, respectively. The acoustic wave is vertically incident from the right side and passes through the unit cell. At the left side, the wave is reflected by the rigid thermal insulation boundary and returns to the outside.

Throughout this work, the characteristics of the reflected long-focus AF are simulated using the COMSOL Multiphysics software. The parameters of the reflected unit cell are $l = 8$ cm, $h = 1$ cm, $d = 0.05$ cm, and $T_0 = 300$ K. We also account for the influence of temperature on the thermal parameters of air, such as thermal conductivity k and thermal capacity C_p, which are expressed as: $k = -0.00227583562 + (1.15480022 \times 10^{-4}) \times T - (7.90252856 \times 10^{-8}) \times T^2 + (4.11702505 \times 10^{-11}) \times T^3 - (4.11702505 \times 10^{-15}) \times T^4$ W m^{-1} K^{-1} and $C_p = 1047.63657 - 0.372589265 \times T + (9.45304214 \times 10^{-4}) \times T^2 - (6.02409443 \times 10^{-7}) \times T^3 + (1.2858961 \times 10^{-10}) \times T^4$ J (kg K)$^{-1}$. The acoustic velocity c and density ρ of air in the reflected unit cell are determined by temperature T and can be calculated using Eqs. (7.1) and (7.2), respectively. The heat flux at the external boundaries is continuous, and thermal convection inside the model is neglected in the simulations. Figure 7.42 shows the reflected phase delays φ of the unit cell at different temperatures. The reflected phase delay covers the entire 2π range for temperatures between 560 and 2450 K (the black shaded region).

Figure 7.43a, b show the reflected intensity distributions caused by the theoretical continuous phase distribution and the TPA, respectively. The reflected acoustic energy forms a long focus on the right side of the lens with minimal diffraction, exhibiting the characteristic Bessel-like beam. Based on Eq. (7.6), the theoretical length of the focus D is calculated as 2.53 m, which agrees well with the simulated

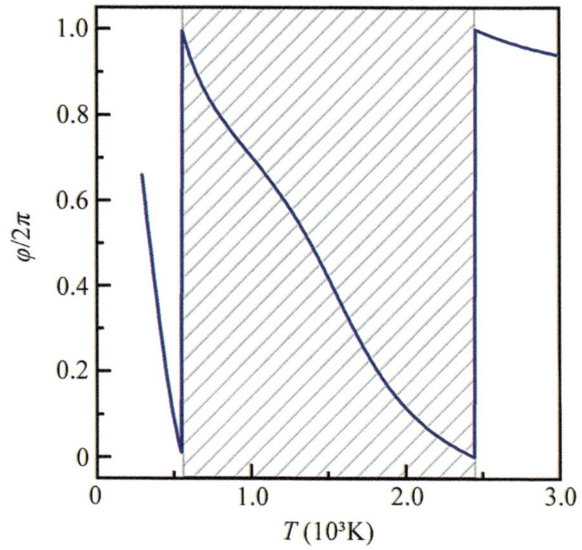

Fig. 7.42 Simulated reflected phase delays of the reflected unit cells ($l = 8$ cm) at different temperatures at 5.0 kHz

result in Fig. 7.43b and is larger than that of previous AF lenses [17–20]. The long-focus AF characteristics from the TPA match the theoretical results in Fig. 7.43a. Additionally, as shown in Fig. 7.43c, the AF effect does not occur in free space, confirming that the long-focus AF effect is achieved using the Bessel-like beam.

To further illustrate the long-focus AF effect, the transverse and longitudinal intensity distributions through the focus (lines I–IV in Fig. 7.43b, c) are simulated, as shown in Fig. 7.44. With the reflected long-focus AF lens, the center of the focus is at (124.1 cm, 0), and the maximum intensity at the center is about 6 times larger than that in free space. As shown in Fig. 7.44a, the intensity along line I (from 0.5 to 2.5 m) exceeds 2.0 Pa^2, demonstrating high long-focus performance. Furthermore, the sound energy is primarily focused at $y = 0$, with very weak diffraction, highlighting the non-diffraction characteristic of the long-focus AF effect.

To demonstrate the bandwidth of the reflected long-focus AF lens, we simulate the reflected intensity distributions at different frequencies, as shown in Fig. 7.45. At the three selected frequencies, the reflected sound energy is focused into a long focus on the right side. The working bandwidth of the lens is about 1.4 kHz, with a fractional bandwidth of 0.29.

As the frequency increases, the long focus shifts to the right, as shown in Fig. 7.46. This occurs because the incident angle and phase distribution of the lens remain constant, and the reflected angle θ_r decreases with increasing frequency, as per

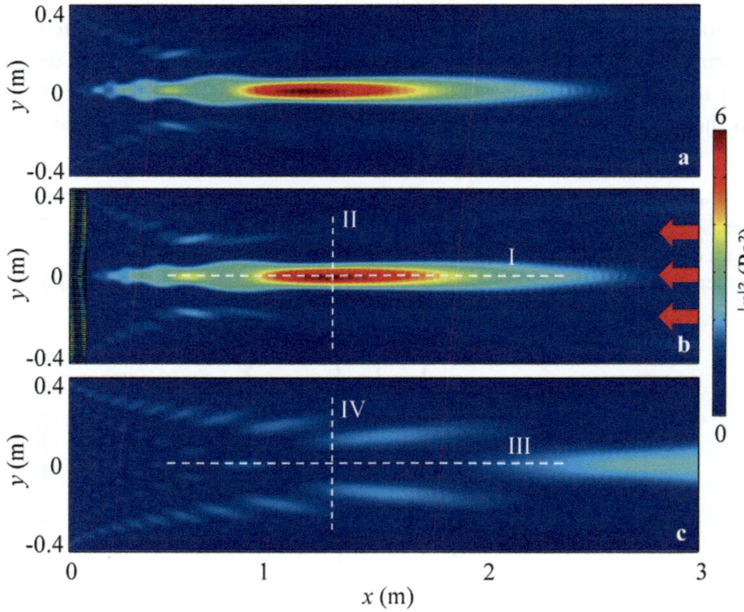

Fig. 7.43 Simulated intensity distributions caused by **a** the theoretical continuous phase distribution and **b** the TPA, and **c** in free space at 5.0 kHz. The red arrows in (**b**) represent the incident acoustic wave

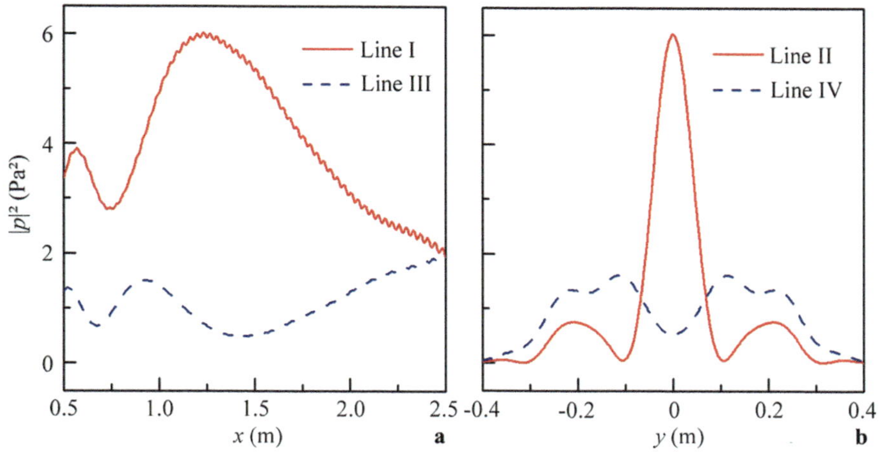

Fig. 7.44 Simulated intensity distributions along **a** lines I and III, and **b** lines II and IV in Fig. 7.43

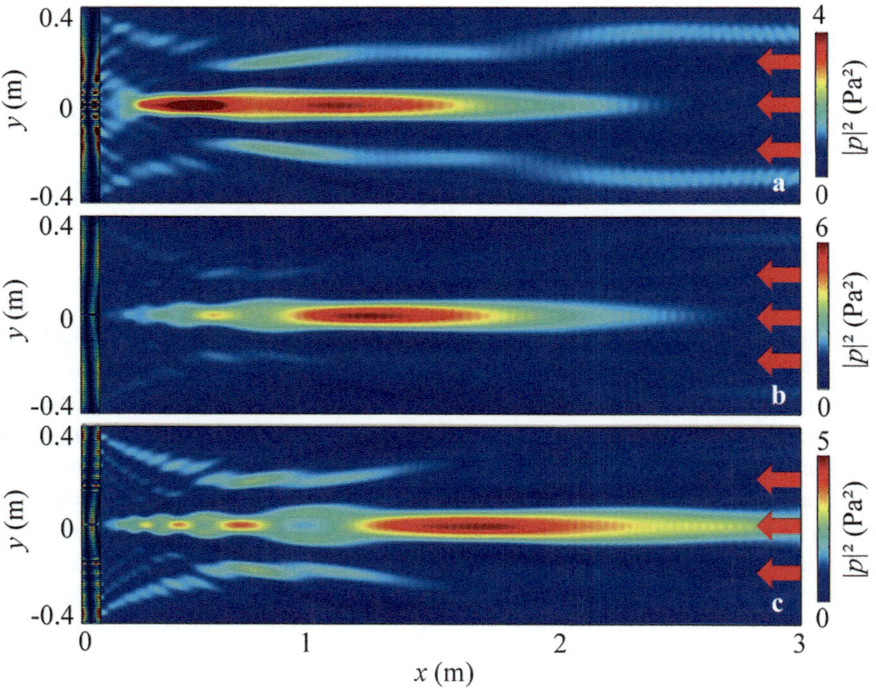

Fig. 7.45 Simulated reflected intensity distributions caused by the reflected long-focus AF lens at **a** 4.2 kHz, **b** 4.9 kHz, and **c** 5.6 kHz

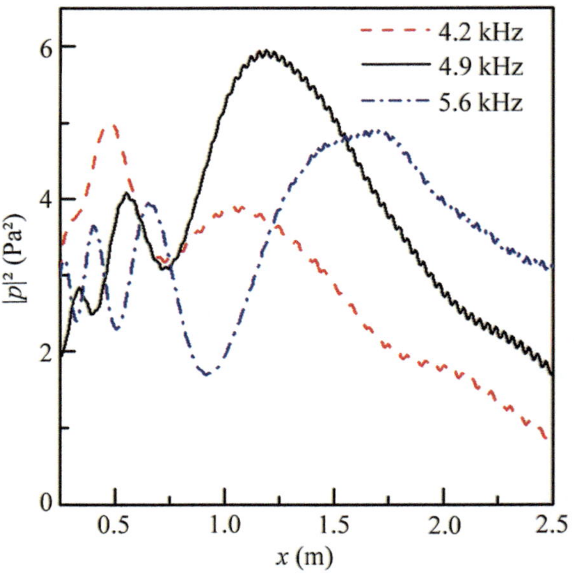

Fig. 7.46 Simulated reflected transverse intensity distributions through the focus at different frequencies

Eq. (2.41), causing the focus to shift rightward. Thus, the reflected long-focus AF lens has broad bandwidth and tunable focus position.

7.4.2 Parameter Analysis

Next, we analyze the influence of the base angle β of the Bessel-like beam on the long-focus AF effect. Figure 7.47 shows the theoretical continuous phase distributions and discrete phase delays of the reflected long-focus AF lenses for $\beta = 6°$, 12°, and 15°, and the simulated reflected intensity distributions caused by the reflected long-focus AF lenses with $\beta = 6°$, 12°, and 15° are presented in Fig. 7.48a–c, respectively. The long-focus AF effect persists across different values of β. As β increases, the length and size of the focus decrease, and the focus shifts to the left, as explained by Eq. (7.6). Thus, the sound energy becomes more concentrated, and the focus moves leftward. Figure 7.49 shows the transverse intensity distributions through the focus for different values of β. As β increases, the intensity at the focus increases, and the focus shifts to the left. This occurs because the refraction angles of the transmitted sound beams increase, causing the interference region to move leftward and shrink. Thus, the sound energy becomes more concentrated, consistent with the results in Fig. 7.48.

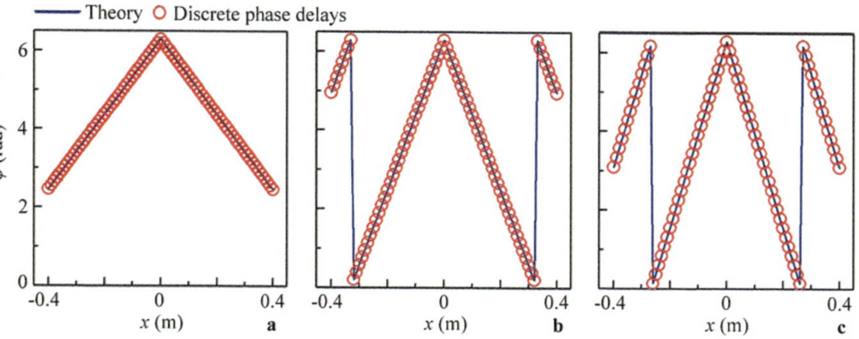

Fig. 7.47 Theoretical continuous phase distributions and discrete phase delays of the reflected long-focus AF lenses with **a** $\beta = 6°$, **b** 12° and **c** 15° at 5.0 kHz

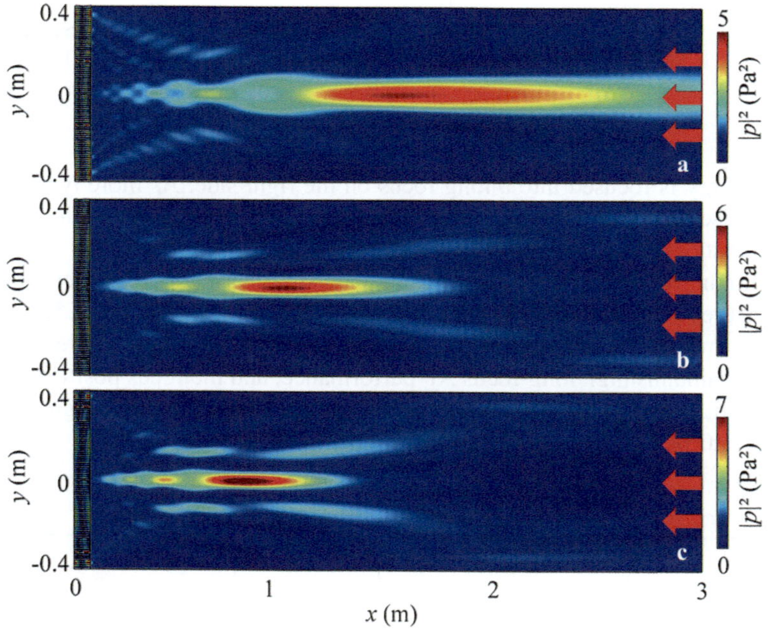

Fig. 7.48 Simulated reflected intensity distributions caused by the reflected long-focus AF lenses with **a** $\beta = 6°$, **b** 12° and **c** 15° at 5.0 kHz

7.4.3 Robustness Demonstration

To demonstrate the robustness of the reflected long-focus AF lens, we design lenses using four, six, and eight types of reflected unit cells with 81 discrete phase delays, as shown in Fig. 7.50a–c. Figure 7.51a–c show the simulated reflected intensity distributions for these cases. The long-focus AF effect persists, and the reflected

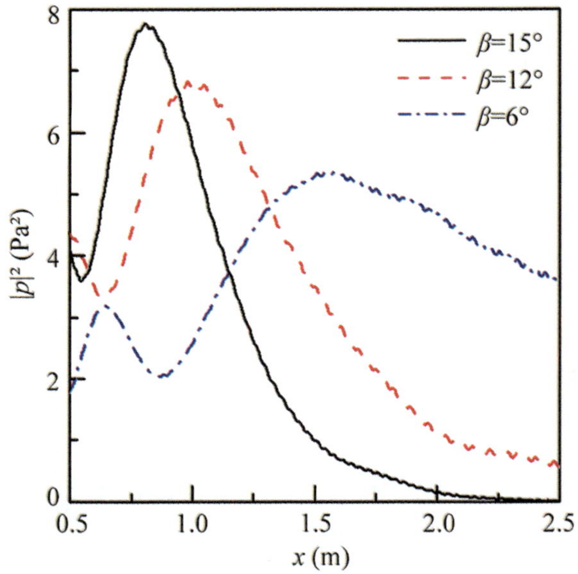

Fig. 7.49 Simulated reflected transverse intensity distributions through the focuses for different values of β at 5.0 kHz

sound energy is focused into a long focus on the right side. As more types of unit cells are used, the maximum intensity at the focus increases slightly, while the focus characteristics remain largely unchanged.

Additionally, we simulate the reflected intensity distributions by increasing the temperatures of each unit cell by 100, 300, and 500 K, as shown in Fig. 7.52 a–c. The other parameters are the same as those in Fig. 7.43b. As the temperature increases, the lens maintains high long-focus AF performance, and the focus position remains unchanged, although the maximum intensity at the focus decreases slightly. These results confirm the high robustness of the reflected long-focus AF lens.

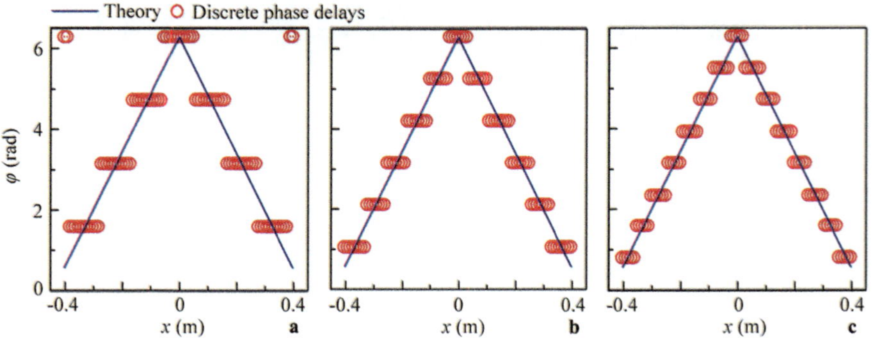

Fig. 7.50 Theoretical continuous phase distributions and discrete phase delays of the reflected long-focus AF lenses with **a** four, **b** six, and **c** eight types of reflected unit cells at 5.0 kHz

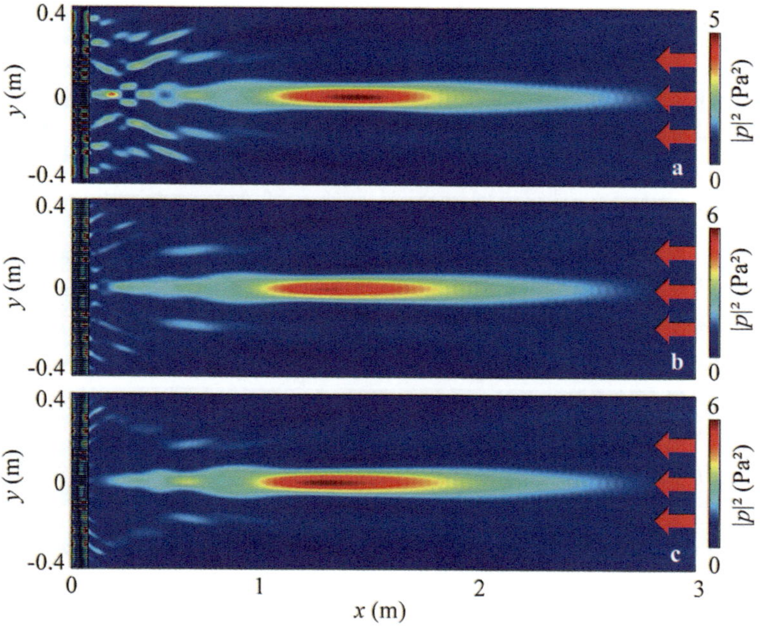

Fig. 7.51 Simulated reflected intensity distributions caused by the reflected long-focus AF lenses with **a** four, **b** six, and **c** eight types of reflected unit cells at 5.0 kHz

7.4.4 Design and Performance of Reflected Long-Focus AF Lens for Cylindrical Sound Source

Based on Eq. (2.41), we design a reflected long-focus AF lens for a cylindrical sound source. As shown in Fig. 7.53, the cylindrical sound source is located at (25 cm, 0), and the reflected sound energy from the lens forms a Bessel-like beam, focusing into a long focus on the right side. The phase distribution $\varphi(y)$ of the TPA is calculated using Eq. (7.5). Figure 7.54a–c show the theoretical continuous phase distributions and 81 discrete phase delays of the reflected long-focus AF lenses for $\beta = 6°$, $12°$, and $15°$, with $L = 0.25$ m.

Figure 7.55a–c show the simulated reflected intensity distributions caused by the reflected long-focus AF lenses with $\beta = 6°$, $12°$, and $15°$, respectively. The reflected sound energy from the cylindrical sound source is focused into a long focus on the right side. As β increases, the focus size decreases, and the focus shifts to the left, consistent with the results in Fig. 7.48.

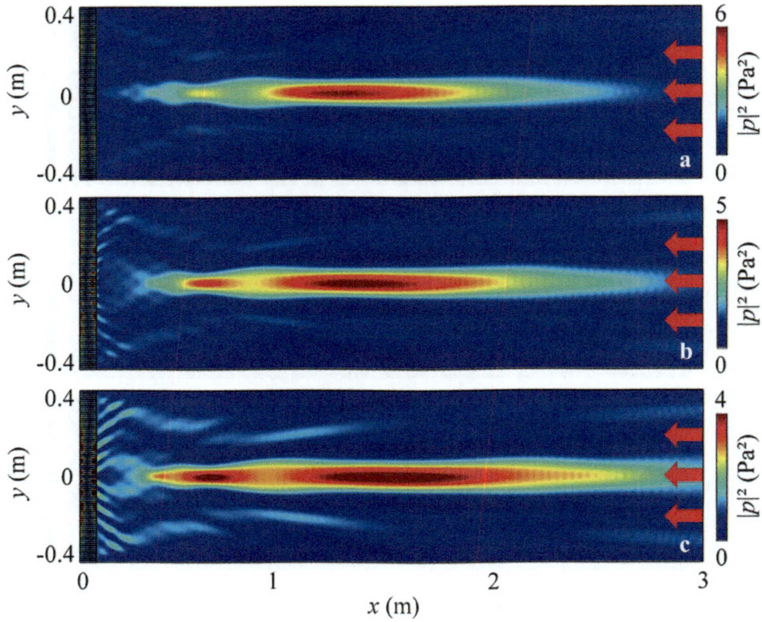

Fig. 7.52 Simulated reflected intensity distributions caused by the reflected long-focus AF lens in Fig. 7.43b by increasing the temperatures of each unit cell by **a** 100 K, **b** 300 K, and **c** 500 K at 5.0 kHz

Fig. 7.53 Schematic of the reflected long-focus AF lens for the cylindrical sound source

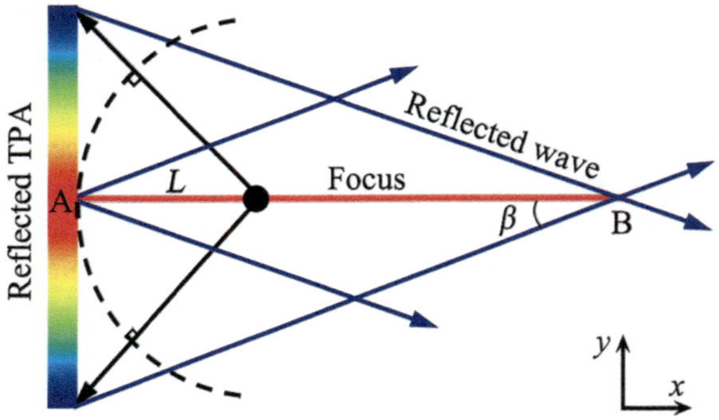

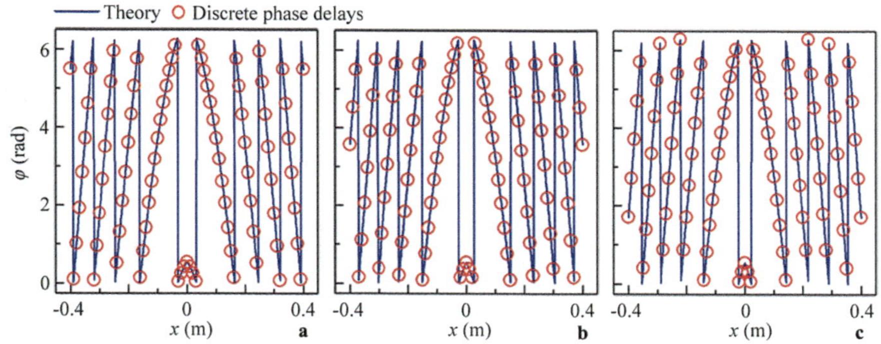

Fig. 7.54 Theoretical continuous phase distributions and discrete phase delays of the reflected long-focus AF lenses with **a** $\beta = 6°$, **b** 12° and **c** 15° for the cylindrical sound source

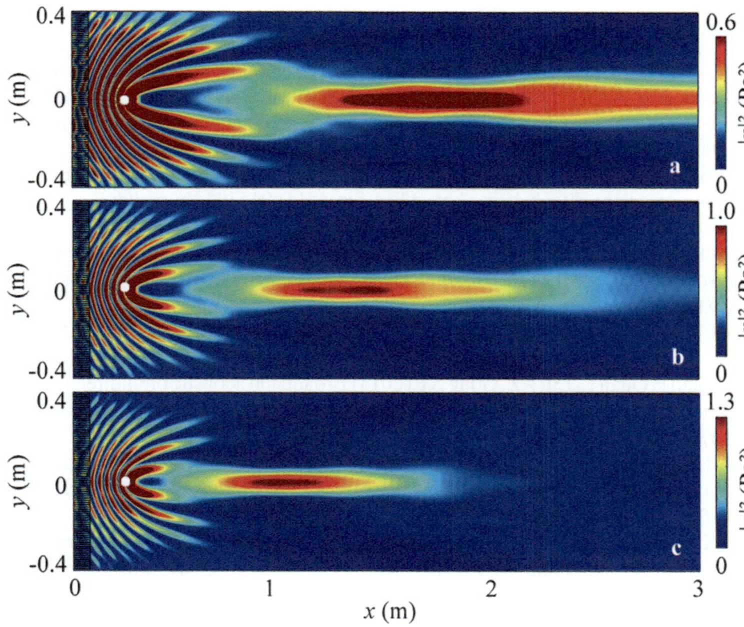

Fig. 7.55 Simulated reflected intensity distributions caused by the reflected long-focus AF lenses with **a** $\beta = 6°$, **b** 12°, and **c** 15° under the excitation of the cylindrical sound source (the white points) at 5.0 kHz

7.5 Conclusion

This chapter investigates AF techniques based on thermoacoustic phase-controlled metamaterials, highlighting three types of AF lenses. First, a self-healing AF lens composed of symmetric TPAs is proposed, demonstrating broadband focusing performance. Second, a long-focus AF lens based on Bessel-like beams is designed, showcasing its long-focus and broadband characteristics. Finally, a reflected long-focus AF lens is introduced, highlighting its potential for long-distance focusing. This research opens new possibilities for the application of thermoacoustic phase-controlled metamaterials in medical ultrasound, directional bird repelling, and acoustic energy harvesting.

References

1. Cai LW, Sánchez-Dehesa J (2007) Microlens array for focusing airborne ultrasound using heated wire grid. Appl Phys Lett 91(18):181915. https://doi.org/10.1063/1.2804564
2. Sánchez-Dehesa J, Angelov MI, Cervera F, Cai LW (2009) Sound control by temperature gradients. Appl Phys Lett 95(20):204102. https://doi.org/10.1063/1.3263949
3. Liu C, Sun HX, Yuan SQ, Xia JP (2016) Broadband acoustic focusing effect based on temperature gradient distribution. Acta Phys Sin 65(4):044303. https://doi.org/10.7498/aps.65.044303
4. Ge Y, Sun HX, Liu C, Qian J, Yuan SQ, Xia JP, Guan YJ, Zhang SY (2016) Acoustic focusing by an array of heat sources in air. Appl Phys Express 9(6):066701. https://doi.org/10.7567/APEX.9.066701
5. García-Chocano VM, Torrent D, Sánchez-Dehesa J (2012) Reduced acoustic cloaks based on temperature gradients. Appl Phys Lett 101(8):084103. https://doi.org/10.1063/1.4747197
6. Qian F, Zhao P, Quan L, Liu XZ, Gong XF (2014) Broadband acoustic omnidirectional absorber based on temperature gradients. Europhys Lett 107(3):34009. https://doi.org/10.1209/0295-5075/107/34009
7. Guan YJ, Sun HX, Xia JP, Yuan SQ (2017) Broadband asymmetric acoustic transmission in a single medium by an array of heat sources. J Phys D Appl Phys 50(16):165102. https://doi.org/10.1088/1361-6463/aa6101
8. Liu C, Xia JP, Sun HX, Yuan SQ (2017) Thermoacoustic focusing lens by symmetric Airy beams with phase manipulations. J Phys D Appl Phys 50(50):505101. https://doi.org/10.1088/1361-6463/aa964a
9. Sun HX, Hou AR, Qian J, Ge Y, Yuan SQ, Guan YJ, Si QR, Liu XJ (2019) Broadband acoustic converging and asymmetric converging based on thermoacoustic phased arrays. J Appl Phys 125(2):024504. https://doi.org/10.1063/1.5055288
10. Hou AR, Gao WT, Qian J, Sun HX, Ge Y, Yuan SQ, Si QR, Liu XJ (2018) Thermoacoustic-reflected focusing lens based on acoustic Bessel-like beam with phase manipulation. Chin Phys B 27(12):124301. https://doi.org/10.1088/1674-1056/27/12/124301
11. Liu C, Sun HX, Yuan SQ, Xia JP, Qian J (2017) Acoustic focusing by thermoacoustic phased array. Acta Phys Sin 66(15):154302. https://doi.org/10.7498/aps.66.154302
12. Gao H, Gu ZM, Liang B, Zou XY, Yang J, Yang J, Cheng JC (2016) Acoustic focusing by symmetrical self-bending beams with phase modulations. Appl Phys Lett 108(7):073501. https://doi.org/10.1063/1.4941992
13. Qian J, Liu BY, Sun HX, Yuan SQ, Yu XZ (2017) Broadband acoustic focusing by symmetric Airy beams with phased arrays comprised of different numbers of cavity structures. Chin Phys B 26(11):114304. https://doi.org/10.1088/1674-1056/26/11/114304

14. Tian Y, Wei Q, Cheng Y, Xu Z, Liu XJ (2015) Broadband manipulation of acoustic wavefronts by pentamode metasurface. Appl Phys Lett 107(22):221906. https://doi.org/10.1063/1.4936762
15. Hao H, Scalo C, Sen M, Semperlotti F (2018) Thermoacoustics of solids: a pathway to solid state engines and refrigerators. J Appl Phys 123(2):024903. https://doi.org/10.1063/1.5006489
16. Sun HX, Chen JH, Ge Y, Yuan SQ, Liu XJ (2018) Broadband and flexible acoustic focusing by metafiber bundles. J Phys D Appl Phys 51(24):245102. https://doi.org/10.1088/1361-6463/aac1aa
17. Li Y, Liang B, Tao X, Zhu XF, Zou XY, Cheng JC (2012) Acoustic focusing by coiling up space. Appl Phys Lett 101(23):233508. https://doi.org/10.1063/1.4769984
18. Li Y, Liang B, Gu ZM, Zou XY, Cheng JC (2013) Reflected wavefront manipulation based on ultrathin planar acoustic metasurfaces. Sci Rep 3:2546. https://doi.org/10.1038/srep02546
19. Li Y, Jiang X, Li RQ, Liang B, Zou XY, Yin LL, Cheng JC (2014) Experimental realization of full control of reflected waves with subwavelength acoustic metasurfaces. Phys Rev Appl 2(6):064002. https://doi.org/10.1103/PhysRevApplied.2.064002
20. Wang WQ, Xie YB, Konneker A, Popa BI, Cummer SA (2014) Design and demonstration of broadband thin planar diffractive acoustic lenses. Appl Phys Lett 105(10):101904. https://doi.org/10.1063/1.4895619

Chapter 8
Acoustic Asymmetric Focusing by Asymmetric Modulation Metamaterials

8.1 Introduction

Asymmetric modulation metamaterials enable multifunctional asymmetric transmission of sound, such as acoustic devices capable of both asymmetric transmission and focusing. These devices not only overcome the limitations of ultrasound imaging and therapy techniques but also enhance the performance of acoustic communication, sonar, and material processing. For example, HIFU, a non-invasive tumor treatment technology, has garnered significant attention. However, to extend the lifespan of HIFU focusing transducers, it is essential to prevent high-intensity reflected sound energy from human tissues from re-entering the instruments. Thus, designing acoustic devices with AAF effects remains a significant challenge.

In this chapter, we numerically and experimentally investigate five types of AAF lenses using asymmetric modulation metamaterials. First, we design an AAF lens composed of dual-layer metasurfaces with different phase distributions [1]. The AAF effect exists in the frequency range of 2.91–3.61 kHz, resulting from asymmetric phase modulation by the dual-layer metasurfaces. Additionally, we can modulate the horizontal and vertical focus positions without altering the metasurface structure and discuss the design and performance of an underwater AAF lens with dual-layer metasurfaces, extending its working band to the ultrasonic frequency range. Second, we propose an AAF lens composed of ten mode-conversion meta-atoms [2]. The AAF effect arises from phase control and the step waveguide of the meta-atoms, respectively. The fractional bandwidth of the AAF effect is 0.19, and the focus length is 0.38λ, demonstrating broadband and subwavelength focusing characteristics. We also discuss the modulation of focus position and the design and performance of an underwater AAF lens in detail. Third, we design an underwater AAF lens consisting of a steel plate with six unit cells [3]. Each unit cell contains two types of slits in the steel plate. The AAF effect exists in the frequency range of 560–592 kHz, resulting from interference enhancement and cancellation caused by the symmetric structure of the lens and the path differences of the two types of slits,

H. Sun et al., *Acoustic Metamaterials*, Microfluidics and Nanophotonics: Science and Engineering 2, https://doi.org/10.1007/978-981-96-8524-0_8

respectively. Fourth, we design an AAF prism composed of 31 phased unit cells [4]. The AAF effect arises from asymmetric phase modulation due to the non-parallel exit interfaces on both sides of the prism. The FLHM of the focus is 0.36λ, indicating subwavelength focusing. Finally, we design a long-focus AAF lens constructed from dual-layer phased arrays [5]. The long-focus AAF effect results from asymmetric phase modulation by the dual-layer phased arrays, with an FLHM of 18.1λ and a fractional bandwidth of 0.14, demonstrating long-focus AAF characteristics. These AAF lenses have practical applications in focused ultrasound therapy and medical ultrasound imaging.

8.2 Acoustic Asymmetric Focusing by Dual-Layer Metasurfaces

8.2.1 Design and Performance of AAF Lens

As shown in Fig. 8.1, we design an AAF lens composed of dual-layer AMs in air [1]. The phase distributions of AM1 and AM2 are symmetric about $y = 0$. The abbreviations LI and RI represent incident waves from the left and right sides of the AAF lens, respectively. For LI (Fig. 8.1a), the transmitted wave through AM1 splits symmetrically into two beams, with the transmitted angles increasing gradually from the edge to the center of AM1. Additionally, all sound energy passing through AM2 converges into a focus on the right side of the lens. For RI (Fig. 8.1b), the transmitted wave through AM2 is rotated 90°, and the sound energy is localized on the surface of AM2, attenuating quickly away from the interface, preventing sound energy from passing through AM1.

To realize AM1 and AM2, we design a phased unit cell (with length $l = 0.505\lambda$ and height $h = 0.1\lambda$) composed of two layers of four Helmholtz resonators and a straight channel with a tunable height h_1, where λ is the sound wavelength (Fig. 8.2). The cavity height h_2 is determined by the pipe height h_1, and the other parameters are $d = \lambda/200$, $h_2 = (h - 4d - h_1)/2$, $l_2 = 0.12\lambda$, $l_3 = 0.023\lambda$, and $l_1 = (l - 4l_3 - d)/4$. The Helmholtz resonators are made of epoxy resin to meet the sound-hard boundary condition. Throughout this work, the performance of the AAF lens is simulated using the COMSOL Multiphysics software, with material parameters as follows: density $\rho = 1180$ kg/m^3 and sound velocity $c = 2720$ m/s for epoxy resin; $\rho = 1.21$ kg/m^3 and $c = 343$ m/s for air. The wavelength λ of the incident acoustic wave is 10 cm (i.e., $f = 3.43$ kHz in air).

Figure 8.3 shows the transmission coefficients and phase delays of the phased unit cells for different values of h_1/h. For $0.1 \leqq h_1/h \leqq 0.8$, most transmission coefficients exceed 0.9, and the phase delay covers the entire 2π range. This allows arbitrary manipulation of acoustic propagation paths by designing desired phase distributions for metasurface structures. As shown in Fig. 8.1, the phase distributions φ_1 and φ_2 of AM1 and AM2 are symmetric about $y = 0$. We calculate the phase distributions of

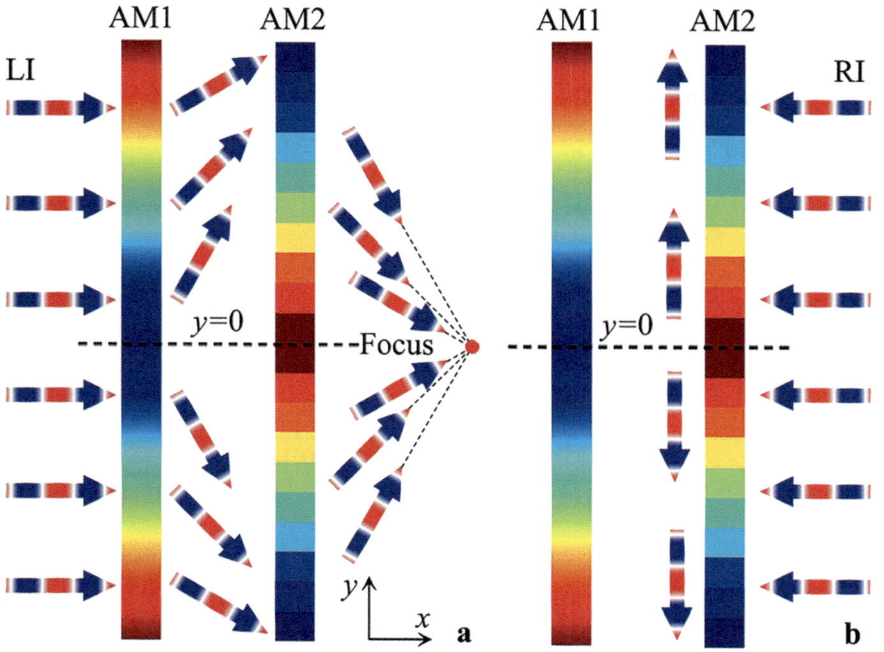

Fig. 8.1 Schematic of acoustic propagation paths through the AAF lens for **a** LI and **b** RI

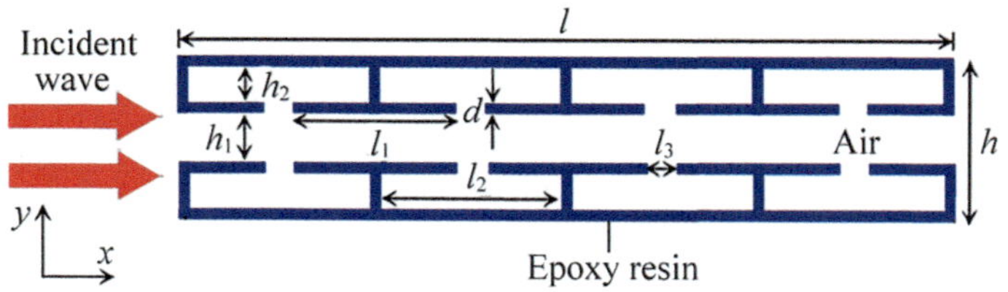

Fig. 8.2 Schematic of the phased unit cell composed of a dual-layer Helmholtz resonator array

both metasurfaces for $y > 0$. For RI, the incident angle θ_i is zero, and the refraction angle θ_t through AM2 is $-\pi/2$. Based on Eq. (2.41), the phase gradient of AM2 is:

$$\frac{1}{k}\frac{\mathrm{d}\varphi_2}{\mathrm{d}y} = 1. \tag{8.1}$$

For LI, the incident acoustic wave passes through both AM1 and AM2, focusing on the right side of AM2. Figure 8.4 shows the acoustic propagation path for LI, where y and y_0 are the coordinates of the incident points on AM1 and AM2, respectively. Based on the geometric relationships of these parameters, we obtain:

$$\tan\theta - y_0 / d_2 = 0, \tag{8.2}$$

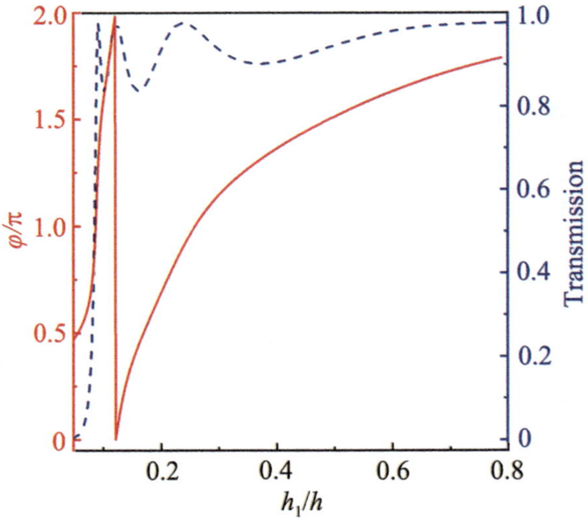

Fig. 8.3 Simulated phase delay (the red solid line) and transmission coefficients (the blue dashed line) of the phased unit cells for different height ratios h_1/h

$$y_0 - y - d_1 \tan\alpha = 0, \tag{8.3}$$

where the distance between AM1 and AM2 is $d_1 = 0.2$ m, and the distance between AM2 and the focus is $d_2 = 0.3$ m. The refraction angles α and θ correspond to the acoustic waves passing through AM1 and AM2, respectively. Based on Eq. (2.41), the phase gradient of AM1 and the relationship between angles α and θ are:

$$\frac{1}{k}\frac{\mathrm{d}\varphi_1}{\mathrm{d}y} = -\sin\alpha, \tag{8.4}$$

$$\frac{1}{k}\frac{\mathrm{d}\varphi_2}{\mathrm{d}y} = \sin\theta + \sin\alpha. \tag{8.5}$$

Using Eq. (8.1), we calculate the phase distribution φ_2 of AM2. The phase gradient $\mathrm{d}\varphi_1/\mathrm{d}y$ is obtained by solving Eqs. (8.2)–(8.5) simultaneously, and the phase distribution φ_1 of AM1 is calculated by integrating $\mathrm{d}\varphi_1/\mathrm{d}y$. Figure 8.5a, b show the theoretically calculated phase distributions (φ_1 and φ_2) of AM1 and AM2, respectively, with 81 discrete phase delays selected to design AM1 and AM2 in the AAF lens. Thus, we design AM1 and AM2 by selecting 81 phased unit cells, with the length of the AAF lens in the y direction being 81 cm.

Figure 8.6a, b show the simulated intensity distributions caused by the AAF lens for LI and RI, respectively, demonstrating a clear AAF effect. For LI (Fig. 8.6a), the transmitted sound energy through AM1 splits symmetrically into two beams, eventually focusing on a point on the right side of AM2. The sound energy is primarily concentrated in the focus, with weaker side lobes. For RI (Fig. 8.6b), the transmitted sound energy is localized on the surface of AM2, attenuating away from the interface,

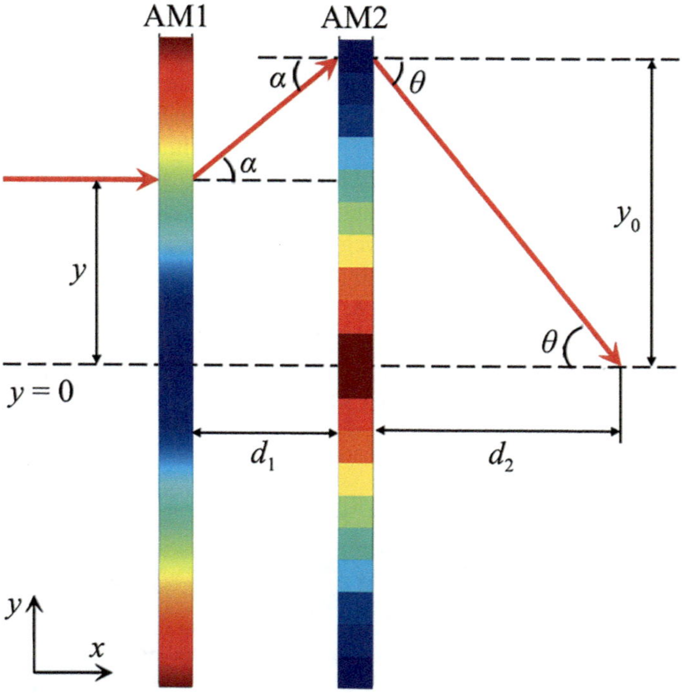

Fig. 8.4 Schematic of the acoustic propagation path for LI

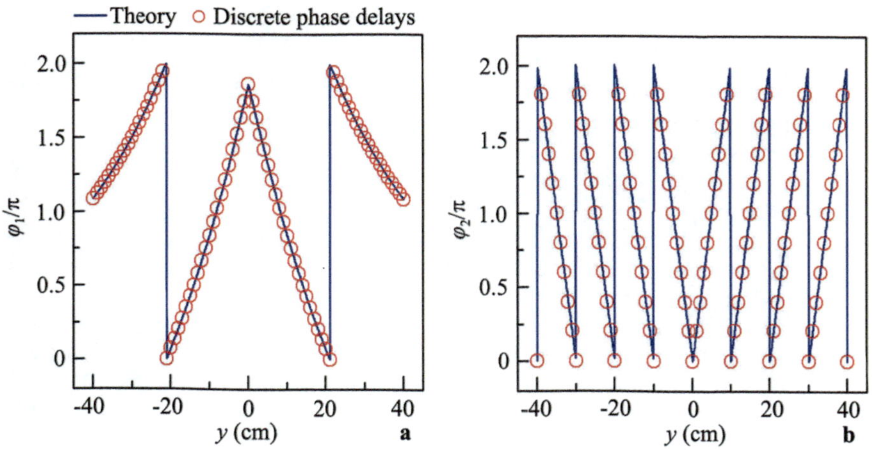

Fig. 8.5 Theoretical continuous phase distributions and 81 discrete phase delays of **a** AM1 and **b** AM2 at 3.43 kHz

preventing the acoustic wave from passing through AM1. Thus, the designed lens exhibits high AAF performance.

To experimentally validate the AAF performance, we measure the intensity distributions in red open rectangles R1 and R2 (9 cm × 16 cm) in Fig. 8.6. The photographs of AM1 and AM2 are shown in Fig. 8.7, and the experimental set-up is displayed in Fig. 8.8. The experiment is conducted in a planar waveguide system (2 m × 2 m × 1 cm) composed of two parallel plates. An anechoic environment is created by placing wedge-shaped foams at the waveguide boundaries. The incident plane wave is generated by a speaker array driven by a power amplifier. Pressure distributions are scanned using a microphone (Brüel & Kjær type-4961) moved by 2D motorized linear stages (Newport: MIN300CC and ILS250CC). Another microphone is fixed toward the speaker array as a reference. Data is recorded by the Brüel & Kjær 3160-A-022 module and analyzed using PULSE LabShop software. The measured and

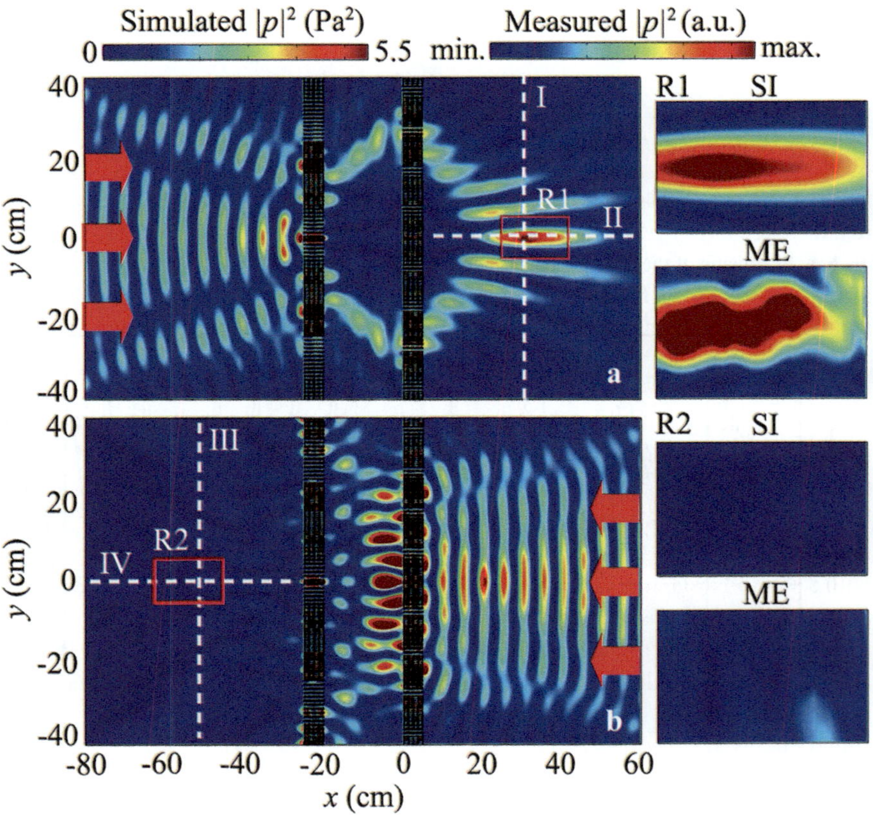

Fig. 8.6 Measured and simulated intensity distributions caused by the AAF lens for **a** LI and **b** RI at 3.43 kHz, and the red arrows represent the incident plane wave. The insets show the measured and simulated intensity distributions in red open rectangles R1 and R2

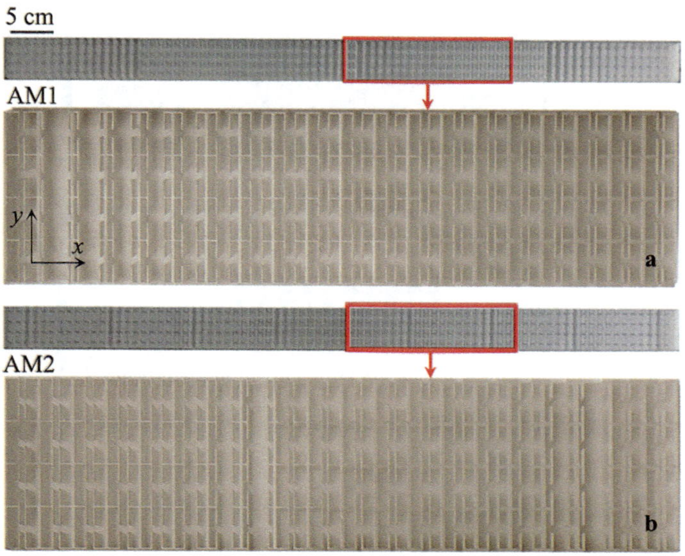

Fig. 8.7 Photographs of the samples of **a** AM1 and **b** AM2

simulated field distributions are shown on the right side of Fig. 8.6. The measured results agree well with the simulations for LI and RI, demonstrating the AAF effect.

Figure 8.9 shows the measured and simulated normalized intensity distributions along lines I–IV in Fig. 8.6. The intensity distributions along lines I and II are significantly higher than those along lines III and IV, and the FLHM of the focus in the y direction (Fig. 8.9a) is 4.0 cm (0.4λ). The measured results match the simulations, indicating that the lens combines AAT and subwavelength AF.

To illustrate the influence of reflected sound energy on the focusing performance of the AAF lens, we design a traditional AF lens composed of a single-layer metasurface with the same length in the y direction for comparison. The phase distribution of the traditional AF lens is shown in Fig. 8.10. Figure 8.11a, b show the simulated total intensity distributions caused by the traditional AF lens and the designed AAF lens with a sound-hard boundary on the right side for LI, respectively. The corresponding reflected intensity distributions in the red dashed rectangles are displayed on the right side. The reflected intensity distributions are obtained by subtracting the total acoustic intensity fields with and without the lenses. The AAF lens exhibits almost no reflected acoustic energy, while the traditional AF lens shows strong reflected sound energy focused on the central region. Thus, the designed AAF lens has low sound reflection, effectively preventing damage to acoustic transducers in HIFU instruments.

Figure 8.12a shows the simulated transmission spectra of the AAF lens for LI and RI. The transmission coefficient is calculated as P_o/P_i, where P_i and P_o represent the integration of the intensity in the same transmitted region without and with the AAF lens, respectively. Notably, in bands I (2.91–3.28 kHz, the red shaded region) and II

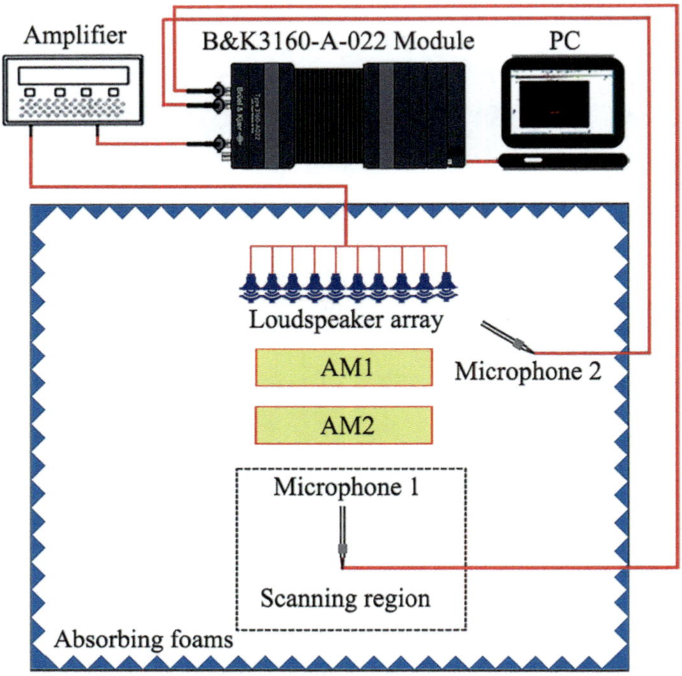

Fig. 8.8 Schematic of the experimental set-up for the AAF lens in Fig. 8.7

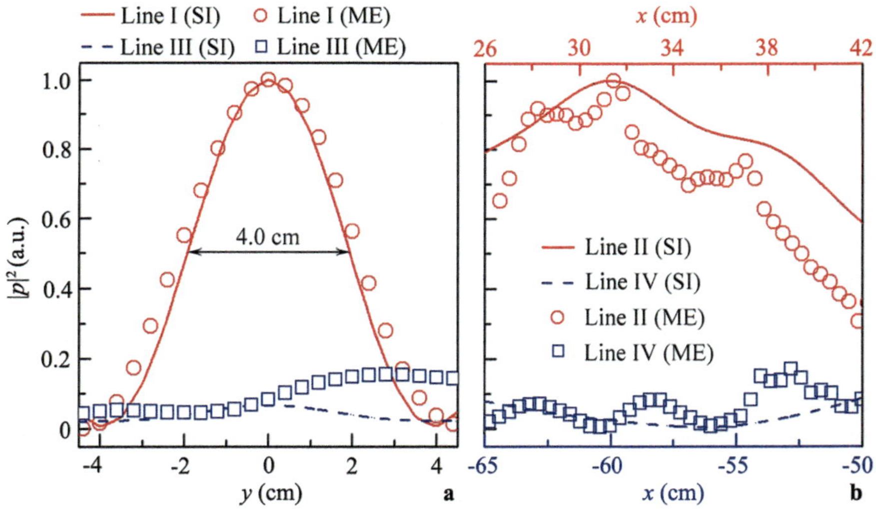

Fig. 8.9 Measured and simulated intensity distributions along **a** lines I and III and **b** lines II and IV in Fig. 8.6

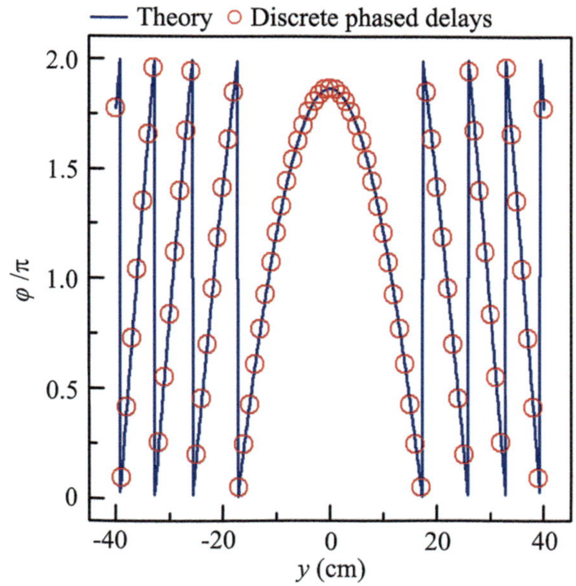

Fig. 8.10 Theoretical continuous phase distribution and 81 discrete phase delays of the traditional AF lens at 3.43 kHz

(3.28–3.61 kHz, the blue shaded region), the lens exhibits a clear characteristic of AAT. However, at certain frequencies in band I, the transmission coefficient for RI is larger, and the AAT effect is weaker compared to that in band II.

Figure 8.12b shows the maximum intensity spectra along lines II and IV in Fig. 8.6a, b. Along line II (LI), the maximum intensities in bands I and II range from 3.0–5.0 Pa^2 and 5.0–7.0 Pa^2, respectively, demonstrating the focusing effect of the AAF lens across broad bandwidths. However, along line IV (RI), the maximum intensities are below 1.0 Pa^2 and 2.0 Pa^2 in bands I and II, respectively, which are significantly lower than those along line II (LI). It is worth noting that the maximum intensities at frequencies with high transmission for RI in band I (Fig. 8.12a) are also below 1.0 Pa^2, indicating that the transmitted sound energy diverges on both sides and does not converge in the central region. Therefore, the harmfulness of the transmitted sound energy for RI at these frequencies is relatively low.

To further verify the bandwidth of the AAF effect, we simulate the intensity distributions caused by the AAF lens at 2.94 and 3.57 kHz. As shown in Fig. 8.13a, c, the acoustic waves can transmit through both metasurfaces and are focused on the right side of AM2. For RI, most of the sound energy cannot pass through AM1, as shown in Fig. 8.13b, d. The results demonstrate that the AAF effect can also be achieved at both edge frequencies, further confirming the broadband characteristic of the AAF lens. This broadband characteristic arises from the fact that the dual-layer metasurfaces are composed of the same type of phased unit cell, and the working bandwidth of the lens is nearly identical to that of the unit cell.

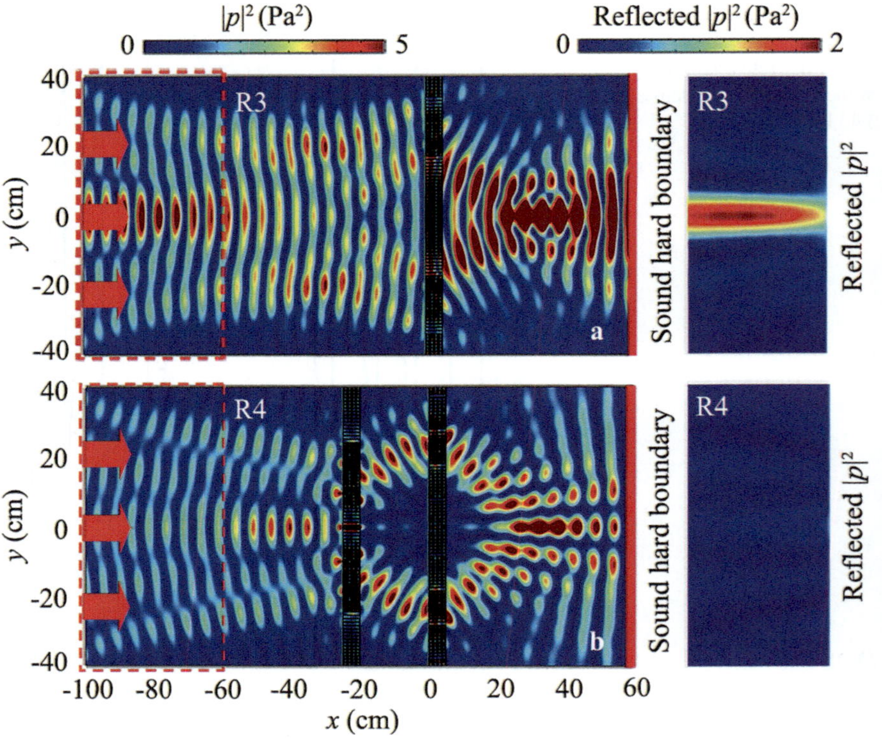

Fig. 8.11 Simulated total intensity distributions caused by **a** the traditional AF lens and **b** the AAF lens with the sound-hard boundaries placed on the right side of the models for LI. The two insets on the right side show the reflected intensity distributions in red open rectangles in R3 and R4

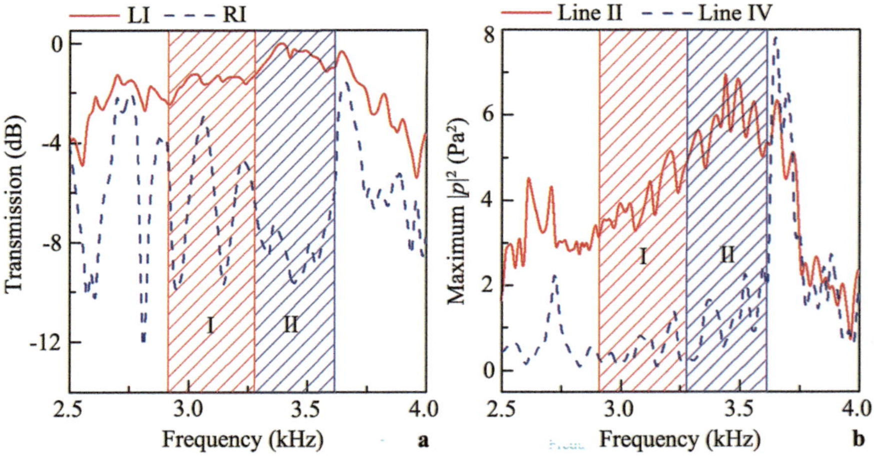

Fig. 8.12 **a** Simulated transmission spectra of the AAF lens for LI and RI, and **b** maximum intensity spectra along lines II and IV in Fig. 8.6 at different frequencies

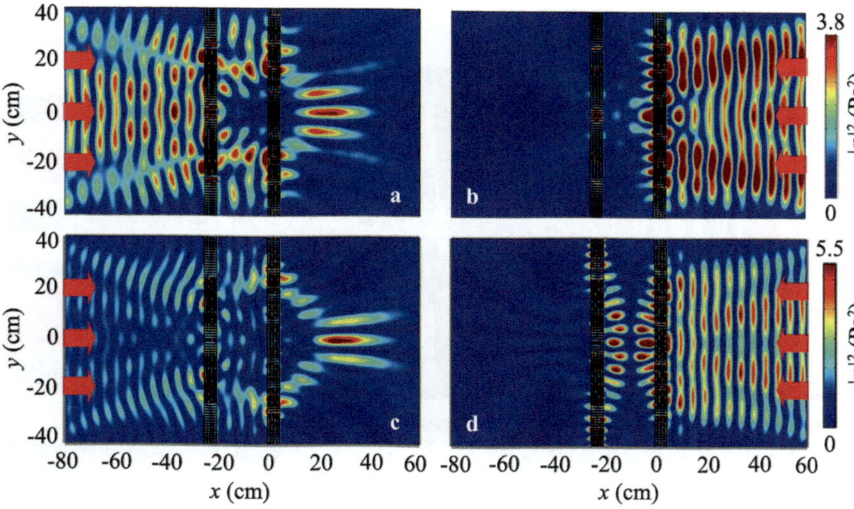

Fig. 8.13 Simulated intensity distributions caused by the AAF lens at **a**, **b** 2.94 kHz and **c**, **d** 3.57 kHz for LI and RI

8.2.2 Performance Modulation

In addition to the broadband AAF effect, we can manipulate the horizontal and vertical positions of the focus by mechanically adjusting the distance d_1 between the metasurfaces and the incident angle, respectively. Figure 8.14 shows the simulated intensity distributions caused by the AAF lens with different values of d_1 for LI, along with the measured intensity distributions presented on the right for comparison. The position of the measured regions (the red open rectangles) is fixed and corresponds to that in Fig. 8.6a.

As shown in Fig. 8.14, the position of AM1 remains unchanged, while AM2 gradually moves to the right. Interestingly, as d_1 increases, the AAF effect persists, and the focus moves gradually to the right. The measured results agree well with the simulated ones for LI. Additionally, most of the sound energy cannot pass through AM1 for RI, as shown in Fig. 8.15. To further illustrate the manipulation of the focus in the x direction, we simulate the intensity distributions along line II (shown in Fig. 8.6a) for different values of d_1, as shown in Fig. 8.16. The focus moves linearly to the right as d_1 increases. Therefore, without changing the structures of AM1 and AM2, the horizontal position of the focus can be controlled by mechanically moving AM2.

Additionally, by adjusting the incident angles of sound, the vertical position of the focus can be controlled. Figure 8.17 shows the intensity distributions caused by the AAF lens with incident angles of $-4°$, $-8°$, and $-12°$ for LI and RI. As shown in Fig. 8.17a–c, for LI, as the incident angle increases, the AF effect remains, and the focus moves downward in the y direction. For RI, most of the sound energy cannot pass through AM1, as shown in Fig. 8.17d–f. Therefore, the vertical position

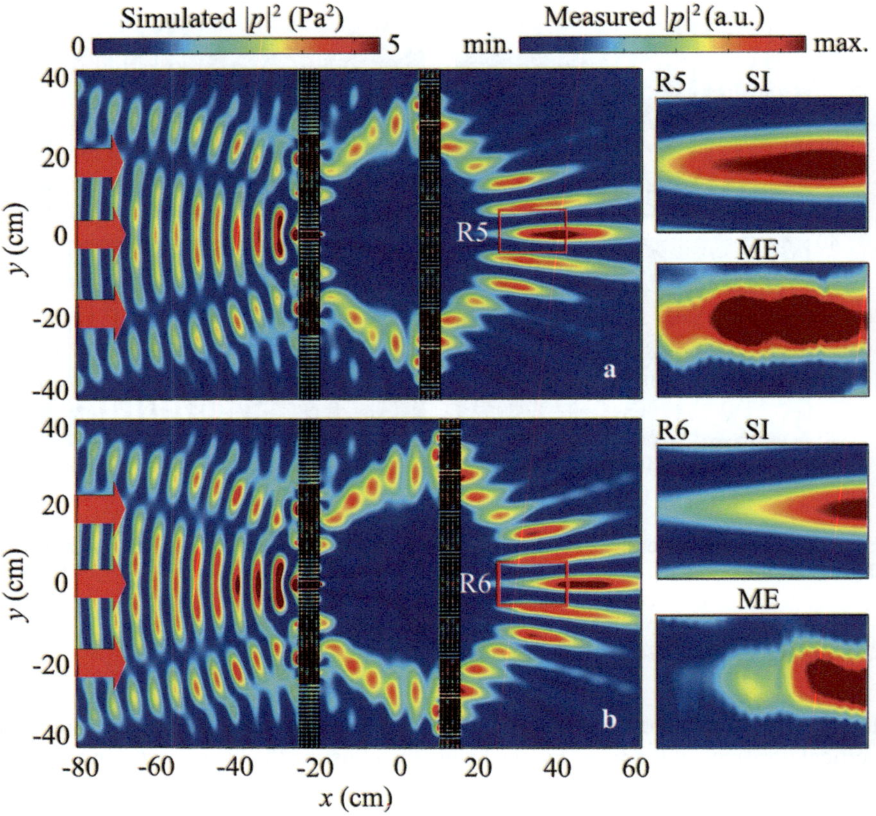

Fig. 8.14 Measured and simulated intensity distributions caused by the AAF lens with **a** $d_1 =$ 25 cm and **b** 30 cm for LI. The four insets on the right side show the measured and simulated intensity distributions in red open rectangles R5 and R6

of the focus can be manipulated by adjusting the incident angle. Given that medical ultrasound often requires precise control of the focus position to scan lesions, the designed AAF lens with tunable focus positions has significant potential in HIFU therapy.

8.2.3 Design and Performance of Underwater AAF Lens

Furthermore, the working bandwidth of the designed AAF lens can be extended to the ultrasonic frequency range. As an example, we design a phased unit cell (with length $l = 0.5\lambda$ and height $h = 0.2\lambda$) made of brass immersed in water, as shown in Fig. 8.18. The structural parameters of the phased unit cell are $l_1 = 0.1l$, $l_2 = 0.225l$, $l_3 = 0.15l$, and $d = 0.1$ cm. The material parameters are density $\rho = 1000$ kg/m^3

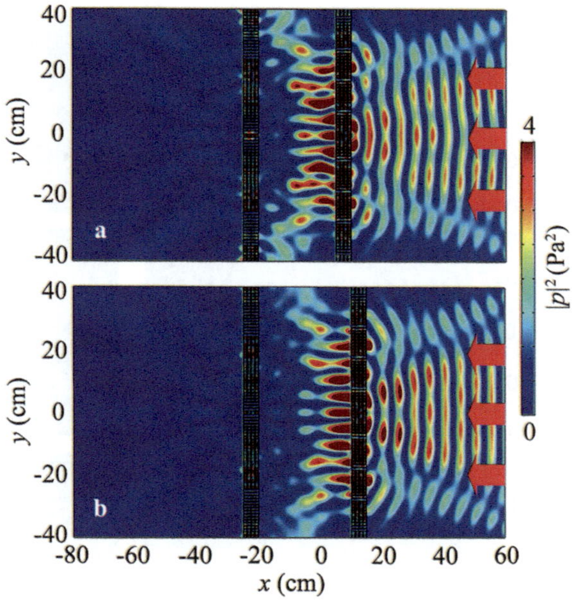

Fig. 8.15 Simulated intensity distributions caused by the AAF lens with **a** $d_1 =$ 25 cm and **b** 30 cm for RI

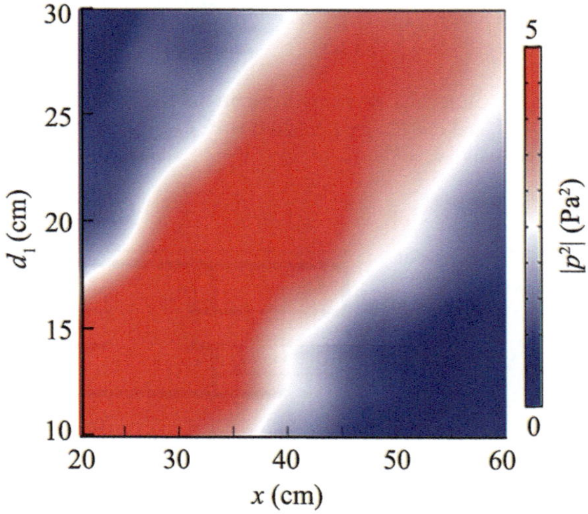

Fig. 8.16 Simulated acoustic intensity distributions along line II in Fig. 8.6a caused by the AAF lenses with different values of d_1

and sound velocity $c = 1483$ m/s for water; and density $\rho = 8800$ kg/m^3, Young's modulus $E = 108.42$ GPa, and Poisson's ratio $v = 0.33$ for brass. Figure 8.19 shows the transmission coefficients and phase delays of the phased unit cells with different values of h_1/h. Most of the transmission coefficients are greater than 0.9, and the phase delays cover the entire 2π range.

Based on the same design concept, we calculate the phase distributions (φ_1 and φ_2) of AM3 and AM4, as shown in Fig. 8.20a, b, respectively. Using 81 discrete phase

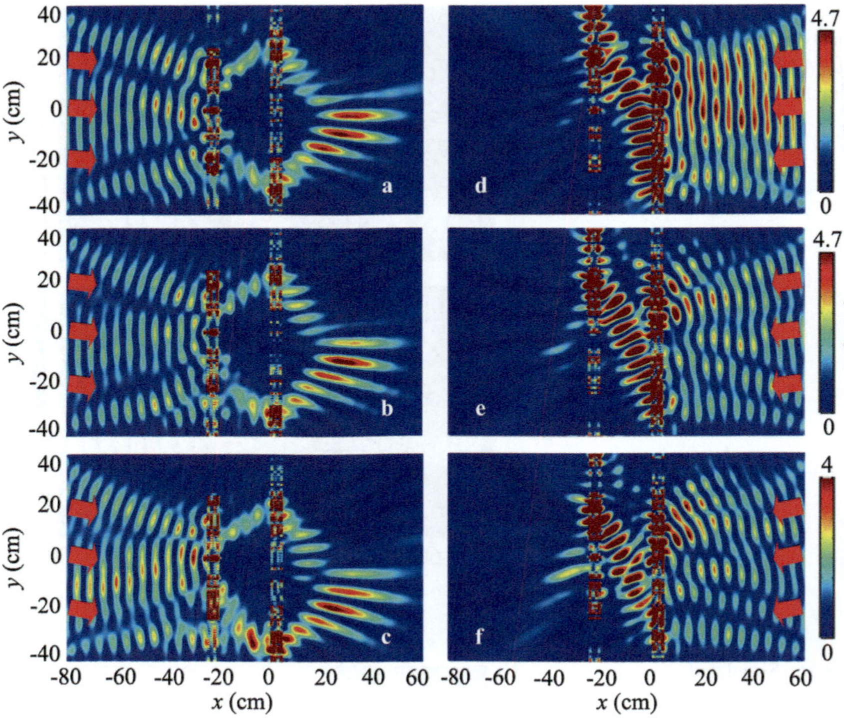

Fig. 8.17 Simulated intensity distributions caused by the AAF lens with incident angles of **a** −4°, **b** −8°, and **c** −12° for LI; and **d** −4°, **e** −8°, and **f** −12° for RI. The red arrows represent the incident plane wave

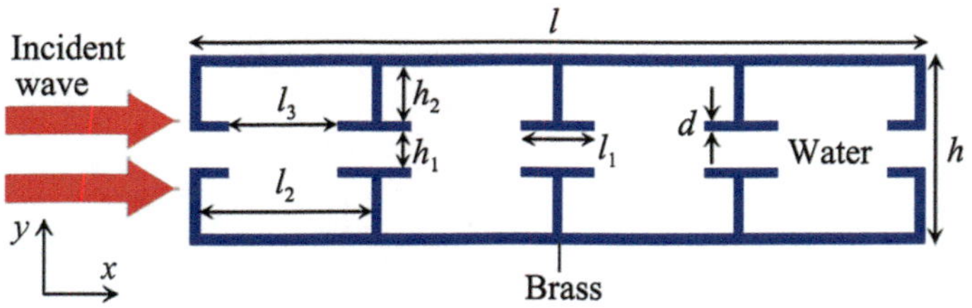

Fig. 8.18 Schematic of the phased unit cell composed of a dual-layer Helmholtz resonator array immersed in water

delays, we design an AAF lens immersed in water, in which the distance between AM3 and AM4 and that between AM4 and the focus are selected as 0.1 m and 0.3 m, respectively.

Figure 8.21a, b show the intensity distributions caused by the AAF lens at 30 kHz for LI and RI, respectively. The transmitted sound energy is focused on the right side of AM4 for LI, while it is prevented from transmitting through AM3 for RI. Therefore,

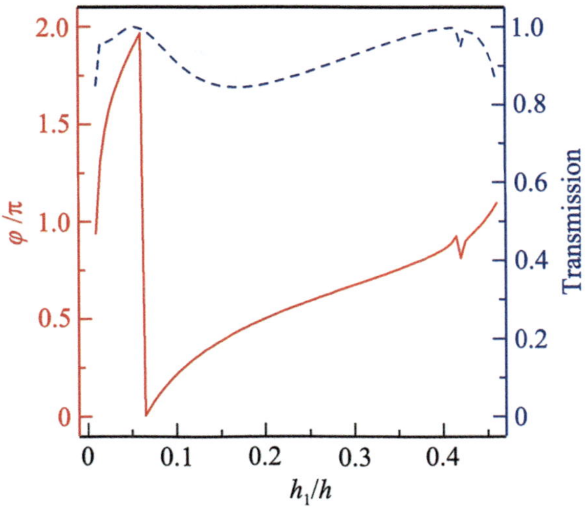

Fig. 8.19 Simulated phase delay (the red solid line) and transmission coefficients (the blue dashed line) of the phased unit cells with different values of h_1/h

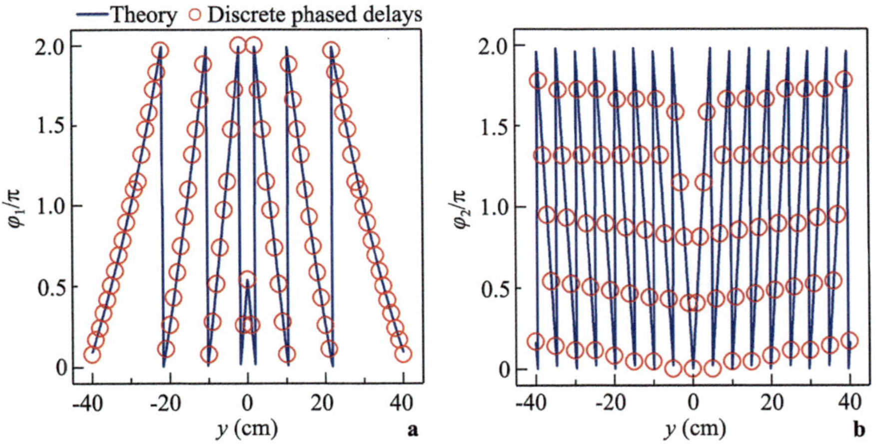

Fig. 8.20 Theoretical continuous phase distributions and 81 discrete phase delays of **a** AM3 and **b** AM4 at 30 kHz

the designed AAF lens can achieve the AAF effect in the ultrasonic frequency range, indicating its potential application in focused ultrasound therapy.

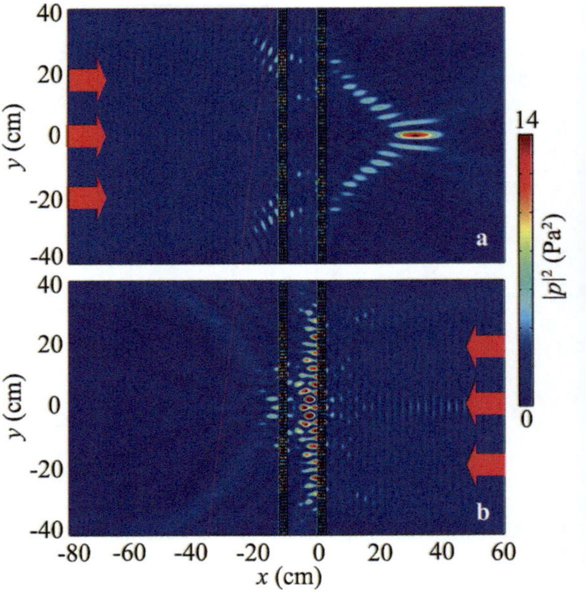

Fig. 8.21 Simulated intensity distributions caused by the AFF lens composed of AM3 and AM4 for **a** LI and **b** RI at 30 kHz

8.3 Acoustic Asymmetric Focusing by Mode-Conversion Meta-Atoms

8.3.1 Design and Performance of Mode-Conversion Meta-Atoms

To design the AAF lens composed of meta-atoms with mode conversion between zero-order and first-order waves, we introduce a phased unit cell (with length $l = \lambda/2$ and width $w = \lambda/10$) composed of four Helmholtz resonators and a straight pipe with a tunable width w_1 (Fig. 8.22) [2]. The parameters are selected as $l_1 = 0.225l$, $l_2 = 0.03l$, and $d = 0.1w$. The Helmholtz resonators are made of epoxy resin to satisfy the sound-hard boundary, and the sound wavelength λ is selected as 5 cm (i.e., $f = 6.86$ kHz in air). Throughout this work, we use the COMSOL Multiphysics software to simulate the characteristics of the AAF lens. In the models, the material parameters are as follows: density $\rho = 1.21$ kg/m^3 and sound velocity $c = 343$ m/s for air; $\rho = 1180$ kg/m^3 and $c = 2720$ m/s for epoxy resin.

Figure 8.23 shows the transmission coefficients and phase delays caused by the phased unit cells with different values of w_1. In the range of 0.75–2.0 mm, the transmission coefficients are greater than 0.75, and their phase delays cover the entire 2π range. Here, we select phased unit cells I and II (the blue open circles) to design the mode-conversion meta-atom, with w_1 values of 1.8 mm and 1.0 mm, respectively. The phase difference between the two unit cells is π, and the transmission coefficients are greater than 0.98. To design an acoustic mode-conversion meta-atom, we select

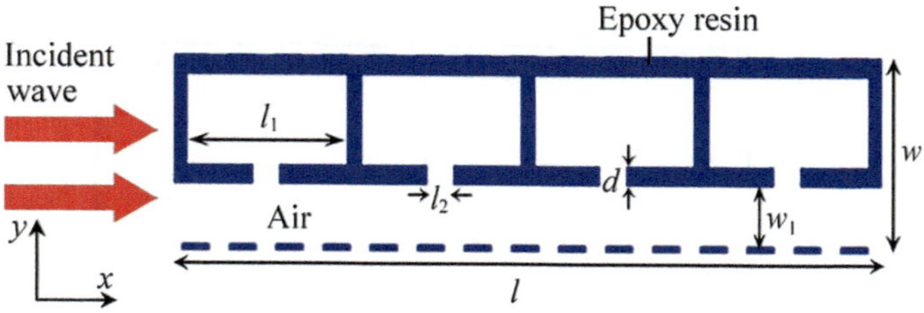

Fig. 8.22 Schematic of the phased unit cell composed of a Helmholtz resonator array

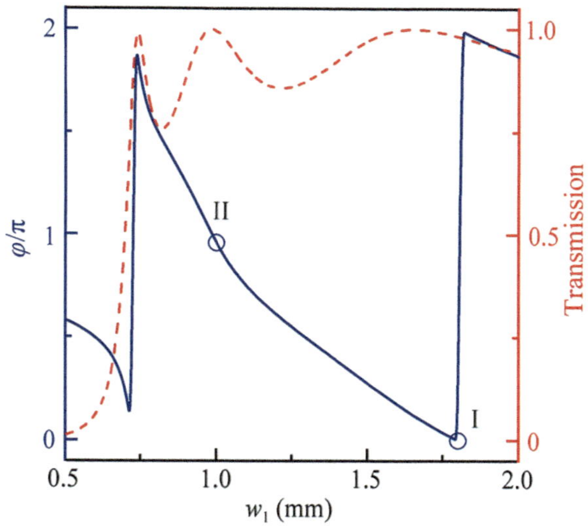

Fig. 8.23 Simulated transmission coefficients (the red dashed line) and phase delays (the blue solid line) caused by the phased unit cells with different values of w_1 at 6.86 kHz

a waveguide with $w_y = 4.2$ cm, which is larger than half the wavelength, allowing the transmission of the first-order wave.

Based on Eq. (2.53), we theoretically calculate the continuous phase distribution of the mode-conversion meta-atom (Fig. 8.24) and select six discrete phase delays to design the meta-atom. Due to the coupling effect between phased unit cells I and II, which have opposite phase delays, we place a solid structure in the middle of the meta-atom to separate the unit cells on both sides, as shown in the inset of Fig. 8.24. Additionally, to achieve AAF, we introduce a step waveguide with a width (w_R) smaller than half the wavelength, placed on the right side of the meta-atom to prohibit the propagation of the first-order wave. The sample of the mode-conversion meta-atom is schematically shown in Fig. 8.25, with parameters $w_L = w_y$, $w_R = 0.46\lambda$, $l_3 = 0.26\lambda$, $l_4 = 0.23\lambda$, and $w_3 = 0.17\lambda$.

Figure 8.26 shows the acoustic propagation paths through the mode-conversion meta-atom, where LI and RI refer to the normal incidence of the zero-order wave from the left and right sides of the meta-atom, respectively. For LI, the converted first-order wave cannot transmit through the step waveguide due to its cutoff

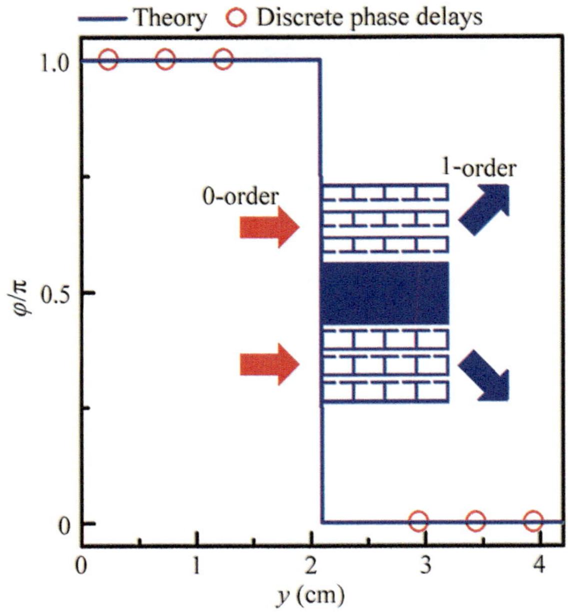

Fig. 8.24 Theoretical continuous phase distribution and discrete phase delays of a meta-atom that can realize mode conversion between zero-order and first-order waves at 6.86 kHz

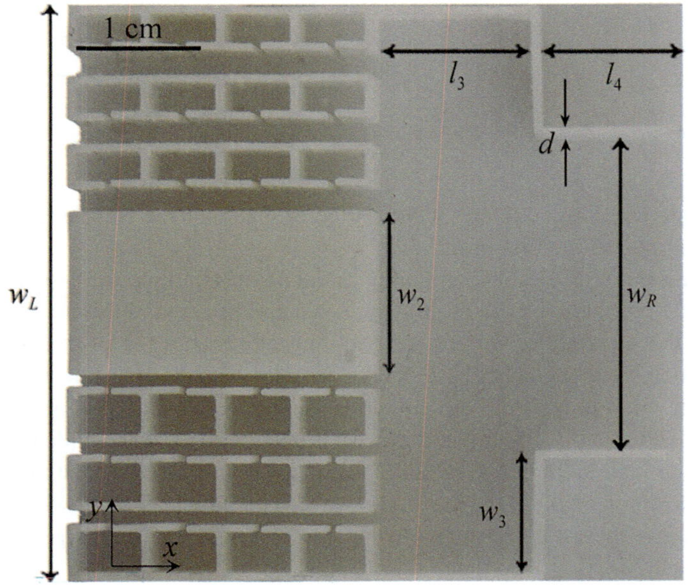

Fig. 8.25 Photograph of sample for the mode-conversion meta-atom with a step waveguide on the right side

frequency. However, for RI, the zero-order wave can pass through the meta-atom and be converted into the first-order wave. The pressure distributions caused by the mode-conversion meta-atom for LI and RI are presented in Fig. 8.27a, b. The zero-order wave can transmit through the meta-atom and propagate as the first-order wave for RI, showing clear characteristics of mode conversion and asymmetric sound transmission.

To quantitatively present the working bandwidth of the meta-atom, we simulate the transmission spectra for LI and RI. As shown in Fig. 8.28a, the meta-atom structure exhibits clear AAT characteristics in bands I (6.36–7.27 kHz, the red shaded region) and II (7.27–7.68 kHz, the blue shaded region). In band I, the transmissions are greater than 0.7 for RI, while those for LI are below 0.4. At 6.86 kHz the transmission coefficients reach 0.98 for RI but only 0.1 for LI, demonstrating high AAT performance. However, in band II, the AAT performance is slightly weaker than in

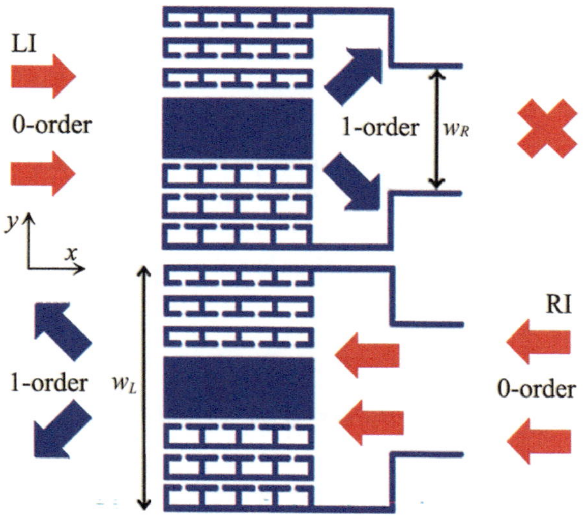

Fig. 8.26 Schematic of acoustic propagation paths through the mode-conversion meta-atom for LI and RI

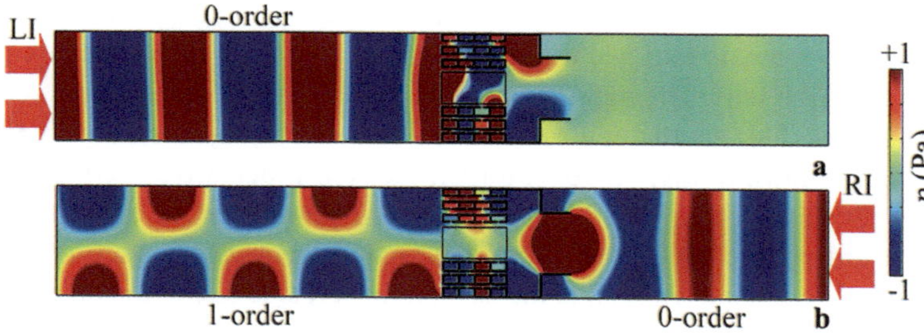

Fig. 8.27 Simulated pressure distributions caused by the mode-conversion meta-atom for **a** LI and **b** RI at 6.86 kHz

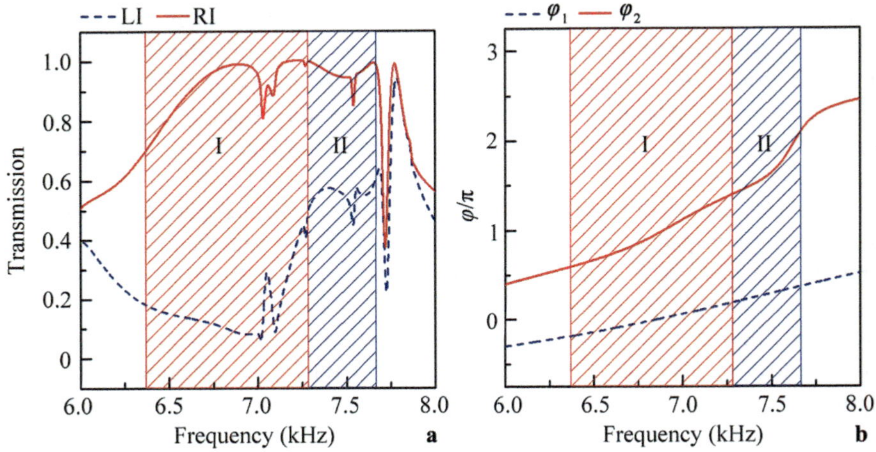

Fig. 8.28 **a** Simulated transmission spectra of the mode-conversion meta-atom for RI (the red solid line) and LI (the blue dashed line), and **b** phase spectra of phased unit cells I (φ_1, the blue dashed line) and II (φ_2, the red solid line)

band I, which arises from variations in the phase difference. To demonstrate this, we simulate the phase spectra of phased unit cells I and II, as shown in Fig. 8.28b. The phase differences between the two unit cells are close to π in band I but slightly larger than π in band II. Therefore, the meta-atom's bandwidth is about 1.32 kHz, with a fractional bandwidth of 0.19.

8.3.2 Design and Performance of AAF Lens

Based on the mode-conversion meta-atom, we design a planar AAF lens [2], which has the advantages of easy integration and a single-layer structure compared to the AAF lens in Fig. 8.1. Using Eq. (6.2), we calculate the phase distribution of the AAF lens, with parameters $e = 15$ cm, $f = 6.86$ kHz $c = 343$ m/s, and $\varphi_0 = \pi/3$. The theoretical continuous phase distribution of the AAF lens is shown in Fig. 8.29.

Based on the theoretical phase distribution and phased unit cells I and II from the mode-conversion meta-atom, we use ten meta-atoms with 60 discrete phase delays to design the AAF lens. The remaining parts are filled with epoxy resin to connect the meta-atoms. The photograph of the AAF lens is displayed in Fig. 8.30a. As schematically shown in Fig. 8.30b, c, for LI, the zero-order wave is converted into the first-order wave, which is prohibited by the AAF lens due to the cutoff frequency of the first-order wave caused by the step waveguide. For RI, the zero-order wave can transmit through the lens and is focused on the left side due to the phase control of the lens by phased unit cells I and II.

Figure 8.31a, b show the measured and simulated intensity distributions caused by the AAF lens for LI and RI, respectively. The sound energy can transmit through the

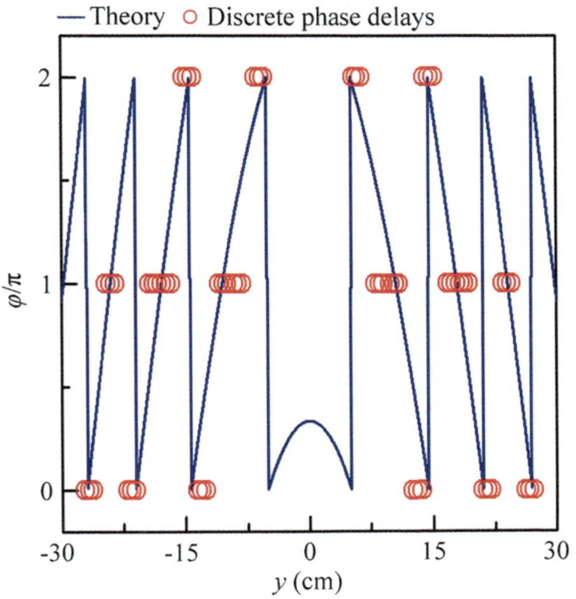

Fig. 8.29 Theoretical continuous phase distribution and 60 discrete phase delays of the AAF lens at 6.86 kHz

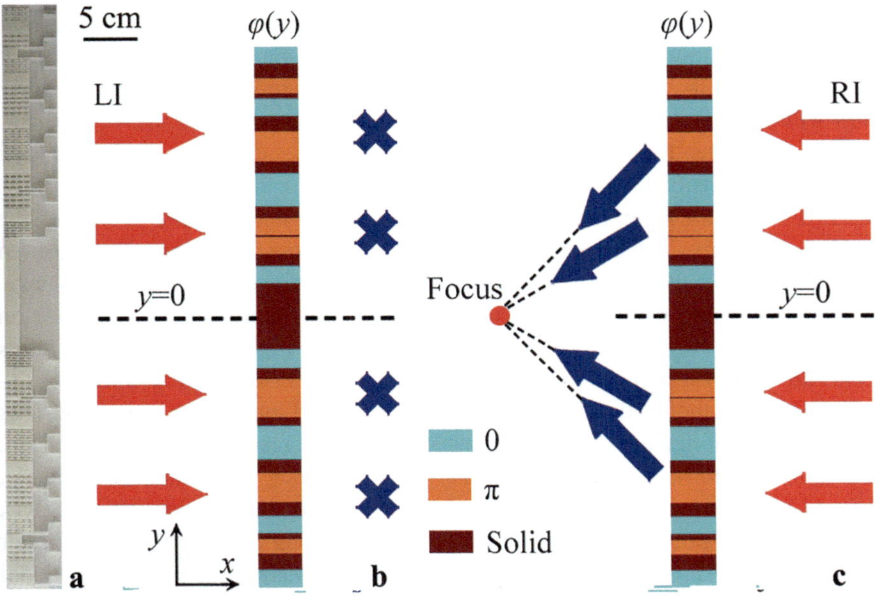

Fig. 8.30 **a** Photograph of the AAF lens composed of ten mode-conversion meta-atoms. Schematic of acoustic propagation paths caused by the AAF lens for **b** LI and **c** RI

lens and is focused on the left side for RI. However, for LI, the incident acoustic wave cannot reach the right side of the lens, demonstrating the characteristic behavior of the AAF lens. It is noted that there are also obvious side lobes on both sides of the focus. This is because the AAF lens is composed of only two types of phased unit cells, and part of the incident sound energy is reflected back by the solid structures. To explain the mechanism of AAF, we simulate the intensity distributions caused by the AAF lens composed of meta-atoms without the step waveguide, with the same parameters as in Fig. 8.31. Figure 8.32a, b show the simulated intensity distributions for LI and RI, respectively. The sound energy can transmit through the lens and is focused for both LI and RI, with focusing characteristics similar to those in Fig. 8.31b. Therefore, the AAT of the lens is attributed to the cutoff frequency of the first-order wave caused by the step waveguide in each meta-atom.

To experimentally demonstrate the AAF effect, we measure the intensity distributions in red open squares R1 and R2 (16 cm × 16 cm) in Fig. 8.31. Figure 8.33 shows the experimental set-up, where the sample is placed in a planar waveguide composed of two parallel plates (dimensions 2 m × 2 m × 1 cm). Wedge-shaped foams are placed at the boundaries of the planar waveguide to create an anechoic environment. The incident plane wave is generated by a loudspeaker array driven by a power amplifier. The pressure distributions are scanned by a microphone (Brüel & Kjær type-4961) moved by a set of 2D motorized linear stages (Newport: MIN300CC and ILS250CC). Another microphone is fixed toward the speaker array to measure the reference acoustic signal. Using the Brüel & Kjær 3160-A-022 module, the pressure amplitude and phase at each position in the scanning region are recorded, and the pressure distribution is retrieved using the PULSE Labshop software.

As shown in the right region of Fig. 8.31, the measured field distributions agree well with the simulated ones for LI and RI, further demonstrating the AAF performance of the lens. Figure 8.34a, b show the measured and simulated results of the normalized intensity distributions along lines I–IV in Fig. 8.31. The intensities along lines III and IV are significantly larger than those along lines I and II. Moreover, the FLHM of the focus in the y direction is only 1.9 cm (about 0.38λ), and the measured and simulated results agree well, demonstrating the performance of the AAF lens.

To quantify the working bandwidth of the lens, we simulate the transmission spectra through the AAF lens for LI and RI, as shown in Fig. 8.35a. The lens exhibits clear AAT behavior in bands I (6.36–7.27 kHz, the red shaded region) and II (7.27–7.68 kHz, the blue shaded region). However, the AAT performance in band II is weaker than that in band I, which is consistent with the results shown in Fig. 8.28a. Figure 8.35b shows the maximum intensity spectra along lines II and IV. Notably, most of the maximum intensities in bands I and II exceed 3 Pa^2 along line IV (RI), while the corresponding values are below 0.5 Pa^2 along line II (LI). This further confirms the AAF effect in bands I and II (6.36–7.68 kHz), indicating a fractional bandwidth of 0.19, which demonstrates the broadband characteristics of the designed AAF lens.

We also investigate the influence of viscous loss on the performance of the AAF lens, as shown in Fig. 8.36. The AAF performance of the lens remains nearly

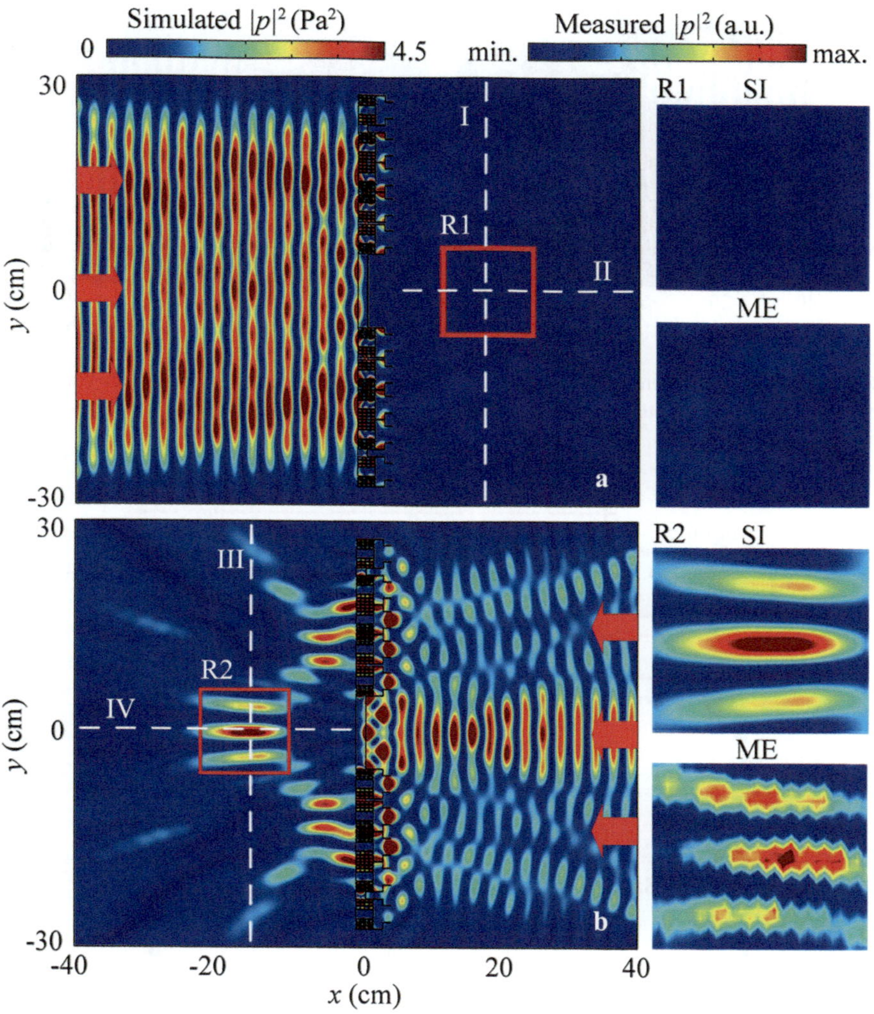

Fig. 8.31 Measured and simulated intensity distributions caused by the AAF lens for **a** LI and **b** RI at 6.86 kHz. The insets on the right side are the measured and simulated intensity distributions in red open squares R1 and R2. The red arrows represent the incident acoustic wave

unchanged when viscous loss is introduced, compared to the case without viscous loss in Fig. 8.31.

To further illustrate this, we simulate the intensity distributions along lines V and VI with and without viscous loss, as shown in Fig. 8.37a, b, respectively. The maximum intensity of the focus decreases by only 2.5% when viscous loss is introduced, indicating that viscous loss has a minimal effect on the AAF lens.

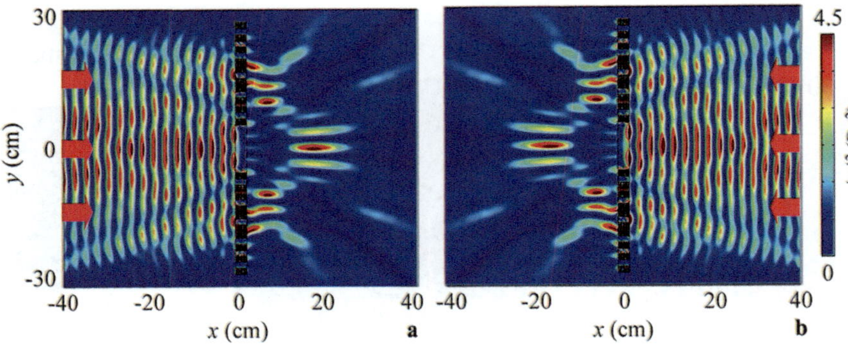

Fig. 8.32 Simulated intensity distributions caused by the AAF lens composed of mode-conversion meta-atoms without the step waveguide for **a** LI and **b** RI at 6.86 kHz. The red arrows represent the incident acoustic wave

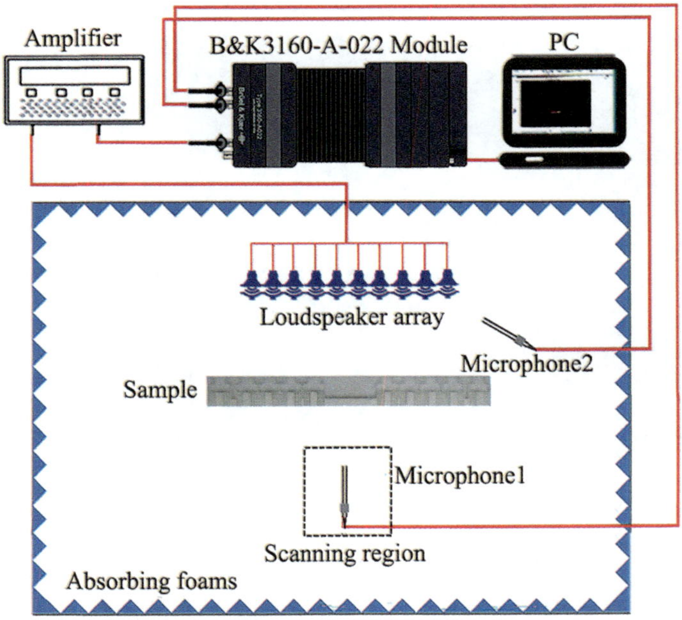

Fig. 8.33 Schematic of the experimental set-up for the AAF lens in Fig. 8.31a

8.3.3 Performance Modulation

The focal length of the AAF lens can be controlled by adjusting the phase distribution. Figure 8.38a, b show the phase distributions of AAF lenses with focal lengths of 20 and 25 cm, respectively.

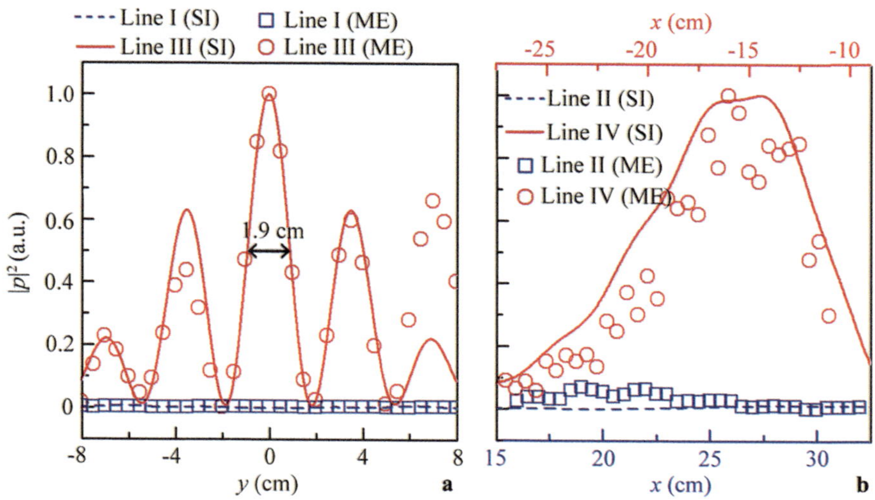

Fig. 8.34 Measured and simulated acoustic intensity distributions along **a** lines I and III and **b** lines II and IV in Fig. 8.31

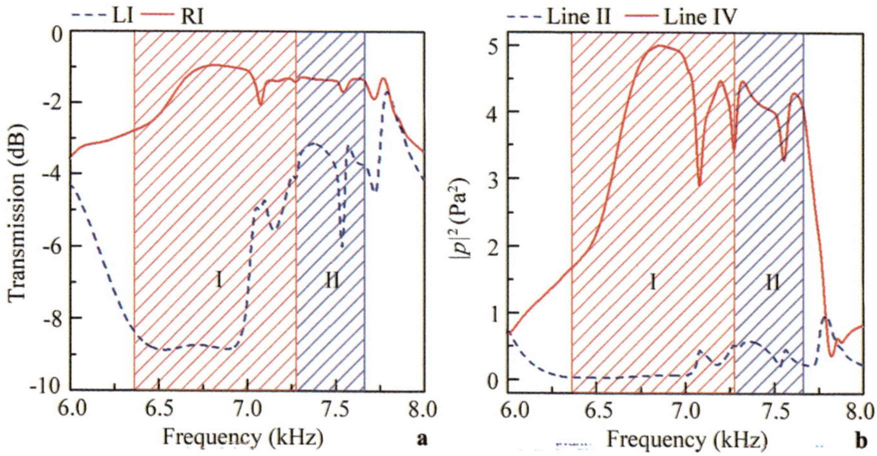

Fig. 8.35 **a** Simulated transmission spectra of the AAF lens for LI and RI. **b** Maximum intensity spectra along lines II and IV in Fig. 8.31

Figure 8.39 shows the simulated intensity distributions caused by AAF lenses with focal lengths of 15, 20, and 25 cm for LI and RI. As shown in Fig. 8.39b, d, and f, the incident wave passes through the AAF lens and is focused on the left side for RI. The focal lengths are 14.6 (Fig. 8.39b), 18.5 (Fig. 8.39d), and 22.5 cm (Fig. 8.39f), respectively, which are close to the theoretical values. However, for LI, the sound energy does not reach the right side of the lens (Fig. 8.39a, c, and e). Therefore, the focal length of the AAF lens can be controlled by adjusting the phase distribution.

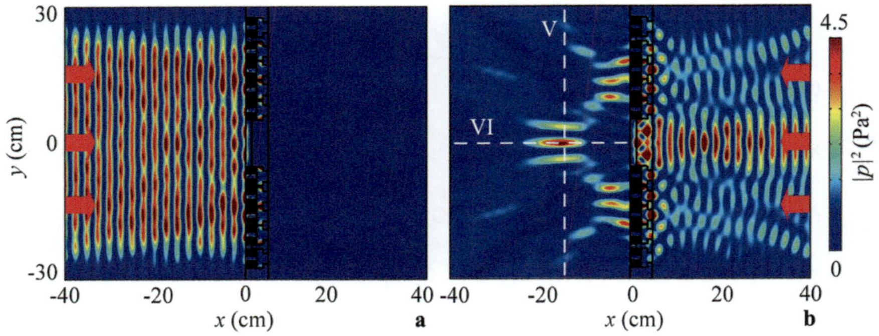

Fig. 8.36 Simulated intensity distributions caused by the AAF lens with viscous loss for **a** LI and **b** RI. The red arrows represent the incident acoustic wave

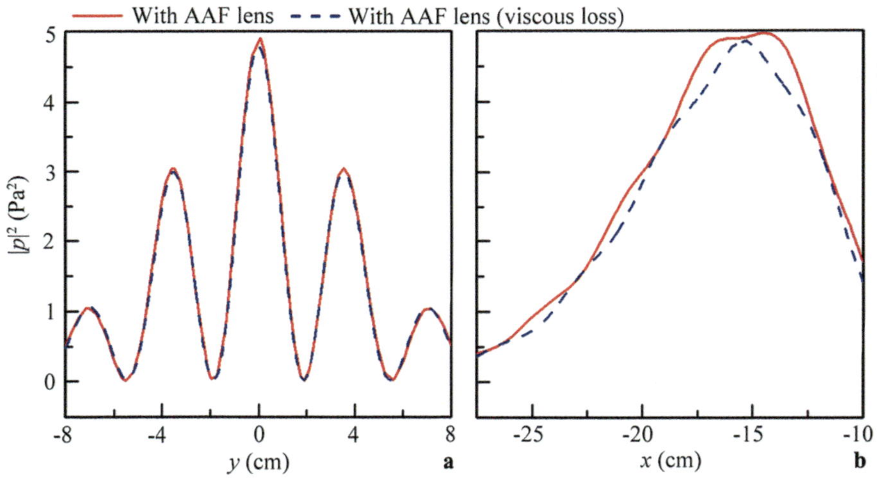

Fig. 8.37 Simulated intensity distributions along lines **a** V and **b** VI in Fig. 8.36b with and without viscous loss

To demonstrate the focusing performance of the AAF lenses with different focal lengths, we simulate the intensity distributions along lines VII, VIII, and IX in Fig. 8.39, as shown in Fig. 8.40. The maximum intensities at these focuses are about 5.5, 5.5, and 3.5 Pa^2 for the lenses with focal lengths of 15, 20, and 25 cm, respectively, indicating high focusing performance across different focal lengths.

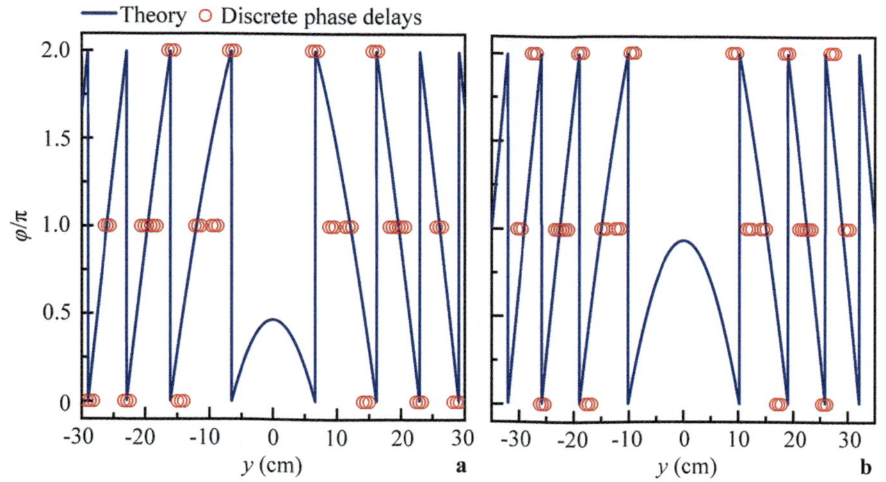

Fig. 8.38 Theoretical continuous phase distributions and 60 discrete phase delays of the AAF lenses with focal lengths of **a** 20 cm and **b** 25 cm at 6.86 kHz

8.3.4 Design and Performance of Underwater AAF Lens

Finally, we extend the working band of the proposed AAF lens to the ultrasonic frequency range by using a type of underwater mode-conversion meta-atom. Figure 8.41 shows an underwater phased unit cell made of tungsten, with the following parameters: $l = 2.5$ cm, $w = 0.5$ cm, $l_1 = 0.5626$ cm, $l_2 = 0.2$ cm, and $d = 0.05$ cm. The material parameters are: density $\rho = 1000$ kg/m^3 and sound velocity $c = 1483$ m/s for water, and $\rho = 19{,}350$ kg/m^3, Young's modulus $E = 411$ GPa, and Poisson's ratio $v = 0.28$ for tungsten. The sound impedance of tungsten is significantly higher than that of water, so a certain thickness of tungsten can satisfy the sound-hard boundary in water. We use the acoustic-solid interaction module of the COMSOL Multiphysics software to simulate the AAF characteristics, with an incident frequency of 30 kHz.

Figure 8.42 shows the transmission coefficients and phase delays of the phased unit cells with different values of w_1. Most of the transmission coefficients exceed 0.9, and the phase delays cover the entire 2π range for w_1 values between 0.04 and 0.33 cm. Based on the same design concept, phased unit cells III and IV, with a phase difference of π, are used to design the underwater mode-conversion meta-atom, with w_1 values of 0.07 cm and 0.29 cm, respectively. The width of the waveguide is set to 4.2 cm to allow the transmission of the first-order wave, and the theoretical continuous phase distribution and discrete phase delays of the meta-atom are the same as those in Fig. 8.24. Figure 8.43 shows the underwater mode-conversion meta-atom, where the step waveguide made of tungsten on the right side prevents the propagation of the first-order wave. The parameters $w_L = 4.2$ cm, $w_2 = 1.2$ cm, $w_r = 2.2$ cm, $w_3 = 0.79$ cm, $l_3 = 0.8$ cm, $l_4 = 2.0$ cm, and $d_2 = 0.21$ cm.

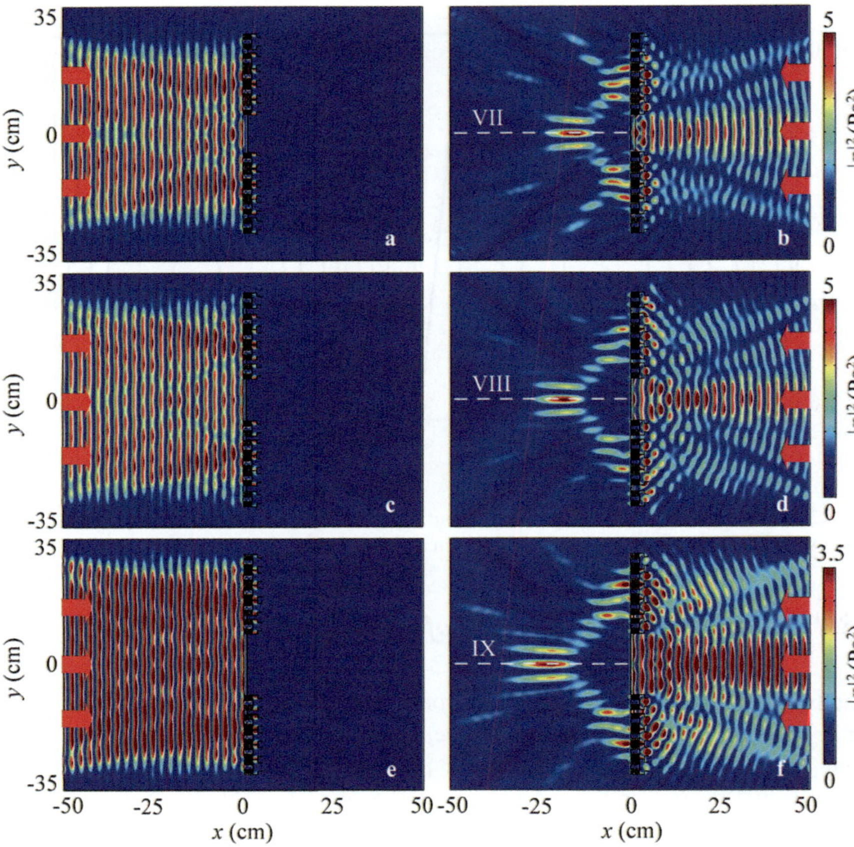

Fig. 8.39 Simulated intensity distributions caused by the AAF lenses with focal lengths of **a**, **b** 15 cm, **c**, **d** 20 cm, and **e**, **f** 25 cm for LI and RI at 6.86 kHz. The red arrows represent the incident acoustic wave

Figure 8.44 shows the pressure distributions caused by the underwater mode-conversion meta-atom for LI and RI at 30 kHz. The zero-order wave passes through the meta-atom and is converted into the first-order wave for RI, but the converted first-order wave cannot pass through the meta-atom for LI. Thus, the underwater meta-atom exhibits mode conversion and AAT characteristics in the ultrasonic frequency range.

To demonstrate the bandwidth of the meta-atom immersed in water, we simulate the transmission spectra for LI and RI, as shown in Fig. 8.45a. In the range of 27.5–31.6 kHz, most of the transmission coefficients exceed 0.8 for RI, while those for LI are below 0.4, showing a clear AAT characteristic. To explain the AAT mechanism, we simulate the phase spectra of phased unit cells III and IV, as shown in Fig. 8.45b. The phase differences of both unit cells are close to π in the black shaded region. Thus,

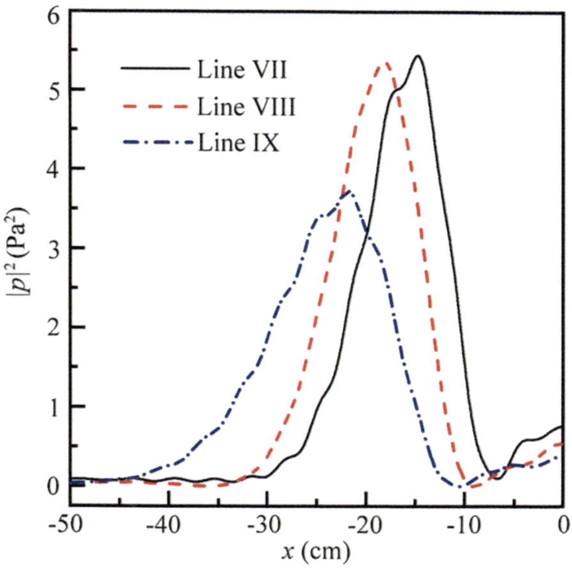

Fig. 8.40 Simulated intensity distributions along lines VII, VIII and IX in Fig. 8.39

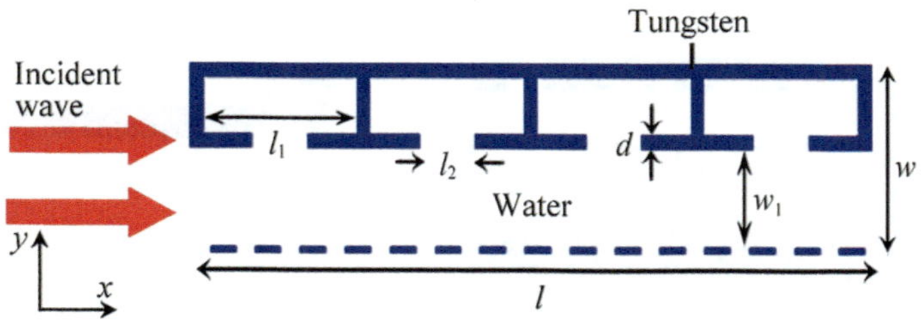

Fig. 8.41 Schematic of the phased unit cell composed of a Helmholtz resonator array immersed in water

the working bandwidth of the meta-atom is 4.1 kHz, with a fractional bandwidth of 0.13.

Based on Eq. (6.2) and the phase distribution of the meta-atom, we calculate the theoretical continuous phase distribution and discrete phase delays of the AAF lens immersed in water, as shown in Fig. 8.46. The parameters e and φ_0 are the same as those in Fig. 8.29. By using ten meta-atoms immersed in water, we design an AAF lens for the ultrasonic frequency range.

Figure 8.47a, b show the simulated intensity distributions caused by the AAF lens for LI and RI at 30 kHz, respectively. The sound energy is focused on the left side of the lens for RI, but the sound energy is blocked by the step waveguide for LI, indicating that the lens can achieve the AAF effect in the ultrasonic frequency range.

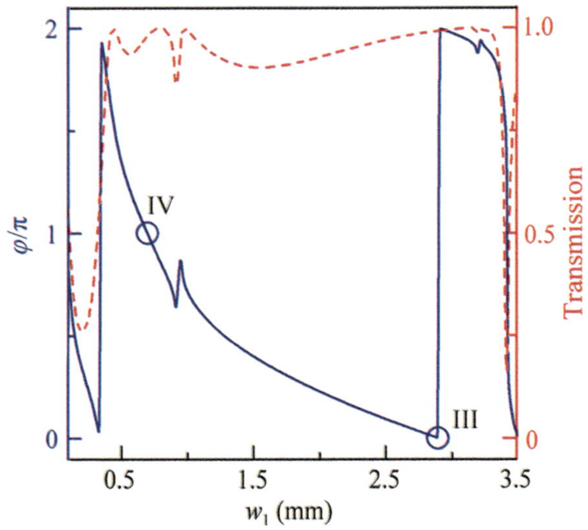

Fig. 8.42 Simulated transmission coefficients (the red dashed line) and phase delays (the blue solid line) caused by the phased unit cells with different values of w_1 at 30 kHz

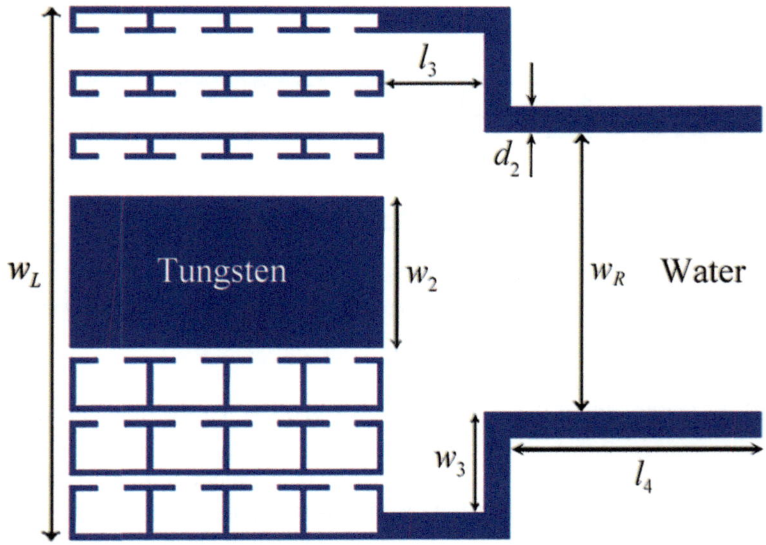

Fig. 8.43 Schematic of the mode-conversion meta-atom immersed in water

To further quantify the AAF performance, we simulate the normalized intensity distributions along lines X-XIII in Fig. 8.47a, b. As shown in Fig. 8.48, the intensities along lines XII and XIII are significantly larger than those along lines X and XI. Additionally, the FLHM of the focus in the y direction is 0.19 cm (0.38λ). Thus,

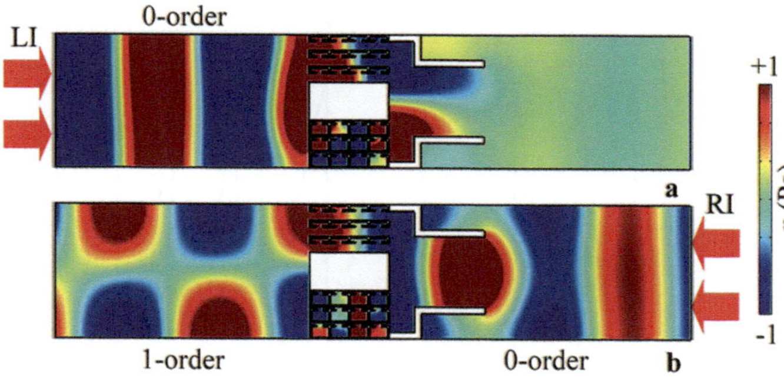

Fig. 8.44 Simulated pressure distributions caused by the underwater mode-conversion meta-atom for **a** LI and **b** RI at 30 kHz

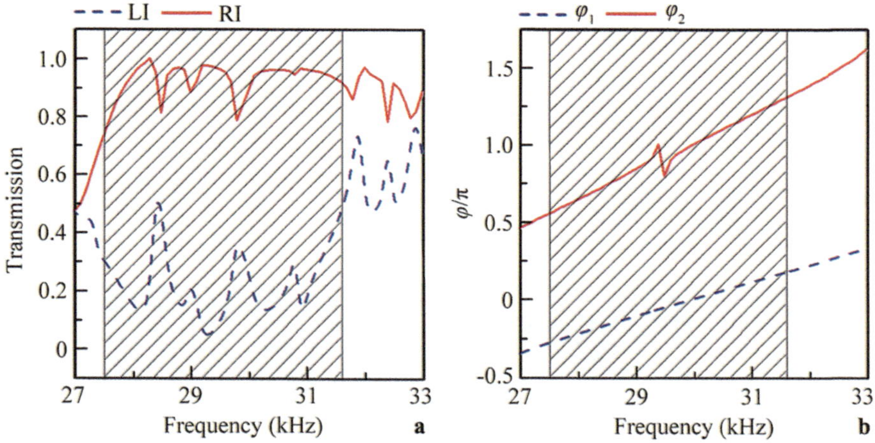

Fig. 8.45 **a** Simulated transmission spectra of the underwater mode-conversion meta-atom for LI and RI. **b** Simulated phase spectra of phased unit cells III (φ_1, the blue dashed line) and IV (φ_2, the red solid line) in Fig. 8.42

the designed AAF lens exhibits high performance in AAT and subwavelength AF, making it highly suitable for focused ultrasound therapy.

Figure 8.49a shows the transmission spectra of the AAF lens immersed in water for LI and RI. The lens exhibits typical AAT characteristics in the range of 27.5–31.6 kHz (the black shaded region). The higher transmission coefficient for LI at certain frequencies may result from coupling resonance between adjacent meta-atoms. To further verify the bandwidth of the AAF lens, we simulate the maximum intensity spectra along lines XI and XIII, as shown in Fig. 8.49b. In the black shaded region, most of the maximum intensities are below 1 Pa^2 along line XI, while the corresponding values exceed 3 Pa^2 along line XIII. Thus, the fractional bandwidth of the AAF lens is 0.13.

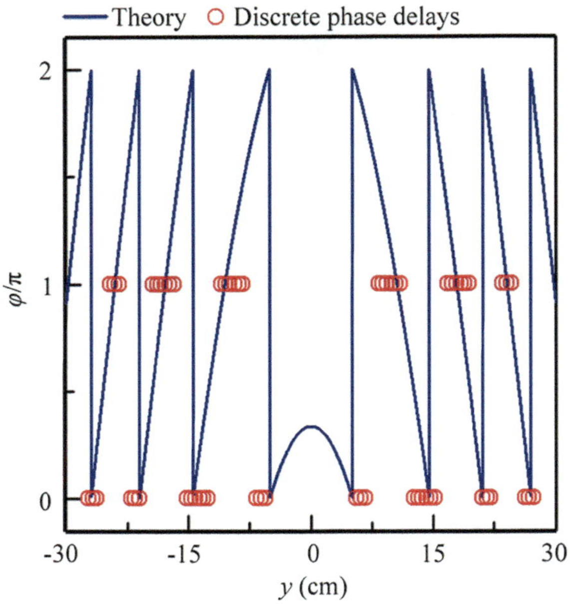

Fig. 8.46 Theoretical continuous phase distribution and 60 discrete phase delays of the AAF lens immersed in water

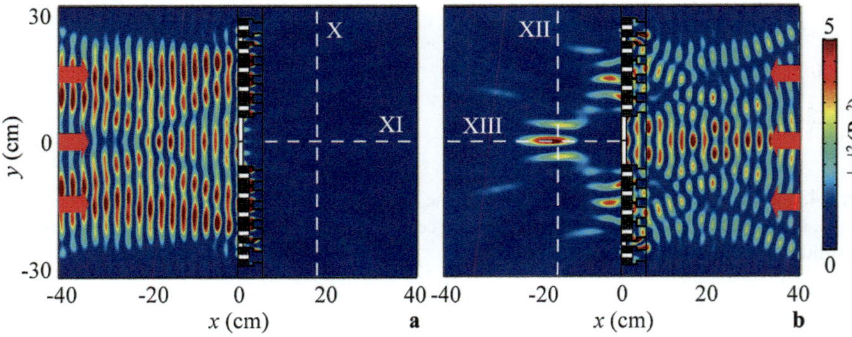

Fig. 8.47 Simulated intensity distributions caused by the underwater AAF lens for **a** LI and **b** RI at 30 kHz

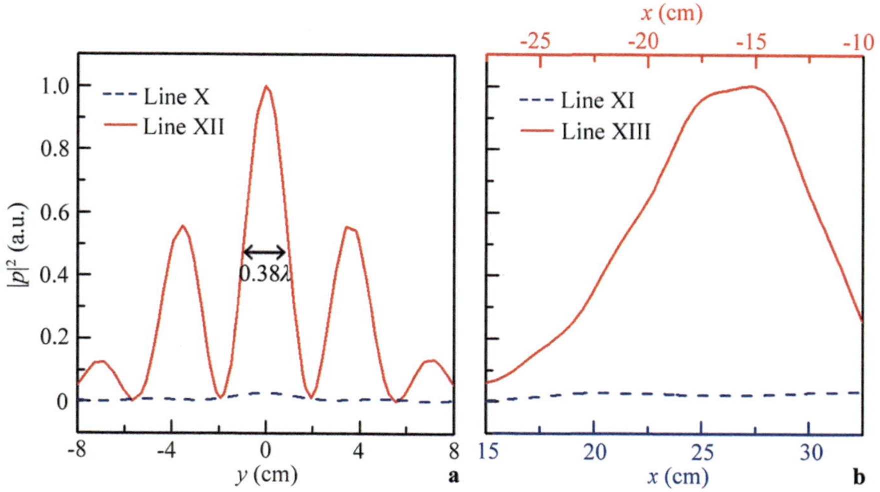

Fig. 8.48 Simulated intensity distributions along **a** lines X and XII and **b** lines XI and XIII in Fig. 8.47

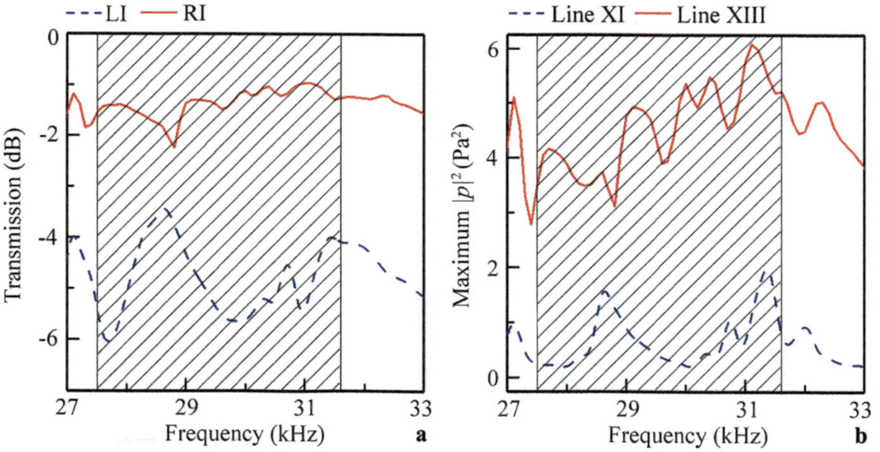

Fig. 8.49 **a** Simulated transmission spectra of the underwater AAF lens for LI and RI. **b** Maximum intensity spectra along lines XI and XIII in Fig. 8.47

8.4 Acoustic Asymmetric Focusing by a Metal Plate with Binary Wave-Path Slits

8.4.1 Design and Performance of AAF Lens

As schematically shown in Fig. 8.50, the designed AAF lens is symmetric about $y=0$ and consists of a steel plate with six unit cells immersed in water [3]. Each unit cell, with a width of a and a thickness of b, contains two types of slits: an inclined slit A with an angle of θ and a horizontal slit B. Both slits have the same width c. The performance of the AAF lens is simulated using the COMSOL Multiphysics software, with the following material parameters: density $\rho = 7800$ kg/m^3, longitudinal wave velocity $c_l = 6100$ m/s, and transversal wave velocity $c_t = 3300$ m/s for steel; and density $\rho = 998$ kg/m^3 and $c_l = 1483$ m/s for water. The acoustic-structure boundary condition is applied to the boundaries of the AAF lens in the simulations.

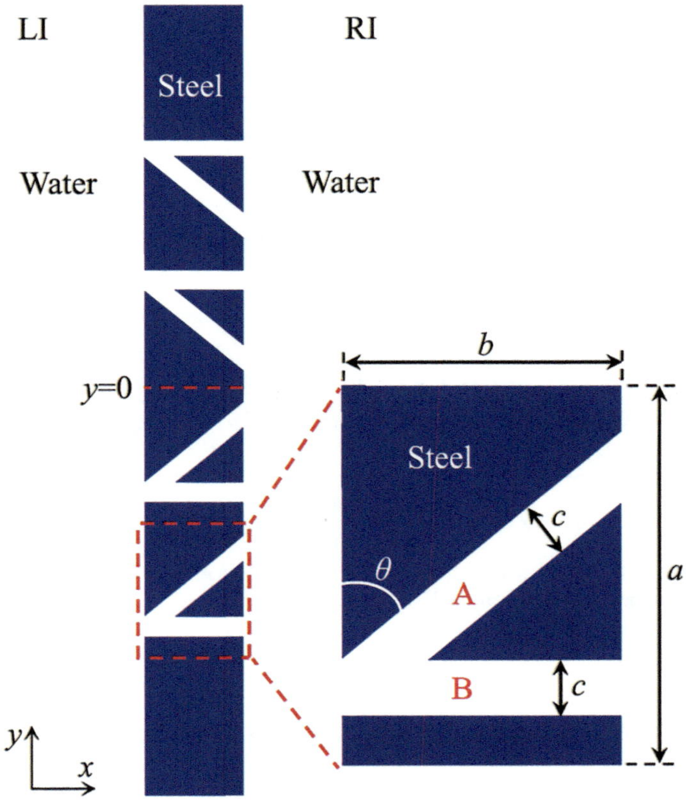

Fig. 8.50 Schematic of the AAF lens. The parameters are $a = 6.8$ mm, $b = 5.0$ mm, $c = 1.0$ mm and $\theta = 50.5°$

Figure 8.51a–f show the intensity distributions caused by the AAF lens at different frequencies, where LI and RI represent incident plane waves emanating from the left and right sides of the lens, respectively. The width and initial amplitude of the incident plane wave are 20 mm and 1.0 Pa, respectively. The acoustic intensity distributions exhibit a clear AAF phenomenon. In the case of LI (Fig. 8.51a, c, and e), strong sound energy is observed in the right region, converging into a line at $y = 0$. Interestingly, the sound energy gradually converges into a point as the incident frequency increases. However, for RI (Fig. 8.51b, d, and f), the acoustic waves are almost entirely reflected, with negligible sound transmission in the left region. Additionally, the AAF effect is obtained over a broader bandwidth (560–592 kHz). The sound intensity distributions at the edge frequencies (560 and 592 kHz) are shown in Fig. 8.52, demonstrating that the designed lens achieves broadband AAF with high performance.

To further demonstrate the AAT characteristic of the lens, the right boundary of the model is set as a sound-hard boundary, and the simulated intensity distribution caused by the lens is shown in Fig. 8.53a. For comparison, the corresponding result with a

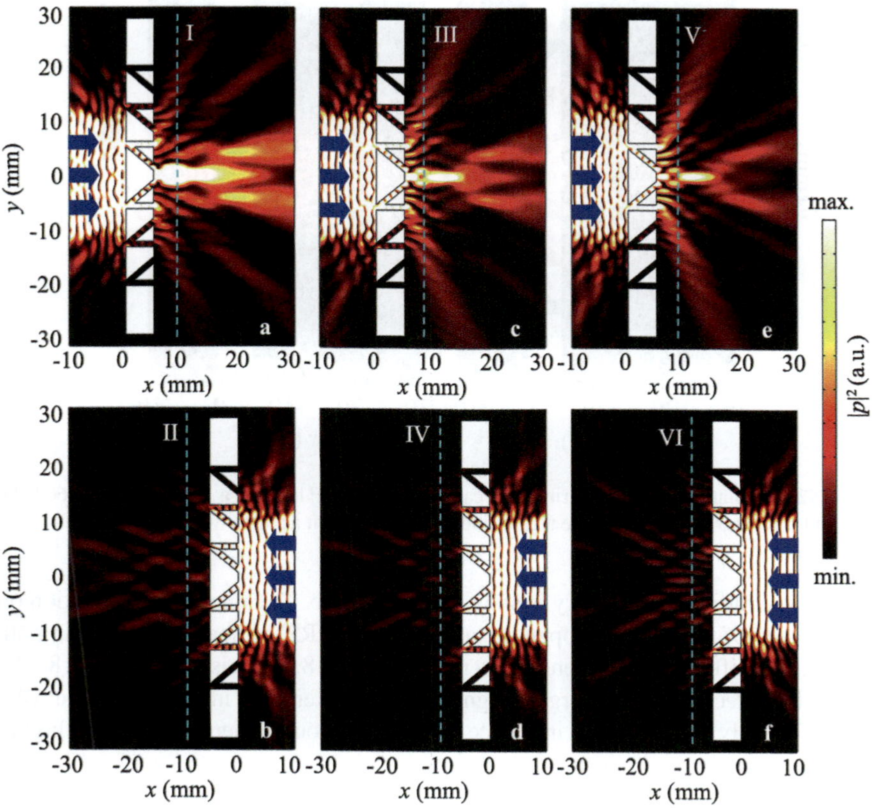

Fig. 8.51 Simulated intensity distributions caused by the AAF lens for **a** LI and **b** RI at 564 kHz; **c** LI and **d** RI at 576 kHz; **e** LI and **f** RI at 588 kHz. The blue arrows indicate the incident acoustic wave

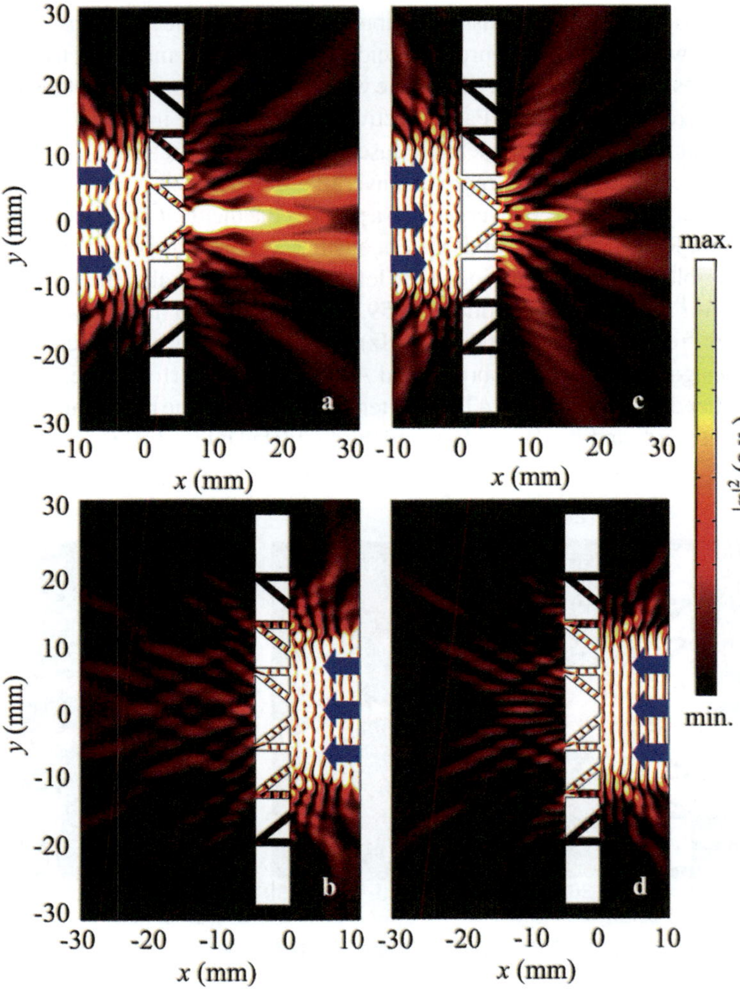

Fig. 8.52 Simulated intensity distributions caused by the AAF lens for **a** LI and **b** RI at 560 kHz; **c** LI and **d** RI at 592 kHz. The blue arrows indicate the incident acoustic wave

plane wave radiation boundary is shown in Fig. 8.53b. The intensity distribution on the left side of the lens (red open rectangles R1 and R2) is nearly identical in both figures. The difference between the left regions in Fig. 8.53a, b is shown in Fig. 8.53c, where the reflected sound energy is significantly weaker. This indicates that the AAF lens effectively mitigates the influence of reflected sound energy, making it highly suitable for designing focusing transducers.

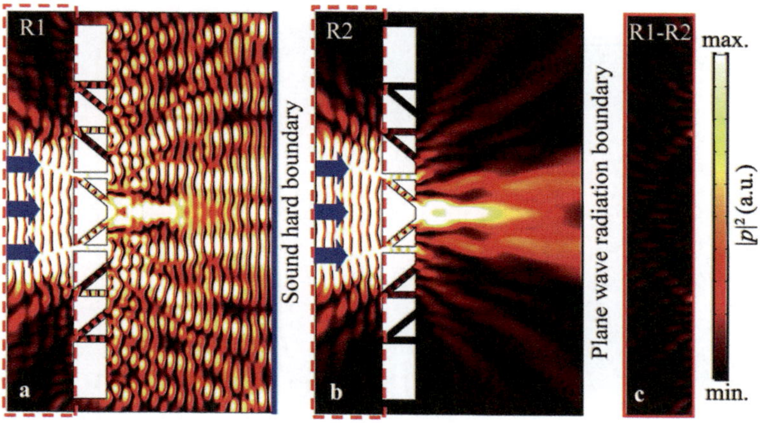

Fig. 8.53 Simulated intensity distributions caused by the AAF lens for LI at 564 kHz, with the right boundary of the model set as **a** sound-hard boundary and **b** plane wave radiation boundary. **c** Difference between the intensity distributions in red open rectangles R1 and R2. The blue arrows indicate the incident acoustic wave

8.4.2 Experimental Measurement

To verify the AAF performance, we experimentally measure the intensity distribution along lines I–VI in Fig. 8.51. The photograph for the left and right sides of the sample are shown in Fig. 8.54a, b, respectively. The sample has a width $w =$ 40.8 mm and length $l = 90.0$ mm, with other parameters matching those in Fig. 8.50. Figure 8.55 shows the experimental set-up, where the sample is placed between two broadband ultrasonic transducers: one as a generator (V318-SU, Olympus, with a central frequency of 0.5 MHz and a diameter of 0.75 inches) and the other as a detector (V323-N-SU, Olympus, with a central frequency of 2.25 MHz and a diameter of 0.25 inches). The set-up is immersed in a water tank. The generating transducer is aligned normally toward the sample at a distance of 20 mm, while the detecting transducer scans along lines I–VI at the same distance as that in Fig. 8.51.

Figure 8.56a–c show the measured intensity distributions caused by the sample along lines I–VI, with corresponding simulations plotted for comparison. As shown in Fig. 8.56a–c, the measured intensities for RI are close to zero, while LI exhibits high intensities, with the maximum intensity occurring at $y = 0$ for the three selected frequencies. This demonstrates the high performance of the AAF effect. The measured results agree well with the simulations, though minor discrepancies exist, particularly for LI in Fig. 8.56b, c. These differences are attributed to the 0.25 inch diameter of the detecting transducer, which limits spatial resolution, and a fabrication accuracy error of ±0.1 mm in the sample.

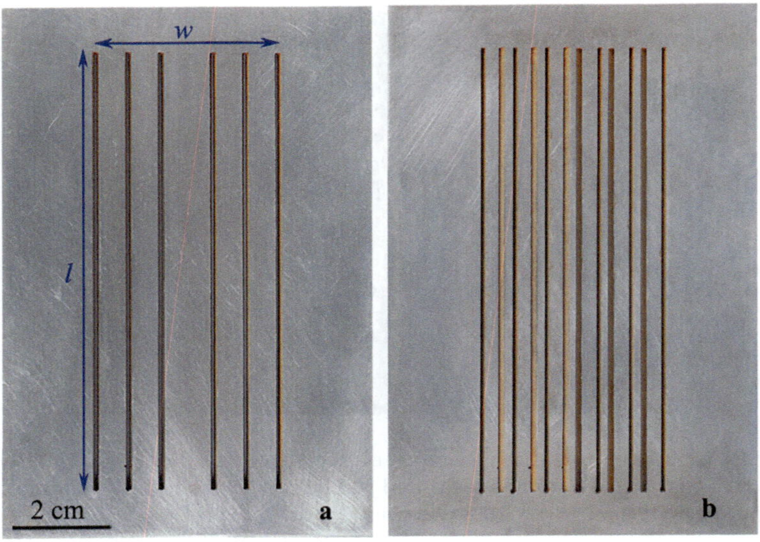

Fig. 8.54 Photographs of the **a** left and **b** right sides of the sample

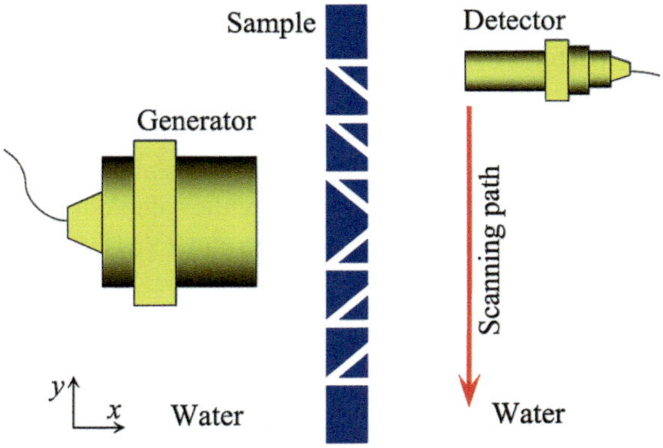

Fig. 8.55 Schematic of the experimental set-up for the AAF lens in Fig. 8.54

8.4.3 Physical Mechanism

To further explore the mechanism of the AAF lens, we simulate the pressure distributions caused by a unit cell of the lens, with an initial incident plane wave amplitude of 1.0 Pa. Figure 8.57a shows the acoustic propagation path in the unit cell for RI, where the incident acoustic wave is divided into two beams passing through slits A and B. The two beams intersect at midpoint O of the red dashed line. The propagation distances in slits A and B are l_1 and l_2, respectively, and the incident plane wave is located on the left side of the model.

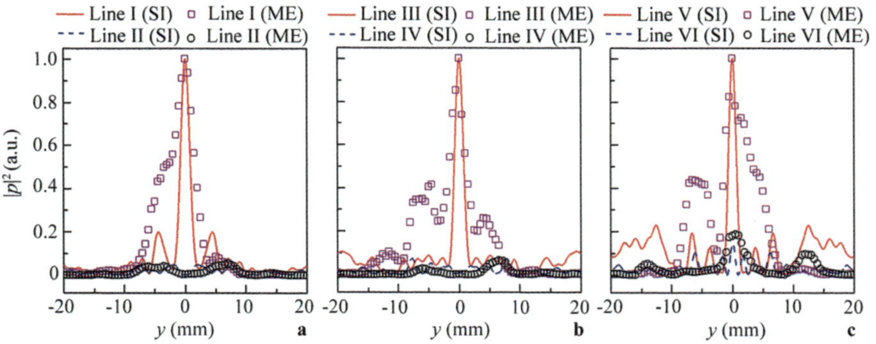

Fig. 8.56 Measured and simulated intensity distributions along **a** lines I and II; **b** lines III and IV; **c** lines V and VI in Fig. 8.51. The intensities are normalized by the maximum value of each figure

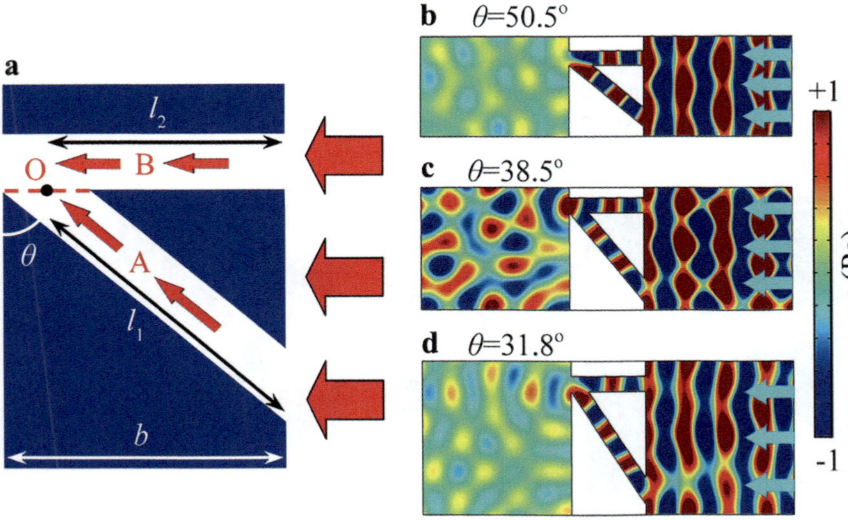

Fig. 8.57 **a** Schematic of the acoustic propagation path for RI. Simulated pressure distributions caused by a unit cell with inclined angles of **b** 50.5°, **c** 38.5°, and **d** 31.8° at 564 kHz for RI. The sky-blue arrows in (**b–d**) indicate the incident acoustic wave

Figure 8.57b presents the pressure distribution caused by the unit cell at 564 kHz, with parameters matching those in Fig. 8.50. Negligible sound energy is observed in the left region, consistent with Fig. 8.57a. This is because the difference Δl between distances l_1 and l_2 is 1.25 mm, close to the incident half-wavelength of 1.31 mm. Thus, interference cancellation occurs at point O, preventing sound energy from passing through the unit cell. Additionally, the half-wavelengths at the edge frequencies of 560 kHz and 592 kHz are 1.32 mm and 1.25 mm, respectively, both close to Δl. Therefore, the prohibition of acoustic waves for RI arises from interference cancellation. To further verify this, we increase Δl by adjusting the inclined angle θ, while keeping other parameters constant. As shown in Fig. 8.57c, d, Δl values of 2.64 mm and 3.96 mm correspond to λ and 1.5λ, respectively. In Fig. 8.57c, the acoustic wave transmits through the unit cell due to interference enhancement, while in Fig. 8.57d, negligible sound energy reaches the left region due to interference cancellation.

Furthermore, we develop a unit cell with two inclined slits, A and B, as shown in Fig. 8.58a. The incident acoustic line sources I_1 and I_2 are located at the input ports of slits A and B, respectively, with phases φ_1 and φ_2 and equal amplitudes. The parameters b, θ, and c match those in Fig. 8.50, so the propagation distances in slits A and B are the same. As shown in Fig. 8.58b, when the initial phases of I_1 and I_2 are the same (the in-phase case, $\varphi_1 - \varphi_2 = 0$), sound energy transmits through the unit cell due to interference enhancement. However, for the out-of-phase case ($\varphi_1 - \varphi_2 = \pi$), the acoustic wave cannot pass through the unit cell due to interference cancellation (Fig. 8.58c). These results further demonstrate the mechanism of the AAT lens.

Figure 8.59a, b show the pressure distributions caused by a single unit cell and the AAF lens at 564 kHz, respectively. The incident plane wave is located at the left surface of the model, and the unit cell parameters match those in Fig. 8.50. As shown in Fig. 8.59a, the acoustic wave passes through slits A and B, and the transmitted

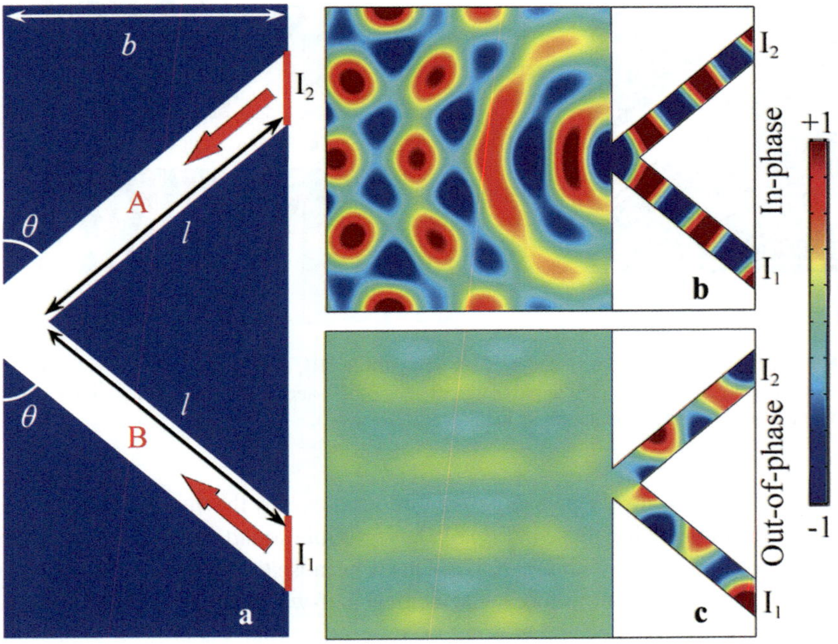

Fig. 8.58 **a** Schematic of the unit cell with two inclined slits. The red lines placed at the input ports of slits A and B represent two acoustic line sources with initial phases φ_1 and φ_2. Simulated pressure distributions through the two inclined slits under the excitation of **b** in-phase ($\varphi_1 - \varphi_2 = 0$) and **c** out-of-phase ($\varphi_1 - \varphi_2 = \pi$) line sources

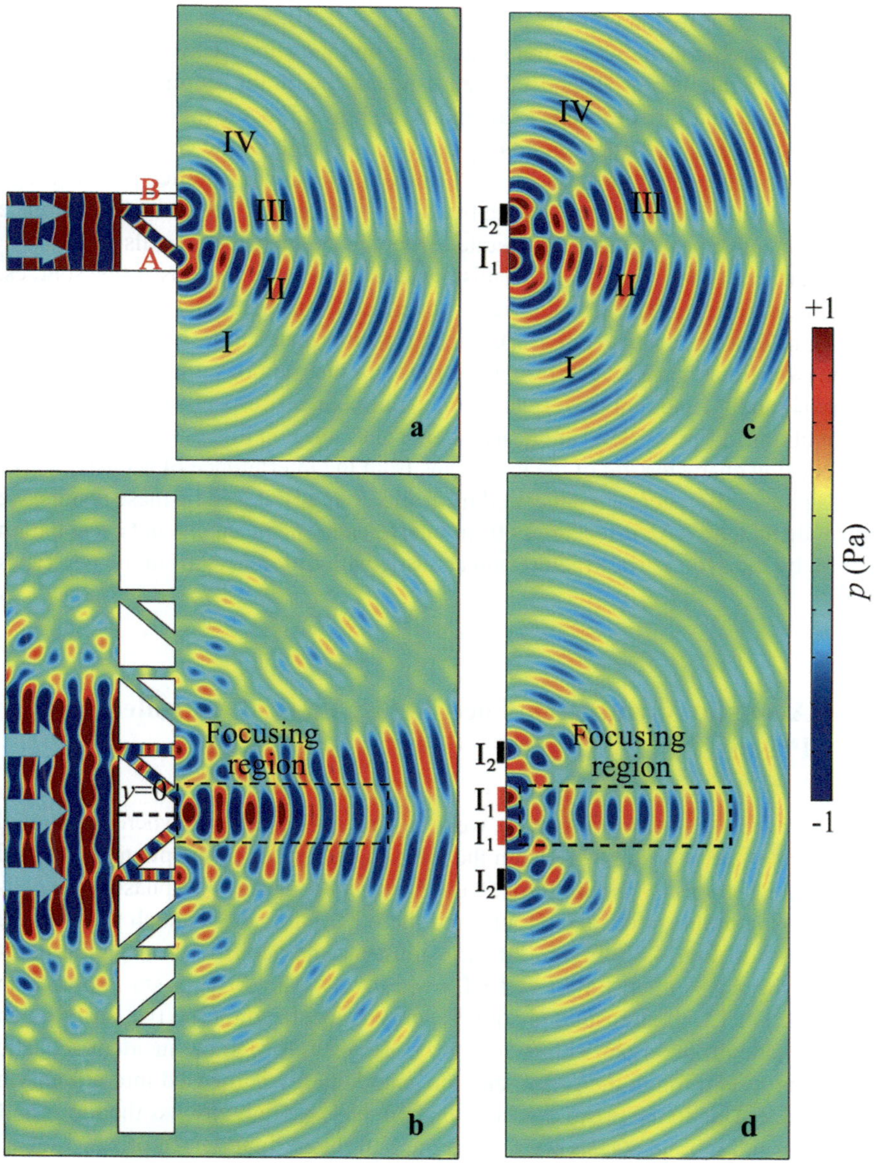

Fig. 8.59 Simulated pressure distributions caused by the **a** unit cell and **b** AAF lens at 564 kHz. Simulated pressure distributions under the excitation of **c** two (I_1 and I_2) and **d** four (two I_1 and two I_2) line sources at 564 kHz. The incident line sources I_1 and I_2 have a phase difference of π

wave in the right region consists of four beams: I, II, III, and IV, resulting from the interference of acoustic waves transmitted through the slits. Beams I and II incline downward, while beams III and IV incline upward, with the amplitude of beam II being larger than those of beams I, III, and IV. In Fig. 8.59b, the transmitted pressure distribution is primarily determined by the two unit cells at the center of the AAF lens, which are symmetric about $y = 0$. Consequently, the transmitted sound energy converges into a line along $y = 0$ (the black open rectangle), due to the interference enhancement of beams I and II generated from the symmetric unit cells.

To further discuss the AAF mechanism, we use two types of acoustic line sources, I_1 and I_2, to represent the acoustic waves generated from slits A and B. I_1 and I_2 have a phase difference of π ($\varphi_1 - \varphi_2 = \pi$), and their widths match those of the output ports of the two slits. Figure 8.59c, d show the pressure distributions under the excitation of two (I_1 and I_2) and four (two I_1 and two I_2) line sources, respectively. The pressure distributions in Fig. 8.59c, d are nearly identical to those in Fig. 8.59a, b. Therefore, the focusing characteristic of the lens stems from the interference of acoustic waves with a phase difference of π generated from slits A and B. Based on these results, we conclude that the AAF effect arises from interference enhancement and cancellation caused by the symmetric structure and acoustic path differences from the two types of slits.

8.5 Acoustic Asymmetric Focusing by Phase-Controlled Prism

As schematically shown in Fig. 8.60, we design an AAF prism [4], where the normal incidence of the acoustic wave from the left and right sides of the prism is denoted as LI and RI, respectively. Similarly, the AAF prism consists of 31 phased unit cells with varying numbers of meta-atoms, and the other parameters match those of the AAT prism in Fig. 4.60. Figure 8.60a, b illustrate the acoustic propagation paths through the AAF prism for LI and RI, respectively. The AAF prism is placed in region II. For LI, the transmitted sound energy is localized on the two surfaces on the right side of the prism with $\theta_{t1} = -90°$. However, for RI, the incident acoustic wave transmits through the prism, and the sound energy is focused into a point on the left side. Due to the prism's structural symmetry, we only discuss the lower half of the prism. Figure 8.60c shows a photograph of the designed AAF prism.

The AF effect is present for RI. Based on Eq. (2.41), the refraction angle θ_{t2} is expressed as:

$$\sin\theta_{t2} = \frac{-x_2}{\sqrt{x_2^2 + e^2}}, \tag{8.6}$$

where the focal length $e = 0.2$ m. Using Eqs. (4.1), (4.4), and (4.5), we theoretically calculate the phase distributions in the y_1 and y_2 directions of the prism, as shown

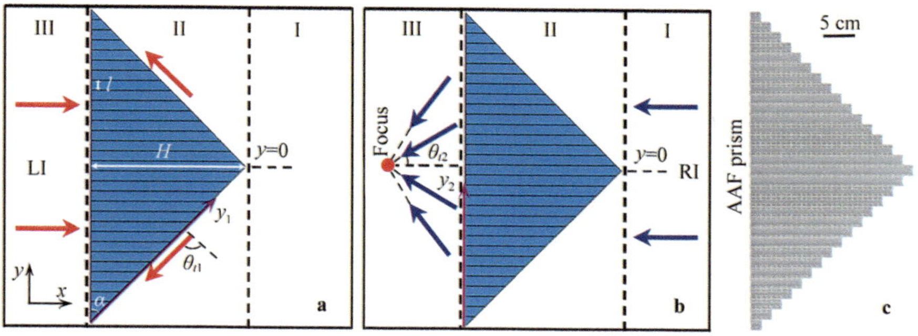

Fig. 8.60 Schematic of acoustic propagation paths caused by the AAF prism for **a** LI and **b** RI, and **c** photograph of the AAF prism

in Fig. 8.61a, b, respectively. Additionally, the sixteen red open circles represent the discrete phase delays of the selected phased unit cells in the y_2 direction. The parameter h_1 of each meta-atom in different phased unit cells and the discrete phase delays of the phased unit cells are presented in Table 8.1.

Figure 8.62a, b show the simulated intensity distributions caused by the AAF prism at 6.66 kHz, demonstrating a clear AAF effect. For RI, the acoustic wave transmits through the prism, and the sound energy is focused into a point on the left side. However, for LI, the transmitted sound energy is very weak. This is because the calculated $\mathrm{d}\varphi_1/k\mathrm{d}y_1$ in the y_1 direction ranges from -0.95 to -1.22, as shown in Fig. 8.63. Based on Eq. (2.41), the maximum refraction angle is calculated as $\theta_{t1} = -72°$ for $\mathrm{d}\varphi_1/k\mathrm{d}x_1 = -0.95$. Moreover, in the range $-1.22 \leq \mathrm{d}\varphi_1/k\mathrm{d}x_1 < -1$, the transmitted acoustic wave is converted into an evanescent surface wave, which cannot reach region I for LI. Therefore, the refraction angle θ_{t1} for most of the sound energy

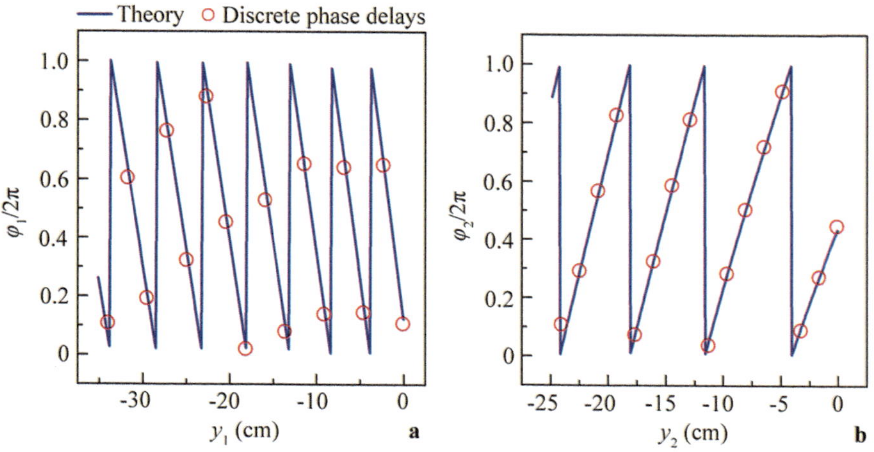

Fig. 8.61 Theoretical continuous phase distributions and sixteen discrete phase delays in the **a** y_1 and **b** y_2 directions for the lower half of the AAF prism at 6.66 kHz

Table 8.1 Selected sixteen discrete phase delays of the phased unit cells for the lower half of the AAF prism

n	1	2	3	4	5	6	7	8	9	10	11	12	13	14	15	16
h_1 (mm)	11	7.7	6.5	6.1	5.9	9.7	8.4	7.7	7.4	10.7	9.7	9	8.6	8.3	8.1	10.5
Phase delay (rad)	0.66	1.83	3.56	5.18	0.45	2.04	3.68	5.25	0.23	1.76	3.09	4.5	5.71	0.59	1.64	2.79

is around −90°, and the remaining part is converted into an evanescent wave, which cannot reach region I for LI. Additionally, the measured intensity distributions in red open squares R1 and R2 are displayed on the right side for comparison, showing good agreement with the simulated results.

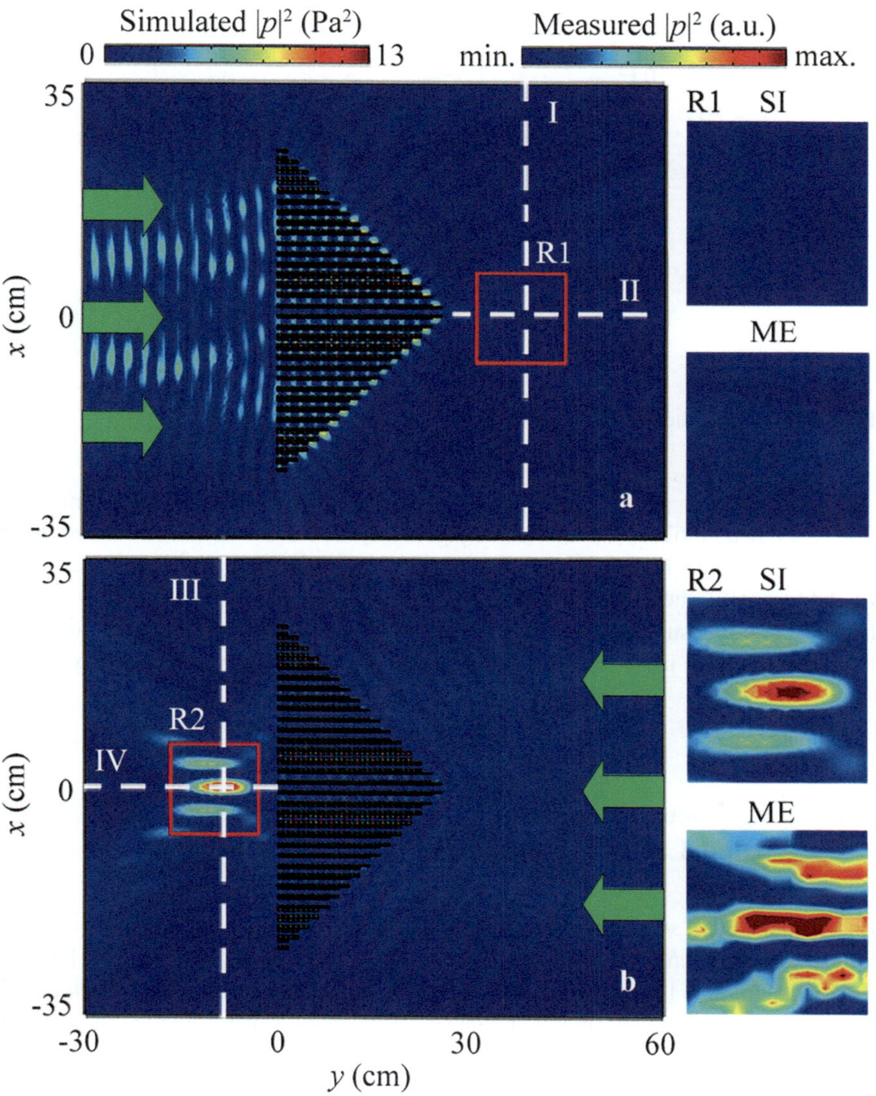

Fig. 8.62 Measured and simulated intensity distributions caused by the AAF prism for **a** LI and **b** RI at 6.66 kHz. The insets on the right side show the measured and simulated intensity distributions in red open squares R1 and R2. The green arrows indicate the incident acoustic wave

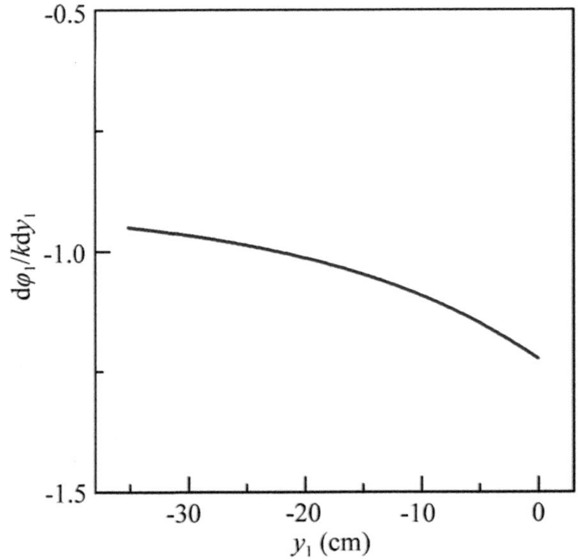

Fig. 8.63 Phase gradient $d\varphi_1/kdy_1$ of the AAF prism in the y_1 direction for LI

Finally, we also present the measured and simulated intensity distributions along lines I–IV (Fig. 8.62a, b), as shown in Fig. 8.64. The measured and simulated intensities along lines III and IV are significantly larger than those along lines I and II. Additionally, the FLHM of the focus in the y direction (Fig. 8.64a) is 1.9 cm (0.36λ). The measured results agree well with the simulations, demonstrating that the designed AAF prism exhibits high performance in subwavelength focusing and AAT.

8.6 Long-Focus Acoustic Asymmetric Focusing by Dual-Layer Thermoacoustic Phased Arrays

Based on the eight selected thermoacoustic phased unit cells in Fig. 7.24, we design a long-focus AAF lens with dual-layer arrays (denoted as TPA1 and TPA2) [5], where the focus length is larger than those of previous AAF lenses, as shown in Fig. 8.65. LI and RI represent the normal incidence of the acoustic wave from the left and right sides of the long-focus AAF lens, respectively. The phase distributions of TPA1 and TPA2 (φ_1 and φ_2) are symmetric about $y = 0$. Therefore, we only discuss the propagation paths and phase distributions of TPA1 and TPA2 in the range $y > 0$.

As shown in Fig. 8.65a, for LI, the acoustic wave passes through TPA1 and is divided into a pair of two beams (symmetric about $y = 0$) with a refraction angle $\beta_1 = -\pi/3$. Additionally, the transmitted sound energy through TPA2 with a refraction angle $\beta_2 = -\pi/6$ forms a Bessel-like beam, exhibiting a long-focus characteristic on

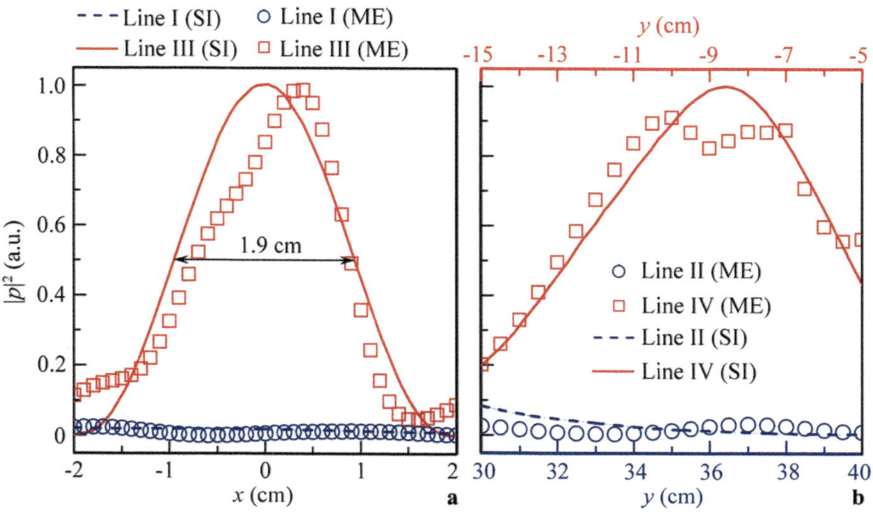

Fig. 8.64 Measured and simulated intensity distributions along **a** lines I and III and **b** lines II and IV in Fig. 8.62. The intensities are normalized by the maximum value of each figure

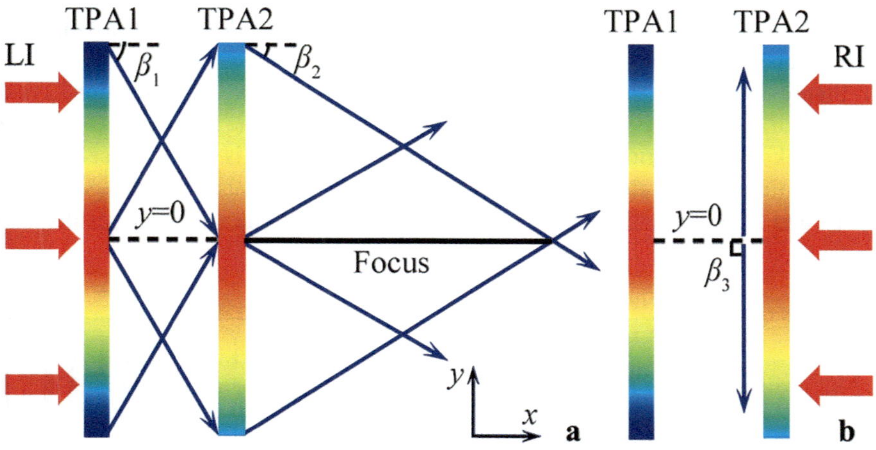

Fig. 8.65 Schematic of acoustic propagation paths through the long-focus AAF lens for **a** LI and **b** RI

the right side of the lens. However, for RI (Fig. 8.65b), the acoustic wave transmits through TPA2 with a refraction angle $\beta_3 = -\pi/2$. Consequently, the sound energy is localized on the surface of TPA2, and no sound energy passes through TPA1 for RI. Based on Eq. (2.41), in the range $y > 0$, the phase gradient of TPA2 is written as:

$$\frac{1}{k_0}\frac{\mathrm{d}\varphi_2}{\mathrm{d}y} = -1. \tag{8.7}$$

The phase gradient of TPA1 is obtained from:

$$\frac{1}{k_0}\frac{\mathrm{d}\varphi_1}{\mathrm{d}y} = -\frac{\sqrt{3}}{2}. \tag{8.8}$$

Figure 8.66a, b show the theoretical continuous phase distributions φ_1 and φ_2 of TPA1 and TPA2, respectively, where the distance d between the two arrays is 24 cm. Similar to Fig. 7.24b, we selected 81 discrete phase delays to design the phase distributions of TPA1 and TPA2 using the eight types of thermoacoustic phased unit cells in Fig. 7.24a.

Figure 8.67a, b show the intensity distributions caused by the long-focus AAF lens for LI and RI, respectively. As shown in Fig. 8.67a, the acoustic wave transmits through TPA1, and the transmitted sound energy is finally focused into a long focus on the right side of TPA2 with weak sound diffraction. However, for RI (Fig. 8.67b), the transmitted sound energy is localized on the surface of TPA2, and thus the acoustic wave cannot pass through TPA1, demonstrating high AAF performance. Figure 8.68 a, b show the intensity distributions along lines I–IV. The intensity distributions exhibit a significant difference between LI and RI, where the maximum intensity at the focus for LI (lines I and II) reaches approximately 3.3 Pa^2, but the intensities for RI (lines III and IV) are below 0.25 Pa^2. Additionally, as shown in Fig. 8.68a, the FLHM of the long focus is approximately 18.1λ, further demonstrating the high performance of the long-focus AAF effect.

Figure 8.69 shows the transmittance spectra caused by the AAF lens for LI and RI. Both transmittance spectra exhibit a clear difference in the range of 4.8–5.5 kHz (the black shaded region), where the transmittance for LI is about twice that for RI. Additionally, we simulate the intensity distributions caused by the AAF lens for LI and RI at 4.9 and 5.3 kHz, as shown in Fig. 8.70. The simulated results also

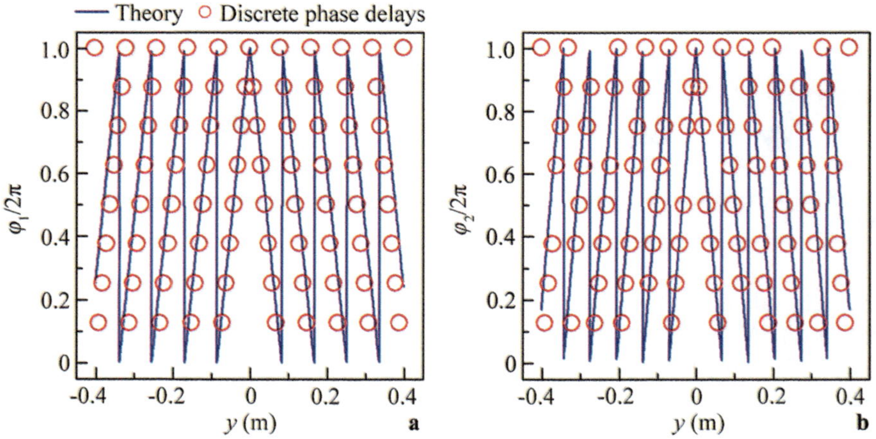

Fig. 8.66 Theoretical continuous phase distributions and 81 discrete phase delays of **a** TPA1 and **b** TPA2 in the long-focus AF lens

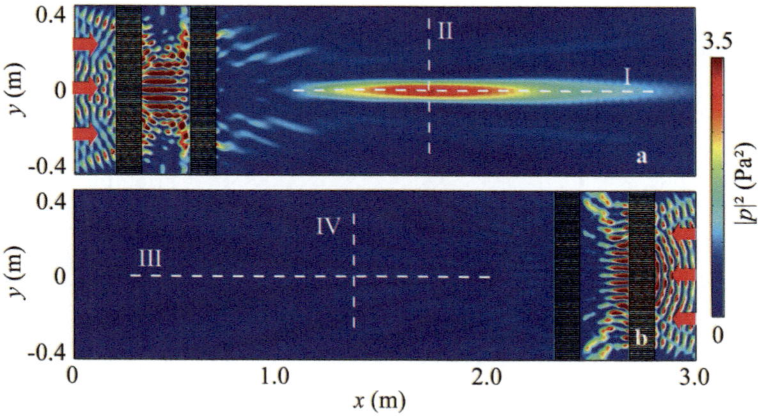

Fig. 8.67 Simulated reflected intensity distributions caused by the long-focus AAF lens for **a** LI and **b** RI at 5.0 kHz

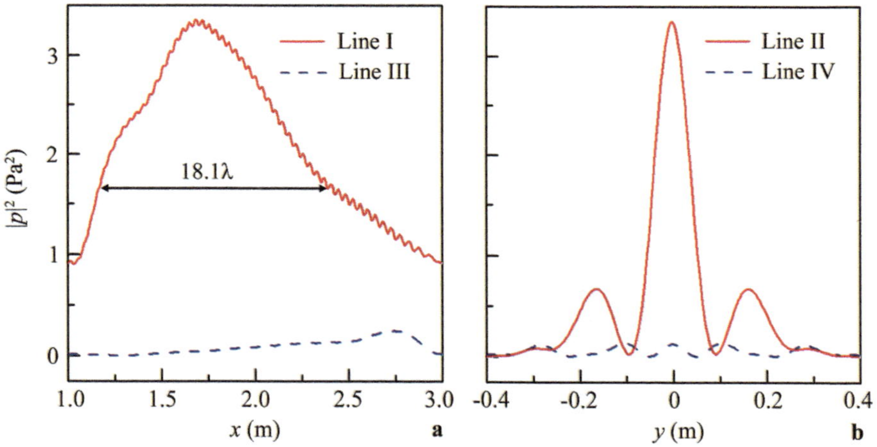

Fig. 8.68 Simulated reflected intensity distributions along **a** lines I and III and **b** lines II and IV in Fig. 8.67

demonstrate high performance of the long-focus AAF effect at both edge frequencies. Therefore, the working bandwidth of the AAF lens is 0.7 kHz, with a fractional bandwidth of 0.14, which is narrower than that of the long-focus AF lens due to the increase in transmittance for RI in the high-frequency region (the blue dashed line). This is because the incident angle and phase distribution of TPA2 remain unchanged. As the frequency increases, the refraction angle gradually decreases based on Eq. (2.41). Consequently, the sound energy cannot be localized on the surface of TPA2 and transmits through TPA1, reaching the left side of the AAF lens.

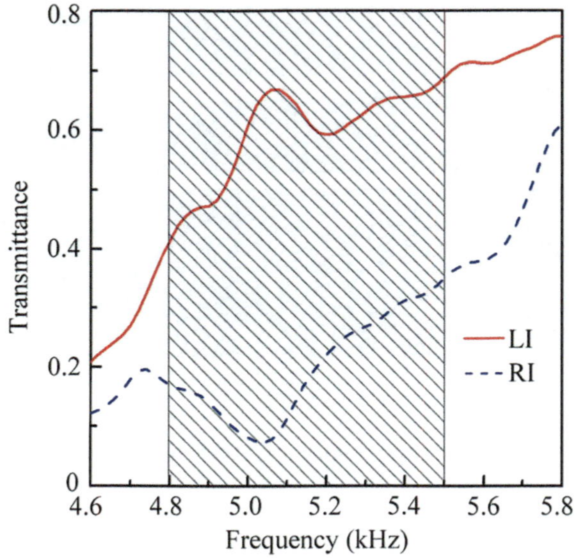

Fig. 8.69 Transmittance spectra of the long-focus AAF lens for LI and RI

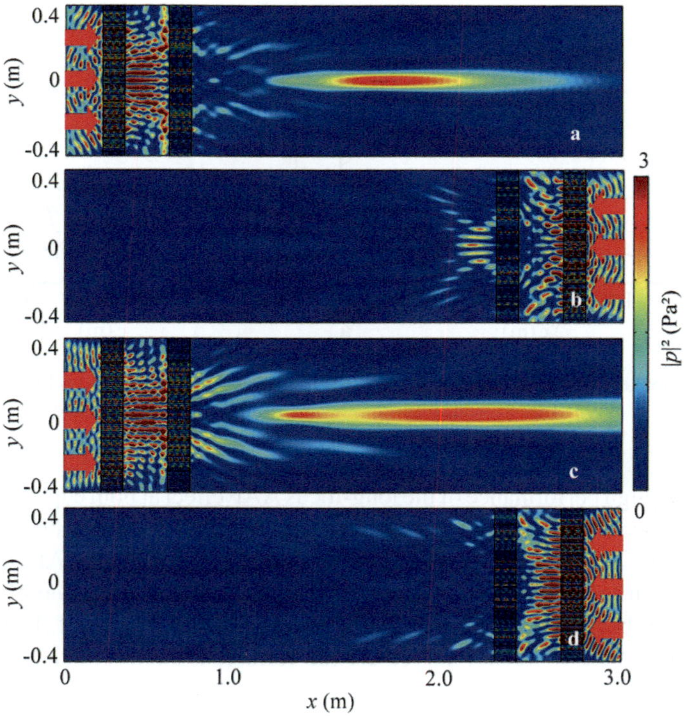

Fig. 8.70 Simulated reflected intensity distributions caused by the long-focus AAF lens for LI and RI at **a**, **b** 4.9 kHz and **c**, **d** 5.3 kHz

8.7 Conclusion

This chapter explores the design and performance of AAF achieved through asymmetric modulation metamaterials. Five types of AAF lenses were designed, including dual-layer metasurface lenses, mode-conversion meta-atom lenses, underwater AAF lenses, acoustic prisms, and long-focus AAF lenses. These lenses demonstrated broadband and subwavelength focusing characteristics through asymmetric phase modulation and acoustic path differences, with notable performance in the ultrasonic frequency range. Both experimental and numerical simulations confirmed their potential applications in focused ultrasound therapy and medical ultrasound imaging.

References

1. Xia JP, Zhang XT, Sun HX, Yuan SQ, Qian J, Ge Y (2018) Broadband tunable acoustic asymmetric focusing lens from dual-layer metasurfaces. Phys Rev Appl 10(1):014016. https://doi.org/10.1103/PhysRevApplied.10.014016
2. Qian J, Wang Y, Xia JP, Ge Y, Yuan SQ, Sun HX, Liu XJ (2020) Broadband integrative acoustic asymmetric focusing lens based on mode-conversion meta-atoms. Appl Phys Lett 116(22):223505. https://doi.org/10.1063/5.0004579
3. Sun HX, Huang YL, Xia JP, Yuan SQ (2017) Asymmetric acoustic convergence in a metal plate with binary wave-path slits. J Phys D: Appl Phys 50(35):35LT02. https://doi.org/10.1088/1361-6463/aa7f7e
4. Wang Y, Xia JP, Sun HX, Yuan SQ, Ge Y, Si QR, Guan YJ, Liu XJ (2019) Multifunctional asymmetric sound manipulations by a passive phased array prism. Phys Rev Appl 12(2):024033. https://doi.org/10.1103/PhysRevApplied.12.024033
5. Sun HX, Hou AR, Qian J, Ge Y, Yuan SQ, Guan YJ, Si QR, Liu XJ (2019) Broadband acoustic converging and asymmetric converging based on thermoacoustic phased arrays. J Appl Phys 125(2):024504. https://doi.org/10.1063/1.5055288

Chapter 9
Three-Dimensional Acoustic Focusing and Asymmetric Focusing by Phase-Controlled Metamaterials

9.1 Introduction

In recent years, the study of AF and AAF based on phase-controlled metamaterials has garnered increasing attention due to its practical applications in various important fields, such as medical ultrasound, nondestructive testing, acoustic communication, and energy harvesting. With the continuous development of phase-controlled metamaterials, it has become feasible to realize several types of high-performance AF and AAF lenses. Typically, these previously demonstrated AF and AAF lenses are 1D systems that achieve AF and AAF effects in 2D space, which poses challenges for practical applications. However, realizing AF and AAF in 3D space using 2D acoustic systems and establishing a theoretical framework for this remains a significant challenge.

In this chapter, we design four types of 2D acoustic lenses using phase-controlled metamaterials and demonstrate AF, AVF, AAF, and AAVF effects in 3D space. First, based on the derived 2D generalized Snell's law, we design four types of 2D reflected AF lenses composed of sixteen types of phased unit cells with a subwavelength thickness of $\lambda/10$ [1]. These lenses achieve point focusing, line focusing in two perpendicular directions, and long-focus focusing in 3D space. Second, we design a reflected long-focus AVF lens composed of sixteen types of phased unit cells [2]. Using the 2D generalized Snell's law, we realize broadband long-focus AVF with the characteristics of acoustic vortex and long-focus focusing in 3D space. The FLHM of the focus and the fractional bandwidth are 9.2λ and 0.28, respectively. Additionally, we discuss the design and performance of the long-focus AVF lens with different topological charges in detail. Third, we propose a 2D AAF lens composed of 44 mode-conversion meta-atoms [3]. The 3D AAF effect arises from the phase distribution of sound focusing and the step waveguide of the meta-atoms. The fractional bandwidth of the AAF lens is 0.14. Furthermore, we study the design and performance of AAF lenses with focal lengths of 10 and 15 cm. Finally, we propose a broadband 2D AAVF lens using three types of mode-conversion meta-atoms [4]. Based on the phase

H. Sun et al., *Acoustic Metamaterials*, Microfluidics and Nanophotonics: Science and Engineering 2, https://doi.org/10.1007/978-981-96-8524-0_9

distributions of focusing and vortex caused by 48 meta-atoms, the 2D AAVF lens with $n = 2$ is observed, which is attributed to the phase control and the step waveguide of the meta-atoms. The fractional bandwidth of the AAVF effect is 0.19. Moreover, by tuning the phase distributions of the lens, we study the design and performance of 2D AAVF lenses with $n = 1$ and 3. The designed 2D lenses can achieve complex sound wavefront manipulations in 3D space, which have potential applications in non-contact particle trapping, medical ultrasound imaging and treatment.

9.2 Three-Dimensional Acoustic Focusing by Reflected Metasurfaces

9.2.1 Design and Performance of Phased Unit Cell

To realize the reflected 2D metasurface with AF based on Eq. (2.50), we design a phased unit cell (Fig. 9.1a) composed of a central straight pipe and two ring-like Helmholtz resonators filled with air and made of epoxy resin [1]. The bottom solid structure serves as a sound-hard boundary. It is noted that the acoustic wave can be significantly delayed after passing through the Helmholtz resonators [5, 6], allowing the reflected phase delays of sound to be flexibly manipulated by tuning the geometrical parameters of the phased unit cell. The cross-sectional sizes of the x–z and y–z planes of the phased unit cells are identical, as shown in Fig. 9.1b. The parameters are selected as $w = 0.1\lambda$, $h = 0.1\lambda$, $d = 0.05\lambda$, and $h_1 = 0.15\lambda$, and the cavity width w_2 is determined by the pipe width w_1. The incident wavelength λ is selected as 10 cm (i.e., $f_0 = 3.43$ kHz in air). The COMSOL Multiphysics software is used to simulate the characteristics of a series of 3D AF effects, with the following material parameters: density $\rho = 1180$ kg/m^3 and sound velocity $c = 2720$ m/s for epoxy resin; $\rho = 1.21$ kg/m^3 and $c = 343$ m/s for air.

Figure 9.2 shows the reflected phase delays of sound caused by the phased unit cells with different values of w_1/w. In the range of $0.1 < w_1/w < 0.8$, the reflected phase delays cover the entire 2π range. Here, we select sixteen values of w_1/w (the red open circles) to design sixteen phased unit cells with equally spaced phase delays from 0 to 2π.

Figure 9.3 shows the reflected pressure distributions caused by the selected sixteen phased unit cells, where the incident acoustic waves are located on the upper side. It is evident that the reflected discrete phase delays caused by the selected sixteen phased unit cells increase gradually from 0 to 2π in steps of $\pi/8$. Due to the identical sizes in both the x and y directions and the deep subwavelength thickness ($\lambda/10$) in the z direction for the phased unit cells, the proposed metasurface structure is an excellent candidate for fabricating 3D acoustic devices with the advantage of easy integration.

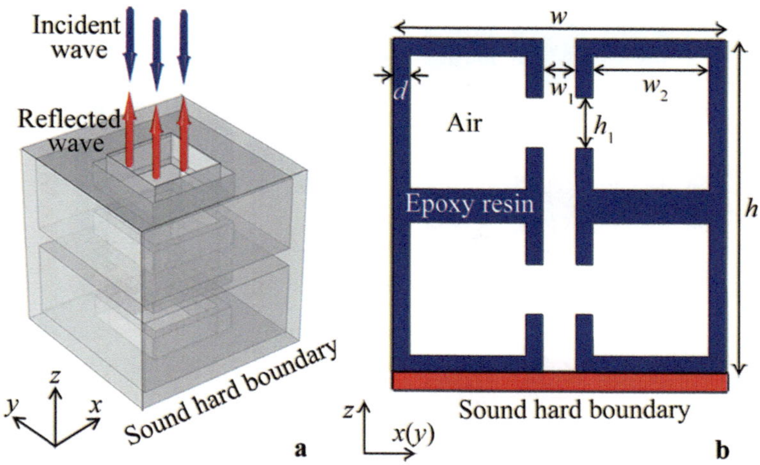

Fig. 9.1 **a** Schematic of the phased unit cell and **b** the cross-section of $x(y)$–z

Fig. 9.2 Simulated reflected phase delays caused by the phased unit cells with different values of w_1/w

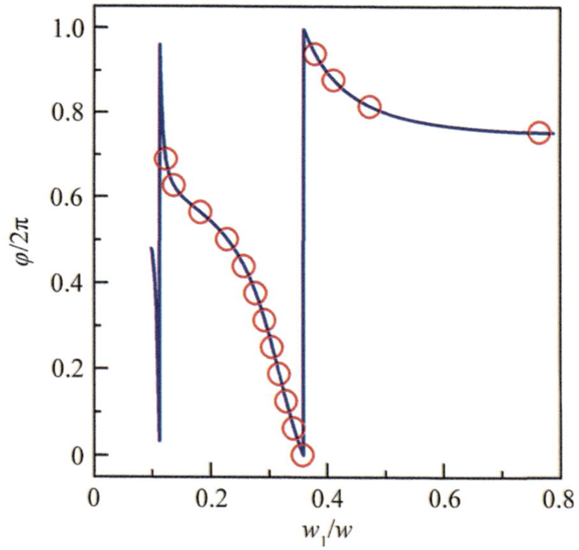

9.2.2 *Design and Performance of Reflected Point AF Lens*

Based on the selected sixteen unit cells and the 2D generalized Snell's law, we design four types of reflected metasurfaces with different AF effects, such as point focusing, line focusing in the x and y directions, and long-focus focusing. Figure 9.4 schematically shows the reflected point-focusing lens [1], where the focus is located at the position $(0, 0, e)$, and e represents the focal length. Thus, the reflected angles θ_r and β_r can be expressed as:

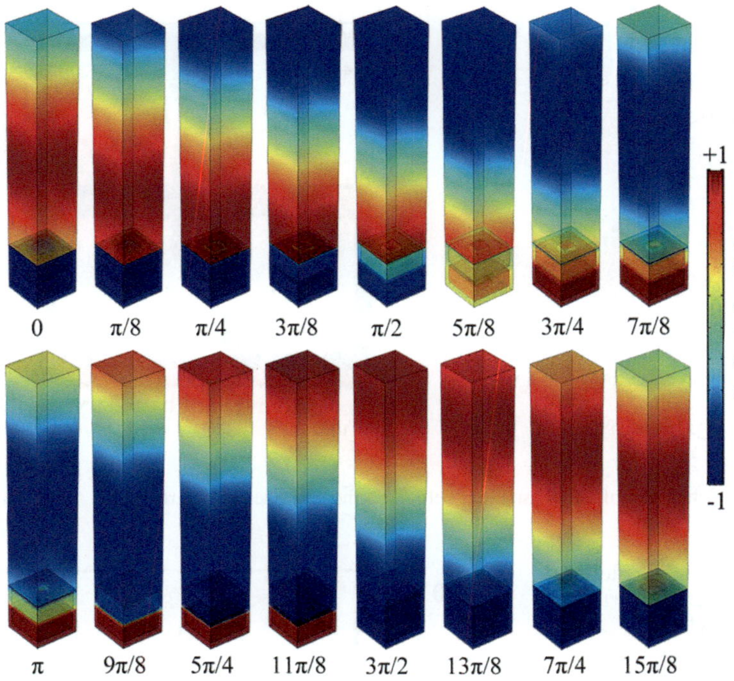

Fig. 9.3 Simulated reflected pressure distributions caused by the selected sixteen quasi-3D unit cells with different values of w_1/w at 3.43 kHz

$$\sin\theta_r = \frac{\sqrt{x^2+y^2}}{\sqrt{x^2+y^2+e^2}}, \tag{9.1}$$

$$\cos\beta_r = -\frac{x}{\sqrt{x^2+y^2}}, \tag{9.2}$$

$$\sin\beta_r = -\frac{y}{\sqrt{x^2+y^2}}. \tag{9.3}$$

Based on Eqs. (9.1)–(9.3), the theoretical phase distribution of the reflected point focusing lens is expressed as:

$$\varphi = -k\sqrt{x^2+y^2+e^2}. \tag{9.4}$$

Here, the focal length $e = 10$ cm, and the size of the metasurface in the x–y plane is 40 cm × 40 cm. Figure 9.5a, b show the theoretical continuous and discrete phase distributions of the reflected point-focusing lens, respectively.

Based on the discrete phase distribution in Fig. 9.5b, we design the reflected point-focusing lens and simulate the reflected 3D intensity distributions caused by the theoretical continuous phase distribution and the reflected point-focusing lens, as

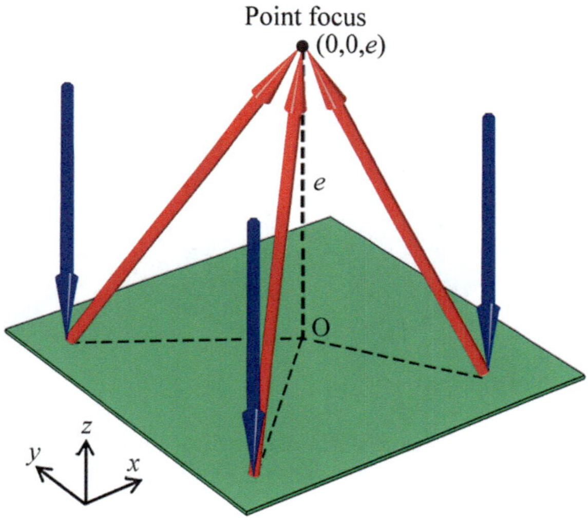

Fig. 9.4 Schematic of the reflected metasurface with point focusing

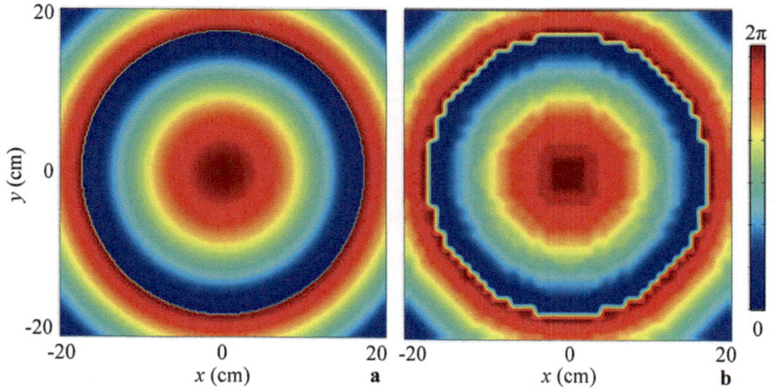

Fig. 9.5 **a** Theoretical continuous and **b** discrete phase distributions of the reflected point-focusing lens

shown in Fig. 9.6a, b, respectively. It is evident that the point-focusing effect caused by the reflected lens agrees well with the theoretical result, and the center position of the focus for both cases is located at (0, 0, 9.5 cm). Therefore, the focal length of the reflected point-focusing lens is 9.5 cm, which is close to the theoretical value of 10 cm. The slight difference in focal length between the two results arises from the limited size of the metasurface.

To quantify the AF performance of the reflected point-focusing lens, we simulate the reflected intensity distributions through the focus in the x and z directions (lines I and II in Fig. 9.6), as shown in Fig. 9.7a, b, respectively. The intensities in Fig. 9.7 are normalized by the maximum intensity of each figure. It is clear that the intensity

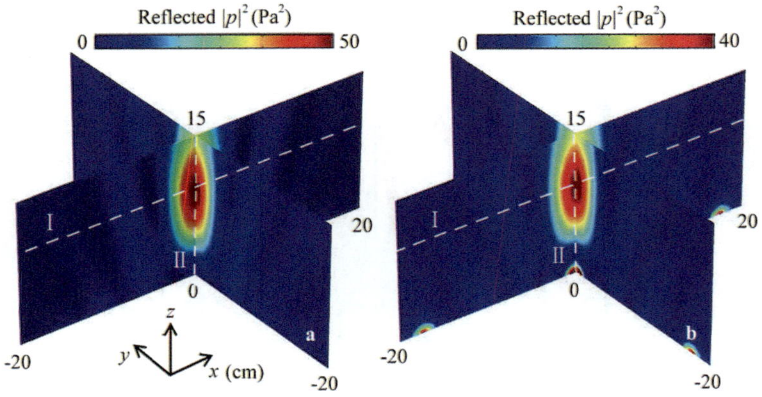

Fig. 9.6 Simulated reflected 3D intensity distributions caused by **a** the theoretical continuous phase distribution and **b** the reflected point focusing lens at 3.43 kHz

distributions agree well for both cases. Additionally, as shown in Fig. 9.7b, the focal length is 9.5 cm, which is close to the theoretical value of 10 cm. Thus, the reflected lens exhibits high performance of point focusing in 3D space.

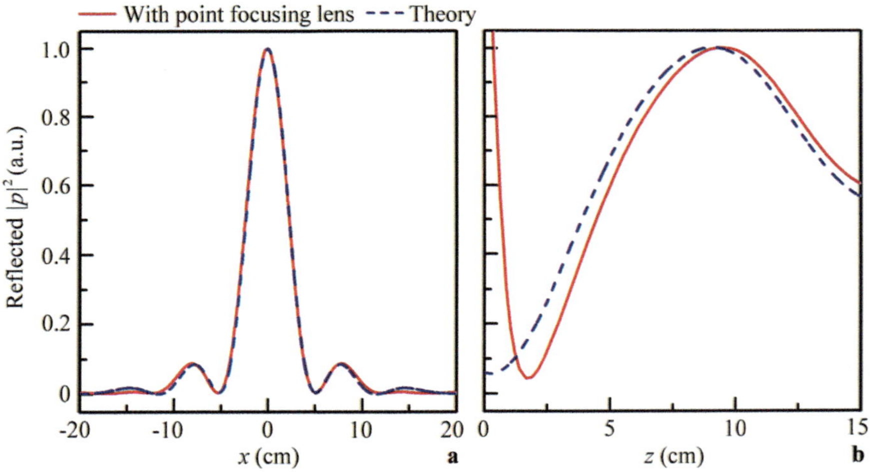

Fig. 9.7 Simulated intensity distributions along lines **a** I and **b** II in Fig. 9.6

9.2.3 Design and Performance of Reflected Line AF Lens

In addition to the reflected point focusing, we design two types of reflected line focusing lenses in the x and y directions [1]. Figure 9.8 presents the reflected metasurface with line focusing along $x = 0$, where the parameters $\sin\theta_r$, $\cos\beta_r$ and $\sin\beta_r$ in Eq. (2.50) are expressed as:

$$\sin\theta_r = \frac{|x|}{\sqrt{x^2 + e^2}}, \tag{9.5}$$

$$\cos\beta_r = -sign(x), \tag{9.6}$$

$$\sin\beta_r = 0. \tag{9.7}$$

Based on Eqs. (2.50) and (9.5)–(9.7), the theoretical phase distribution of the reflected line-focusing lens satisfies:

$$\varphi = -k\sqrt{x^2 + e^2}. \tag{9.8}$$

Here, the focal length e and the size of the metasurface are the same as those in Fig. 9.4, and the theoretical continuous and discrete phase distributions of the reflected line-focusing lens are displayed in Fig. 9.9a, b, respectively. It is evident that the phase distribution is independent of the parameter y, which agrees with Eq. (9.8).

Figure 9.10a, b show the reflected 3D intensity distributions caused by the theoretical continuous phase distribution and the reflected line-focusing lens, respectively. Note that the reflected sound energy is primarily focused on the line $x = 0$ and is very weak in other positions. The characteristics of the reflected line-focusing lens (Fig. 9.10b) agree well with the theoretical results (Fig. 9.10a). Additionally, the

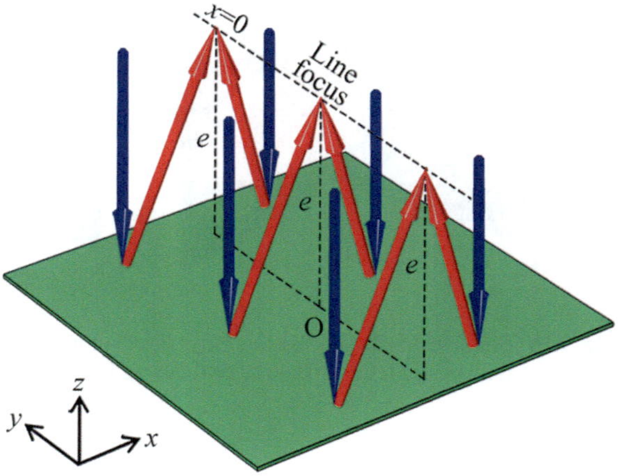

Fig. 9.8 Schematic of the reflected metasurface with line focusing along $x = 0$

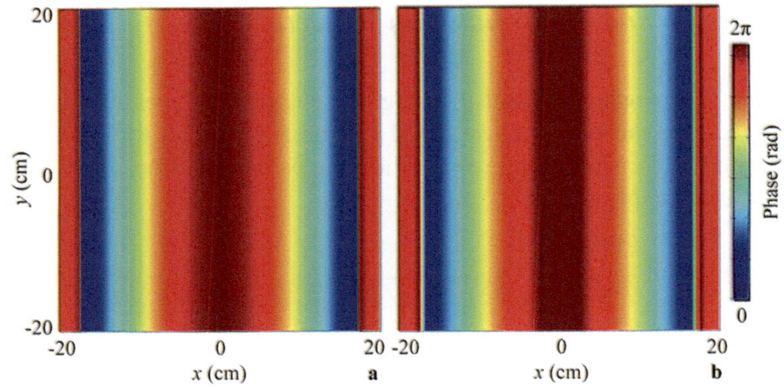

Fig. 9.9 **a** Theoretical continuous and **b** discrete phase distributions of the reflected line focusing lens along $x = 0$

reflected intensity distributions through the focus in the x, y, and z directions (lines III, IV, and V in Fig. 9.10a, b) are presented in Fig. 9.11. By comparing the intensity distributions in the x and y directions, it is clear that the reflected sound energy is mainly focused on the line $x = 0$ (Fig. 9.11a), and the intensity is greater than 0.8 in the range from $y = -10$ to 10 cm (Fig. 9.11b), indicating that the reflected sound energy is focused along the line $x = 0$. Moreover, as shown in Fig. 9.11c, the focal length is close to the theoretical value of 10 cm, further demonstrating the performance of the reflected line focusing in 3D space.

Next, we design a reflected metasurface with line focusing along $y = 0$, as shown in Fig. 9.12. In this case, $\sin\theta_r$, $\sin\beta_r$, and $\cos\beta_r$ are expressed as:

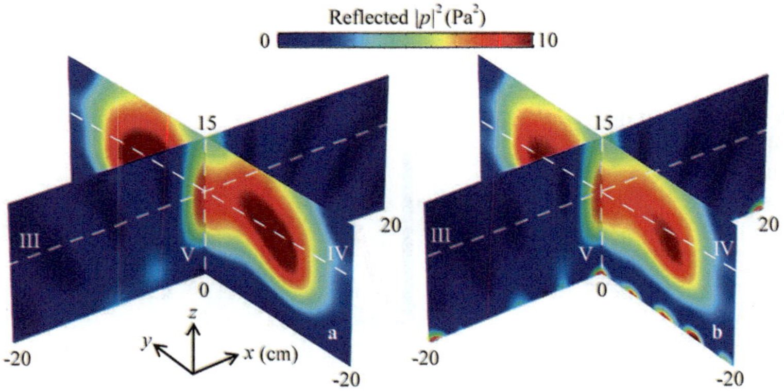

Fig. 9.10 Simulated reflected 3D intensity distributions caused by **a** the theoretical continuous phase distribution and **b** the reflected line focusing lens along $x = 0$ at 3.43 kHz

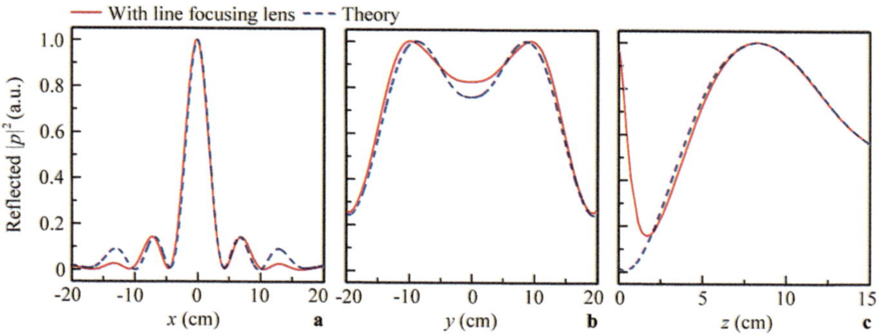

Fig. 9.11 Simulated intensity distributions along lines **a** III, **b** IV and **c** V in Fig. 9.10

$$\sin\theta_r = \frac{|y|}{\sqrt{y^2+e^2}}, \tag{9.9}$$

$$\cos\beta_r = 0, \tag{9.10}$$

$$\sin\beta_r = -sign(y). \tag{9.11}$$

Based on Eqs. (2.50) and (9.9)–(9.11), the theoretical phase distribution of the reflected line-focusing lens along $y = 0$ can be written as:

$$\varphi = -k\sqrt{y^2+e^2}. \tag{9.12}$$

Based on Eq. (9.12), we calculate the theoretical continuous and discrete phase distributions of the reflected line AF lens, as shown in Fig. 9.13a, b, respectively. Figure 9.14a, b show the reflected intensity distributions caused by the theoretical

Fig. 9.12 Schematic of the reflected metasurface with line focusing along $y = 0$

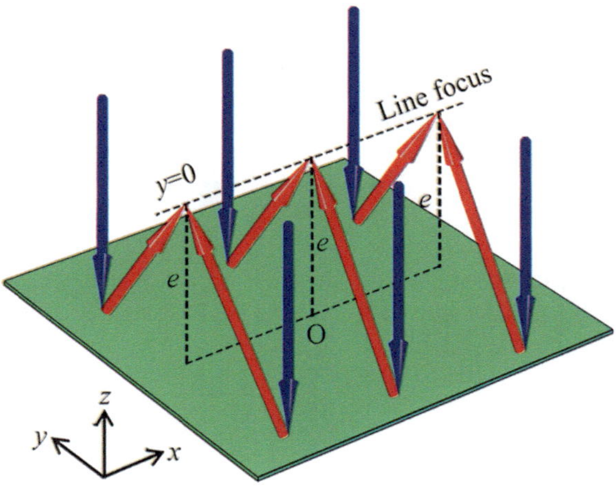

continuous phase distribution and the reflected line-focusing lens, respectively, where the line focusing occurs along $y = 0$. The reflected intensity distributions through the focus along the x, y, and z directions (lines VI, VII, and VIII in Fig. 9.14) are shown in Fig. 9.15a–c, respectively. It is evident that the intensity distributions for both cases agree with each other. Additionally, the intensity distributions in the x and y directions (Fig. 9.15a, b) are opposite to those in Fig. 9.11a, b, while the result in Fig. 9.15c is almost identical to that in Fig. 9.11c, demonstrating the characteristics of the reflected line focusing along $y = 0$.

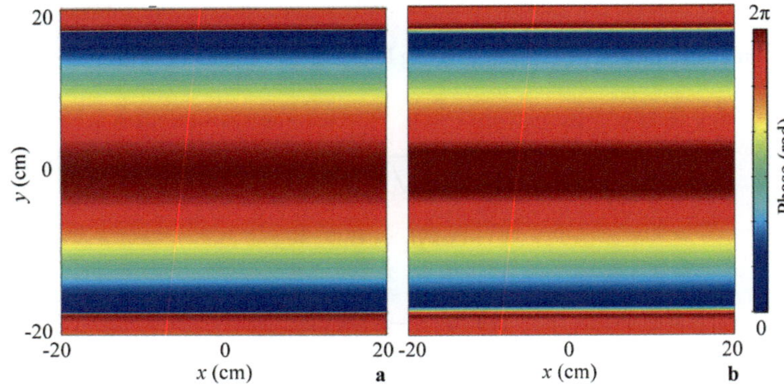

Fig. 9.13 **a** Theoretical continuous and **b** discrete phase distributions of the reflected line-focusing lens along $y = 0$

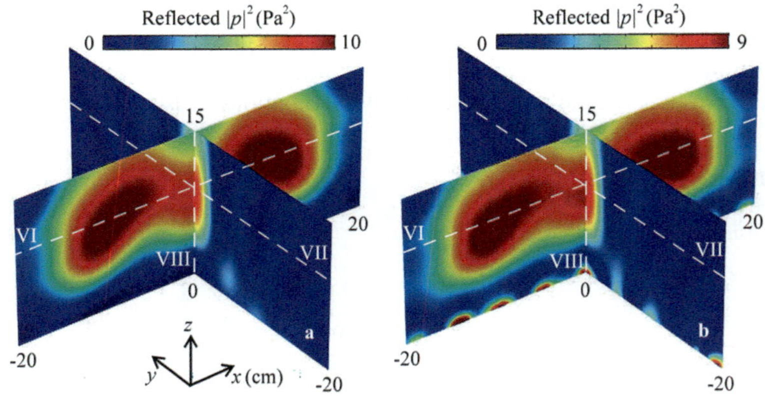

Fig. 9.14 Simulated reflected 3D intensity distributions caused by **a** the theoretical continuous phase distribution and **b** the reflected line focusing lens along $y = 0$ at 3.43 kHz

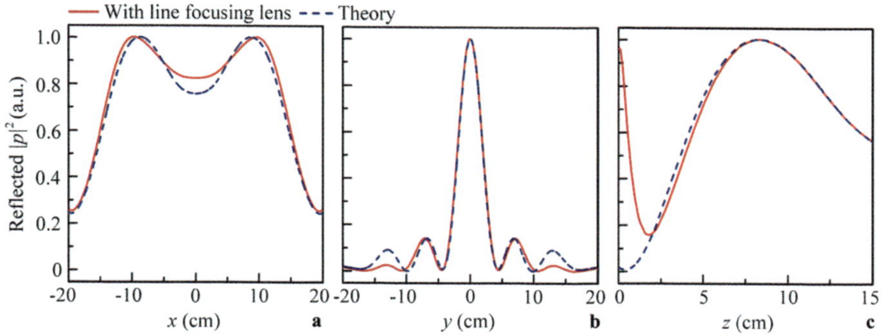

Fig. 9.15 Simulated intensity distributions along lines **a** VI, **b** VII, and **c** VIII in Fig. 9.14

9.2.4 Design and Performance of Reflected Long-Focus AF Lens

Finally, we design a reflected metasurface with a Bessel beam [1], which exhibits the characteristic of long-focus focusing. As shown in Fig. 9.16, the parameter α is the base angle of the Bessel beam. Using Eq. (2.50), the theoretical phase distribution of the reflected long-focus focusing lens in the x–y plane is expressed as:

$$\varphi = -k \sin\alpha\sqrt{x^2 + y^2}. \tag{9.13}$$

Here, the base angle $\alpha = 16°$, and the size of the metasurface is the same as that in Fig. 9.4. Figure 9.17a, b present the theoretical continuous and discrete phase distributions of the reflected long-focus focusing lens, respectively. Figure 9.18a,

Fig. 9.16 Schematic of the reflected metasurface with a non-diffracting Bessel beam

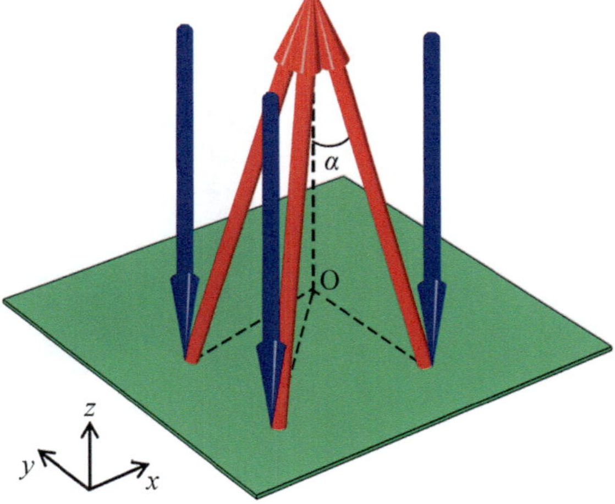

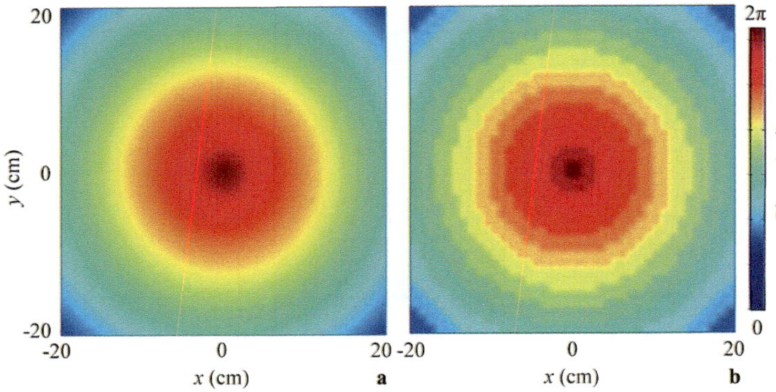

Fig. 9.17 **a** Theoretical continuous and **b** discrete phase distributions of the reflected long-focus focusing lens

b show the reflected 3D intensity distributions caused by the theoretical continuous phase distribution and the reflected long-focus focusing lens, respectively. It is evident that a non-diffracting Bessel beam is formed by the reflected long-focus focusing lens (Fig. 9.18b), and the sound energy propagates along the z direction with a long focus, indicating that the reflected lens exhibits high performance in long-focus AF. Moreover, the reflected intensity distributions through the long focus in the x and z directions (lines IX and X in Fig. 9.18) are displayed in Fig. 9.19. Both types of results are nearly identical. Additionally, the intensity is high at the long focus in the z direction, indicating that high performance of the long-focus AF effect in 3D space has been achieved.

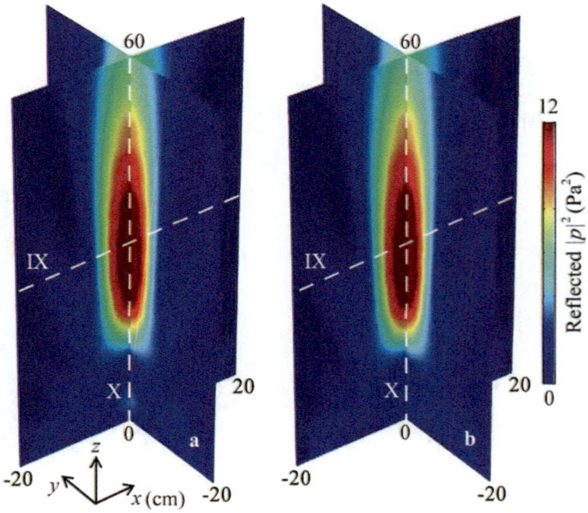

Fig. 9.18 Simulated reflected 3D intensity distributions caused by **a** the theoretical continuous phase distribution and **b** the reflected long-focus focusing lens at 3.43 kHz

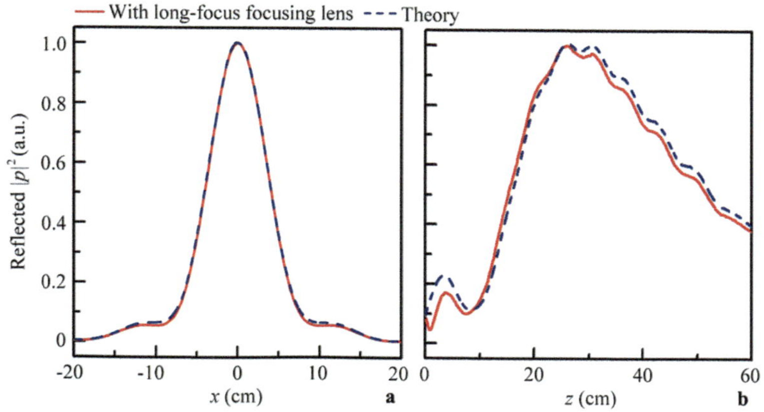

Fig. 9.19 Simulated intensity distributions along lines **a** IX and **b** X in Fig. 9.18

9.3 Three-Dimensional Long-Focus Acoustic Vortex Focusing by Reflected Metasurfaces

9.3.1 Design and Performance of Phased Unit Cell

As schematically shown in Fig. 9.20, we propose a reflected phased unit cell composed of a cuboid solid with a groove on the upper side. The cross-sections of the x–z and y–z planes of the phased unit cell are identical, and the structural parameters are $w = 1.0$ cm, $h = 2.5$ cm, and $d = 0.1$ cm. The groove depth h_0 is tunable, allowing for the modulation of reflected phase delays of sound. The incident wavelength λ of sound is selected as 3.8 cm (i.e., $f = 9.0$ kHz in air). Throughout this work, the characteristics of long-focus AVF are simulated using the COMSOL Multiphysics software, with the following material parameters: density $\rho = 1180$ kg/m^3, longitudinal wave velocity $c_l = 2720$ m/s, and transversal wave velocity $c_t = 1460$ m/s for epoxy resin; $\rho = 1.21$ kg/m^3 and sound velocity $c = 343$ m/s for air.

Figure 9.21 shows the reflected phase delays caused by the phased unit cells with different values of h_0. It is evident that, in the range of $0.1 < h_0 < 2.4$ cm, the theoretical reflected phase delays cover the entire 2π range. Here, we select sixteen different values of h_0 (the red open circles) to achieve equally spaced phase delays from 0 to 2π in steps of $\pi/8$.

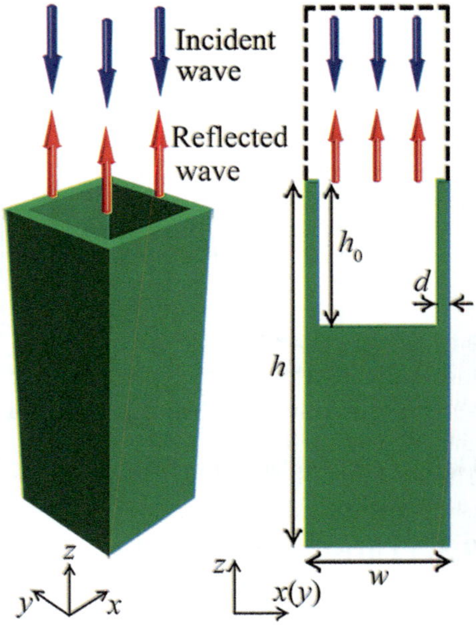

Fig. 9.20 Schematic of the phased unit cell and its cross-section of $x(y)$–z

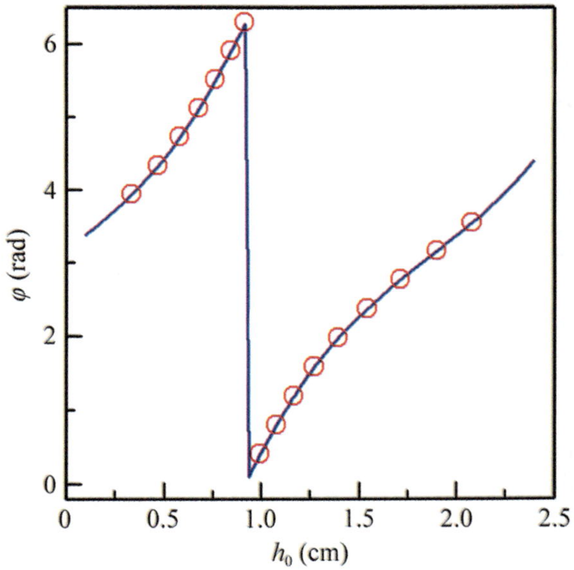

Fig. 9.21 Reflected phase delays (the blue solid line) caused by the phased unit cells with different values of h_0. The red open circles represent the sixteen phase delays from 0 to 2π in steps of $\pi/8$

9.3.2 Design and Performance of Reflected Long-Focus AVF Lens

Based on the selected sixteen phased unit cells, we design a planar reflected metasurface with long-focus AVF. Figure 9.22 shows the normal incidence of an acoustic wave on the reflected metasurface, where the sound reflections form a long vertex focus. To design this, we introduce the phase distribution of a vortex beam in free space, which is expressed as:

$$\phi_s(\theta) = n\theta, \tag{9.14}$$

where n is the topological charge of the vortex beam [7], and θ is the azimuth angle, ranging from 0 to 2π in the x–y plane. To simultaneously achieve acoustic vortex and long-focus focusing in free space, we introduce the phase distribution of a non-diffracting Bessel beam. Based on Eqs. (2.50) and (9.14), the theoretical phase distribution of the reflected long-focus AVF lens in the x–y plane is written as:

$$\phi = -k \sin\alpha\sqrt{x^2 + y^2} + n\theta, \tag{9.15}$$

where $k = 2\pi f/c$ is the wave number in air, and c and f are the sound velocity and frequency, respectively. α is the base angle of the Bessel beam, and the parameter θ satisfies $\tan\theta = x/y$. Here, the parameters α and n are selected as 15° and 1, respectively, and the size of the reflected lens is 40 × 40 cm^2 in the x–y plane. Figure 9.23a, b show the theoretical continuous and discrete phase distributions of the reflected long-focus AVF lens in the x–y plane, respectively. It is worth noting

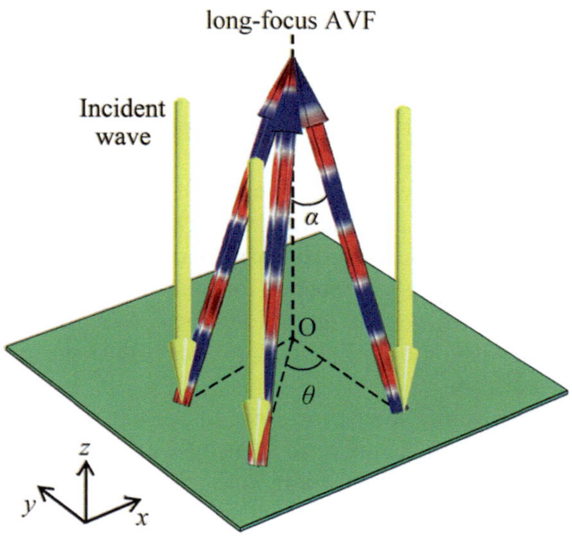

Fig. 9.22 Schematic of the reflected metasurface with long-focus AVF

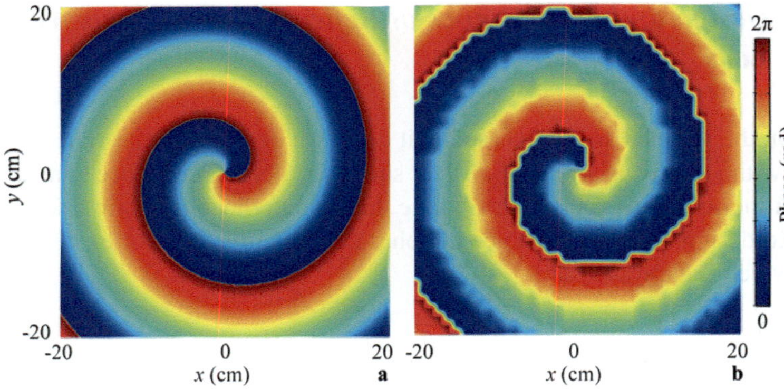

Fig. 9.23 **a** Theoretical continuous and **b** discrete phase distributions of the reflected long-focus AVF lens with $n = 1$

that the phase distribution exhibits the typical characteristics of a vortex beam with $n = 1$, and both types of phase distributions agree with each other.

Based on the two types of phase distributions, we design the reflected long-focus AVF lens composed of the sixteen types of phased unit cells. Figure 9.24a shows the 3D reflected intensity distribution caused by the reflected long-focus AVF lens. It is evident that the sound reflection is concentrated into a long hollow focus in the z direction, covering the range of 28–72 cm ($|p|^2 > 5.0$ Pa^2). On the other hand, the sound energy at the center of the long focus is almost zero, showing the typical characteristics of both Bessel and vortex beams.

To further illustrate the characteristics of the long-focus AVF, we display the reflected intensity and phase distributions of the long focus at five selected cross-sections of $z = 30$, 40, 50, 60, and 70 cm, as shown in Fig. 9.24b, c, respectively. It is evident that, as z increases, a concentric ring-shaped region appears at the center of each cross-section (Fig. 9.24b), consistent with Fig. 9.24a. Additionally, the sound energy in the ring-shaped region ($|p|^2 > 6.0$ Pa^2) is much larger than that at the center of the five cross-sections ($|p|^2 \approx 0$) due to the existence of a phase singularity (Fig. 9.24c). The phase distributions in Fig. 9.24c exhibit the typical characteristics of a vortex beam with $n = 1$, further verifying the characteristics of long focus and acoustic vortex in 3D free space. In addition to the aforementioned results, we simulate the intensity and phase distributions caused by the theoretical continuous phase distribution for comparison, as shown in Fig. 9.24d–f. It is evident that the characteristics of the long-focus AVF match well with those in Fig. 9.24a–c, further demonstrating the performance of the long-focus AVF caused by the reflected metasurface.

To quantify the performance of the long-focus AVF, we display the simulated longitudinal and transverse intensity distributions through the long focus along lines I–III in Fig. 9.24a, as shown in Fig. 9.25. It is evident that two sound energy peaks ($|p|^2 \approx 12.0$ Pa^2) exist on both sides of $x = 0$, while the sound intensity at $x = 0$ is close to zero (Fig. 9.25a). Additionally, as shown in Fig. 9.25b, the intensities along

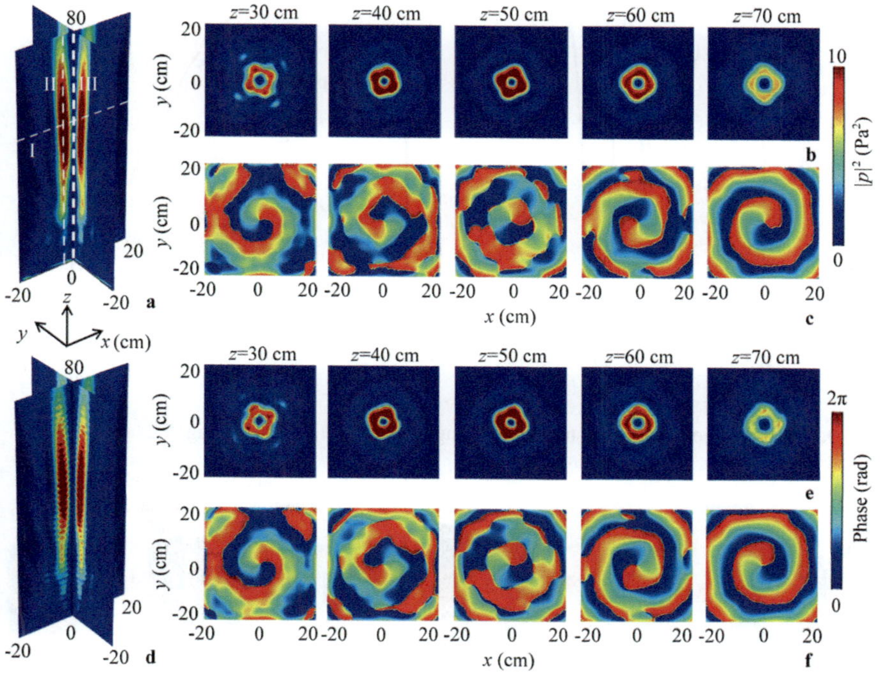

Fig. 9.24 **a** Simulated 3D reflected intensity distribution caused by the reflected long-focus AVF lens at 9.0 kHz, and its corresponding reflected **b** intensity and **c** phase distributions at the five selected cross-sections of $z = 30$ cm, 40 cm, 50 cm, 60 cm, and 70 cm. **d** Simulated 3D reflected intensity distribution caused by the theoretical continuous phase distribution, and its corresponding reflected **e** intensity and **f** phase distributions at the five cross-sections

line II are much larger than those along line III ($|p|^2 \approx 0$), which arises from the existence of the phase singularity in line III. Moreover, the FLHM of the focus along line II reaches 35 cm (9.2λ), showing the typical characteristic of long-focus AVF. Therefore, we further verify high performance of both long-focus AF and vortex in 3D space caused by the reflected metasurface.

Next, we discuss the working bandwidth of the reflected long-focus AVF lens. Figure 9.26 shows the intensity and phase distributions caused by the long-focus AVF lens at different frequencies, with other parameters remaining unchanged. It is noted that, as the frequency increases, the position of the long focus gradually moves along the $+z$ direction. However, the characteristics of the long-focus AVF remain the same, consistent with those in Fig. 9.24. Moreover, we simulate the intensity and phase distributions caused by the long-focus AVF lens at both edge frequencies (8.0 and 11.0 kHz), as shown in Fig. 9.27. It is evident that the field distributions exhibit clear characteristics of reflected long-focus AVF at both edge frequencies. Therefore, the working bandwidth of the long-focus AVF lens reaches 3.0 kHz, and its fractional bandwidth is 0.28, demonstrating a broadband characteristic.

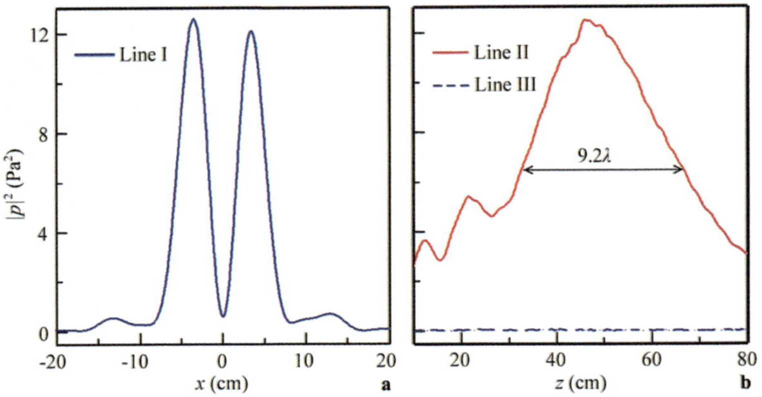

Fig. 9.25 Simulated intensity distributions along lines **a** I, **b** II and III in Fig. 9.24a

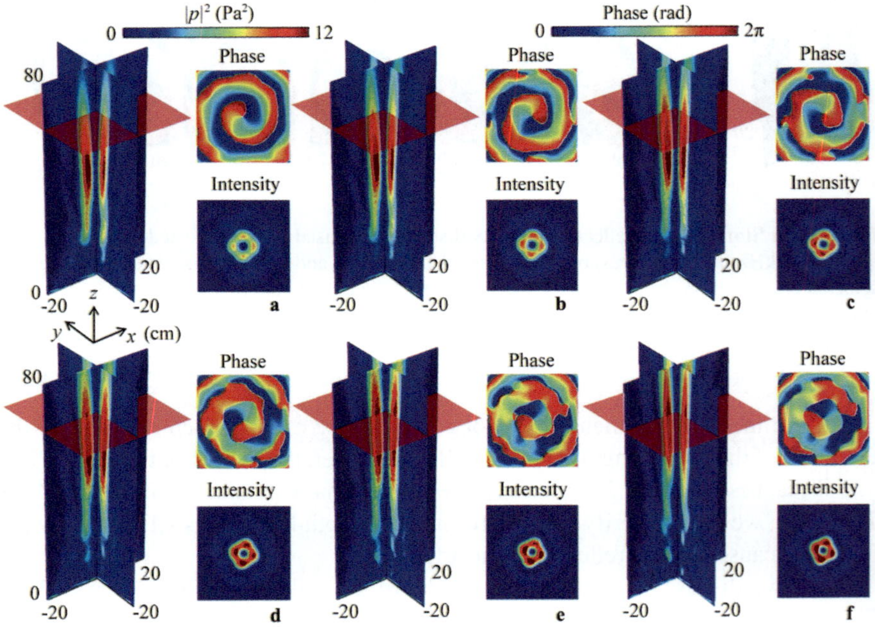

Fig. 9.26 Simulated 3D reflected intensity distributions caused by the reflected long-focus AVF lens at **a** 8.2 kHz, **b** 8.7 kHz, **c** 9.2 kHz, **d** 9.7 kHz, **e** 10.2 kHz, and **f** 10.7 kHz. Insets on the right side represent the phase and intensity distributions at the cross-section of $z = 60$ cm (the red translucent planes)

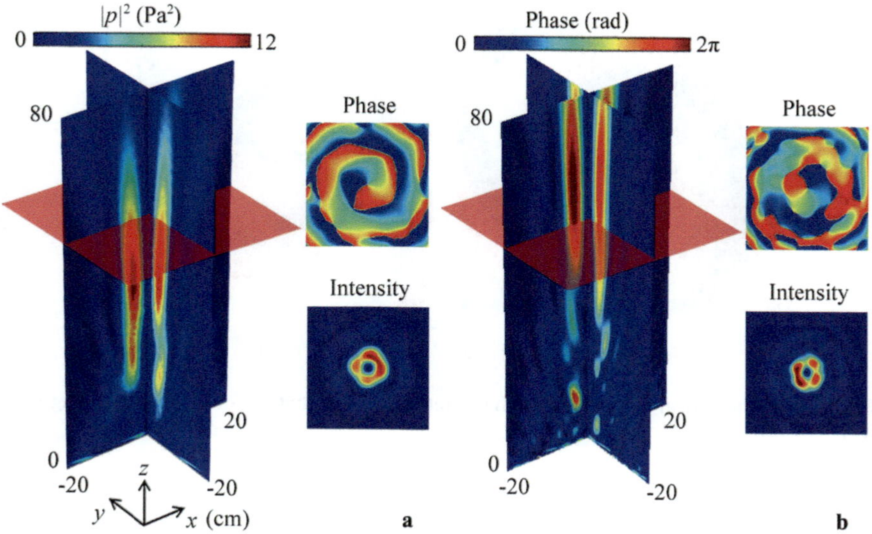

Fig. 9.27 Simulated 3D reflected intensity distributions caused by the reflected long-focus AVF lens at **a** 8.0 kHz and **b** 11.0 kHz. Insets on the right side represent the phase and intensity distributions at the cross-section of $z = 55$ cm (the red translucent planes)

9.3.3 Design and Performance of Reflected Long-Focus AVF Lens with n = *2*

Furthermore, based on Eq. (9.15), we design a reflected long-focus AVF lens with $n = 2$. The theoretical continuous and discrete phase distributions are shown in Fig. 9.28a, b, respectively. Figure 9.29 shows the simulated intensity and phase distributions caused by the reflected long-focus AVF lens with $n = 2$. It is evident that the sound reflections are concentrated into a long hollow focus with a height of 9λ, and the characteristics of the long-focus AVF are similar to those of the reflected lens with $n = 1$ in Fig. 9.24. However, as shown in Fig. 9.29b, the phase delay changes from 0 to 2π twice around a circle at the three cross-sections of $z = 20$, 40, and 60 cm, which differs from that in Fig. 9.24. This phenomenon is also clearly displayed in Fig. 9.30.

9.3.4 Design and Performance of Underwater Reflected Long-Focus AVF Lens

Finally, we design a type of reflected long-focus AVF lens immersed in water. Figure 9.31 schematically shows the $x(y)$–z cross-section of the reflected phased unit cell, in which the parameters are the same as those in Fig. 9.20. The material

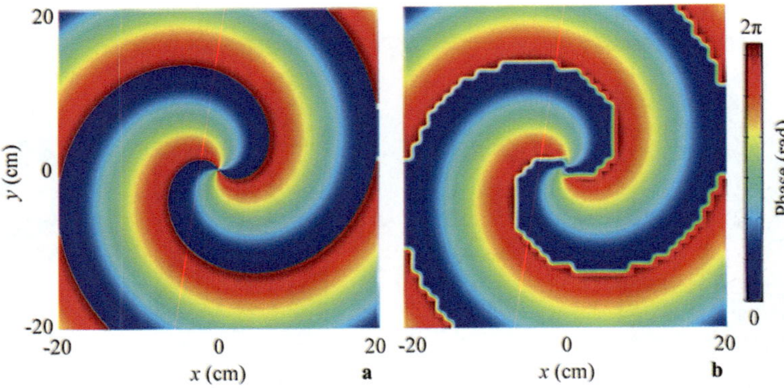

Fig. 9.28 **a** Theoretical continuous and **b** discrete phase distributions of the reflected long-focus AVF lens with $n = 2$

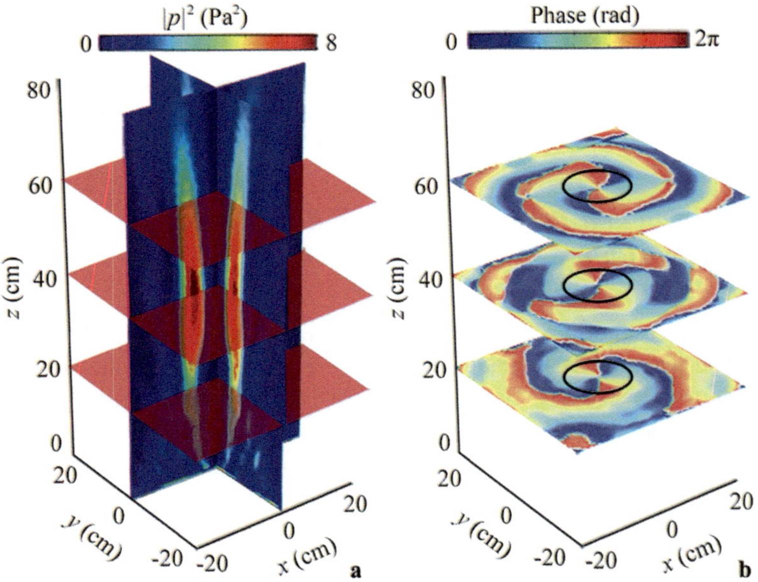

Fig. 9.29 **a** Simulated 3D reflected intensity distribution caused by the reflected long-focus AVF lens with $n = 2$ at 9.0 kHz, and its corresponding reflected **b** phase distribution at the cross-sections of $z = 20$ cm, 40 cm, and 60 cm

parameters used in the simulations are density $\rho = 1000$ kg/m^3 and sound velocity $c = 1490$ m/s for water; and $\rho = 8800$ kg/m^3, Young's modulus $E = 108.42$ GPa, and Poisson's ratio $\nu = 0.33$ for brass. Figure 9.32 shows the reflected phase delays of sound caused by the phased unit cells with different values of h_0, and its theoretical reflected phase delays can cover the whole 2π range in the range of $1.0 < h_0 <$

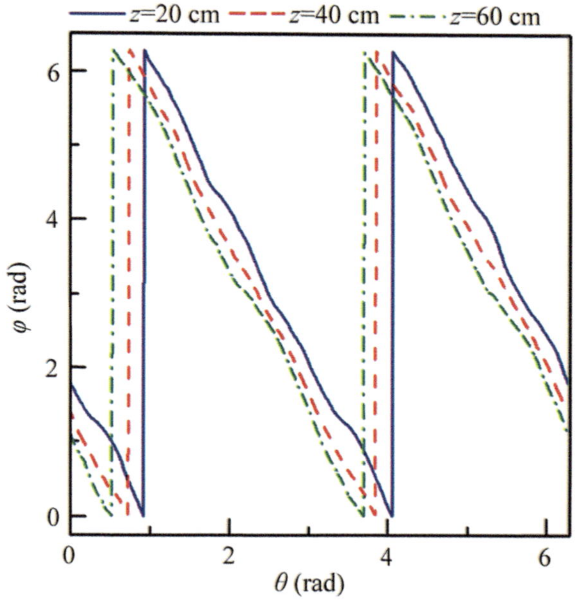

Fig. 9.30 Simulated phase delays around in a circle with a radius of 6 cm (the black open circles in Fig. 9.29b) with different values of θ at the three cross-sections

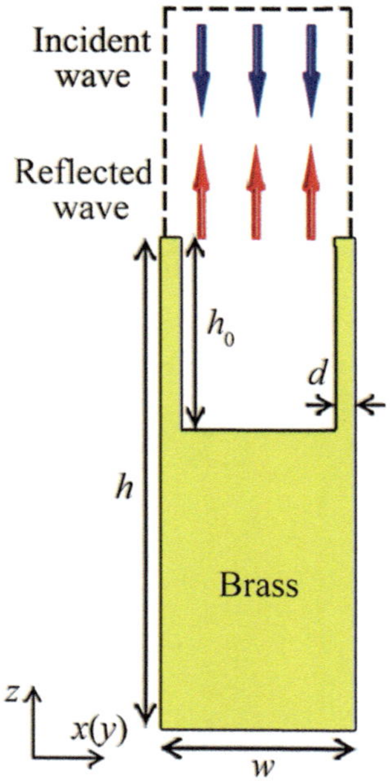

Fig. 9.31 Schematic of the $x(y)$–z cross-section of a phased unit cell made of brass immersed in water

2.0 cm. Here, we select sixteen discrete values of h_0 (the red open circles) to realize an equally spaced phase delay from 0 to 2π with a step of $\pi/8$.

Fig. 9.32 Reflected phase delays caused by the phased unit cells with different values of h_0 at 30 kHz

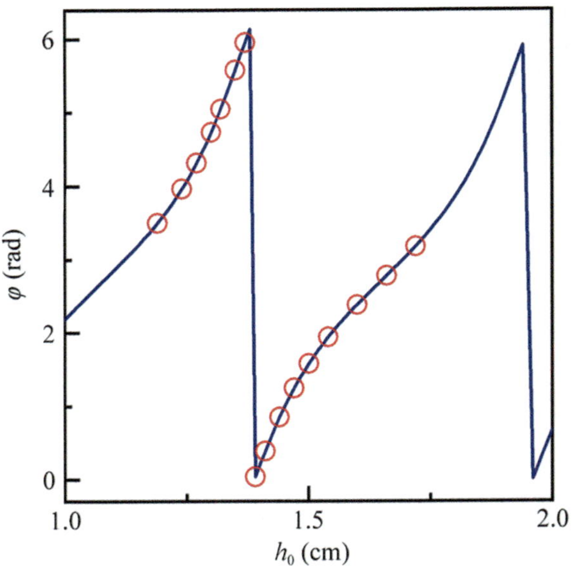

Fig. 9.33 Schematic of the reflected metasurface with long-focus AVF immersed in water

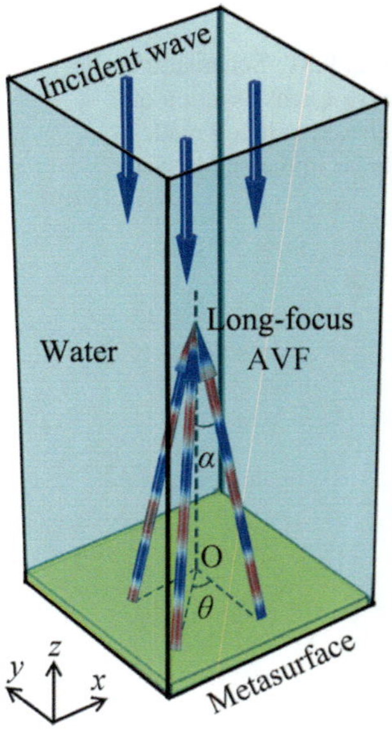

Figure 9.33 schematically shows the reflected metasurface with long-focus AVF immersed in water. Here, the size of the metasurface is the same as that in Fig. 9.22, and its theoretical continuous and discrete phase distributions in the x–y plane are shown in Fig. 9.34a, b, respectively.

Figure 9.35 shows the simulated intensity and phase distributions caused by the reflected long-focus AVF lens immersed in water. We can see that there exist typical

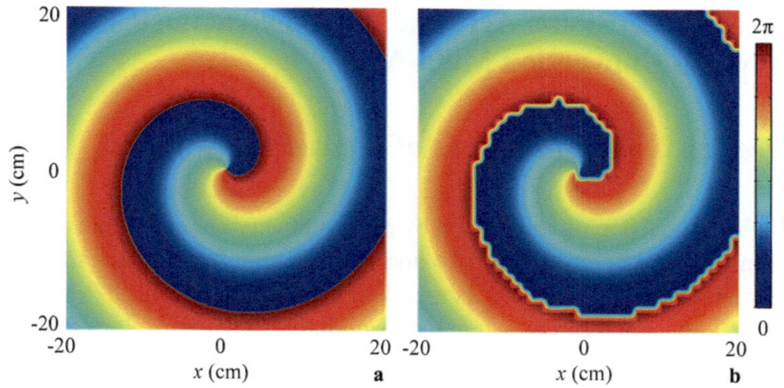

Fig. 9.34 **a** Theoretical continuous and **b** discrete phase distributions of the reflected long-focus AVF lens at 30 kHz

characteristics of the acoustic vortex and the long focus in the range of 30–120 cm ($|p|^2 > 2.5$ Pa^2). The reflected long-focus AVF lens is important for realizing advanced in-vivo acoustic tweezers as it can penetrate deeper to the lesion location, which can be potentially applied to broken bones removing, thrombus ablation, and nerve stimulation.

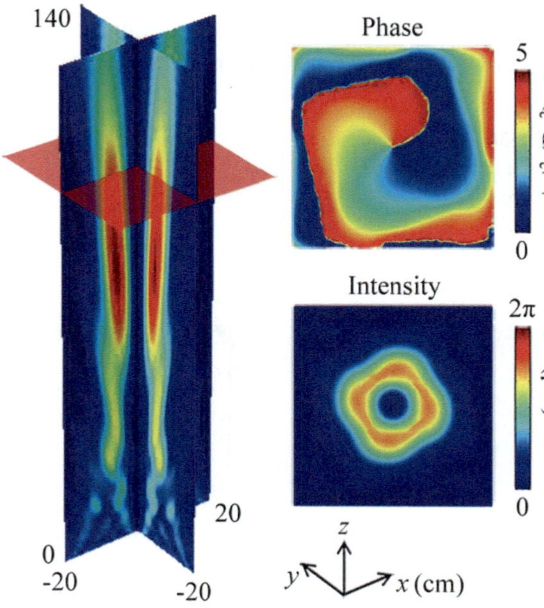

Fig. 9.35 Simulated 3D reflected intensity distributions caused by the reflected long-focus AVF lens in water at 30 kHz. Insets on the right side represent the phase and intensity distributions at the cross-section of $z = 110$ cm (the red translucent plane)

9.4 Three-Dimensional Acoustic Asymmetric Focusing by Mode-Conversion Meta-Atoms

9.4.1 Design and Performance of Phased Unit Cell

To realize the 2D AAF lens constructed by mode-conversion meta-atoms, we design a phased unit cell [3] composed of two arrays of Helmholtz resonators (Fig. 9.36). The parameters are selected as $h = 3$ cm, $w = 2$ cm, $h_2 = 0.55$ cm, $d_1 = 0.1$ cm, and $d_2 = 0.2$ cm, with w_1 and h_1 being tunable to modulate the phase delay of sound. The sound wavelength λ is selected as 6 cm (i.e., $f = 5716$ Hz in air). Throughout this work, the propagation characteristics of sound are simulated using the COMSOL Multiphysics software. In the simulations, the 2D AAF lens is made of epoxy resin to satisfy the sound-hard boundary condition, and the material parameters of air are selected as $\rho = 1.21$ kg/m^3 and $c = 343$ m/s.

Figure 9.37a, b show the simulated phase delays and transmittances caused by the phased unit cells with different values of w_1 and h_1, respectively. It is evident that, when both parameters w_1 and h_1 are in the range of 0.2–0.5 cm, the phase delay of sound can cover the entire 2π range, and most of the transmittances exceed 0.8. Here, we select two phased unit cells (the black open circles labeled "I" and "II") to design the mode-conversion meta-atom. The phase difference between the two unit cells is π, and the corresponding transmittances are 0.96 and 0.99. Figure 9.38 shows the total pressure distributions through the phased unit cells I and II excited by zero-order waves with the same initial phases. The transmitted pressure distributions

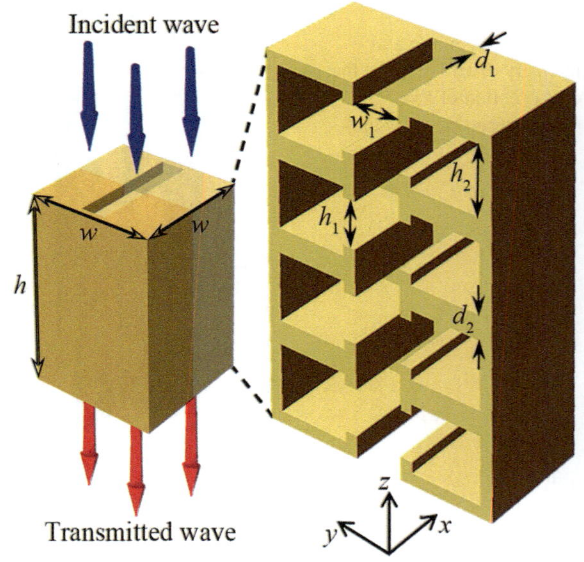

Fig. 9.36 Schematic of the phased unit cell and its cross-section

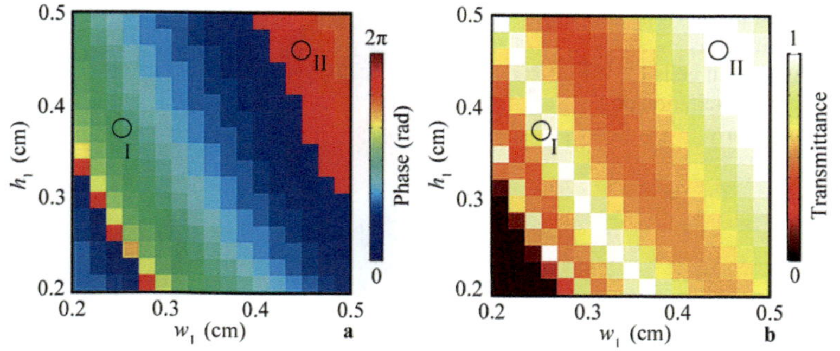

Fig. 9.37 **a** Simulated phase delays and **b** transmittances caused by the phased unit cells with different values of w_1 and h_1 at 5716 Hz. The two black open circles represent phased unit cells I ($w_1 = 0.24$ cm and $h_1 = 0.38$ cm) and II ($w_1 = 0.46$ cm and $h_1 = 0.48$ cm)

clearly exhibit high transmission and opposite phase delays caused by the phased unit cells I and II.

9.4.2 *Design and Performance of Mode-Conversion Meta-Atom*

Next, we discuss the mode conversion between the zero-order ($n_x = 0$) and first-order ($n_x = 1$) waves in the x direction, where the order of both waves in the y direction is selected as $n_y = 0$. As shown in Fig. 9.39, we place a mode-conversion meta-atom in a straight waveguide with widths w_x and w_y in the x and y directions, respectively. When the zero-order wave impinges on the meta-atom along the z direction, the sound energy transmits through the meta-atom and is converted into the first-order wave in the x direction. Based on Eq. (2.53), we obtain the theoretical phase distribution of the mode-conversion meta-atom, as shown in Fig. 9.40a.

Based on the theoretical continuous phase distribution in Fig. 9.40a, we select four discrete phase delays (two pairs of "0" and "π") to design a mode-conversion meta-atom, as shown in Fig. 9.40b. Figure 9.41 schematically shows the mode-conversion meta-atom composed of two pairs of phased unit cells I and II, with a cross-shaped solid structure placed at the middle of the meta-atom to separate the unit cells on both sides. The phase distribution of the meta-atom is the same as that in Fig. 9.40b. Additionally, we design a step waveguide placed on the upper side of the meta-atom. When the width of the step waveguide is smaller than or equal to $\lambda/2$, the first-order wave cannot pass through it, enabling the realization of AAT. By selecting the parameters $w_B = w_x = 5$ cm, $w_T = 3$ cm, $l_3 = l_4 = 1.5$ cm and $w_3 = 1$ cm, we design the mode-conversion meta-atom with both phase control and AAT functionalities. Figure 9.42a schematically shows the acoustic propagation paths through the mode-conversion meta-atom excited by the zero-order wave. The normal incidence of sound from the top and bottom sides of the meta-atom is defined as top incidence (TI) and

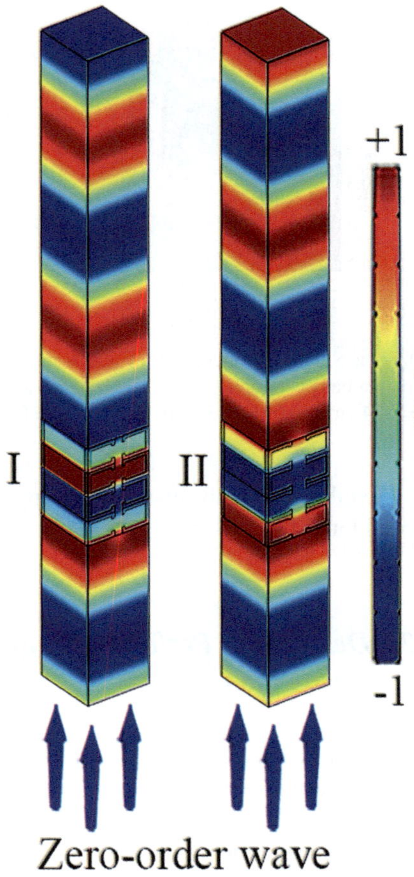

Fig. 9.38 Simulated pressure distributions through phased unit cells I and II under the excitation of the zero-order wave

bottom incidence (BI), respectively. For TI, the zero-order wave transmits through the meta-atom and is converted into the first-order wave in the x direction. However, for BI, the converted first-order wave inside the step waveguide cannot pass through a narrow channel on the upper side due to the limitation of its cutoff frequency. To verify this, we simulate the pressure distributions caused by the meta-atom under the excitation of the zero-order wave for TI and BI, as shown in Fig. 9.42b. The corresponding pressure distributions further demonstrate the acoustic propagation paths in Fig. 9.42a, showing typical characteristics of mode conversion and AAT for the designed mode-conversion meta-atom.

Moreover, we simulate the transmittance spectra caused by the mode-conversion meta-atom for TI and BI. As shown in Fig. 9.43a, in the range of 5.38–6.31 kHz, the transmittances are greater than 0.7 for TI but lower than 0.2 for BI, demonstrating high performance of AAT. This phenomenon is attributed to the stable phase differences of both phased unit cells. To verify this, we simulate the phase difference spectrum of the phased unit cells I and II, as shown in Fig. 9.43b. It is evident that the phase

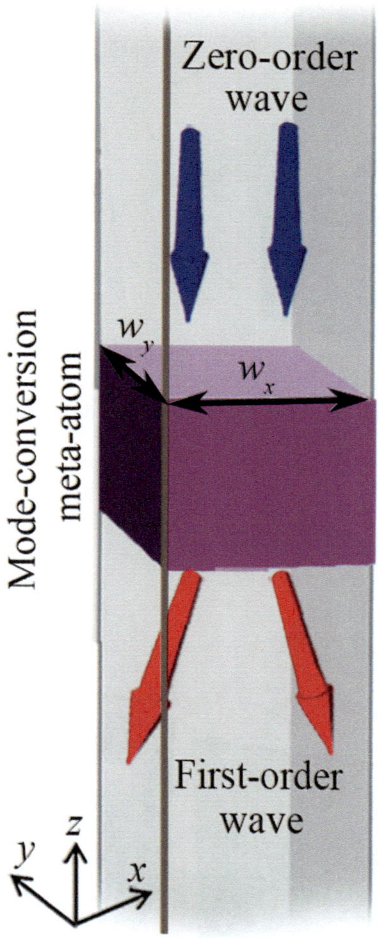

Fig. 9.39 Schematic of the conversion between the zero-order and first-order waves through a mode-conversion meta-atom in a straight waveguide

difference is around π in the black shaded region (5.38–6.31 kHz), indicating the stable phase difference of both phased unit cells. Therefore, the working bandwidth of the mode-conversion meta-atom is 0.93 kHz, and the corresponding fractional bandwidth reaches 0.16.

9.4.3 Design and Performance of AAF Lens

As schematically shown in Fig. 9.44, we design a 2D AAF lens based on mode-conversion meta-atoms [3]. For TI of a zero-order wave (Fig. 9.44a), the sound energy

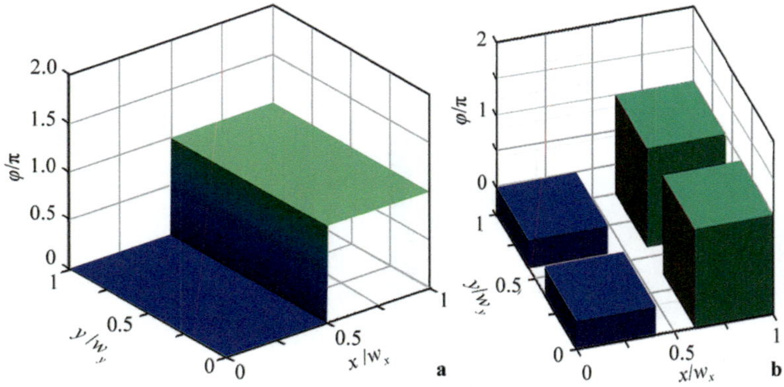

Fig. 9.40 **a** Theoretical continuous and **b** discrete phase distributions of the mode-conversion meta-atom

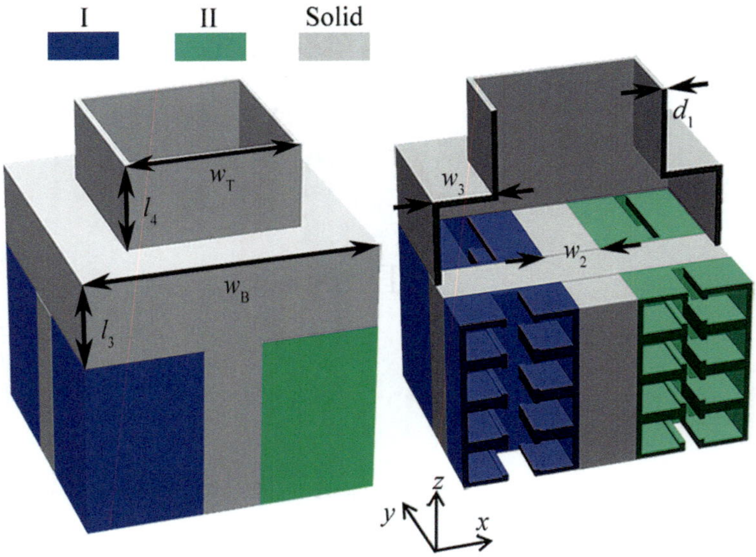

Fig. 9.41 Schematic of the mode-conversion meta-atom composed of two phased unit cells I and II and a step waveguide on the upper side

transmits through the lens and is focused on the bottom side. For BI of the zero-order wave (Fig. 9.44b), the sound energy cannot pass through the lens, demonstrating the characteristic of AAF in 3D space.

Based on Eq. (2.50), the theoretical phase distribution of the 2D AF lens can be expressed as:

$$\phi(x, y) = -k_0\sqrt{x^2 + y^2 + e^2}, \tag{9.16}$$

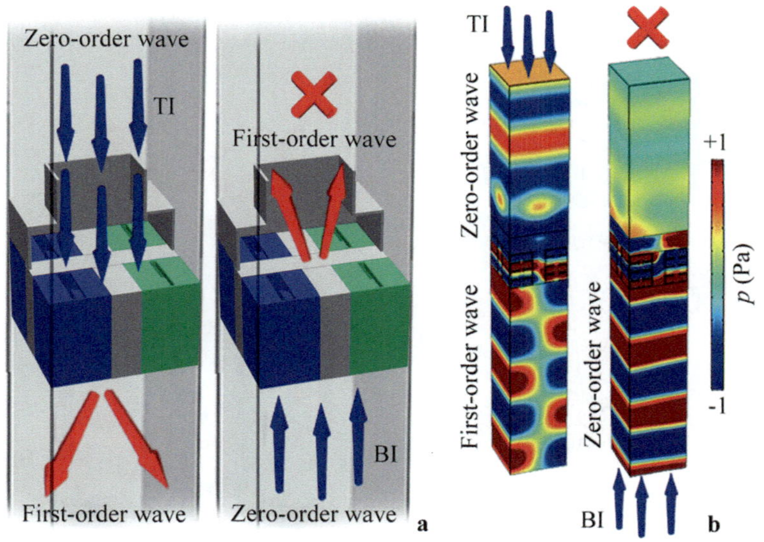

Fig. 9.42 **a** Schematic of acoustic propagation paths for TI and BI. **b** Simulated pressure distributions through the mode-conversion meta-atom under the excitation of the zero-order wave for TI and BI at 5716 Hz

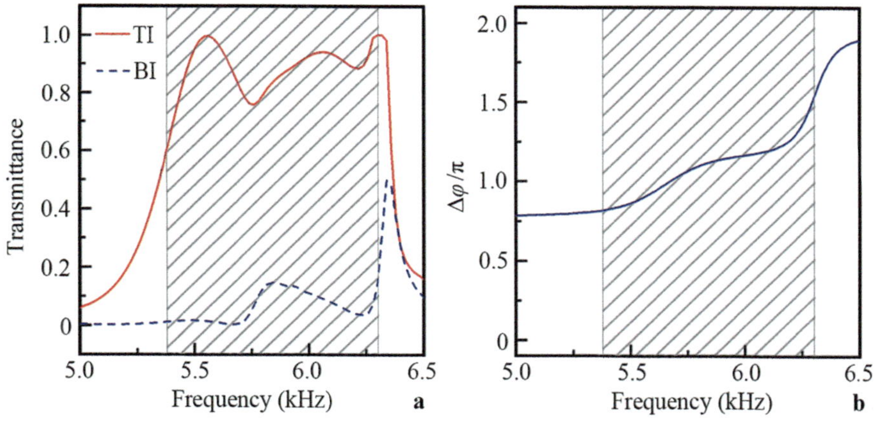

Fig. 9.43 **a** Simulated transmittance spectra through the mode-conversion meta-atom excited by the zero-order wave for TI and BI. **b** Simulated phase difference spectrum of phased unit cells I and II. The black shaded regions (5.38–6.31 kHz) in (**a**, **b**) represent the working band of the mode-conversion meta-atom

where the focal length $e = 20$ cm. Figure 9.45a shows the theoretical continuous phase distribution of the 2D AAF lens. Based on the theoretical continuous phase distribution and the phased unit cells I and II in the mode-conversion meta-atom, we use 44 meta-atoms with 176 discrete phase delays to design the 2D AAF lens (Fig. 9.45b). The other parts are filled with epoxy resin to connect adjacent structures.

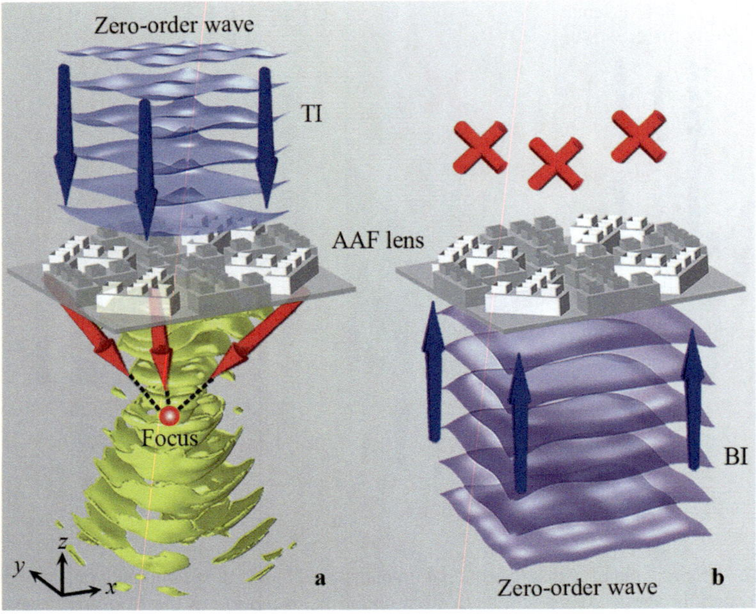

Fig. 9.44 Schematic of acoustic propagation paths through an AAF lens under the excitation of the zero-order wave for **a** TI and **b** BI

The photographs of the designed 2D AAF lens, as well as its upper and bottom surfaces, are shown in Fig. 9.46.

Next, we conduct an experiment to measure the intensity distributions caused by the AAF lens. Figure 9.47a, b present the schematic and photograph of the experimental set-up, respectively. The incident zero-order wave is generated from a loudspeaker array composed of 36 loudspeakers driven by a power amplifier. A set of motorized linear stages moves microphone 1 (Brüel & Kjær type-4961) to detect

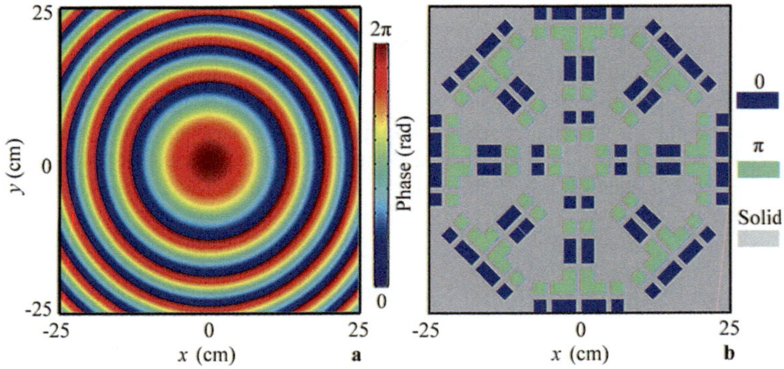

Fig. 9.45 **a** Theoretical continuous and **b** discrete phase distributions of the AAF lens

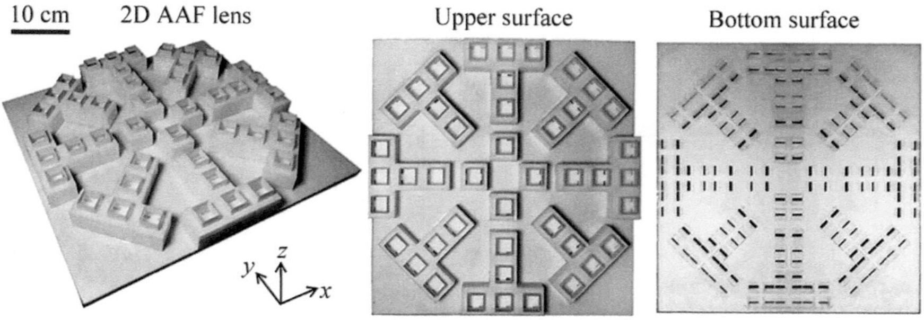

Fig. 9.46 Photographs of the AAF lens composed of 44 mode converters, and the upper and bottom surfaces

sound signals at each position in a square scanning region, and microphone 2 is used as a reference. The software Pulse Labshop is used to record the sound amplitude and phase at each position in the scanning region.

Figure 9.48a, b show the intensity distributions caused by the AAF lens for TI and BI, respectively. The bottom insets show the measured and simulated intensity distributions in regions R1 and R2 for TI, and R3 and R4 for BI. It is observed that the sound energy transmits through the lens for TI and is focused on the bottom side, while the transmitted sound energy is very weak for BI. The measured results agree well with the simulated ones, demonstrating the AAF effect of the lens. It is noted that the effects of AF and AAT are attributed to the phase modulation of the lens and the cutoff frequency of the first-order wave for the step waveguide. To verify this, we simulate the intensity distributions caused by the lens composed of meta-atoms without the step waveguide for TI and BI, as shown in Fig. 9.49a, b, respectively. It is evident that the sound energy transmits through the lens and is focused for TI and BI. The focusing characteristics for both cases are the same, further verifying the mechanisms of both AF and AAT.

To further demonstrate the performance of the AAF lens, we measure and simulate the normalized intensity distributions along lines I–IV in Fig. 9.48, as shown in Fig. 9.50a, b. It is observed that the intensities along lines I and II are greater than those along lines III and IV, and the focal length is 17 cm (Fig. 9.50a), which is close to the theoretical value of 20 cm. Additionally, the FLHM of the focus in the x direction is 3.4 cm, equal to 0.57λ (Fig. 9.50b). The measured and simulated results are consistent with each other, demonstrating the AAF effect of the lens.

Finally, to present the working bandwidth of the AAF lens, we measure and simulate the intensity distributions caused by the lens for TI and BI at 5650, 5900 and 6150 Hz, which are shown in Fig. 9.51. We observe that the AAF effect still exists at the three selected frequencies, and the measured results match well with the simulated ones.

In addition, we simulate the transmittance spectra caused by the lens for TI and BI, and the maximum intensity spectra along line I, which are presented in Fig. 9.52. As shown in Fig. 9.52a, in the black shaded region (5.47–6.31 kHz), the transmittances

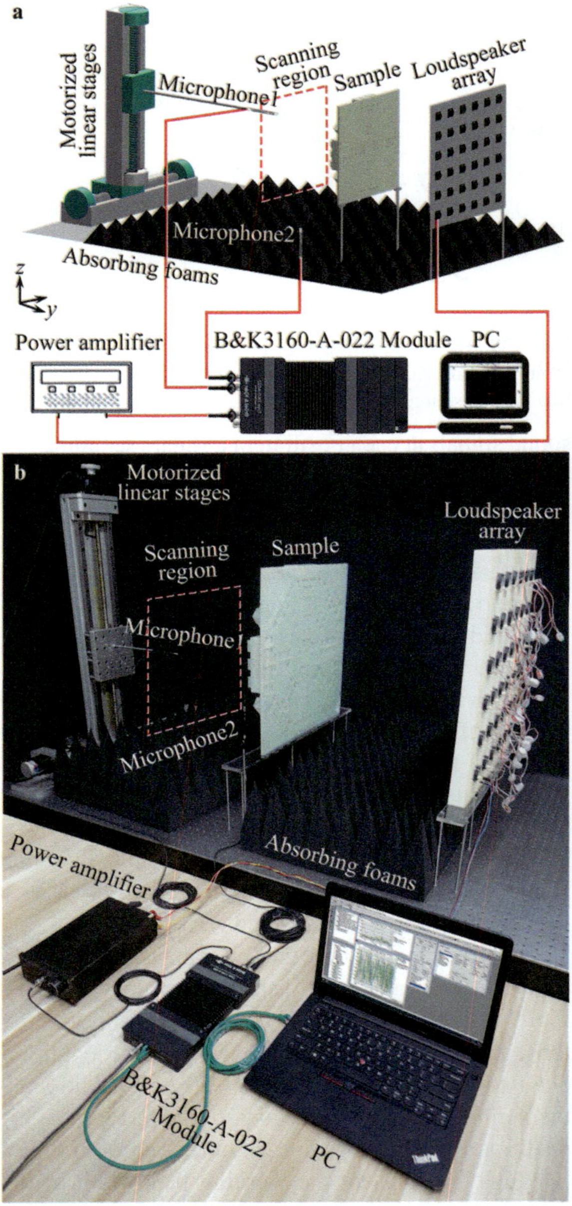

Fig. 9.47 **a** Schematic and **b** photograph of experimental set-up

are larger than 0.6 for TI, but are lower than 0.2 for BI, showing high performance of AAT for the lens. Moreover, the maximum intensity along line I is larger than 4.0 Pa^2 in the black shaded region (Fig. 9.52b), demonstrating high performance of AF for the lens. Therefore, the working bandwidth of the 2D AAF lens is 0.84 kHz,

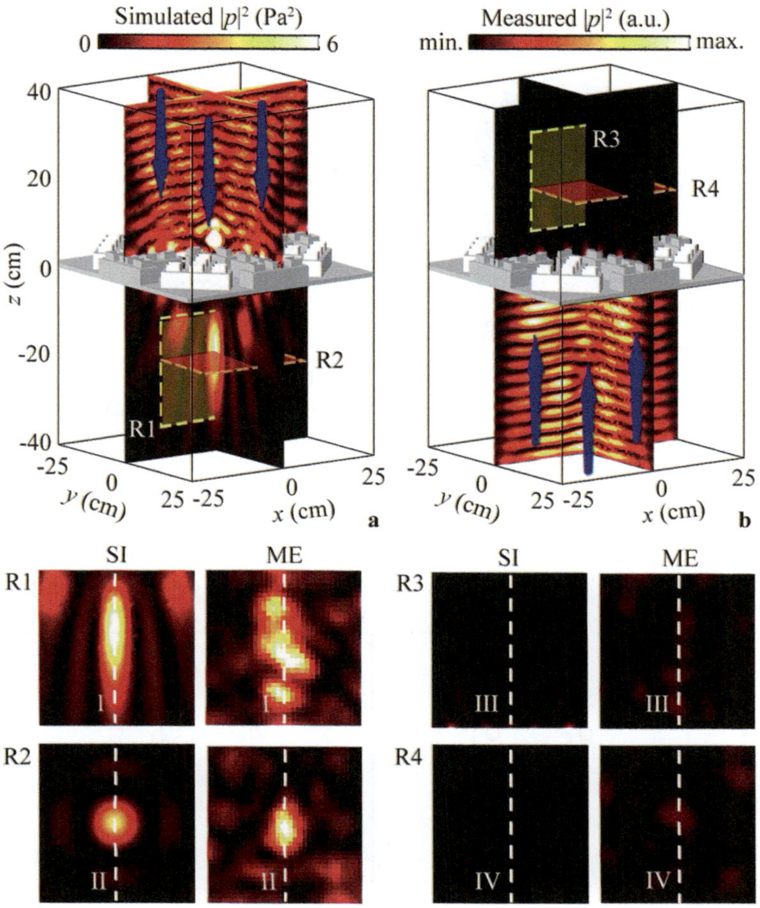

Fig. 9.48 Measured and simulated intensity distributions caused by the AAF lens for **a** TI and **b** BI at 5716 Hz. The blue arrows represent the incident zero-order wave, and the bottom eight insets represent the measured and simulated intensity distributions in regions R1–R4

and the fractional bandwidth is 0.14, showing a broadband characteristic of the 2D AAF lens.

9.4.4 Performance Modulation

In addition to the AAF lens with $e = 20$ cm, we design another two types of AAF lenses with the focal lengths of 10 and 15 cm. Figure 9.53 shows the theoretically continuous and discrete phase distributions of the 2D AAF lenses with the focal lengths of 10 and 15 cm, respectively.

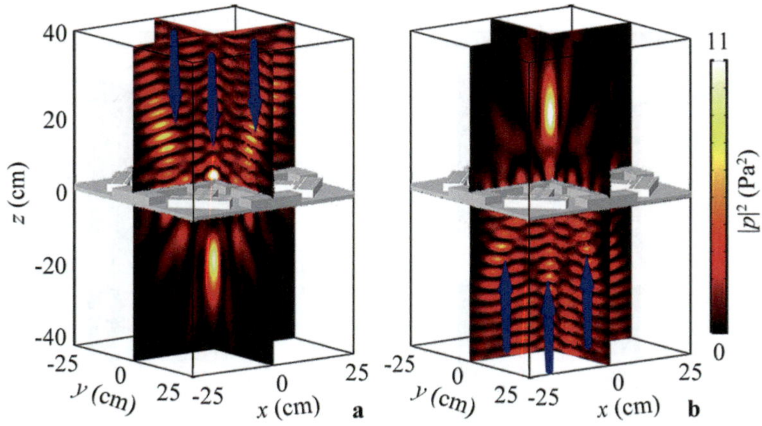

Fig. 9.49 Simulated intensity distributions caused by the AAF lens composed of the meta-atoms without the step waveguide for **a** TI and **b** BI at 5716 Hz. The blue arrows represent the incident zero-order wave

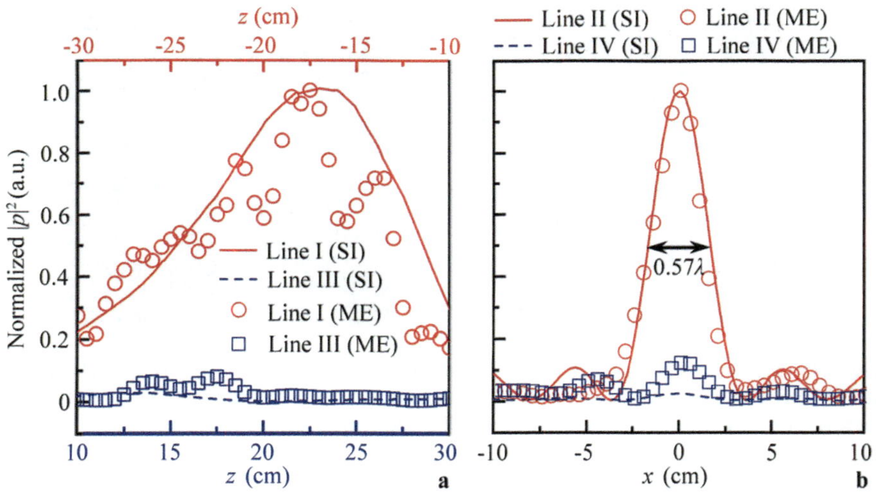

Fig. 9.50 Measured and simulated intensity distributions along **a** lines I and III and **b** lines II and IV in Fig. 9.48. The intensities are normalized by the maximum value of each figure

Figure 9.54 shows the simulated intensity distributions caused by the AAF lenses with the focal lengths of 10 and 15 cm for TI and BI. The middle insets are the intensity distributions in regions R11–R14. As shown in Fig. 9.54a, c, the zero-order wave can transmit through the lens, and is focused at the bottom side for TI. However, for BI (Fig. 9.54b, d), the zero-order wave cannot pass through the lens.

To further present the focusing performance of the AAF lenses, we simulate the intensity distributions along the line I in Fig. 9.48, and the lines V and VI in Fig. 9.54,

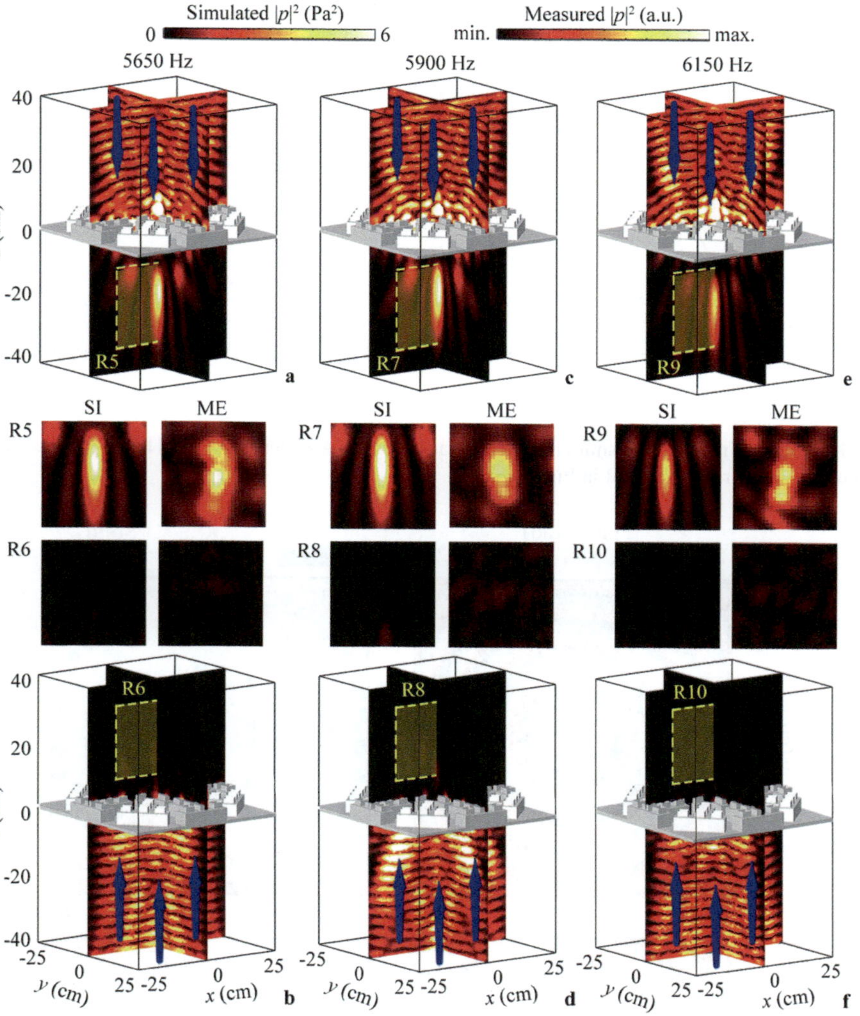

Fig. 9.51 Measured and simulated intensity distributions caused by the 2D AAF lens at **a**, **b** 5650 Hz, **c**, **d** 5900 Hz, and **e**, **f** 6150 Hz for TI and BI. The blue arrows represent the zero-order wave, and the middle twelve insets are measured and simulated intensity distributions in R5–R10

which are shown in Fig. 9.55. We can see that the maximum intensities at the focuses are about 6.9, 11.7, and 13.5 Pa^2, and the focal lengths are about 17.0, 11.1, and 13.1 cm, which are close to the theoretical values of 20, 10 and 15 cm. Therefore, we can design the AAF lenses with different focal lengths by selecting different phase distributions.

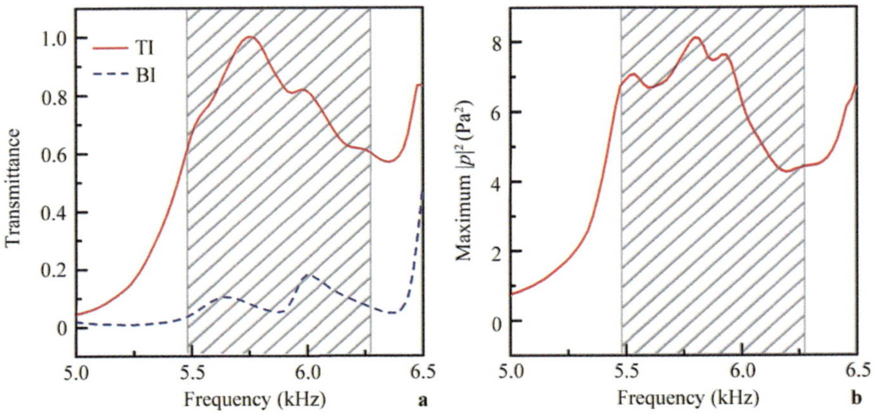

Fig. 9.52 **a** Simulated transmittance spectra caused by the AAF lens for TI and BI, and **b** maximum intensity spectra along line I in Fig. 9.48a

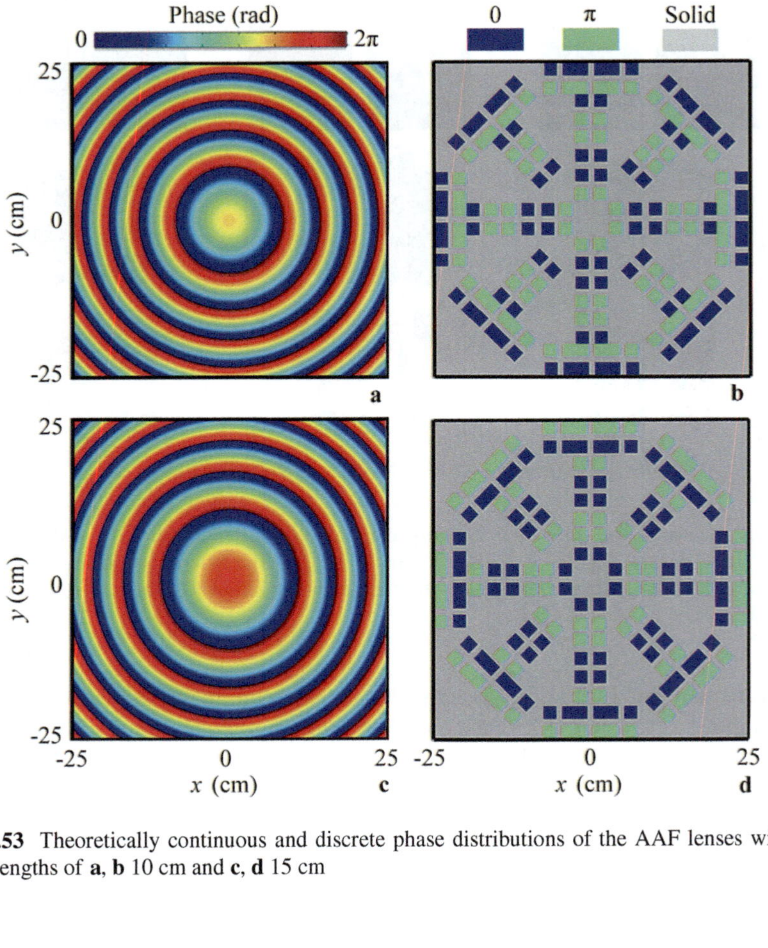

Fig. 9.53 Theoretically continuous and discrete phase distributions of the AAF lenses with the focal lengths of **a**, **b** 10 cm and **c**, **d** 15 cm

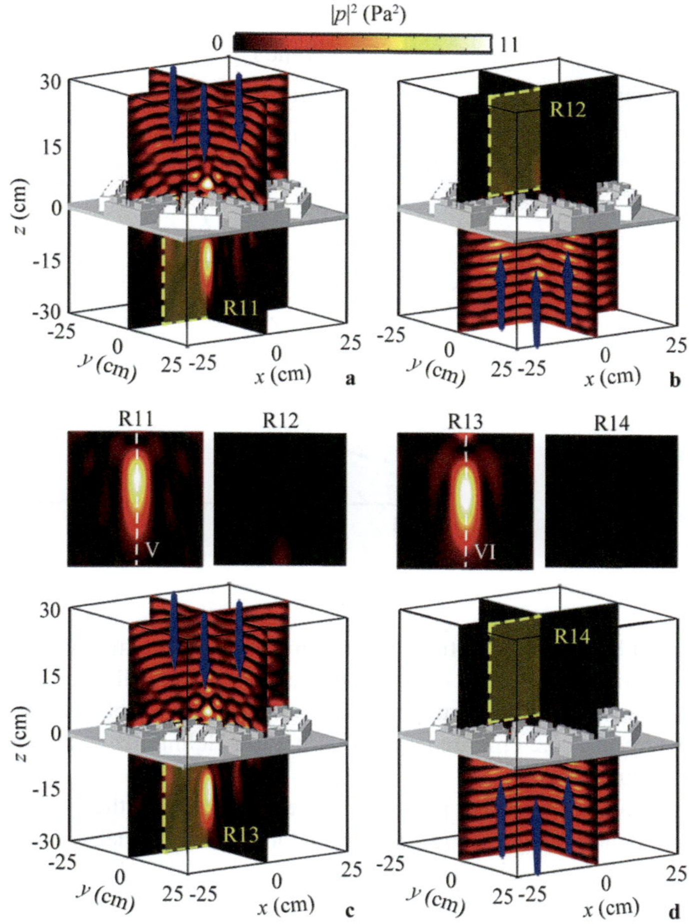

Fig. 9.54 Simulated intensity distributions caused by the AAF lenses with focal lengths of **a**, **b** 10 cm and **c**, **d** 15 cm for TI and BI. The blue arrows represent the zero-order waves. The middle four insets show the intensity distributions in regions R11–R14

9.5 Three-Dimensional Acoustic Asymmetric Vortex Focusing by Mode-Conversion Meta-Atoms

9.5.1 Design and Performance of Mode-Conversion Meta-Atom

Figure 9.56 shows the photograph of the designed 2D AAVF lens [4] with a radius $r = 25.8$ cm, composed of a central solid cylinder and twelve fan-shaped structures (the red shaded region). Each fan-shaped structure (the upper inset) is constructed by four mode-conversion meta-atoms. The 2D AAVF lens is made of epoxy resin to satisfy

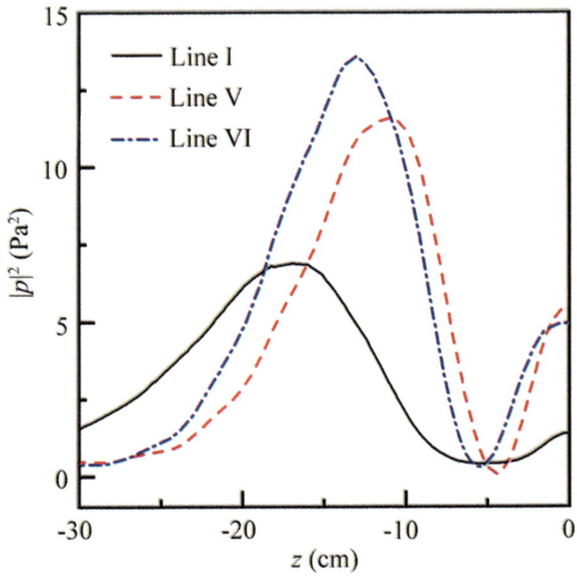

Fig. 9.55 Simulated intensity distributions along line I in Fig. 9.48, and along lines V and VI in Fig. 9.54

the hard sound boundary condition. As shown in Fig. 9.56b, the mode-conversion meta-atom consists of two pairs of out-of-phase phased unit cells (denoted as A and B) and a step waveguide on the upper side. Both phased unit cells (with length $l =$ 4 cm and width $w = 1.2$ cm) are composed of eight Helmholtz resonators on both sides and a central channel.

To design the out-of-phase phased unit cells, we adjust the volume of each Helmholtz resonator by tuning the width w_1 of the central channel, thereby realizing different resonant frequencies and phase delays of sound. Based on two pairs of out-of-phase phased unit cells, we can achieve the conversion between the zero-order and first-order waves. It is noted that the zero-order wave can pass through the waveguide with an arbitrary width, while the first-order wave can only transmit through the waveguide with a width larger than half the wavelength. Here, the selected width w_4 of the step waveguide is smaller than half the wavelength of sound, which prohibits the propagation of the first-order wave, enabling the AAT effect. Throughout this work, we use COMSOL Multiphysics software to simulate the performance of AAVF. The structural parameters of the mode-conversion meta-atom are selected as $w_2 = 1.3$ cm, $w_3 = 4.8$ cm, $w_4 = 2$ cm, $l_1 = 0.9$ cm, $l_2 = 0.15$ cm, $l_3 = 1.5$ cm and $d = 0.08$ cm, The sound wavelength λ is 5 cm (i.e., $f_0 = 6.86$ kHz in air). The material parameters of air are $\rho = 1.21$ kg/m^3 and $c = 343$ m/s in the simulations.

Figure 9.57 shows the simulated transmission coefficients and phase delay spectra through the phased unit cells with different values of w_1. Here, we select six phased unit cells with equally spaced phase delays from 0 to 2π, and their transmission coefficients are greater than 0.9. Based on these six phased unit cells, we design three types of binary-phase mode-conversion meta-atoms, denoted as I, II, and III, as shown in Fig. 9.58.

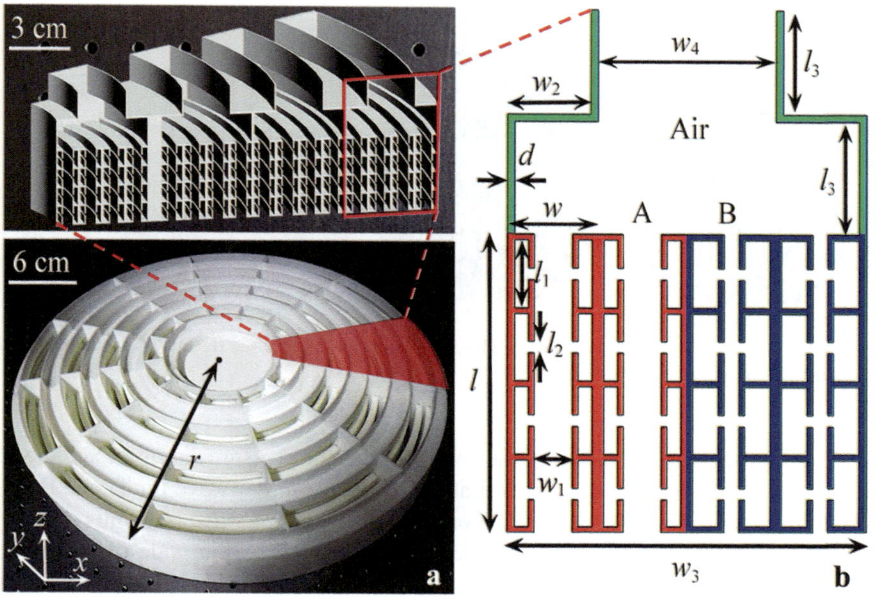

Fig. 9.56 **a** Photograph of the 2D AAVF lens composed of a central solid cylinder and twelve fan-shaped structures with four mode-conversion meta-atoms. **b** Schematic of the mode-conversion meta-atom composed of two pairs of out-of-phase phased unit cells (A and B) and a step waveguide on the upper side

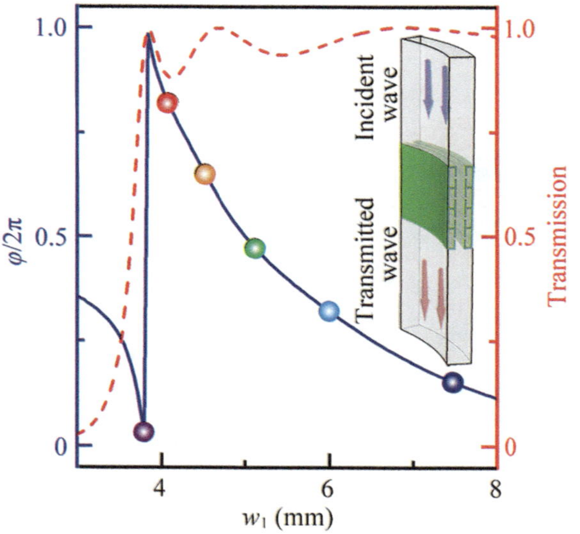

Fig. 9.57 Simulated phase delay (the blue solid line) and transmission coefficients (the red dashed line) of the phased unit cells with different values of w_1 at 6.86 kHz. The phase delays caused by the selected six phased unit cells (the color points) from the left to the right are 0, 5π/3, 4π/3, π, 2π/3, π/3, with corresponding values of w_1 being 0.384 cm, 0.406 cm, 0.45 cm, 0.51 cm, 0.6 cm and 0.75 cm, respectively

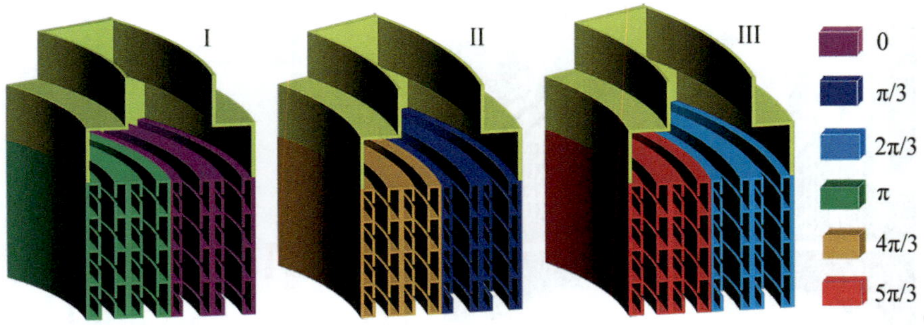

Fig. 9.58 Schematic of three types of mode-conversion meta-atoms I, II, and III constructed by the selected six phased unit cells in Fig. 9.57

To demonstrate their characteristics, we simulate the pressure distributions caused by the mode converters under the excitation of the zero-order wave for TI and BI, as shown in Fig. 9.59a, b, respectively. For TI, the zero-order wave transmits through the step waveguide and phased unit cells, reaching the bottom side of the mode converter as the converted first-order wave. However, for BI, the zero-order wave is converted into the first-order wave by the phased unit cells but cannot pass through the step waveguide, as the opening width of the step waveguide is smaller than half the wavelength of sound. Additionally, we simulate the transmittance spectra of meta-atoms I, II, and III for TI and BI, as shown in Fig. 9.60. The difference in transmittance coefficients between TI and BI is greater than 5 dB in the black shaded regions, demonstrating high performance of AAT for each meta-atom.

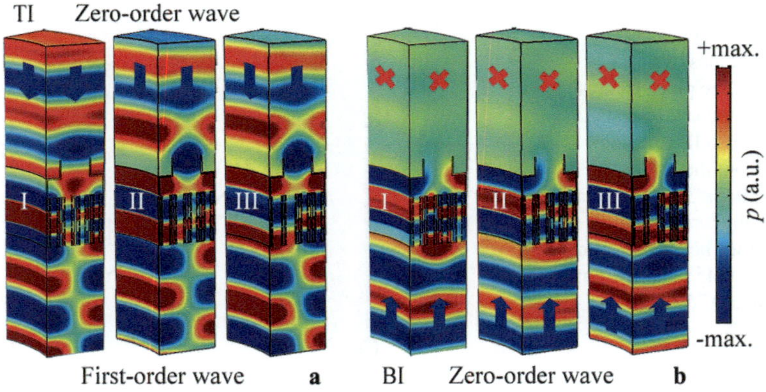

Fig. 9.59 Simulated pressure distributions caused by mode-conversion meta-atoms I, II, and III for **a** TI and **b** BI at 6.86 kHz

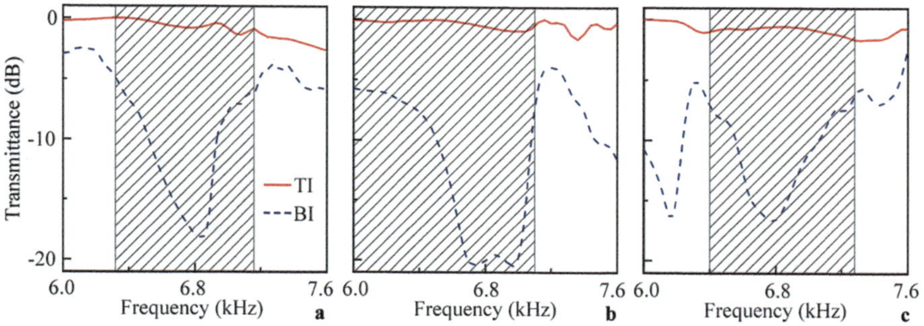

Fig. 9.60 Simulated transmittance spectra caused by mode-conversion meta-atoms **a** I, **b** II, and **c** III for TI and BI. The working bands (the black shaded regions) are 6.32–7.16 kHz, 6.00–7.10 kHz, and 6.40–7.28 kHz for meta-atoms I, II and III, respectively

9.5.2 *Design and Performance of AAVF Lens with* n = 2

Figure 9.61 schematically shows the acoustic propagation paths through a 2D AAVF lens, where TI and BI represent the normal incidence of acoustic waves from the upper and bottom sides of the lens, respectively. For TI, the acoustic wave transmits through the lens and forms a vortex focus on the bottom side. However, for BI, the acoustic wave cannot transmit through the lens, demonstrating the characteristic of AAVF. To theoretically design this, we introduce the phase distribution of a vortex beam in free space as follows:

$$\varphi_{\mathrm{v}}(\theta) = n\theta, \tag{9.17}$$

where n is the topological charge of the vortex beam [7], and θ is the azimuth angle. Here, by introducing the AVF, the theoretical phase distribution of the AAVF lens in

Fig. 9.61 Schematic of acoustic propagation paths through the 2D AAVF lens for TI and BI

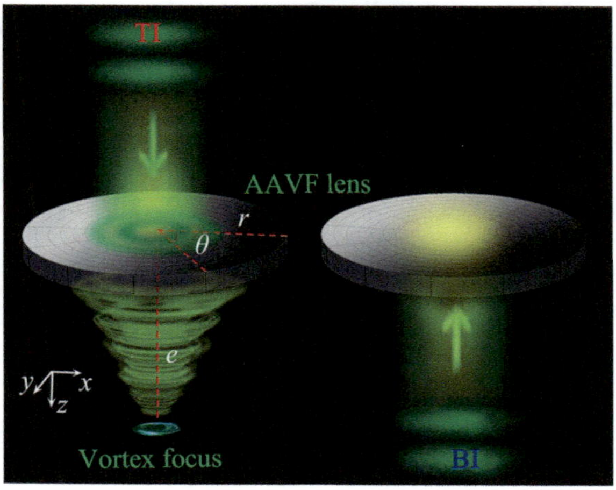

the x–y plane based on Eq. (2.41) can be written as:

$$\varphi = -k\sqrt{r^2 + e^2} + n\theta, \tag{9.18}$$

where $k = 2\pi f/c$ is the wave number in air, c and f represent the sound velocity and the frequency, respectively. The parameters $r = \sqrt{x^2 + y^2}$ and e are the radial length and the focal length of the lens, respectively, and the azimuth angle θ satisfies $\tan\theta = x/y$. Based on Eq. (9.18), we can achieve the AVF effect.

Using the three mode-conversion meta-atoms, we design a broadband 2D AAVF lens with $e = 20$ cm and $n = 2$. Based on Eq. (9.18), we theoretically calculate the continuous phase distribution of the lens in the x–y plane (Fig. 9.62a), which exhibits the characteristic of a vortex beam with $n = 2$. Based on the theoretical continuous phase distribution, we use 48 meta-atoms to design the AAVF lens. However, gaps smaller than the size of the meta-atoms inevitably exist, so we fill these gaps with epoxy resin (the light gray solid in Fig. 9.62b) to connect adjacent meta-atoms. As shown in Fig. 9.62b, the phase distribution of the AAVF lens is consistent with that in Fig. 9.62a. Therefore, we can realize the AAVF lens based on the theoretical continuous phase distribution, and the transmitted vortex focusing can be achieved through phase control.

Figure 9.63a, b show the simulated intensity distributions through the 2D AAVF lens for TI and BI, respectively. For TI (Fig. 9.63a), the sound energy passes through the lens and is focused into a narrow hollow cylinder, with the sound energy at the center being almost zero, demonstrating the characteristic of AVF. However, for BI (Fig. 9.63b), the sound energy cannot transmit through the lens, showing the characteristic of AAVF. This is because the converted first-order wave cannot pass through the step waveguide of each mode converter. Furthermore, we simulate the intensity distributions caused by the lens composed of meta-atoms without the step waveguide, as shown in Fig. 9.64. The acoustic wave transmits through the lens and

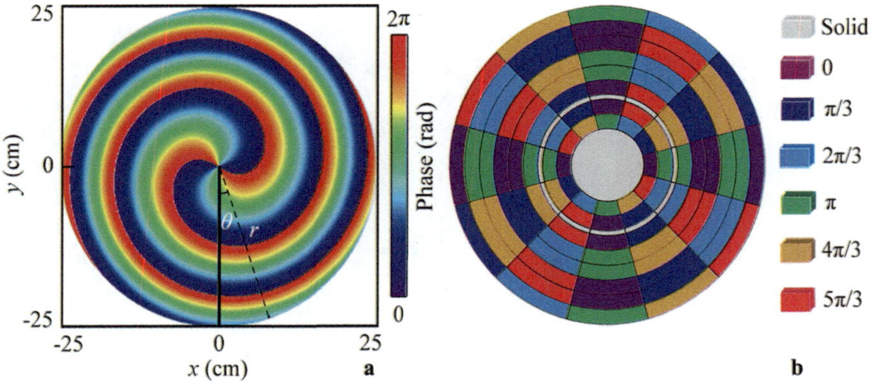

Fig. 9.62 **a** Theoretical continuous and **b** discrete phase distributions of the 2D AAVF lens in the x–y plane at 6.86 kHz. The designed 2D AAVF lens in (**b**) is composed of 48 mode-conversion meta-atoms

forms a vortex focus for both TI and BI, with the characteristics of AVF for TI and BI being the same as those for TI in Fig. 9.63a, further demonstrating that the AAT effect is determined by the step waveguide.

To demonstrate the AAVF performance of the lens, we conduct an experiment to measure the intensity and phase distributions caused by the lens in R1 and R2 for TI, and R3 and R4 for BI, as shown in Fig. 9.63. Figure 9.65 schematically shows the experimental set-up, where the AAVF lens sample is placed in 3D free space. The incident plane wave is generated from a loudspeaker driven by a power amplifier, and the distance between the loudspeaker and the sample is 1 m. Additionally, we use two 0.25 inch-diameter microphones to measure the pressure amplitude and phase distributions in the scanning region. Microphone 1 detects sound signals in the scanning region moved by a set of 2D motorized linear stages, and microphone 2 is used as a reference.

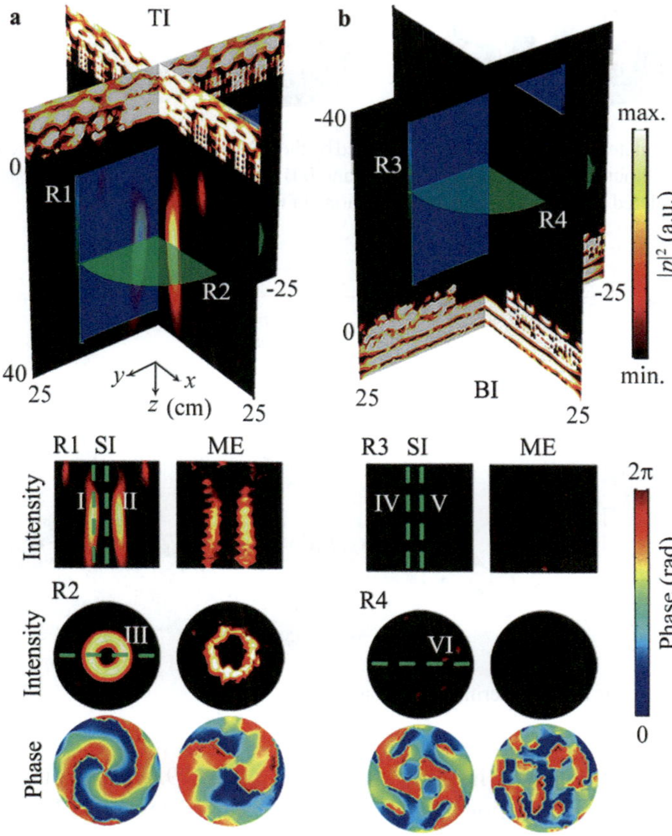

Fig. 9.63 Measured and simulated intensity distributions caused by the AAVF lens for **a** TI and **b** BI at 6.86 kHz. The twelve insets at the bottom show the intensity and phase distributions caused by the AAVF lens in R1 (y–z plane at $x = 0$) and R2 (x–y plane at $z = 20$ cm) for TI, and R3 (y–z plane at $x = 0$) and R4 (x–y plane at $z = -20$ cm) for BI

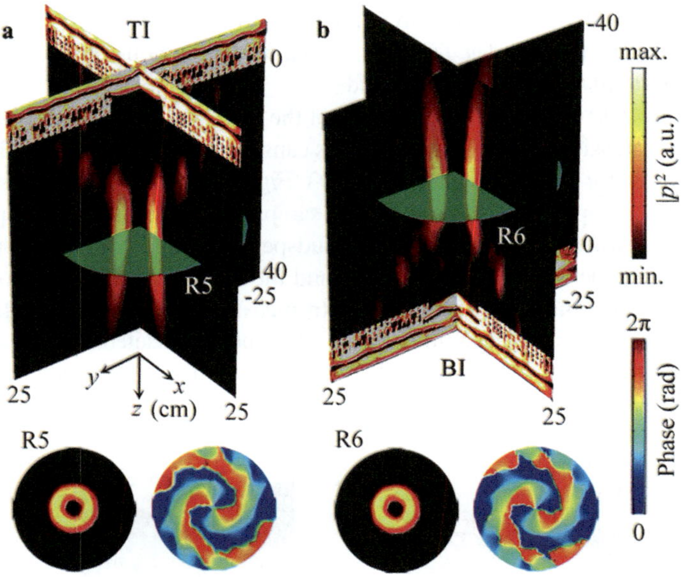

Fig. 9.64 Simulated intensity distributions caused by the AAVF lens composed of mode-conversion meta-atoms without the step waveguide for **a** TI and **b** BI at 6.86 kHz. The four insets at the bottom show the simulated intensity and phase distributions in R5 for TI and R6 for BI

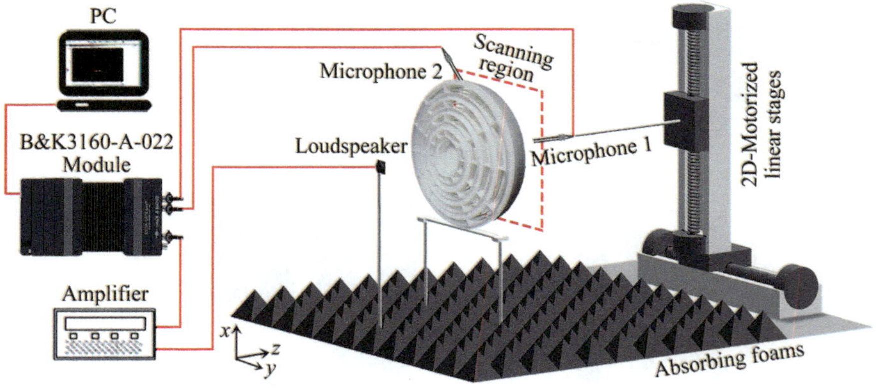

Fig. 9.65 Schematic of the experimental set-up

As shown in the twelve insets at the bottom of Fig. 9.63, the sound energy is focused into two focal regions in R1 and a ring-shaped region in R2. The phase distribution in R2 exhibits the characteristic of a vortex beam with $n = 2$ for TI. The measured results agree well with the simulated ones. Additionally, no sound energy passes through the lens in R3 and R4, and the vortex characteristic is absent, further experimentally verifying the performance of the 2D AAVF lens.

Figure 9.66 shows the measured intensity distributions along lines I-VI in Fig. 9.63, with the simulated results presented for comparison. As shown in Fig. 9.66a, the focus is located at $z = 20$ cm along line I, and the intensities along line I are much greater than those along line II for TI due to the phase singularity. Moreover, as shown in Fig. 9.66b, two acoustic energy peaks exist on both sides of $x = 0$ along line III, with the intensity at $x = 0$ being close to zero, demonstrating the characteristic of AVF. For BI (Fig. 9.66c, d), the intensities along lines IV–VI are close to zero, indicating that the sound energy cannot pass through the lens. The measured intensity distributions agree with the simulated ones, experimentally demonstrating the performance of the lens.

Finally, we experimentally demonstrate the bandwidth of the 2D AAVF lens. Figure 9.67a–d show the simulated 3D intensity distributions caused by the 2D AAVF lens for TI and BI at four selected frequencies. As the frequency increases, the focal length gradually increases for TI, and the characteristics of the focus are the same as

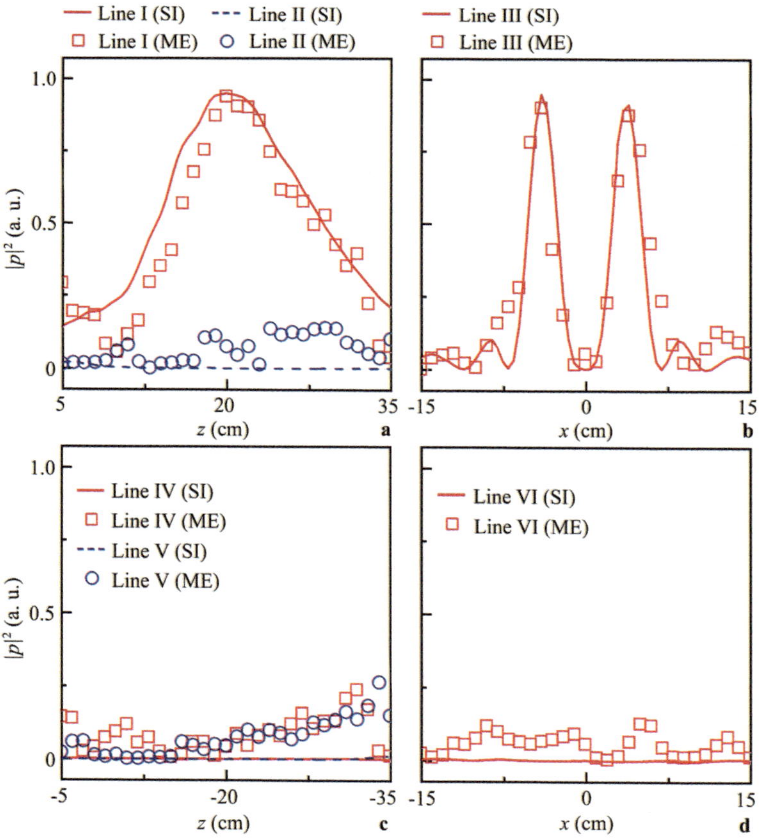

Fig. 9.66 Measured and simulated intensity distributions along **a** lines I, II, and **b** III in Fig. 9.63a, and along **c** lines IV, V, and **d** VI in Fig. 9.63b

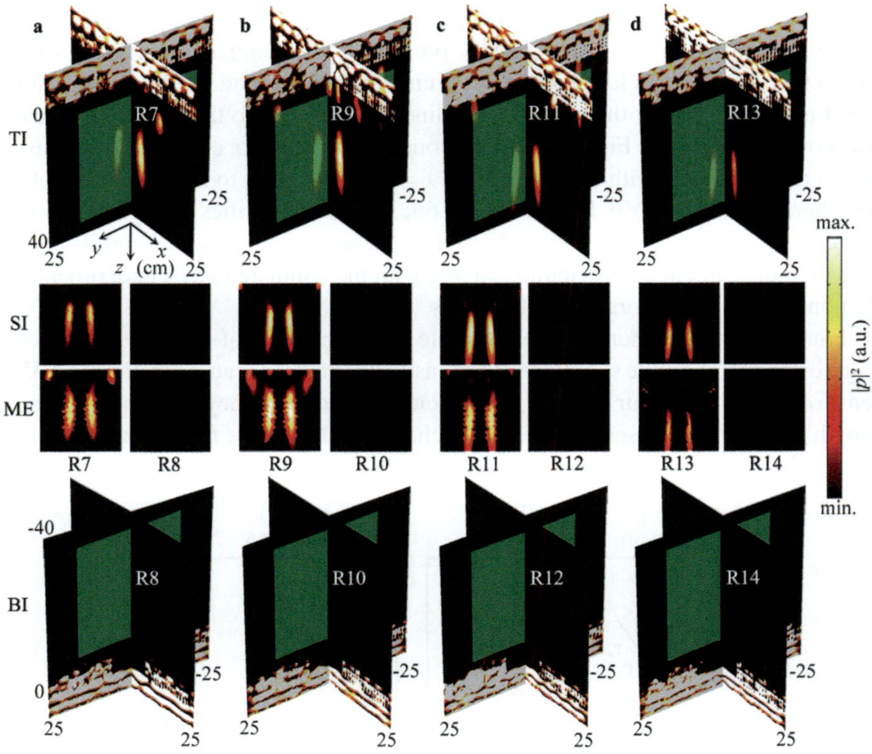

Fig. 9.67 Measured and simulated intensity distributions caused by the 2D AAVF lens for TI and BI at **a** 6.2 kHz, **b** 6.6 kHz, **c** 7.0 kHz, and **d** 7.4 kHz

those in Fig. 9.63a. Additionally, the acoustic wave cannot pass through the lens for BI. The measured and simulated results in R7–R14 at these frequencies are presented in the middle region, with the measured results agreeing well with the simulated ones. The measured results demonstrate that the bandwidth of the 2D AAVF lens exceeds 1.2 kHz, and the fractional bandwidth is greater than 0.19, showing the broadband characteristic of the lens. Moreover, we simulate the contrast ratio of the AAVF lens, as shown in Fig. 9.68. The simulated contrast ratio is greater than 0.8 in the range of 6.2–7.4 kHz, indicating high-performance AAVF of the designed lens.

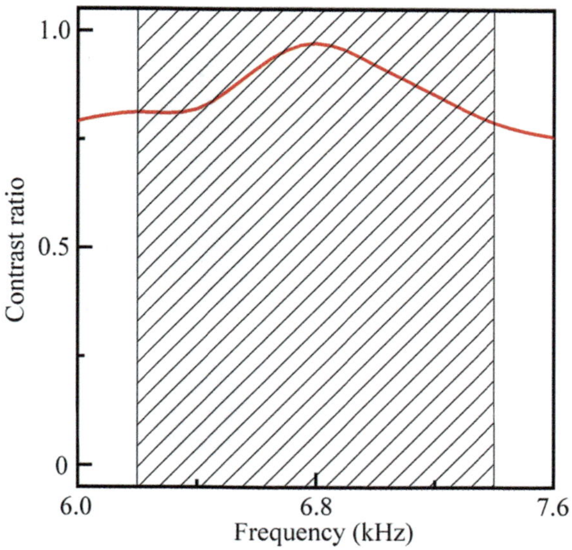

Fig. 9.68 Simulated contrast ratio of the 2D AAVF lens. The black shaded region (6.2–7.4 kHz) represents the working band

9.5.3 Design and Performance of AAVF Lenses with n = *1 and 3*

In addition to the lens with $n = 2$, we designed two types of 2D AAVF lenses with $n = 1$ and 3. Figure 9.69a, b show the simulated intensity distributions caused by the 2D AAVF lenses with $n = 1$ and 3 for TI and BI, respectively. It shows that sound energy can pass through both lenses and is focused into the shape of a narrow hollow cylinder for TI. However, for BI, the sound energy cannot reach the upper side of the lens. These results are similar to those produced by the lens with $n = 2$, as shown in Fig. 9.63.

In regions R15 and R17 for TI, the phase distributions complete one full cycle (from 0 to 2π) around a circle for the 2D AAVF lens with $n = 1$, and three full cycles for the lens with $n = 3$. Additionally, as n increases, the radius of the AAVF beam gradually increases, which provides a means for manipulating and screening particles of different sizes. In regions R16 and R18 for BI, no sound energy passes through the lens, and the vortex characteristic is absent. This demonstrates the high performance of AAVF for lenses with $n = 1$ and 3.

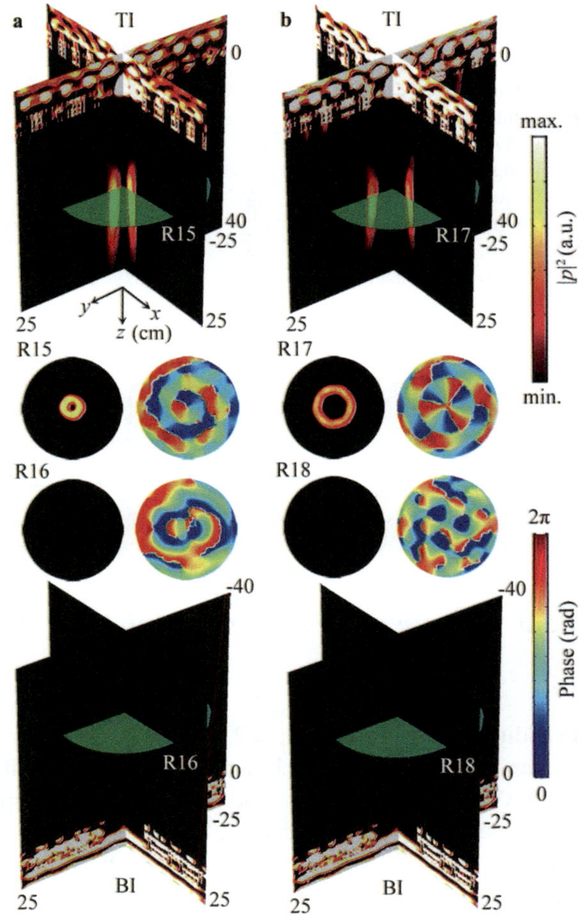

Fig. 9.69 Simulated intensity distributions generated by the 2D AAVF lens with **a** $n = 1$ and **b** $n = 3$ for TI and BI. The middle eight insets show the simulated intensity and phase distributions in R15 and R17 for TI, and in R16 and R18 for BI

9.6 Conclusion

This chapter presents 3D focusing and asymmetric focusing using phase-controlled acoustic metamaterials. Four types of 2D acoustic lenses were designed, achieving point focusing, line focusing, long-focus focusing, and vortex focusing. By using the 2D generalized Snell's law, these lenses demonstrated complex sound wavefront manipulation in 3D space. Experimental and simulation results highlighted their potential applications in non-contact particle trapping, medical ultrasound imaging, and therapy, showcasing their versatility in advanced acoustic manipulation.

References

1. Qian J, Wang Y, Yuan SQ, Sun HX, Liu XJ (2019) Reflected acoustic wavefront manipulation by an ultrathin metasurface based on three-dimensional generalized Snell's law. Appl Phys Express 12(9):094001. https://doi.org/10.7567/1882-0786/ab3492
2. Wang Y, Qian J, Xia JP, Ge Y, Yuan SQ, Sun HX, Liu XJ (2021) Acoustic Bessel vortex beam by quasi-three-dimensional reflected metasurfaces. Micromachines-Basel 12(11):1388. https://doi.org/10.3390/mi12111388
3. Yin JL, Zou HY, Wang Y, Zhao KQ, Ge Y, Yuan SQ, Sun HX (2023) Three-dimensional acoustic asymmetric focusing by mode-conversion structure. J Appl Phys 134(8):084504. https://doi.org/10.1063/5.0162326
4. Wang Y, Zou HY, Lu YJ, Gu S, Qian J, Xia JP, Ge Y, Sun HX, Yuan SQ, Liu XJ (2023) Two-dimensional acoustic asymmetric vortex focusing lens by binary-phase mode converters. Prog Electromagn Res 177:127–137. https://doi.org/10.2528/PIER23042502
5. Li Y, Jiang X, Liang B, Cheng JC, Zhang LK (2015) Metascreen-based acoustic passive phased array. Phys Rev Appl 4(2):024003. https://doi.org/10.1103/PhysRevApplied.4.024003
6. Xia JP, Zhang XT, Sun HX, Yuan SQ, Qian J, Ge Y (2018) Broadband tunable acoustic asymmetric focusing lens from dual-layer metasurfaces. Phys Rev Appl 10(1):014016. https://doi.org/10.1103/PhysRevApplied.10.014016
7. Jia YR, Wei Q, Wu DJ, Xu Z, Liu XJ (2018) Generation of fractional acoustic vortex with a discrete Archimedean spiral structure plate. Appl Phys Lett 112(17):173501. https://doi.org/10.1063/1.5026646

Chapter 10
Conclusions, Challenges, and Future Prospects

10.1 Summary of Key Findings and Contributions

This monograph provides a comprehensive exploration of acoustic metamaterials, focusing on their ability to manipulate sound in innovative ways. It begins with an introduction to the fundamental concepts and historical development of acoustic metamaterials, followed by detailed discussions on a variety of applications, including AAT, AF, and AAF. Key findings include the successful design and experimental demonstration of multiple systems for AAT and AF in both underwater and airborne environments. These AAT and AF systems use advanced theoretical frameworks such as the EM theory, generalized Snell's law, and mode conversion theory, enabling precise control over sound propagation. The monograph also highlights the development of broadband and ultra-broadband AF lenses using phase-controlled and thermoacoustic metamaterials, which have significant potential in medical ultrasound, nondestructive testing, and acoustic communication. One of the major contributions of this work is the demonstration of AAF and 3D sound manipulation using acoustic metamaterials, opening new possibilities for applications in focused ultrasound therapy, acoustic encryption, and non-contact particle trapping. The key findings and contributions of this monograph are listed as follows:

1. Fundamental theories and modeling methods. Beginning with the establishment of foundational theories, this work delves into the principles necessary for understanding and designing acoustic metamaterials. These include the interference theory, EM theory, FE method, generalized Snell's law, mode conversion theory, and band theory. These frameworks provide the methodological support required to analyze the effects of AAT, AF, and AAF, providing a robust theoretical basis for exploring the underlying physical mechanisms.
2. Underwater acoustic metamaterials. Focusing on underwater environments, the design and experimental demonstration of several AAT systems are explored. Composed of brass plates and periodic gratings, these systems exhibit AAT effects across multiple frequency bands below 1500 kHz. The capability to

H. Sun et al., *Acoustic Metamaterials*, Microfluidics and Nanophotonics: Science and Engineering 2, https://doi.org/10.1007/978-981-96-8524-0_10

control sound propagation paths, asymmetric transmission, and isolation in water holds significant implications for underwater acoustic communication, sonar detection, and medical ultrasound imaging and treatment.

3. Airborne acoustic metamaterials. Shifting to airborne applications, the design of AAT systems is examined, particularly for use in architectural acoustics and mechanical engineering. Systems such as broadband AAT tunnels/windows and multi-channel AAT lenses demonstrate the potential for creating open structures with AAT effects. Their fractional bandwidths underscore their broadband characteristics, making them highly suitable for a variety of practical applications.
4. Ring-shaped metamaterials. The exploration of AF systems using ring-shaped metamaterials reveals their ability to leverage rich eigenmodes and resonant cavity structures to achieve perfect AF with high intensity. These systems show promise in applications such as underwater signal detection, medical ultrasound treatment, and acoustic communication.
5. Phase-controlled metamaterials. Highlighting advancements in AF lenses, phase-controlled metamaterials enable precise shaping of acoustic wavefronts. Composed of phased unit cells with varying parameters, these lenses exhibit broadband and ultra-broadband characteristics. Their flexibility and robustness make them ideal for applications in acoustic communication, energy harvesting, and medical ultrasound.
6. Thermoacoustic phase-controlled metamaterials. The use of thermoacoustic phase-controlled metamaterials, which manipulate sound based on temperature gradients, is also investigated. These materials facilitate the creation of broadband AF lenses with self-healing and long-focus characteristics. Potential applications include medical ultrasound, directional bird repelling, and acoustic energy harvesting.
7. Asymmetric focusing and 3D sound manipulation. Concluding with advanced applications, the design of AAF and 3D focusing lenses using acoustic metamaterials is presented. These systems demonstrate the ability to achieve complex sound wavefront manipulations in 3D space, with potential applications in focused ultrasound therapy, acoustic encryption, and non-contact particle trapping.

10.2 Future Research Directions

While the advancements in acoustic metamaterials presented in this monograph have significantly expanded our ability to manipulate acoustic waves, the field continues to face challenges that require further exploration. As we look to the future, it is essential to identify key research directions that can address these challenges and unlock new possibilities for innovation. The following sections outline several promising areas for future investigation, ranging from the development of broadband and tunable devices to the integration of advanced computational tools and the exploration of 3D sound manipulation. By focusing on these areas, researchers can push the boundaries of acoustic metamaterials, enabling transformative applications in fields such as medical ultrasound, communication, and environmental acoustics.

1. Broadband and ultra-broadband acoustic devices. Future research should prioritize the development of broadband and ultra-broadband acoustic metamaterials for applications in AAT, AF, and AAF. Current acoustic metamaterials often operate within narrow frequency bands, limiting their practical utility in real-world scenarios. By optimizing multi-resonant structures and phased unit cells, researchers can create devices capable of functioning across a wide frequency range. For instance, broadband AAT systems could enable more efficient underwater communication by reducing interference from other sound sources, while ultra-broadband AF lenses could improve medical ultrasound imaging by capturing a wider range of frequencies. Additionally, these advancements could enhance noise control in architectural acoustics, allowing for more effective soundproofing in buildings. The challenge lies in balancing the trade-offs between bandwidth, structural complexity, and fabrication costs. Future work could explore the use of advanced materials, such as gradient-index metamaterials, to achieve broader bandwidths without compromising performance. By addressing these challenges, broadband and ultra-broadband acoustic devices could become more versatile and widely applicable in fields ranging from medical ultrasound to telecommunications.
2. Integration of active and tunable metamaterials. The integration of active and tunable metamaterials represents a significant future direction for acoustic metamaterials, particularly in applications involving AAT, AF, and AAF. Unlike passive metamaterials, which have fixed properties, active and tunable metamaterials can dynamically adjust their acoustic characteristics in response to external stimuli, such as temperature changes, mechanical deformation, or electric and magnetic fields. This adaptability could enable real-time control of sound propagation, allowing for reconfigurable acoustic lenses, tunable sound barriers, and dynamic noise cancellation systems. For example, in medical ultrasound therapy, tunable metamaterials could adjust their focusing properties to target different tissue depths without requiring physical adjustments to the device. Similarly, in architectural acoustics, active metamaterials could adapt to changing noise conditions, providing optimal sound insulation in real-time. However, challenges remain in designing materials that can reliably and efficiently respond to external stimuli while maintaining structural integrity. Future research could explore the use of smart materials, such as shape-memory alloys or piezoelectric materials, to create robust and responsive acoustic devices. By overcoming these challenges, active and tunable metamaterials could revolutionize the field of acoustics, enabling more flexible and adaptive applications.
3. 3D acoustic wavefront manipulation. Expanding acoustic metamaterials into 3D space is a critical future direction for achieving complex wavefront manipulations, such as 3D focusing, vortex generation, and asymmetric transmission. Most current AAT, AF, and AAF systems are limited to 2D space, which restricts their practical applications. By designing multi-layered or multi-dimensional

phased arrays, researchers can create 3D acoustic metamaterials capable of precise control over sound waves in all spatial dimensions. For instance, 3D AF systems could improve medical imaging by providing more accurate and detailed images of internal organs, while 3D AAF systems could enhance focused ultrasound therapy by targeting tumors with greater precision. Additionally, 3D acoustic vortex generation could enable non-contact particle trapping, allowing for the manipulation of small objects in fields like microfluidics and biotechnology. However, designing 3D acoustic metamaterials presents significant challenges, including increased structural complexity and fabrication difficulties. Future research could explore advanced manufacturing techniques, such as 3D printing, to create intricate 3D structures with subwavelength precision. By addressing these challenges, 3D acoustic wavefront manipulation could open up new possibilities in medical imaging, underwater communication, and industrial applications.

4. Robustness and environmental adaptability. Enhancing the robustness and environmental adaptability of acoustic metamaterials is essential for their practical deployment in real-world applications. Current AAT, AF, and AAF systems are often sensitive to environmental factors such as temperature, humidity, and mechanical stress, which can degrade their performance over time. Future research should focus on developing materials and structures that can withstand varying environmental conditions while maintaining their acoustic properties. For example, thermoacoustic metamaterials, which manipulate sound based on temperature gradients, could be designed to operate reliably in both high and low-temperature environments. Additionally, self-healing materials could be incorporated into acoustic metamaterials to repair damage caused by mechanical stress or wear. These advancements would make acoustic devices more reliable for outdoor and industrial applications, such as directional sound transmission in harsh environments or underwater acoustic communication in varying ocean conditions. Furthermore, robust and adaptable acoustic metamaterials could improve the longevity and performance of medical ultrasound devices, reducing the need for frequent maintenance or replacement. By addressing these challenges, researchers can create acoustic metamaterials that are not only highly effective but also durable and adaptable to a wide range of environmental conditions.
5. Multi-functional acoustic devices. The development of multi-functional acoustic devices that combine AAT, AF, and AAF capabilities in a single system is a promising future direction. Currently, most acoustic metamaterials are designed for a single function, such as focusing sound or blocking it in one direction. However, many real-world applications require devices that can perform multiple tasks simultaneously. For example, a multi-functional acoustic device could focus sound in one direction while blocking it in another, making it ideal for applications like medical ultrasound therapy, where precise sound focusing and isolation are both critical. Similarly, in architectural acoustics, multi-functional devices could provide both soundproofing and directional sound transmission, improving the acoustic environment in buildings. The challenge lies in designing structures that

can achieve multiple acoustic effects without compromising performance. Future research could explore the use of hybrid metamaterials, which combine different types of acoustic structures to achieve multi-functionality. Additionally, advanced optimization techniques could be used to balance the trade-offs between different functions, ensuring that the device performs well in all desired applications. By developing multi-functional acoustic devices, researchers can simplify the design of acoustic systems, reduce costs, and expand their practical applications in fields like medical ultrasound, communication, and noise control.

6. Machine learning and optimization for design. Leveraging machine learning and optimization algorithms to accelerate the design and optimization of AAT, AF, and AAF systems is a key future direction. Designing acoustic metamaterials with complex structures often requires extensive trial and error, which is time-consuming and computationally expensive. Machine learning algorithms can analyze large datasets of acoustic properties and structural parameters, identifying optimal designs more efficiently than traditional methods. For example, machine learning could be used to predict the performance of different metamaterial configurations, allowing researchers to quickly identify the most promising designs for further testing. Additionally, optimization algorithms can fine-tune structural parameters to achieve specific acoustic effects, such as broadband focusing or asymmetric transmission. These tools could significantly reduce the time and cost of developing advanced acoustic metamaterials, enabling rapid prototyping and innovation in the field. Furthermore, machine learning could be used to explore new design possibilities that may not be immediately apparent through conventional methods. By integrating machine learning and optimization into the design process, researchers can create more efficient and effective acoustic devices, paving the way for new applications in medicine, communication, and energy harvesting. This approach could also facilitate the development of customized acoustic solutions for specific applications, further expanding the potential of acoustic metamaterials.

10.3 Conclusion and Outlook

In conclusion, this monograph provides a comprehensive overview of the current state of research in acoustic metamaterials, highlighting their potential to revolutionize the field of acoustics. The key findings and contributions presented in this book demonstrate the ability of acoustic metamaterials to manipulate sound waves in innovative ways, enabling new applications in medical ultrasound, nondestructive testing, acoustic communication, and more. However, significant challenges remain, and further research is needed to fully realize the potential of these materials. By addressing these challenges and exploring new directions, the field of acoustic metamaterials can continue to evolve, offering innovative solutions to longstanding problems in acoustics and beyond. The future of acoustic metamaterials is bright, and the possibilities for new discoveries and applications are vast.

GPSR Compliance

The European Union's (EU) General Product Safety Regulation (GPSR) is a set of rules that requires consumer products to be safe and our obligations to ensure this.

If you have any concerns about our products, you can contact us on ProductSafety@springernature.com

In case Publisher is established outside the EU, the EU authorized representative is:

Springer Nature Customer Service Center GmbH
Europaplatz 3
69115 Heidelberg, Germany

Batch number: 10704305

Printed by Printforce, the Netherlands